LOGIC SYNTHESIS AND VERIFICATION ALGORITHMS

LOGIC SYNTHESIS AND VERIFICATION ALGORITHMS

by

Gary D. Hachtel
University of Colorado

Fabio Somenzi
University of Colorado

KLUWER ACADEMIC PUBLISHERS
Boston / Dordrecht / London

Distributors for North America:
Kluwer Academic Publishers
101 Philip Drive
Assinippi Park
Norwell, Massachusetts 02061 USA

Distributors for all other countries:
Kluwer Academic Publishers Group
Distribution Centre
Post Office Box 322
3300 AH Dordrecht, THE NETHERLANDS

Consulting Editor: Jonathan Allen, Massachusetts Institute of Technology

Library of Congress Cataloging-in-Publication Data

A C.I.P. Catalogue record for this book is available
from the Library of Congress.

Printed on acid-free paper.

Printed in the United States of America

To:

Linda, *Jordan*, and *Kira*,

and

Chiara and *Laura*.

Contents

List of Figures

List of Tables

Preface

Genesis of the Book

This book grew from courses taught at the University of Colorado (Boulder) and at the Universidad Politecnica de Madrid, Spain. As the title suggests, we were motivated by two disparate objectives. First, the VLSI CAD group at Boulder was given the responsibility for teaching a course which satisfied the ABET requirement for an upper division algorithms and discrete mathematics course in a EE or ECE curriculum. Hence we started looking for an appropriate book, and taught trial courses from various books including [241], [162], and [190]. While each of these books had their individual strengths, there were always significant areas that were neglected.

Second, logic synthesis has matured as a field to the point of almost universal designer acceptance and is used in every major IC design/production house worldwide. Further, the younger field of formal verification, perhaps spurred on by the infamous Pentium bug, appears to be following a trajectory very much like that taken by logic synthesis over the last decade.

Consequently, we wanted an orderly integration of modern developments in logic synthesis and formal verification, into the traditional subject matter of Switching and Finite Automata Theory. This clearly eliminated texts like [241], [190], and [146]. The book that came closest to our requirements was Kohavi's book [162]. Although this text was excellent and long lived, it is now outdated, since it does not deal many modern developments in discrete mathematics that were significant to bringing VLSI CAD to its current advanced state.

Thus we decided to occupy the niche previously filled by the Kohavi book [162] and supplement the coverage with recent theoretical developments most significant to the emergence of automatic synthesis and verification tools during the nineties.

As an example of the aforementioned integration, consider the problem of formally verifying that two distinct FSMs are or are not equivalent. The solution of this problem for problems of practical size was due to the efforts of Coudert, Madre, and McMillan [70, 194]. The solution rested on ingenious interweaving of BDDs (Binary Decision Diagrams) and the state equivalence theory covered in [162]. Boolean function manipulation using BDDs evolved from the work of Bryant in 1986 [47], and is not covered in previous comparable textbooks.

Other examples of recent theoretical advances which had profound effects on the development of automatic synthesis tools are :

1. Complexity theory and the development of "optimal complexity" algorithms like Tarjan's strong components algorithm;
2. The unate recursive paradigm, and its various applications in the industry standard ESPRESSO program for logic minimization [37, 39, 250];
3. Branch and bound algorithms with lower bound pruning that has proved to be exceptionally powerful for such disparate logic synthesis applications as two-level logic minimization and state minimization of Finite State Machines [228];
4. The Kerneling theory of Brayton and McMullen, [38], which led to efficient and widely adopted algebraic methods for restructuring a circuit so minimization techniques could be more powerfully applied;

5. The PODEM algorithm for automatic test pattern generation, [116], and the learning/implication heuristic used to make it widely applicable, [122];
6. The development and deployment of the theory of don't care conditions, [15], [204];
7. The deployment of BDD-based symbolic processing as discussed above along with modified ATPG techniques which made sequential synthesis and ATPG practical for large circuits, [63];

It was very difficult for a VLSI CAD group to teach switching theory without covering these subjects. Consequently, we wanted a book which met ABET requirements and had the above developments woven into the fabric of the theory.

Equally important was the emergence over the last decade of widely applicable public domain logic synthesis and verification tools, such as ESPRESSO, MIS, SIS, SMV, BOLD, as well as commercial tools such as SYNOPSIS' design compiler, The CADENCE, VIEWLOGIC, and MENTOR tool suites, and IBM's BOOLEDOZER package. Many of the solved and unsolved (an answer book is available) problems are based on using easily available tools. All tools used in the book can be obtained from anonymous FTP from the following web sites:

IC.berkeley.edu,
monk.colorado.edu.

Also available from the latter site are PERL5 scripts which implement most of the discrete math algorithms, like longest path analysis, determinization of nondeterministic automata, minimization of completely specified FSMs, et cetera.

Thus we wanted the book to reflect the concurrent evolution of switching and automata theory, and of VLSI Logic Synthesis and Verification, over the last decade, while also providing the necessary background in Boolean algebras and discrete mathematics.

Our objective from the onset was a senior course, with enough depth to be of interest to first year graduate students as well. It is quite likely that any student who takes a job in the still burgeoning (at publication time INTEL was still planning two $1B fabrication sites) semiconductor industry will use one or more of these tools.

Some students may eventually be involved in the design of the tools themselves. A persistent focus of the book is "hands-on" relation of actual design tools to the theoretical material. Whenever possible, the algorithms covered in the text are the subject of one or more problems based on the use of available synthesis programs. An appendix describes the details of the installation and use of the software at our institution.

Problems. The book contains a large collection of solved problems. This is because we generate new homework problems every semester, as part of the development of the course. We also have another equally large collection of solved problems which will go into the Instructor's Manual. We plan to maintain this manual in the Colorado web site quoted above. Thus it should continue to grow over the years as a resource for all users of the book.

Many solved problems require the use of software tools. Much of the assigned work involves the use of Berkeley's sequential/combinational synthesis program SIS-1.2 [250] and CMU's formal verification program SMV [194]. Whenever possible, the algorithms covered in the text are the subject of one or more problems based on the use of synthesis programs.

Itineraries

It would be difficult for undergraduates at most institutions to cover the entire book in one semester. We describe here some sample itineraries for efficaciously traversing the book. The "Logic Design Option" skips Chapters 6-9, and possibly 3 (our experience is that students find this chapter appealing). The "Algorithms and Discrete Mathematics Option skips Chapter 2, 5, 8, 12, and 13. Finally, the graduate "Introduction to Synthesis and Verification Option" takes well prepared and motivated graduate students through the whole text.

Convention When theorems have obvious proofs, or are well known theorems from the literature whose proofs are outside the scope of the present text, we omit the proof. In the latter case we give a citation in which the proof may be found.

Notation

Notation Types		
font	environment	type
bold	text	Keywords, Eg., in algorithms and procedures.
slanted	text	emphasis (instead of italic)
upper case (small)	text	acronyms, abbreviations
bold lower (italic)	math	vectors, or scalar functions
bold lower (normal)	math	sequences (strings, runs, tapes, etc., sets of vectors.
i, j, k, l, m, n, p, q	math	indices and integer quantities
upper case	math	sets, or characteristic functions of sets, or BDDs of same
upper case	math	graphs (ordered set of 2 or more sets)
upper case	math	
upper case	math	characteristic functions of relations (sets)
bold upper	math	matrices, characteristic functions of sets (Only when it is necessary to make a distinction between a set and its characteristic function)
small caps	math	algorithm, function, operator, and procedure names, as well as commands, file names, etc.
calligraphic	math	Finite state machines and automata (ordered sets of 4 or more sets), languages (sets of strings) Relations, Sets of Sets.
bold calligraphic	math	Sets of languages, such as regular languages.

Pseudo-Code Conventions

We shall adopt the following further notation for pseudo-code description of algorithms and formal procedures. The beginning and end of the statement blocks of procedures, or **if** or **else** blocks, or of **for**, **foreach**, **while**, or other loop or branching structures are denoted by matching braces, as in the C/C++ programming languages. Comments are either denoted by /* ... */ as in C, or are set to the far right of pseudo-code statements without such delimiters when the context is clear. Semicolons will be used as statement separators only — generally they are omitted at the end of a line of pseudo-code.

Acknowledgments

We acknowledge the impact of a career's worth of association with Bob Brayton.

We acknowledge the impact of a previous book-experience with Bob Brayton, Alberto Sangiovanni-Vincentelli, and Curt McMullen, and of an aborted (so far) book attempt with Bob, Alberto, and Rick Rudell (Rick McGeer may yet save that book). In both of these former endeavors, Carl Harris is to be recognized for his gentle, enlightened, and persistent persuasion. The help of John Hayes was valuable in giving us early feedback on book direction.

We acknowledge the many contributions of University of Colorado students and associates, whose work is featured in this book — to name just a few: Karen Bartlett, Chris Morrison, Reily Jacoby, Hyunwoo Cho, Sehwoong Jeong, June Rho. In-Ho Moon helped by proofing several chapters of the last draft.

We are also indebted in this regard to many other colleagues and friends — among others, Richard Newton, Srinivas Devadas, Kurt Keutzer, Aart DeGeus, Randy Bryant, Louise Trevillyan, and Giovanni DeMicheli were particularly impactful.

We acknowledge the impact of numerous "summer seminars", first at IBM Yorktown, and then rotating between Boulder, Berkeley, and Stanford.

The indefatigable help of our administrative assistants Karen Schneider and Ruth Major must also be acknowledged.

Penultimately, we acknowledge the support (over the last decade) of the National Science Foundation, in particular that of our friend and mentor, Bob Grafton, who helped us see the dawning of a new era in the advent of Computer Aided Verification.

Finally, we acknowledge the patience and appetite for learning shown by our undergraduate students in enduring some "very rough" drafts.

Part I

Introduction

The Zen of Painting

> A Zen Master, intoning that "The journey of 1000 miles begins with a single step", had sent two adepts to paint the great chalk cliffs. After years of patient endurance, the rythmic swing of the brush had opened their minds to the beauty and drama of nature. But in the the tenth year, the first adept noticed that the second was "painting" with an empty hand, so he called out: "where's your brush?". After a goodly pause, the second adept's voice floated back in the morning breeze: "Sure does.".

The moral of the story is that every repititive task can be automated. For example, doing the dishes, printing this book, or designing an advanced microprocessor. By reading this book, you take the first step toward joining the great journey of design automation. Any sufficiently advanced technology is indistinguishable from magic, and the magic of Design Automation is literally shaping our future (as well as making a lot of people fabulously wealthy).

Roadmap of Part I Part I is the "big picture" part of this book. In Chapter 1, we briefly describe how the revolution of design automation has enabled the semiconductor industry to design microprocessor chips containing on the order of 10^7 transistors. At the time of writing (1996), an "advanced" CAD workstation had a billion bytes of memory, and could do a billion operations per second. Fifteen years ago, the operative numbers were a million bytes of memory, and a million operations per second. In 1996, the IBM "Deep Blue" computer managed to beat world champion Gary Kasparov (perhaps the best player who ever lived) in one game of a match (Kasparov still won the match handily, but a blow had been struck).

After briefly sketching design mehtodologies, we characterize the IC design process as one of optimal tradeoff of competing design goals:

1. The area on the chip required by the circuit being designed;
2. The critical path delay of the circuit;
3. The testability of the circuit, measured in stuck-at fault coverage;
4. The power dissipated by the circuit.

Unfortunately, an engineer designing a chip is like a politician pleasing his political consitutencies: an ultra-small chip is going to be too slow, and an ultra-fast chip is going to be too big. The art of automated design is finding optimal tradeoffs. To paraphrase an infamous politican: "You can please part of the circuit all of the time, or all of the circuit part of the time, but you can't please all of the circuit all of the time".

In this chapter we also introduce such disparate subjects as CMOS implementation of logic functions, complexity of algorithms, and graphs. This in turn enables us to discuss efficient algorithms for critical path delay analysis and test generation. All of these notions will be recurring themes in the sequel — our goal is to develop theory and intuition for design automation.

In Chapter 2, we give a "quick tour" of the design of a simple decoder circuit. In this chapter, we contrast two approaches to design optimization. The first approach is "optimization by clever tricks", as might be performed by an experienced designer. The second is "design with VLSI CAD tools". Like the match between Kasparov and Deep Blue, a good designer will always win in the competition between these two approaches. But good tools on powerful computers do pretty well, and are far more cost effective when applicable, because "time to market" is the dominant cost factor.

Chapter 1

Introduction

This book is about the theoretical underpinnings of VLSI (Very Large Scale Integrated circuit) CAD tools for logic synthesis and verification. Historically, the field of VLSI CAD has been driven by the inexorable push of microchip technology, as more or less accurately described by Moore's law: "microchips will gain a factor of two in speed every 1.5 years". This speed improvement is basically due to shrinking of feature sizes (improvement due to architectural innovation is added on top of this). Of course, decreased size means more transistors per chip. Consequently, the size of the chip component designed by one individual or team grows correspondingly. VLSI CAD tools must therefore cope with "Large Scale".

We begin this first chapter with two sections which briefly characterizes this technology, and the opportunities and challenges it presents. We then discuss the design styles engendered by the push of technology. We conclude with some sections which give an overview of the specific sub-area of VLSI CAD known as "Logic Synthesis and Verification", including some basic concepts which will be useful throughout the sequel.

1.1 VLSI: Opportunity and Challenge

This text is concerned with the design of complex digital systems. Digital circuits are used in computers, communication equipment, compact disc players, and so on. It is possible today to put several million transistors on a single integrated circuit (IC) or chip (for example the high-end INTEL Pentium chips). This Very Large Scale Integration (VLSI) is made possible by two *technologies*: Manufacturing technology and Design technology.

1.1.1 Manufacturing Technology

The most common indicator of the degree of sophistication of a manufacturing process is the so-called *minimum feature size*. It indicates the minimum achievable length of a MOS transistor channel. A typical value of the minimum feature size for an advanced IC manufacturing process in 1996 is $0.25\mu m$.

Another important indicator is the number of interconnection layers available on the chip. An advanced IC process may have three metal layers, that is, interconnections and power and ground lines can use three layers that communicate through *vias*. A metal line on a chip is somewhat larger than the minimum transistor and in case of multiple metal layers, it increases with the distance from the substrate. Nevertheless, vast resources, both in terms of active devices (transistors) and interconnects are made available to the chip designer by today's manufacturing technology. High speed is also possible, thanks to the reduced capacitances associated with these small devices. Clock frequencies in excess of 200 MHz have been achieved for large ICs.

A complex system is typically composed of several integrated circuits. In recent years there has been a rapid evolution in the fields of packaging and printed wiring boards. These advances have made it possible to design, for instance, today's notebook computers.

1.1.2 Design technology

The design of a system with several million parts is an arduous task. It is even more challenging because of the increasing concern with time-to-market and quality considerations. The *revenue life* of a product is the time during which a product generates most of its revenues. The electronic industry is characterized by extremely short and steadily shrinking revenue lives. Whereas ten years ago a revenue life of 3–5 years was normal, today many products enjoy less than one year of revenue life.

It is clear that in such a scenario, delaying the launch of a new product may have a serious impact on its profitability. Several studies have shown that a large overrun in product cost has a much smaller impact on profitability than a relatively modest delay in development.

It is therefore necessary to automate as much as possible design, in order to increase designer productivity, eliminate trivial mistakes, help eliminate more subtle mistakes that may cause a design not to work right the first time.

Over the last three decades, Computer Aided Design (CAD) tools have gradually been adopted for various tasks of the design process. Initially, computers have been used to perform mundane chores. In time, however, more creative design activities have been delegated to CAD programs. Today, activities that are routinely performed by CAD programs include many forms of verification (geometrical and electrical design rule checking, several types of simulation, etc.) as well as various aspects of design synthesis (generation of masks from layout, generation of layouts for modules like custom RAMs or ALUs, automatic placement and routing, logic synthesis and optimization). Computers are also used for design entry, data and design management.

The adoption of tools has influenced the way circuits are designed in various ways. In some cases some restrictions on what was an acceptable design had to be imposed, so that some tools may be used. In other cases new tools have opened up new possibilities, like the one of exploring several design alternatives, that were precluded as long as the design was done manually. The advent of design data bases and design management tools has contributed to the development of design disciplines that help cope with the complexity of large circuits. Finally, in some cases, entirely new design styles have emerged, that are largely defined by the tools that are used.

The body of knowledge that is at the foundation of the CAD tools constitutes the *design technology*. As the previous examples suggest, it may be divided according to domains like process design, physical design, electrical design, logic design, high-level design, design management, and so on. We shall focus primarily on logic design and we shall examine the techniques and tools that go under the name of logic synthesis.

Logic synthesis has a relatively long history as a scientific discipline, being born approximately at the same time as logic design. However, the systematic use of computer programs to do logic design is more recent. About fifteen years ago, IBM pioneered extensive use of logic design tools when PLA-based design became popular. In the early eighties, proprietary tools began to be used also for other styles of designs. Finally, in the late eighties, CAD companies began selling logic synthesis tools and logic synthesis became mainstream.

1.1.3 Why VLSI

The investments associated with the manufacturing of VLSI circuits are very high. A production line today may easily cost in excess of $500M. Even higher costs are predicted for the future. Design costs are also high. Commercial software for a VLSI design workstation is typically worth much more than the workstation itself. Some tools cost hundreds of thousands of dollars per copy. However, the design of a complex microprocessor may require hundreds of man-years of labor, so that in spite of the high cost of tools, labor is still the predominant cost of development.

It is then reasonable to ask why VLSI is successful or even viable. There are several reasons. First of all, VLSI has created new markets. As already mentioned, notebook computers simply could not exist without VLSI. The same is true of digital audio and personal communications equipment.

In spite of the high investments, there is also an economic advantage in VLSI since it exploits in the best possible way the benefits of mass production. Once the masks are made, VLSI circuits are obtained by a process of replication that takes time largely independent of the number of transistors on the chip and the number of chips on the wafer. By contrast, assembly costs are closely related to the number of parts. Reducing the number of parts is therefore beneficial and VLSI technology plays a major role in it.

Speed is another important factor in promoting the use of VLSI components and in pushing the advancement of the technology. Smaller components imply smaller capacitances and therefore faster circuits. Intra-chip communications are also faster than inter-chip communications, so that reducing the number of parts also helps in that respect. This consideration has been important to the success of RISC (reduced instruction set computer) architectures. Reducing the number of parts has also a beneficial effect on the reliability of a system.

Finally, the use of VLSI technology helps protect investments in design, since it is more difficult to copy an integrated circuit than, say, a board.

1.2 VLSI Processes

We briefly review the most common processes used to fabricate integrated circuits. Our purpose is just to provide some context for our work. Hence we shall only emphasize those aspects that are relevant to synthesis.

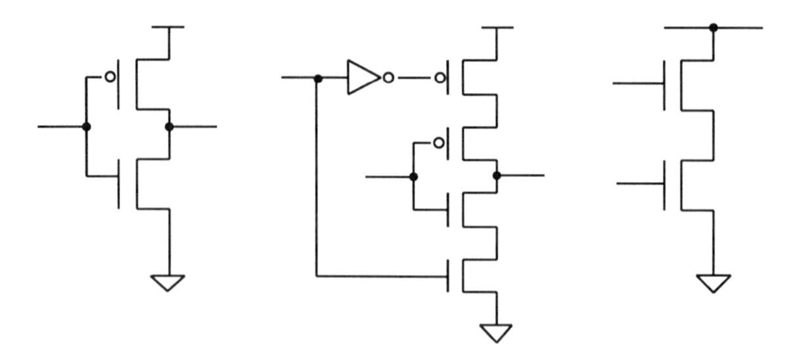

Figure 1.1: MOS gates.

A first distinction is based on the substrate material. By far the most common material is silicon. Gallium Arsenide is used for very high speed applications, but it is much more expensive and has more mechanical problems.

Another distinction is between bipolar processes and Metal-Oxide-Semiconductor (MOS) processes. The distinction obviously refers to the types of transistors fabricated on the chips. Bipolar technologies like Emitter Coupled Logic (ECL) tend to be faster than MOS technologies, but they also tend to consume more power and to be less dense. There are also mixed bipolar-MOS processes, notably BICMOS processes, that allow the fabrication of both BJTs and MOSFETs on the same chip.

Among the MOS processes, today Complementary MOS (CMOS) is the only one we consider. Initially considered too slow and not too dense, it has gradually become the dominating technology for VLSI circuits. The main reason is the very low static power consumption. Some CMOS logic gates are shown in Figure 1.1. On the left a simple inverter is shown. In the center an inverting tri-state driver (the second input switches the mode of operation from inverting or high-impedance), and on the right an open-drain NAND gate (used in pre-charged design styles — the line at the top represents a net connecting a number of these open drain NANDs to a common pullup resistor (not shown)).

Every large manufacturer has typically a certain number of CMOS processes, adapted to different uses. Some CMOS processes differ in their architecture (e.g., p-well vs. n-well); others have additional steps required for special applications (like fabrication of EEPROMs or of mixed analog-digital circuits). Finally, several generations of processes typically coexist, with the more recent and advanced processes being used for the newest and more aggressive designs. However, in what follows all these differences will seldom be important and we shall customarily refer to CMOS, without any further qualification.

1.3 Design Styles

1.3.1 Design Decomposition

A fairly typical approach to the design of large digital systems is a design decomposition which separates the control functions of the design from the data processing functions. It is not uncommon that different design methods are then applied to the separate parts. For instance, in a microprocessor, the design of the data path (ALUs and register files) is approached differently from the design of the instruction decoder or the cache controllers. The data path may be pipelined, for example, and resource allocation decisions must be made on the basis of space-time tradeoffs. Typically such design steps are made at a high level of abstraction in which data path components are regarded simply as functional blocks, with space dimensions and pin delays as attributes.

The controller, on the other hand, may first be designed at a high level of abstraction as an FSM, an acronym for **Finite State Machine** . FSMs are introduced in Chapter 7, and are often represented by a directed (cyclic graph). Synthesis methodology for FSMs is treated in Chapter 8. After such steps as state encoding and state minimization, the controller design may be synthesized as combinational logic with latch-controlled feedback cycles.

In addition, the design of a complex circuit is typically carried out by a team. This results in further design decomposition of both the data path and the controller into sub-designs. A block diagram of the overall system is then developed which defines the interfaces between the blocks designed by different people/teams. Often this block diagram is expressed in an HDL (Hardware Description Language), such as VHDL [109, 54, 135] or Verilog[10].

Defining the top-level block diagram of a circuit goes under the name of *high-level* or *architectural* design. In the sequel, we shall assume that this phase of the design has been carried out already—either manually or automatically—and we shall concentrate on the succeeding phases. It should be clear, however, that architectural design has a large impact on the outcome of the entire design process.

VHDLs are examples of textual descriptions of digital systems. We can also employ graphical means of representing the design decomposition. We shall see examples of both in the sequel. It is typical of a real-life design system to support both ways of representation, for each has its own advantages.

A legion of vlsi CAD tools have been developed to aid in executing the synthesis methodology described above. These tools may be classified according to their associated levels of abstraction, as follows.

The Path to Silicon

1. **Behavioral Synthesis**
   - Resource Allocation [191],
   - Pipelining [210, 209],
   - Control Flow Parallelization [264, 265],
   - Communicating Sequential Processes
   - Partitioning [17]
2. **Sequential Synthesis**
   - Register Movement and Retiming [174, 173, 182]
   - State Minimization [117]
   - State Assignment (Encoding) [93, 87, 84, 198],

- Synthesis for Testable FSM's [86];
- State Machine Verification [71, 72, 262, 60]

3. **Logic Synthesis**

- Extraction of combinational logic from hardware description languages [240, 68];
- Two-level (PLA) minimization [37],
- Algebraic Decomposition [38, 36, 236, 268],
- Multilevel Logic Minimization [15, 39, 123],
- Synthesis for Multifault Testability [128],
- Test Generation via Minimization [15, 151];
- Technology Mapping [158, 85],
- Timing Optimization [82, 120, 39].

4. **Technology Mapping**

- Mapping to Library of Logic Gates[158, 85],
- Timing Optimization [82, 120, 39].

5. **Physical Design Synthesis**

- Cell Placement
- Routing
- Fabrication
- Engineering Changes

Logic Synthesis tools provide the VLSI chip design team with two otherwise non-existent capabilities: (a) automatic translation of high-level language descriptions into logic designs, and (b) automatic optimization of chip area, speed, and testability. Thus, designers can take advantage of logic synthesis tools as key parts of a top down synthesis methodology, or for optimization assistance with manually created logic. The advantages of such an automatic synthesis methodology in VLSI design are clear. They include reduced design time, reduced probability of design error, and higher-quality designs because more effort is focused at a higher-level. However, the use of computer-aided synthesis tools provide an increase in designer productivity only if designs of acceptable quality are produced. In the sequel we will show how and why these tools were developed, and why we expect that synthesis tools can improve on design quality available from manual design.

1.3.2 Logic (Circuit) Design Styles

VLSI circuits are used for various applications. It is customary to divide them into *standard parts* and *Application Specific ICs* (ASICs) depending on whether they are used in many applications or are designed to perform a specific function in one system.

Standard parts tend to be produced in higher volumes (many millions) and to have a longer revenue life. Typical examples of standard VLSI parts are microprocessors, computer peripherals like disk controllers, memory chips, programmable digital signal processors, and field programmable gate arrays (FPGAs).

ASICs are sometimes built in just a few units (that could be the case of a space-borne circuit) and sometimes in hundreds of thousands of units. Some ASICs are designed to the specifications of a single user, for the needs of a particular system; others are designed in such a way that many systems can be built around them. An example of the latter could be a chip-set used by many PC manufacturers. A common feature of ASICs is that development costs and times tend to be predominant.

Because of the differences in the economic factors, standard parts and ASICs tend to be designed differently. When designing a microprocessor, it makes sense to spend more time to reduce the area and increase performance than when designing the typical ASIC.

We shall distinguish three major styles of design:

- **Full Custom Design:** Every circuit part is especially optimized for the purpose it must serve in the design.
- **Semi-Custom Design:** The circuit is designed by assembling pre-designed and pre-characterized sub-circuits. Manufacturing may use a pre-diffused substrate.
- **Programmed Design:** The design is obtained by programming a standard part. Some circuits may be programmed only once (by blowing fuses or anti-fuses), while others may be programmed an unlimited number of times.

There are no clear boundaries between these design styles. For instance, many circuits are designed with a mixture of full custom and semi-custom styles. The full custom approach is used for the critical parts. Another way of combining these approaches is to create a library of sub-circuits specifically tuned for a project. Typically, dynamic or pre-charged design styles are reserved to full custom chips.

Within the semi-custom approach, one may distinguish several sub-styles. There are circuits that are called *gate arrays* that are made of pre-diffused cells. Each cell can be customized to a specific function (e.g., 2-input NAND gate) by metalization. Interconnections are also obtained by laying out metal wires on two or more layers. Since the cells are known in advance, they can be accurately characterized. The designer does not need to know all the details of the cells to successfully work with them. The fact that the circuit is pre-diffused means that the fabrication turn-around time is shorter and the cost of the masks is lower.

Figure 1.2 presents a typical six-transistor cell. The black dots indicate possible contact points. How to obtain a three-input NAND gate from the cell is shown in Figure 1.3. The dotted lines indicate metal connections. More complex circuits like a flip-flop may be obtained by customizing more than one cell. It is also possible to have different topologies for the basic cell, with more or less transistors and connected differently.

There are two major variants of gate arrays: Channeled and channel-less. In the former, the programmable cells are grouped in rows that are divided by *routing channels*. Figure 1.4 presents the conceptual view of a channeled gate array. The shaded

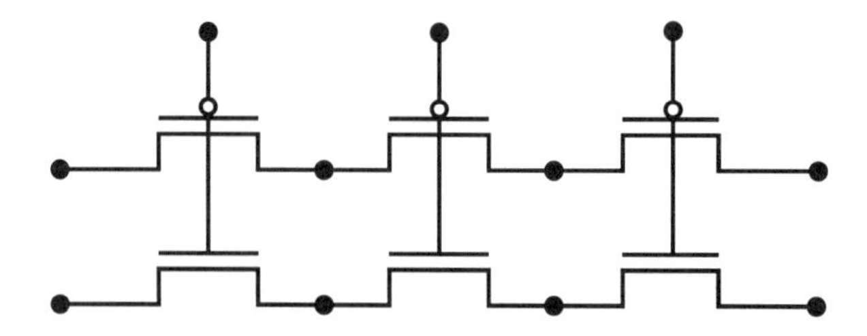

Figure 1.2: A six-transistor gate array cell.

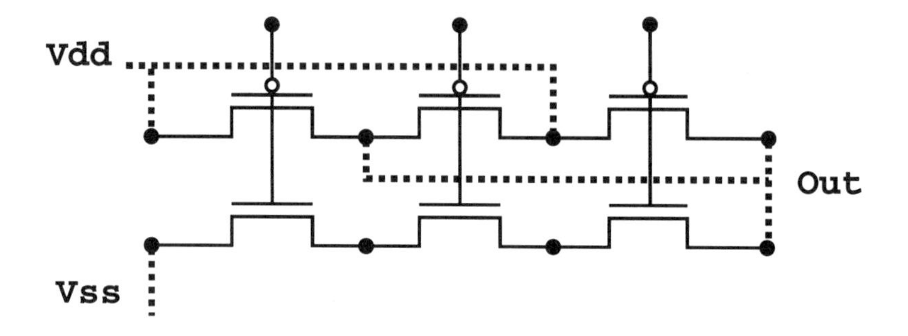

Figure 1.3: A three-input NAND gate obtained from the cell of Figure 1.2.

squares are the input/output pads, while the shaded horizontal strips represent the rows of basic cells. The space between two rows is called *routing channel* and is reserved for the interconnections. In the channel-less gate arrays, also called *sea-of-gate* arrays, cells are available on the entire surface of the chip. Interconnects use the space left by unused cells, as well as on top of the cells (thanks to the multiple routing layers). Channel-less architectures provide a better utilization of the chip area and are used for most recent families of large gate arrays.

Non pre-diffused semi-custom designs are often called *standard cell* circuits. This comes from the use of a pre-designed library of cells. Though fabrication turn-around time is larger than for gate arrays, standard cell designs are still considerably faster to design than full custom circuits. In the simplest cases, the organization of a standard

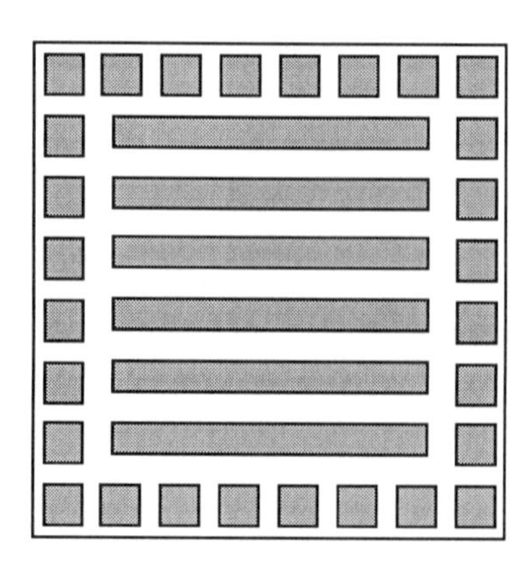

Figure 1.4: Organization of a channeled gate array.

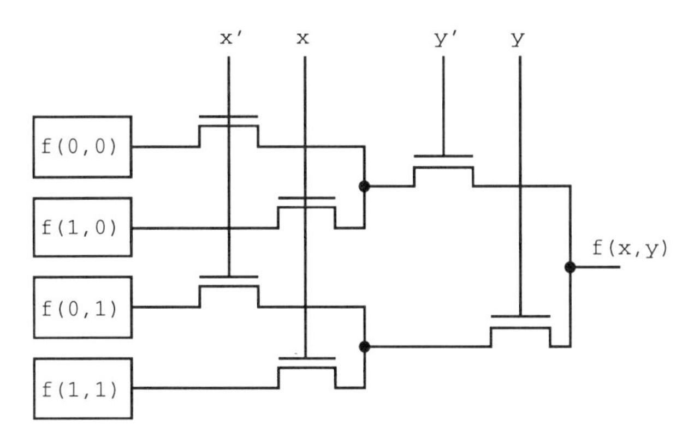

Figure 1.5: Two-input look-up table for FPGAs (Field Programmable Gate Arrays).

cell chip resembles that of a channeled gate array, with the exception that the height of the routing channels is not fixed. In more complex schemes, a chip may contain several rows of standard cells and some large blocks, like memories and data paths, implemented in different styles.

Typically, semi-custom designs are automatically placed and routed. This means that once the circuit is detailed to the level of a netlist of gates, the rest of the process is automatic. This allows IC design to be performed by people that are knowledgeable about electronic design, but not about IC design specifically. This separation of competence is one of the reasons of success of the semi-custom design style.

There are several types of field programmable devices. The simplest, from a functional standpoint, are the non-volatile memories (PROMs, EPROMs,...). Programmable logic devices (PLDs) is a name commonly given to circuits based on programmable function, but rather fixed interconnections. They typically implement two-level functions. Finally, in field programmable gate arrays (FPGAs), both the cells and the interconnections are programmable. We concentrate in our discussion on FPGAs.

Field programmable gate arrays belong to two categories depending on whether they can be reprogrammed or not. To the first category belong devices that store their configurations in static RAM (SRAM). The program is downloaded to the device every time the power is turned on. Those devices that can be programmed only once are typically based on *antifuses*, i.e., open circuits that can be closed by applying a sufficiently high voltage. Although less flexible, antifuse-based devices are denser and faster than SRAM-based counterparts.

We discuss here one possible cell architecture for SRAM-based FPGAs, called look-up table (LUT). In Figure 1.5, the principle of the look-up table is illustrated. We take as example a LUT that can implement all two-input combinational functions. The boxes on the left are SRAM cells. Each stores one value of the function. The inputs to the cell control the switches that route one of the values to the output. A typical LUT-based cell may contain more than one LUT, each with three to five inputs, one or two latches, and miscellaneous logic.

Semi-custom designs are those that have been most amenable to logic synthesis, together with field programmable devices. This is because the design is mostly carried out at the logic level and good characterization data is available. Therefore in the sequel we shall mostly deal with them.

1.4 Overview of Optimal Logic Synthesis

The benefits of automating the logic design process are lost if the result does not meet its area, speed, or power constraints, while optimizing the design tradeoffs as well as an expert designer could. Therefore, a critical aspect of automatic logic synthesis is the optimization problem of deriving a high-quality design from the initial specification. The accepted optimization criteria for multi-level logic are to **Minimize** some convex function of:

1. **Area** occupied by the logic gates and interconnect;
2. the **Critical Path Delay** of the longest path through the logic;
3. the **Degree of Testability** of the circuit, measured in terms of the percentage of faults covered by a specified set of test vectors, for an appropriate fault model (Eg., single stuck faults, multiple stuck faults, etc.);
4. **Power** consumed by the logic gates.

This minimization is to be performed while simultaneously satisfying upper or lower bound constraints placed on these physical quantities. While humans are superbly equipped to solve such problems once they are clearly formulated, the sheer mass of detail in VLSI designs makes the use of synthesis tools imperative.

Delay constrained area minimization has an immediate effect on practical microchip design — the design has to physically fit on the chip, which typically is on the order of 1cm square. We note also that area minimization has an economically important impact on yield, because net yield is known to decrease exponentially with the size of the chip. Alternatively, algorithms for area-constrained delay minimization are often used to maximize performance[1].

Another criterion which is increasingly important, especially for mobile technologies such as laptop and palmtop computers, and cellular phones, is to minimize the power of the final circuit. The area, delay, and power of a design before layout are estimated using models which predict the effects of physical design based on the cells and nets in the final design. Often power is minimized by just making the design slower, but transformations exist which reduce power while not increasing delay.

Another important part of integrated circuit design is the manufacturing test which determines if a fabricated chip works as expected. A connection (wire) is untestable (or redundant) if replacing the connection with a constant value does not affect the functionality of the circuit. The most typical fault model is the so-called stuck-at fault model. In this model, a connection in the circuit is testable if there exists a set of inputs for which at least one circuit output changes value when the connection is forced to logical 1 or 0.

If a connection is not testable, it is called redundant. Aside from the observation that a smaller circuit usually results from removing a redundant connection, there is

[1] In the sequel, we will use this term to mean the inverse of critical path delay

the further problem that redundancies interfere with the production-line testing of the integrated circuit. Therefore, another goal for logic synthesis is to produce designs with no redundancies.

Synthesis tools are also in an ideal position to tackle the testing problem. These tools create and alter designs, so that they can easily modify them for testability. It makes sense for synthesis tools to insert the special structures needed for test, if any, and then optimize the whole design, including the test structures, for minimum speed and area penalty. Similarly, it is logical for synthesis tools to produce vectors that are inherently integrated with a design's test structures, and to minimize the vector set required to test the design. Thus the marriage of synthesis and test can achieve a fully automated test solution, and serves to move test forward in the design cycle. The testing/testability issues are explored in Chapter 11.

The design of the optimal circuit which meets all of these constraints is a difficult problem due to the tremendous number of potential solutions for even a small set of logic equations. The size of VLSI circuits makes logic synthesis for VLSI a difficult optimization problem, which usually requires automated tools for practical designs. As we shall see, graph algorithms and models, Boolean algebras and switching theory, optimization theory, and the theory of finite state machines and finite automata, provide the mathematical tools needed to design and efficiently utilize computer aided design tools for automatic logic synthesis, testing, and verification.

1.4.1 Area-Time Tradeoff Curves

Synthesis tools give the designer a means of exploring the optimal tradeoff curve relating the multiple objectives of chip area and performance. Typically the optimal tradeoff curve is a convex function of Area A and Delay τ, such as a hyperbola, as shown in Figure 1.6. Usually there are delay constraints ($\tau \leq \tau_C$), usually determined by marketing forecasts, and area constraints ($A \leq A_C$), due to the size of the chip and the Area budget of the circuit being designed. Thus designs in the shaded area of Figure 1.6 are *feasible* (that is, acceptable), and designs lying on the tradeoff curve are optimal. By definition, it is impossible to realize designs lying below or to the left of the optimal tradeoff curve. Optimal tradeoffs are discussed briefly here, and given a more detailed treatment in Figure 4.1.

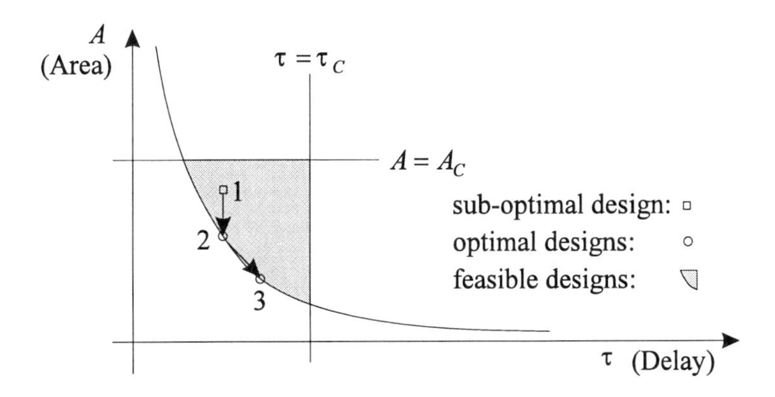

Figure 1.6: Area-Delay tradeoff curves.

Given a delay specification, synthesis for optimal Area would produce a design lying on the intersection of the delay constraint line $\tau = \tau_C$ and the optimal tradeoff curve. Designs with Area smaller than this one would have Delay which exceeds the specification. Similarly, given an area specification, synthesis for optimal Delay would produce a design lying on the intersection of the Area constraint line $A = A_C$ and the optimal tradeoff curve. Designs with Delays smaller than this one would have Area which exceeds the specification.

In the figure, feasible, but suboptimal designs are represented by a small square, whereas optimal designs are represented by a small circle. Thus Design 1 is feasible, but suboptimal, while Designs 2 and 3 are both feasible and optimal. We stress that there usually are many optimal designs, and it is up to the designer to specify his priority of tradeoff. A design move from Design 1 to Design 2 is a tradeoff-free move, which decreases area without sacrificing delay. In contrast, the design move from Design 2 to Design 3 is an optimal tradeoff move, which gives up some delay for a small decrease in area.

The ability to make such design transformations automatically was a key factor in enabling synthesis tools to pervade the microchip design process. Although performance is usually the top design priority, sometimes this type of move is necessary so that the entire design fits on a single chip. If the delay specification is not violated, such a move is desirable and acceptable.

1.4.2 The Technology Independent View — A Bit-Serial Full Adder Circuit

A crucial aspect of any optimization problem is the accuracy of the cost function used. In logic synthesis, there is no guarantee that truly minimizing some cost function based on area estimation would result in a truly minimum size for the fabricated chip. Logic synthesis tools have typically identified two types of cost models, each with their own assortment of synthesis tools. The first is called a **technology independent cost model**, in which the area cost estimate is taken as the number of logic symbols, called **literals** in a given set of logic formulas. The literal count is defined as the number of literals on the right hand side of the formulas in the given set. A simple example is $f = ac$, $c = ab' + a'b$, for which the literal count $2 + 4 = 6$.

The technology independent view has proven useful for improving the structure, or overall architecture of the final circuit, and can thus be regarded as a more *global* type off optimization. In contrast, the technology independent view has proven its worth in tuning the design by well motivated *local* optimizations.

Typically, the delay estimator in the technology independent cost model is the length of the longest dependency chain in the given set of formulas. In this view, the dependency chain is thought of as the "critical path" of the formulas, and its length as the "critical path delay". In the above example f depends on c and a, while c depends on b and a, so the length is 2. Note a and b are inputs in the sense they do not appear on the left hand side of any formula.

We demonstrate the basic ideas of technology independent logic synthesis with discussion of the bit-serial adder circuit of Figure 1.7. In this synchronous sequential circuit, there is combinational logic comprised of a 1-bit full adder, with a latch storing the value of the adder's carry output. By definition, a full adder adds two logic variables x_i and y_i and a carry input c_i, according to the following logic equations:

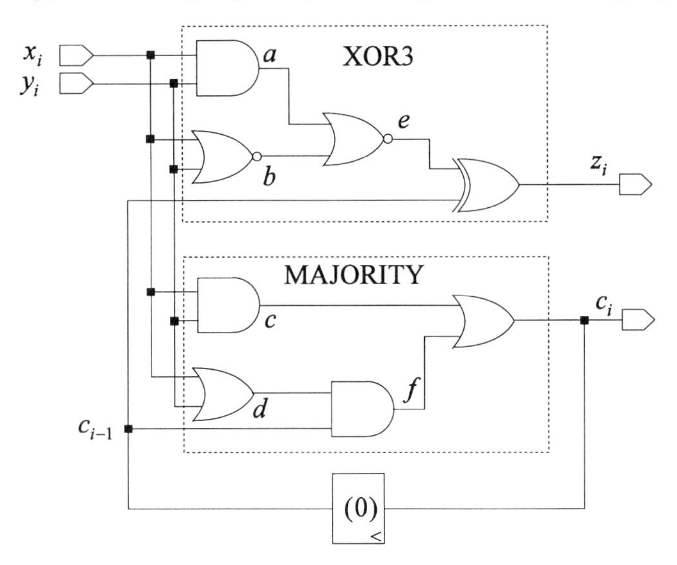

Figure 1.7: Bit-serial adder circuit.

$$
\begin{aligned}
z_i &= \text{XOR3}(x_i, y_i, c_{i-1}) &= \text{XOR2}(\text{XOR2}(x_i, y_i), c_{i-1}) \\
c_i &= \text{MAJORITY}(x_i, y_i, c_{i-1}) &= x_i y_i + c_{i-1}(x_i + y_i),
\end{aligned}
$$

where $\text{XOR}(x_i, y_i) = \overline{x}_i y_i + x_i \overline{y}_i = (xy + (x+y)')'$[2]. Here the i subscripts represent ticks on an implicit clock. In the figure, the XOR3 and MAJORITY subcircuits are enclosed in dashed boxes. This representation is typical of a specification, which might have been originally expressed in an HDL.

Often such a compact representation is inefficient in area or delay or both when mapped into Silicon, so a logic synthesis too might further decompose the XOR3 and MAJORITY subcircuits into 2-input logic gates, as shown in the figure, and in the following set of Boolean equations.

$$
\begin{aligned}
a &= x_i y_i, \\
b &= x_i' y_i', \\
e &= a' b', \\
z_i &= e c_{i-1}' + e' c_{i-1}, \\
c &= x_i y_i, \\
d &= x_i + y + i, \\
f &= d c_{i-1}, \\
c_i &= c + f.
\end{aligned}
$$

Here we count 18 literals on the right hand sides, and a longest dependency chain of 3, so in this example, the "critical path delay" is 3, corresponding to the length of any of the three critical paths $a - e - z_i$, $b - e - z_i$, or $d - f - c_i$.

Observe that each logic signal is associated with exactly one logic gate — for example, signal a is associated with the AND2 gate of the XOR3 subcircuit. Thus it is customary to refer to that gate as "gate a".

[2] Note that we use $\overline{x}$ and x' interchangeably to denote complement.

Note that this simple design is not optimal from a chip area viewpoint. For example, gate (formula) c may be deleted from the MAJORITY subcircuit if signal c of the formula $c_i = c+f$ is replaced by signal a, so that $c_i = a+f$. Note his optimization move would save 2 literals, but does not affect the critical path delay. Thus, from the technology independent view, the above formulas are not optimal, since area can be decreased without increasing delay.

1.4.3 The Technology Dependent View — Technology Mapping

In the technology dependent view, each formula is mapped into one or more logic gates in a pre-designed set of gates called a **technology library**. Gates in the library have a highly optimized, pre-defined path to Silicon, so that the area and delay parameters are known much more accurately. Algorithms for technology mapping are discussed in Chapter 13.

A typical result of technology mapping is shown by the following set of equations.

signal		formula		gate	transistors
g	$=$	$(x_i y_i)'$	$=$	NAND(x_i, y_i)	4,
a	$=$	g'	$=$	NOT(g)	2,
b	$=$	$(x_i + y_i)'$	$=$	NOR(x_i, y_i)	4,
e	$=$	$(a+b)'$	$=$	NOR(a, b)	4,
z_i	$=$	$(ec_{i-1} + ec_{i-1})'$	$=$	XOR(e, c_{i-1})	8,
d	$=$	b'	$=$	NOT(b)	2,
h	$=$	$(dc_{i-1})'$	$=$	NAND(d, c_{i-1})	4,
f	$=$	h'	$=$	NOT(h)	2,
j	$=$	$(a+f)'$	$=$	NOR(a, f)	4,
c_i	$=$	j'	$=$	NOT(j)	2.

Note that all the formula are negative in the sense of having an outer complementation. Thus the corresponding library gates have a transistor count of 2 for each literal in the formula (see Section 1.7 for a discussion of the XOR count). Note that 36 transistors are required, and if we divide this by 2, we get a count of 18 equivalent literals. This happens to be identical to the original literal count of 18. Note that although we did some optimizations, we also added 4 inverters.

These formulas correspond to the mapped circuit of Figure 1.8. The logic sharing between the XOR3 subcircuit and the MAJORITY subcircuit is one of the primary objectives of logic synthesis.

In the mapped circuit, the critical path delay[3] of the circuit can still be estimated by the greatest number of logic levels on any path between any input and any output (not counting the pentagonal input and output buffers). In the mapped circuit, this *critical path length* is 6 corresponding to the length of the critical path $x_i - b - g - e - h - f - j - c_i$. Note that the critical path length is measured through the combinational logic, regarding the latches as an open circuit.

Note that technology dependent delay estimate is 6, which is twice the estimate obtained in the technology independent view. Typically, the technology dependent

[3]Typically, technology mapping tools estimate critical path delay by more sophisticated measures, including fanout load effects and/or specific pin-to-pin delays in terms of gate size parameters.

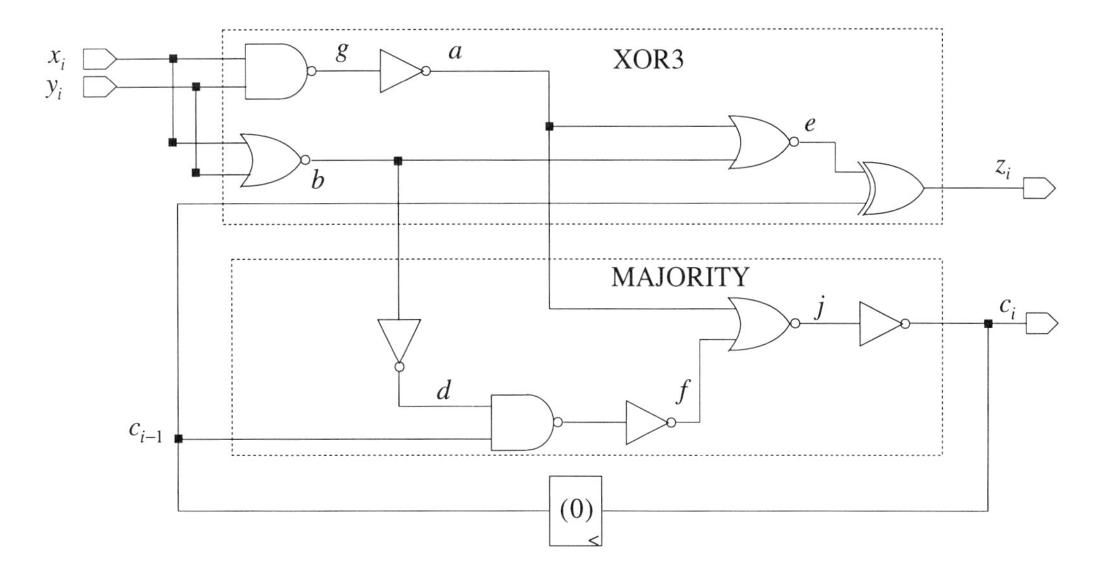

Figure 1.8: Bit-serial adder circuit after technology mapping.

delay estimate is more accurate, which indicates a relative strength of the technology dependent view.

1.4.4 Testing — Is What I Fabricated What I Wanted?

The simple example discussed above may also be use to illustrate some basic concepts of testing. The most common fault model in VLSI test is the so-called *stuck-at fault*. In this model, *two* versions of the circuit under test are simultaneously considered: the specified circuit, as shown in Figure 1.7, and a faulted circuit, for example that of Figure 1.9. In the faulted circuit, one specified logic signal, in this case a of the

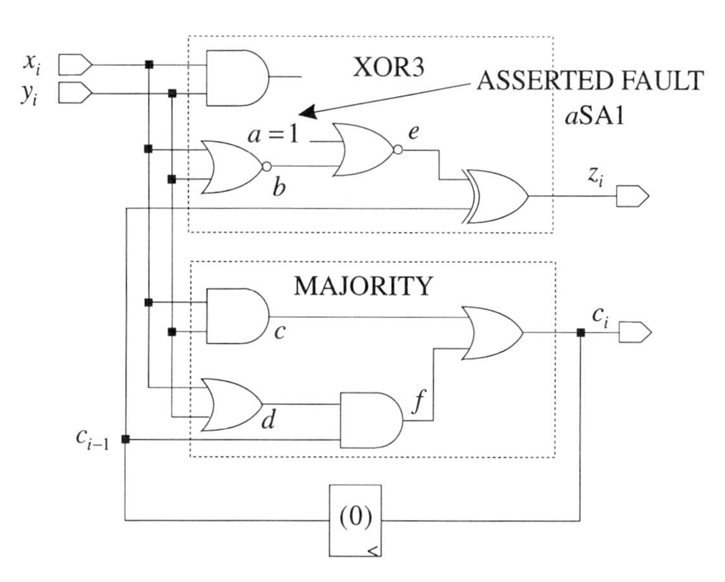

Figure 1.9: Bit-serial adder circuit with fault aSA1 asserted.

XOR3 subcircuit, is connected to logical 1, as shown in the figure. This simulates the possibility that on the manufactured chip, the metal wire implementing this logical connection has, through some manufacturing fault, come in contact with the wire supplying the high voltage power supply. The fault thus simulated is referred to as aSA1, a mnemonic for "a stuck-at 1".

Similarly, if a were inadvertently connected to logical 0, we would be modeling an "a stuck-at 0" fault, denoted aSA0. Typically a single test

We shall use G and F superscripts to distinguish between the "Good Machine" (that is, the specified circuit) and the "Fault Machine" (the same circuit, but with the fault asserted, as in Figure 1.9). Thus the set of tests for an asserted fault can be written as follows. We first note that z_i is the only circuit output affected by the asserted fault, and that the input vector (x_i, y_i) is a test for aSA1 if and only if logic expression $T^F(x_i, y_i) = 1$, where

$$\begin{aligned} T^F &= z_i^G \oplus z_i^F \\ &= (c_{i-1}^G \oplus e^G) \oplus (c_{i-1}^F \oplus e^F) \\ &= (c_{i-1}^G \overline{e}^G + \overline{c}_{i-1}^G e^G) \oplus (c_{i-1}^F \overline{e}^F + \overline{c}_{i-1}^F e^F). \end{aligned} \tag{1.1}$$

We then note since there is no logic path from the fault to c_{i-1}, we have $c_{i-1}^G = c_{i-1}^F = c_{i-1}$, so that

$$\begin{aligned} T^F &= (c_{i-1}\overline{e}^G + \overline{c}_{i-1}e^G)\overline{(c_{i-1}\overline{e}^F + \overline{c}_{i-1}e^F)} + \overline{(c_{i-1}\overline{e}^G + \overline{c}_{i-1}e^G)}(c_{i-1}\overline{e}^F + \overline{c}_{i-1}e^F) \\ &= (c_{i-1}\overline{e}^G + \overline{c}_{i-1}e^G)(c_{i-1}e^F + \overline{c}_{i-1}\overline{e}^F) + (c_{i-1}e^G + \overline{c}_{i-1}\overline{e}^G)(c_{i-1}\overline{e}^F + \overline{c}_{i-1}e^F) \\ &= c_{i-1}\overline{e}^G e^F + \overline{c}_{i-1}e^G\overline{e}^F + c_{i-1}e^G\overline{e}^F + \overline{c}_{i-1}\overline{e}^G e^F \\ &= (c_{i-1} + \overline{c}_{i-1})(e^G \oplus e^F) \\ &= (e^G \oplus e^F). \end{aligned} \tag{1.2}$$

This result reflects the fact that an XOR gate is "always sensitized". In this case, this means that the existence of a test for the fault aSA1 does not depend on the value of the carry in input c_{i-1}.

To get to the second line of the equation, we used the fact that $\overline{(c_{i-1}\overline{e}^F + \overline{c}_{i-1}e^F)} = (c_{i-1}\overline{e}^F + \overline{c}_{i-1}e^F)$, and to get the last identity, we used $(c_{i-1} + \overline{c}_{i-1}) = 1$. (The basis for all such identities is given in Chapter 3, which deals with Boolean Algebras.)

Each specific stuck fault creates a new logic function $e^F(x_i, y_i, c_{i-1})$, but the good machine is invariant, so for any asserted fault we always have $e^G = x_i\overline{y}_i + \overline{x}_i y_i$. A **test vector** for a given fault F is a triple of inputs $t = (t_1, t_2, t_3) = (x_i, y_i, c_{i-1})$ such that $T^F(x_i, y_i, c_{i-1}) = 1$. For any such input, we know that the good machine and fault machine will have at least 1 differing output value. We say that the test vector t **covers** the fault F.

Typically a single test vector t will cover many stuck faults. In the example of Figure 1.9, it can be verified that each of the vectors $(0, 1, 0)$ and $(1, 0, 0)$ cover both the faults aSA1 and bSA1. The key point in efficient testing is to find a small set of test vectors which cover every possible stuck fault in a given circuit.

Of course there are many other fault models which can occur during chip design and fabrication. For example, multiple stuck faults could conceivably occur, as could stuck open faults, bridging faults (shorts), etc. However, it has been observed empirically that if a test set that covers all single stuck faults is applied to a given circuit, it is highly likely that they will cover not only the single stuck faults, but a wide variety of other faults as well. Such topics are discussed in detail in Chapter 12.

Now let us consider how to test for the single stuck fault aSA1. In the Good Machine, we have $e^G = x_i\overline{y}_i + \overline{x}_i y_i$ and in the Fault Machine (aSA1) we have $a^F = 1 \Rightarrow e^F = 0$. (Note a 1 value on the NOR gate input is controlling in the sense that it causes the output e^F to be 0, independent of the value of b.) Hence the set of tests for aSA1 is just the set of $(x_i y_i)$ pairs $\{(0,1),(1,0)\}$ that satisfy

$$T^{a\mathrm{SA1}}(x_i, y_i) = (e^G \oplus e^F) = (e^G \oplus 0) = e^G(x, y) = x_i \oplus y_i = 1.$$

Note that for either of these inputs, we have $a^G = 0$, which is a necessary condition for testability of this fault, since $a^F = 1$.

It may similarly be verified that for the fault aSA0, we have $a^F = 0 \Rightarrow e^F = \overline{b}^F = (x_i + y_i)$. Again $e^G = x_i\overline{y}_i + \overline{x}_i y_i$, so that the set of tests for aSA0 is just the singleton set of $(x_i y_i)$ pairs $\{(1,1)\}$ that satisfy

$$T^{a\mathrm{SA0}}(x_i, y_i) = (e^G \oplus e^F) = (x_i\overline{y}_i + \overline{x}_i y_i \oplus (x_i + y_i)) = x_i y_i = 1.$$

Again, note that the necessary condition $x_i y_i = 1 \Rightarrow a^G = 1$, is satisfied by this test.

Finally, note that the test vectors derived for the two stuck faults are valid independent of the logic value of the carry in signal c_{i-1}. The symbol (0) in the latch of Figures 1.7, 1.8, and 1.9 is meant to indicate that the initial value on the latch is logical 0. Since each of the three test vectors identified above are independent of c_{i-1}, they are guaranteed to work for $c_{i-1} = 0$.

Although this circuit is a sequential circuit, the specific faults discussed above propagate immediately to the output, where they are observable. In many more realistic cases, the fault may take two or more input vectors on successive clock cycles to reach the output. For example, suppose that a fault, say dSA1, occured in the MAJORITY subcircuit, and that the carryout was not an output (that is, was not observable). In this case, a second and third input pair would be necessary to propagate the fault to the output z_i, which we demonstrate as follows.

Initially, $c_{i-1} = c_0 = 0$. Thus on the first clock cycle, $c_1^F = (x_1 y_1) = c_1^G$, since the initial latch value $c_0 = 0$ is controlling for the output OR gate of the majority subcircuit. Thus all that can be done on the first cycle is to generate a carry-out, which requires $x_1 = y_1 = 1$ so the first input pair must be 11.

On the second cycle, we want $c_2^G \neq c_2^F =$, and with dSA1 and $c_1 = 1$, which implies $c_2^F = 1$. This will be true if and only if $c_2^G \oplus c_2^F = (x_2 + y_2) \oplus 1 = \overline{x}_2\overline{y}_2 = 1$. Thus, the second input pair must be 00, which implies that $c_2^G = 0$.

On the third cycle, we have a carry in of $c_2^G = 0$ in the good machine, but $c_2^F = 1$ (in the fault machine). In this case, $T_3^F = e_3^G \oplus e_3^F = (x_3 \oplus y_3) \oplus \overline{x_3 \oplus y_3} = 1$. That is, T_3^F is the tautology, or true for all assignments of its inputs. It thus follows that any input pair will suffice as the 3rd pair in a 3-pair sequence. Thus the tests for dSA1 are $\{(11,00,00),(11,00,01),(11,00,10),(11,00,11)\}$.

1.4.5 Graph Models and Finite State Machines

We can regard this circuit as a so-called FSM (Finite State Machine — see Chapter 7). Since the circuit has only 1 latch, the machine has $2^1 = 2$ states, depending on whether a 1 or a 0 is stored in the latch as the value of c_{i-1}. The associated FSM is shown in Figure 1.10

By convention the latch initially provides a carry-in of 0, so the initial state of the machine (designated by the thick arrow) is the left state. From this state, if and only if both of the two inputs x_i and y_i are 1 does the machine make $c_i = 1$, thus causing a transition to the right state. This provides a carry-in of $c_{i-1} = 1$ on the next clock cycle. Similarly, if the machine is in the $c_{i-1} = 1$ state, it returns to the initial state if

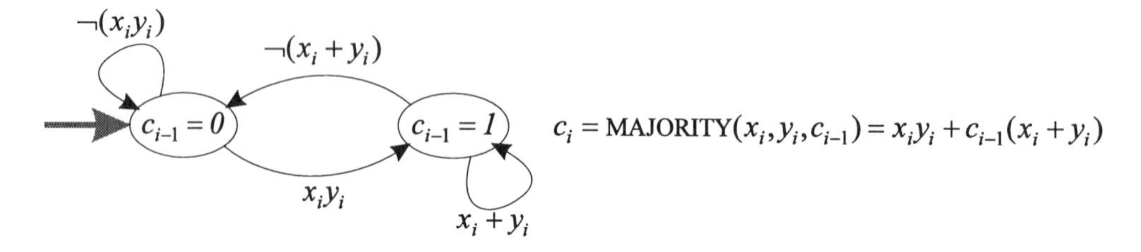

Figure 1.10: Finite State Machine for Majority Circuit.

and only if both of the two inputs x_i and y_i are 0 (else we would again have $c_i = 1$).

Tools for optimizing FSMs, and for testing whether two FSMs are equivalent, are discussed in Chapter 7. Also discussed there are related topics such as the theory and design of Finite Automata.

We have just seen how an FSM can be specified as labeled directed graph. Throughout the sequel we shall have extensive recourse to the theory of such graphs. In the interest of self-containment, we briefly review some of this theory here. Graph theory and models are among the most fundamental tools of discrete mathematics, and are featured in texts introductory to that subject [190, 241]. Consequently, most significant CAD algorithms are based on graph models. Graph models are key in the definitions of Finite State Machines (often abbreviated FSM) and Finite Automata (Deterministic Finite Automata are often abbreviated DFA). State Machines and Finite Automata are, in turn the most fundamental models in the specification, design, synthesis, and verification of sequential digital systems. Indeed, any sequential circuit can be modeled as a single FSM, consisting of combinational logic and latches (also called registers, or flip flops).

We develop in this section enough background in graphs to enable a reasonably self-contained treatment of Finite Automata and State Machines. We shall see that graph models play a predominant role in the discourse.

We refer the reader to [99, 241, 69], for a more comprehensive treatment of relevant aspects of Graph Theory. Graphs provide a visual and intuitive framework for formulating design problems. Further, it is relatively easy to prove theorems in this framework, and there is a vast literature available on graph theory. Here we limit our treatment to a discussion of (1) the decomposition of a graph into components based on connectivity, and (2) algorithms for graph traversal based on depth first search and breadth first search, which is fundamental to the synthesis and verification of logic circuits.

Depth first search is used in many recursive situations, for example, in the "collapsing" phase of multilevel logic synthesis (to be discussed below). Breadth first search is key to many shortest and longest path algorithms and is used in many

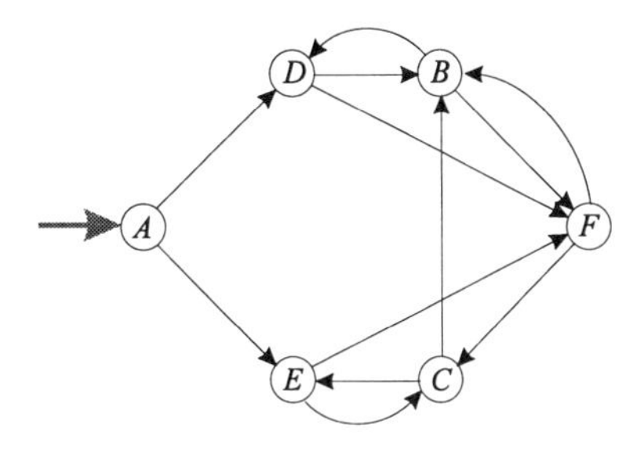

Figure 1.11: A simple directed graph.

aspects of FSM synthesis and verification, as well as in timing optimization of combinational circuits. These search algorithms will be discussed in the context of FSMs in Chapter 7. In this chapter, we shall limit our attention to a single application: finding the longest path in a graph. This problem corresponds directly to that of computing the delay of a logic circuit.

A **graph** is defined as a pair $G = (V_G, E_G)$, where V_G is a set of **vertices** and E_G is a set of edges, sometimes called arcs. An **edge** $e = (u, v)$ is a member of the edge set E_G of graph G, so of course $u \in V_G$, $v \in V_G$. Note that an edge denoted (u, v) is an ordered set if the edges have directivity, and an unordered set $\{u, v\}$ if they do not. Vertices u and v are called the **tail vertex** and **head vertex** (respectively) of the edge (u, v) from u to v. A graph with undirected edges is called an **undirected graph**, or just a graph. If all its edges are directed, a graph is called a **digraph**, and if it has both directed and undirected edges it is called a **mixed graph**.

An example of a directed graph $G = (V_G, E_G)$ is given in Figure 1.11. Here

$$\begin{aligned} V_G &= \{A, B, C, D, E, F, G\} \\ E_G &= \{(A, E), (A, D), (B, F), (B, D), (C, E), (C, B), \\ &\quad (D, F), (D, B), (E, C), (E, F), (F, B), (F, C)\}. \end{aligned}$$

In this graph, the broad arrow is not an edge, but an indicator of where to start a graph search.

A **path** is a sequence of connected edges, and a simple path is one in which no vertex appears either twice as a head vertex or twice a tail vertex of the edges of the path (that is, a simple path doesn't cross itself). A **cycle** is a closed path. A simple cycle is a cycle in which every vertex appears exactly once as both a head vertex and tail vertex. In Figure 1.11, $((A, D), (D, F))$ is a simple length-2 path from A to F, and $((A, D), (D, B), (B, F))$ is a length-3 path from A to F. Similarly, $((B, F), (F, C), (C, B))$ is a length-3 cycle from B to F and back to B. When the context is clear, we will denote the path $((A, D), (D, B), (B, F))$ by the simpler (A, D, B, F). The path (A, D, B, D, F) is an example of a non-simple path from A to F.

If a digraph (directed graph) $G = (V, E)$ has no cycles it is called a **DAG (Directed Acyclic Graph)**.

1.4.6 Successors and Predecessors

It is often useful to denote the existence of an edge $(u, v) \in E$ in a digraph $G = (V, E)$ by $u \rightarrow v$, and the existence of a path from u to v in G by $u \xrightarrow{*} v$. Here the $*$ is used to indicate a transitive, rather than direct, relation between u and v. We shall use the notation $\xrightarrow{*} \subseteq V \times V$ to denote the set of pairs of vertices (u, v) such that the graph contains a path from u to v[4].

In the statement of algorithms and procedures it is useful to have convenient notation for certain vertex subsets

Definition 1.4.1 *In a directed graph* $G = (V, E)$, *we define*

$$u_0 \rightarrow = \{v | (u_0, v) \in E\}.$$

as the set of **successor** *vertices of* u_0. *Symmetrically, we define*

$$\rightarrow v_0 = \{u | (u, v_0) \in E\}.$$

as the set of **predecessor** *vertices of* v_0.

Similarly, if $u_0 \xrightarrow{*} v_1$ in a directed graph $G = (V, E)$, then G contains a path from node u_0 to node v_1. Thus we say that $v_1 \in u_0 \xrightarrow{*}$, that is, v_1 is a **transitive successor** of u_0 and $u_0 \in \xrightarrow{*} v_1$, that is, u_0 is a **transitive predecessor** of v_1. For example, for the graph of Figure 1.11, $A \rightarrow = \{D, E\}$, and $A \xrightarrow{*} = \{B, C, D, E, F\}$ and $\xrightarrow{*} F = \{A, B, C, D, E, F\}$. The extended edge relation $E^*(u, v)$ thus derived from a given edge relation $E(u, v)$ is called the **transitive closure** of E.

As in Figure 1.10, a graph may have a label $x_{u,v}$ associated with each edge, In this case the graph is called an edge labeled graph, and is denoted $G = (V, E, X)$, where the label set is in one to one correspondence with the edge relation E. As discussed above, we denote an edge from u to v by $u \rightarrow v$, and if u, v is labeled with a symbol x, we denote it by $u \xrightarrow{x} v$, and say that v is an x-successor of v. As we shall see in Section 7.6, edge labeled graphs are essential to the discussion of finite state machines and automata. A graph may also have a label w_v associated with each vertex, in which case it is denoted $G = (V, W, E)$. Vertex-labeled graphs are essential to the discussion of Moore machines.

In the context of graphs (V, E) associated with logic circuits, if a vertex $g \in V$ represents a logic gate, then its successors are often called the **fanouts** of g and its predecessors are often called the **fanins** of g. In the circuit of Figure 1.12, the fanins of gate 9 are gates 6 and 8, and the fanouts of buffer gate 2 are gates 4, 5, and 6.

A vertex v for which $\rightarrow v = \emptyset$ (that is, v has no predecessors) is called a **source** vertex. A vertex u for which $u \rightarrow = \emptyset$ is called a **sink** vertex.

1.5 Graph Algorithms and Complexity

1.5.1 Complexity

In most cases CAD tools are developed by modeling a physical problem by some graph analysis or optimization problems. There shall be numerous examples of such

[4]Often $\xrightarrow{*} \subseteq V \times V$ is called the **transitive closure** of the edge relation $E = \rightarrow \subseteq V \times V$.

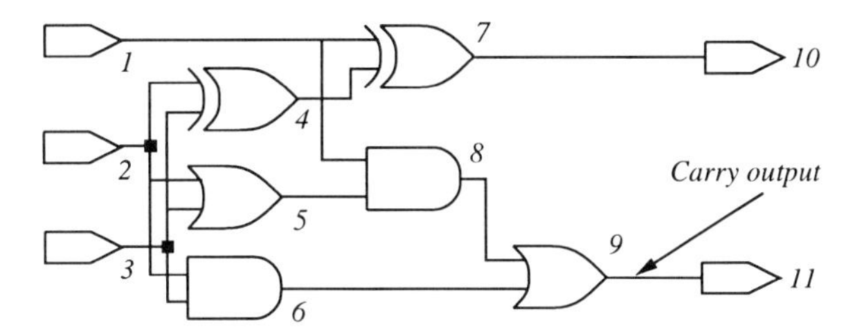

Figure 1.12: Logic Graph of 1-bit full adder. The gate outputs are the vertices of the graph and the nets connecting gate outputs to gate inputs are the edges of the graph.

a modeling process throughout the sequel. If one has to design robust and efficient CAD tools, it is useful to learn how to compute, or at least estimate, the way the running time of specified graph algorithms depends on the size of the data input. VLSI CAD tools focus on problems of very large size. Let us denote problem size by n, and running time by a function $T(n)$. Then we can say that programs for which $T(n) = c_0 + c_1 n$, (linear functions) or depends linearly on n or at worst $T(n) = c_0 + c_1 n \log(n)$(log-linear functions) are tractable. Running times which have a quadratic or stronger dependence on n are usually considered intractable.

Unfortunately it is prohibitively difficult to compute $T(n)$ exactly for most algorithms of interest. Thus we are forced to develop a notion of **asymptotic complexity** of algorithms, which applies when the exact form of $T(n)$ is not known. The idea is to show that there exist known functions $g(n)$ for which either $T(n) \leq g(n)$ or $T(n) \leq g(n)$ holds for all sufficiently large n.

Although both upper and lower bounds for best, average, and worst case estimates of $T(n)$ are all of interest, we shall focus in general on "tight" asymptotic *upper bounds* of *worst case* behavior. This has the advantage of giving a guarantee that there is no input data that will cause the algorithm run longer than the estimate. This relieves the user of the algorithm from the responsibility of making educated guesses about running time. If the bounds are tight enough, the upper bound will provide a good estimate of actual running time as well. As we shall see, this is possible for many, but not all, algorithms of interest.

Further, for some algorithms, the worst case corresponds to fairly typical behavior. For example, once we define Boolean functions in the Chapter on Boolean Algebras, we shall see that for some data representations, the problem of checking to see if the function is the constant 0 function takes time exponential in the number of inputs. We shall see that this question needs to be asked quite frequently in logic synthesis and verification.

We organize this section as follows. First we treat some simple examples for which good estimates of the function $T(n)$ can be computed in closed form. Then we formalize the notion of asymptotic complexity, and develop some notation that can be used throughout the book to characterize the complexity of the numerous presented algorithms.

Line	Procedure SET_CARTESIAN_PRODUCT(G, H) {	Ops	Times/call	
			Best	Worst
1	$m = \|G\|;\ n = \|H\|$	c_1	1	1
2	$P = \emptyset$	c_2	1	1
3	**for**$(i = 1, 2, \ldots, m)\{$	$c_3 m$	1	1
4	**for**$(j = 1, 2, \ldots, n)\{$	$c_4 n$	m	m
5	$P = P \cup (G_i \cap H_j)$	c_5	mn	mnq
	}			
	}			
6	**return**(P)	c_6	1	1
	}			

INPUTS: G and H are sets.

SET_CARTESIAN_PRODUCT gives the product $P = G \times H$ of sets G and H. The elements e of G and H are sets of cardinality $|e|$, where $1 \leq |e| \leq q, \quad \forall e \in G \cup H$.

Output P has at most $mn = |G| \times |H|$ elements, each of which is a set of q or fewer elements.

Figure 1.13: Procedure for Intersecting 2 sets of sets.

1.5.2 Computing the Product of Sets of Sets

We begin by studying the complexity of Procedure SET_CARTESIAN_PRODUCT of Figure 1.13, which takes as inputs the sets G and H, and computes the product $P = G \times H$ of sets G and H. The elements e of G and H are sets of cardinality $|e|$, where $1 \leq |e| \leq q, \quad \forall e \in G \cup H$. By definition each element p in the product P is formed from the set intersection $e_g \cap e_H$, where $e_G \in G$ and $e_H \in H$. Thus $|p| \leq q$. Since it is possible that $e_g \cap e_H = \emptyset$ (meaning e_G and e_H have no elements in common) and P, we know that has at most $mn = |G| \times |H|$ elements, each of which is a set of q or fewer elements.

We can compute $T(n)$ for this algorithm for the best and worst cases of input data. A way to organize this computation is illustrated in the "step table" given at the right of the algorithm. Here the inputs, G and H, are sets of q-element sets. That is, each member of G or H is a set of cardinality q. The procedure simply looks at each element (G_i, H_j) of the Cartesian product of G and H, intersects G_i and H_j, and adds the resulting element $G_i \cap H_j$ to the product set P.

It is instructive to consider how the entries of the "step table" are obtained. First, consider Line 2 of SET_CARTESIAN_PRODUCT. This statement initializes a set P of no elements. The exact number of machine instructions required for this depends on implementation details, but is clearly independent of m, n, or q. Further, it is clear that this statement is executed exactly once per call to SET_CARTESIAN_PRODUCT. Thus, referring to the table entries, we see that the "frequency" of executing this statement is 1, and the operations count per execution is some constant c_2. Looking at the **for** loop structure of Figure 1.13, we see that the Line 3 is executed exactly m times per call, and the operations count per execution is some constant c_3. Line 4 is similar.

Line 5 shows a distinction between best and worst case behavior. In the best case, every element e in the sets G and H is a set with just one element, whereas in the worst case, every e has q elements. Since Line 5 is executed mn times per call, Line 5 contributes c_5mn operations in the best case and c_5mnq operations in the worst case. With this approach, we can estimate the running time $T(n)$ in the best case by

$$T(m,n,q) = c_1 + c_2 + c_3m + c_4mn + c_5mn + c_6,$$

and in the worst case

$$T(m,n,q) = c_1 + c_2 + c_3m + c_4mn + c_5mnq + c_6,$$

1.5.3 Longest Paths

We now treat a practical algorithm that is often used to compute the critical path delay of a VLSI circuit. We shall first discuss the qualitative behavior of the algorithm, and then study its running time.

Procedure LONGEST_PATH of Figure 1.14 is applicable only to directed acyclic graphs. The operations counts on the right of the figure will be discussed later. The algorithm is based on traversing the graph from nodes in I, meanwhile iteratively increasing a lower bound on the length λ_v of a longest path from any node $u \in I$ to each node $v \in V$. At the end of the traversal, the lower bound has become exact for all nodes reachable by paths from I.

An auxiliary array array D_v is defined at all times to be the number *incoming edges* to node $v \in V$ which have not yet been traversed. Once D_v becomes 0, it can be shown that we have, implicitly, traversed all directed paths to v, and at this point the lower bound has become exact. The notation $\rightarrow v$, stands for the set of nodes in the fanin (that is, its predecessors — see Section 1.4.6 for the formal definition of fanin and fanout) of v, and is used to initialize D_v.

LONGEST_PATH accepts as input an directed acyclic graph $G = (V, E)$ with edge lengths L, as well as a set of "initial" nodes (often these nodes have no incoming edges). When this algorithm is used to monitor project completion, the 4th argument, *spec*, denotes a specified limit on how long the longest path should be. Returned is a longest path $\pi = (v_1, \ldots, v_n)$ and an array of longest path lengths λ_v, $\forall v \in V$. The path π satisfies $v_1 \in I$, and $(v_i, v_{i+1}) \in E, \quad i = 1, \ldots, n$.

Once all incoming edges are traversed by LONGEST_PATH, the lower bound becomes exact, and λ_v has converged to the longest path length. Each time this occurs (Line 9), the relevant node is appended to Q by the function QUEUE. On each pass through the **while** loop of Line 4, the node at the head of the queue is taken off and assigned as the currently active node (Line 5). Updates to λ_a and D_a are made in the **foreach** loop of Line 6.

In Line 11, the length λ^* of the longest path to any node is computed, and used in Line 12 to select one node v^* for which $\lambda_{v^*} = \lambda^*$. The selected node is then passed in Line 13 as an argument to Subprocedure BACK_TRACE, which actually traces out a path of maximum length, based on the fact that the array λ of maximum path lengths from any source node to all $v \in V$ is now known. The 6th argument, with value (*spec* $- \lambda^*$), denotes the amount by which the path is "too long".

Line	**Procedure** LONGEST_PATH$(V, E, L, I, spec)$ {	Ops	Times/call	
			Best	Worst
0	$n = \|V\|;\ m = \|E\|;\ q = \|I\|$	c_0	1	1
	for$(v \in V)$ {			
1	$\lambda(v) = 0$	c_1	n	n
2	$D_v = \|{\to}v\|$	c_2	n	n
	}			
3	$Q = I$	$c_3 q$	1	1
4	**while**$(Q \neq \emptyset)$ {	c_4	n	n
5	$v =$DEQUEUE(Q)	c_5	n	n
6	**foreach**$(a \in v{\to})$ {	c_6	m	m
7	$\lambda_a = \max(\lambda_a, (\lambda_v + L_{v,a}))$	c_7	m	m
8	$D_a = D_a - 1$	c_8	m	m
9	**if**$(D_a = 0)$ QUEUE(Q, a) }	c_9	m	m
10	}	c_{10}	n	n
11	$\lambda^* = \max_{v \in V}\{\lambda_v\}$	$c_{11} n$	1	1
12	$v^* =$SELECT1(V, λ^*)	$c_{12} n$	1	1
13	$\pi =$BACK_TRACE$(V, E, L, \lambda, v^*, (spec - \lambda^*))$	c_{13}	1	1
14	**return**(π, λ) }	c_{14}	1	1

Input: A directed graph $G = (V, E)$, an edge length vector, L.

Output: π, a longest weighted path from any source vertex to any sink vertex, and the length, λ_{v^*} of π.

Figure 1.14:

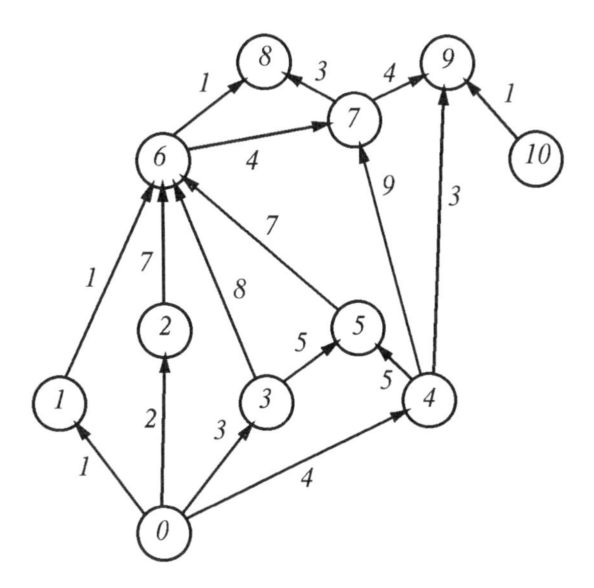

Figure 1.15: A weighted directed acyclic graph.

Subprocedure BACK_TRACE (discussed below) accepts as inputs the graph data V, E, L, as well as v^* and the array λ. The output of BACK_TRACE is called a **critical path**, that is, a longest directed path. Implementations of BACK_TRACE vary, but in our version, discussed below, only one path is traced.

The actions of LONGEST_PATH are illustrated in the data trace given in Table 1.1 for the example of Figure 1.15. We assume here that the **foreach** loop of Line 6 processes the fanout $v\!\!\rightarrow$ of the active node v in *ascending* lexicographical order, the final values of $slack_v$ produced by Procedureback_trace are:

In this example, 11 passes through the **while** loop are required. At the **foreach** loop (Line 6), the quantities (v, λ_v) and the fanout list $v\!\!\rightarrow$ are given in the table. At the end of the each pass (Line 10), the values of (λ_a, D_a) are given for each node $a \in V$.

On subsequent passes only the quantities which actually change are given. Thus the sparsity of the table is representative of the efficiency of the algorithm.

1.5.4 Backtracing

Especially when applied to logic graphs, the complete picture of delay evaluation often involves contains 2 separate arrays of timing data, with one entry in both arrays for each node in the graph. The first array is sometimes called *ArrivalTime*. Assuming that the arrival times of the input nodes I are 0, this array corresponds directly to the "longest path length" array $\lambda(v)$ computed by LONGEST_PATH. The second array is often called called *RequiredTime*. This is an external specification given for designated nodes of the graph, as in logic circuits or PERT charts[5].

These concepts will be elaborated in the Chapter on technology mapping, since sophisticated delay analysis is usually done only in that setting.

[5]PERT is a project management technique, and stands for Performance Evaluation and Review Technique

Line	v, λ^* 6	$v\rightarrow$ 6	(λ_a, D_a) 10										
k			0	1	2	3	4	5	6	7	8	9	10
1	0/0	1,2,3,4	0/0	1/0	2/0	3/0	4/0	0/2	0/4	0/2	0/2	0/3	0/3
2	10/0	9										1/2	
3	1/1	6							2/3				
4	2/2	6							9/2				
5	3/3	5,6						8/1	11/1				
6	4/4	5,7,9						9/0		13/1		7/1	
7	5/9	6							16/0				
8	6/16	7,8								20/0	17/1		
9	7/20	8,9									23/0	24/0	
10	8/23	$\emptyset$											
11	9/24	$\emptyset$											
	final	λ:	0	1	2	3	4	9	16	20	23	24	0

Table 1.1: Data trace for Procedure LONGEST_PATH

In Procedure LONGEST_PATH, only one *RequiredTime* value is specified, namely, the upper limit *spec* on the length of the longest path. Thus $RequiredTime_v = \lambda^*$, and, implicitly, $RequiredTime_v = \infty, \quad \forall v \neq v^*$.

In Procedureback_trace, instead of computing $RequiredTime_v$ for all nodes, we compute the "slack" at each node $v \in V$, denoted $slack_v$. Roughly speaking, the slack at a node v is the difference between $RequiredTime_v$ and the $ArrivalTime_v$. Negative or zero slacks indicate criticality. Formally, we give the following definition.

Definition 1.5.1 *The* **slack** *of an edge (a, v) is the slack of v plus the difference between the length of the longest path to v and the longest path to v through (a, v). In formula*

$$slack_{a,v} \quad = \quad slack_v + (\lambda_v - (\lambda_a + L_{a,v})).$$

Here λ_v is the length of the longest path to v, and $(\lambda_a + L_{a,v})$ is the length of the longest path to v that passes through the edge (a, v). The slack of a node u is be defined as the minimum of its fanout edge slacks, so

$$slack_a \quad = \quad \min_{v \in a\rightarrow} slack_v.$$

In Procedureback_trace, $slack_{v^*}$ is initialized to the value $(spec - \lambda^*)$, when called by LONGEST_PATH. The slack values of all nodes other than v^* are initialized to λ^*, and then selectively updated by the formula

$$slack_a \quad = \quad slack_v + (\lambda_v - (\lambda_a + L_{a,v})). \tag{1.3}$$

We can use this simpler formula when, as in Procedureback_trace, we trace only a single critical path.

Procedureback_trace is quite similar in operation to Procedure LONGEST_PATH. Again a FIFO priority QUEUE Q is employed. The main difference is that for each active node v, as soon as a new 0-slack node a is encountered in the backward traversal of the graph, a is put on the end of Q and the **foreach** loop is exited by the **break** statement.

```
Line   Procedure BACK_TRACE(V, E, L, λ, v*, slack*) {
 0         foreach(v ∈ V) slack_v = λ_v*
           slack_v* = slack*
           π = (v*)
           QUEUE(Q, v*)
 1         while(Q ≠ ∅) {
               v =DEQUEUE(Q)
 2             foreach(a ∈ →v) {
 3                 slack_v = slack_v + (λ_v − λ_a − L_a,v)
                   if(slack_a = 0) {
 4                     QUEUE(Q, a)
 5                     π = (a, π)
 6                     break
                   }
               }
           }
           return(π, slack)
       }
```

Input: A directed graph $G = (V, E)$, an edge length vector, L.

Output: π, a longest weighted path from any source vertex to any sink vertex, and the length, λ_{v^*} of *pi*.

Figure 1.16:

Note that the computed critical path always propagates to the first encountered critical node in the fanin. Consequently, if the input parameter $slack^*$ comes with value 0, then $slack_v$ will always have value 0 in the above equation. Note further that Procedureback_trace has a **break** statement at Line 6. This means that a soon as a new critical node is encountered, we break out of the enclosing **foreach** loop (Line 2), and then start a new pass through the **while** loop of Line 1.

Note nodes which do not lie on a critical path may not be updated by this formula, and will still have their initialized slack values (of λ^*) after completion of Procedure-back_trace. Also note that the final values of $slack_v$ will depend on the order in which the nodes in the fanin $\rightarrow v$ are processed in Line 2.

An edge (a, v) is **critical** if it connects two nodes a and v with slack value 0. Consider the example of Figure 1.15, for which we had $v^* = 9$, and $\lambda^* = 24$. Assuming that the **foreach** loop of Line 2 processes the fanin of v (that is, its predecessors — see Section 1.4.6 for the formal definition of fanin and fanout) in *ascending* lexicographical order, the final values of $slack_v$ produced by Procedureback_trace are:

v	10	9	8	7	6	5	4	3	2	1	0
$slack_v$	24	0	24	0	0	0	0	5	7	14	0

Note that the fanin nodes $4, 7, 10$ of node 9 are processed first, so $a = 4$ the first time Line 3 is executed, and at that point its $slack_4 = 0 + (24 - 4 - 3) = 17 > 0$, so Line 4 is not executed on this path. However, on the second pass through Line 3, we get $slack_7 = 0 + (24 - 20 - 4) = 0$. Thus node 7 is identified as critical and put on both the path π and the queue Q. The ensuing **break** statement breaks the **foreach** loop, so node 10 is not processed, and retains its initial slack setting of $slack_{10} = 24$.

On the next iteration, 7 is active, and $slack_4$ is further reduced to $0+(20-4-9) = 7$, and then node 6 is identified as critical, and so on. At completion, the nodes on the critical path are

$$\pi = (0, 4, 5, 6, 7, 9),$$

and are precisely those for which $slack_v = 0$. Only non-critical nodes $1, 2, 3, 10$ have had their slack adjusted downward from their initial values of 24.

1.5.5 Complexity of Computing the Longest Path

We now discuss the complexity of Procedure LONGEST_PATH. The key point in the design of this algorithm is that every edge is traversed exactly once. That is, assuming that there is actually a path from a node in I to every node $v \in V$, it must happen that every node is put onto the queue exactly once, and taken off exactly once. Further, every fanout edge of every v is thus processed in Lines 7, 8, and 9 exactly once.

It is somewhat tricky to analyze the number of operations expended in Line 6. This statement is executed n times, and each time it executes, it performs $c_6|v\rightarrow|$ operations. Summing over $v \in V$, this is $(c_6/n)\sum_v |v\rightarrow| = c_6(m/n)$ operations for each of the n passes through Line 6, or $c_6 m$ operations total. Note this result obtains no matter what topology the specified graph has. Thus, excluding internal operations in the call to Subprocedure BACK_TRACE,the running time for this algorithm can be written

$$T(m, n, q) = (c_0+c_{13}+c_{14})+c_3 q+(c_1+c_2+c_4+c_5+c_{10}+c_{11}+c_{12})n+(c_6+c_7+c_8+c_9)m.$$

The best case running time for this algorithm depends on what happens at Line 6. Assuming that the graph is connected, the best case occurs when each gate has only one fanout[6]. Thus $m = n - 1$, so Line 6 is executed $n - 1$ times exactly. Lines 7-9 are similar. Thus we have in the best case $m = n - 1$, hence $n = m + 1$, and we have the following expression for $T(m, n, q)$.

$$\begin{aligned} T &= (c_0 + c_{13} + c_{14}) + c_3 q + \\ &\quad (c_1 + c_2 + c_4 + c_5 + c_{10} + c_{11} + c_{12})(m+1) + \\ &\quad (c_6 + c_7 + c_8 + c_9)m, \\ &\geq (c_1 + c_2 + c_4 + c_5 + c_{10} + c_{11} + c_{12} + (c_6 + c_7 + c_8 + c_9))m. \ \forall m > 0. \end{aligned}$$

To establish this lower bound linear function of n, we simply dropped the constant terms as well as the term proportional to q.

Further, since for all directed graphs, we have $n \leq m+1$, and since we also know that $q \leq n$, we also have

$$\begin{aligned} T(m,n,q) &= (c_0 + c_{13} + c_{14}) + c_3 q + \\ &\quad (c_1 + c_2 + c_4 + c_5 + c_{10} + c_{11} + c_{12})n + \\ &\quad (c_6 + c_7 + c_8 + c_9)m, \\ &\leq (c_0 + c_{13} + c_{14}) + \\ &\quad (c_1 + c_2 + c_3 + c_4 + c_5 + c_{10} + c_{11} + c_{12})n + \\ &\quad (c_6 + c_7 + c_8 + c_9)m, \\ &\leq (c_0 + c_{13} + c_{14})(m+1) + \\ &\quad (c_1 + c_2 + c_3 + c_4 + c_5 + c_{10} + c_{11} + c_{12})(m+1) + \\ &\quad (c_6 + c_7 + c_8 + c_9)m, \\ &\leq 2(\textstyle\sum_{i=0}^{i=14} c_i)m, \quad \forall m > 1 \end{aligned}$$

To get to the first inequality we just use $q \leq n$. For the second, we use $n \leq m + 1$. For the third we note that $m + 1 \leq 2m$, for all $m \geq 1$.

Thus this algorithm has optimal complexity, in the sense that it must at least read its input, which will take on the order of m operations, and completes its entire task in the same order of operations.

In the next section, we formalize the notion "on the order of", and show that the above inequalities may be expressed in more compact notation as

$$T(m,n,q) = \Omega(m), \quad \text{and} \quad T(m,n,q) = O(m),$$

respectively.

1.6 Asymptotic Complexity (or just complexity)

By definition, problems solved with VLSI CAD algorithms have "**V**ery **L**arge **S**cale." This requires both the developers and users of VLSI software tools to pay close attention to how the computational resource requirements (e.g. cpu time, cpu storage, operations counts, etc.) depend on problem size. Computer users are often surprised to find that program A, which was significantly faster than program B for "small

[6] In this case the graph is a tree, for which $m = n - 1$.

instances" of problem T, becomes orders of magnitude slower than B for "very large" instances of T. The purpose of this section is to establish a meaningful definition for the size of a problem instance, and meaningful ways to talk about how cpu resource requirements of VLSI Algorithms depend on this size.

The "size" of a problem instance is a parameter (or set of parameters) of the problem input which determines things like the size of arrays in the corresponding application programs, and/or the frequency of execution of **for** loops, **while** loops, etc. For example, **procedure** LONGEST_PATH of Section 1.5.3 has as input a graph $G = (V, E)$ for which the number of nodes is $n = |V|$. It would be useful to know the function $\zeta(n)$, which gives the *exact* operations count of algorithms of interest, just as we computed $T(m, n, q)$ for LONGEST_PATH. Unfortunately, the task of finding such a function for most VLSI CAD algorithms of practical interest is usually prohibitively difficult. Further, the exact count itself is not so much of interest as its qualitative dependence on problem size n. Thus algorithm designers, [161], have developed the concept of the asymptotic complexity of a program.

1.6.1 Worst Case Asymptotic Upper Bound Complexity

Given that an algorithm has to be applied to VLSI (Very Large Scale Integration) CAD applications, it must be efficient and robust for the case of large problem instances. For instance, the control logic on the INTEL P6 chip probably has on the order of 10^5 logic gates. The design of CAD tools for such large applications requires the development of algorithms that are robust and efficient for large scale applications. This has in turn necessitated the development of some notion of the behavior of an algorithm for asymptotically large problem instances. Roughly speaking, what is needed is some way to say "for large n, the algorithm will perform at most on the order of $g(n)$ operations". The notation adopted in the literature for this statement is "the algorithm is $O(g(n))$" [161, 69].

Speaking more formally, we define $O(g(n))$ to be a set, or "class" of functions (where n is some parameter of the "size" of a problem instance). A function $F(n)$ needn't be completely specified in order to prove that it is in the set $O(g(n))$. All that is required is to prove that $F(n)$ is asymptotically bounded from above by a known function $g(n)$, according to the following definition.

Definition 1.6.1 *(Asymptotic Complexity) A function $F(n)$ is in the set $O(g(n))$ if and only if there exist positive constants c_O and n_O such that*

$$F(n) \leq c_O g(n), \quad \forall n \geq n_O \; . \tag{1.4}$$

In words, this means that $F(n)$ is asymptotically bounded from above by a linear function of $g(n)$.

Similarly, a function $F(n)$ is in the set $\Omega(g(n))$ if and only if there exist positive constants c_Ω and n_Ω such that

$$F(n) \geq c_\Omega g(n), \quad \forall n \geq n_\Omega \; . \tag{1.5}$$

In words, this means that $F(n)$ is asymptotically bounded from below by a linear function of $g(n)$.

■

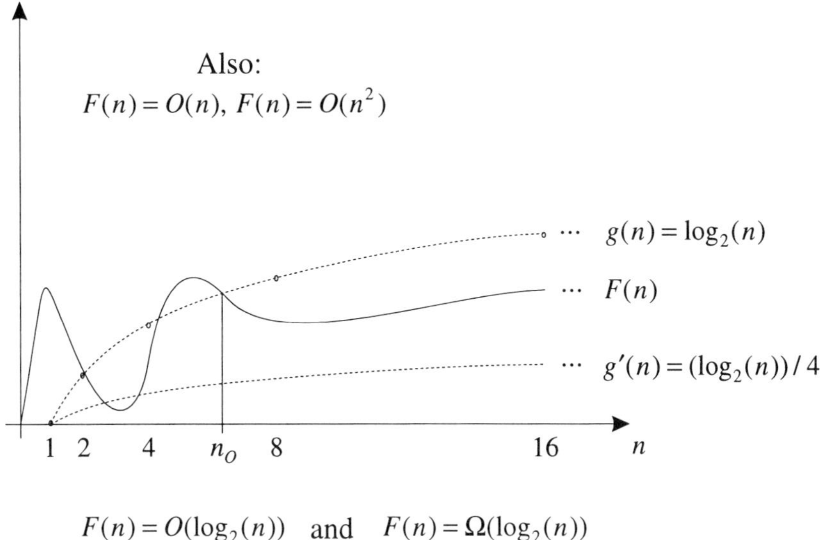

Figure 1.17: A function $F(n)$ in the set $O(\log_2(n))$ and also in the set $\Omega(\log_2(n))$.

This definition is illustrated, for the case $g(n) = \log_2(n)$, in Figure 1.17. Note that after some initial oscillations, $F(n)$ settles down and grows slowly as a function of n. Further, for all n such that $n > 6.5$, $F(n) < g(n)$ (here $c_O = 1$, $n_O = 6.5$). Thus $F(n)$ is in the set $O(\log_2(n))$. The equivalent notations $F(n) = O(\log_2(n))$, and $F(n)$ is $O(\log_2(n))$ carry the same meaning. Similarly, $F(n) = \Omega(\log_2(n))$, since for all $n > n_\Omega = n_O = 6.5$, $F(n) > (\log_2(n))/4$ (here $c_\Omega = 1/4$, $n_\Omega = 6.5$).

Finally, note that $F(n) = O(\log_2(n))$ implies that $F(n) = O(n)$ and $F(n) = O(n^2)$. Similarly, $F(n) = \Omega(n^2)$ implies that $F(n) = \Omega(n)$ and $F(n) = \Omega(\log_2(n))$.

Thus the "big-O" [161] notation gives a way to express the fact that $g(n)$ is an asymptotic upper bound to $F(n)$, to within a constant factor, and similarly for the "big-Ω" notation. $O(1)$ can be regarded as a conceptual set of asymptotically *constant* functions, whose *value* is not specified. Consequently, if $F(n) = O(1)$, all we know is that their exist constants c_O and n_O such that the function F is bounded from above by the constant c_O (c_O is independent of n) for all $n > n_O$.

Some small examples will help clarify the definition:

Example:

$$\begin{array}{lll} 3n+2 = O(n), & \text{since } 3n+2 \leq 4n, & \forall n \geq 2\ (c_O = 4,\ n_O = 2), \\ 3n+2 = O(n^2), & \text{since } 3n+2 \leq 3n^2, & \forall n \geq 2\ (c_O = 3,\ n_O = 2), \\ 3n^2-3 = O(n^2), & \text{since } 3n^2-3 \leq 3n^2, & \forall n \geq 0\ (c_O = 3,\ n_O = 0), \\ 6(2^n)+n^2 = O(2^n), & \text{since } 6(2^n)+n^2 \leq 7(2^n), & \forall n \geq 4\ (c_O = 7,\ n_O = 4). \end{array}$$

Note that it is unnecessary to find the smallest constants c_O and n_O that satisfy the definition. Any convenient values which establish the inequalities of Equations 1.4 and 1.5 suffice.

Note also that the the values of the constant coefficients in the equation for $F(n)$ are similarly inconsequential:

$$
\begin{aligned}
10^{999} &= O(1) \ , \\
3n + 10^{999} & \quad O(n) \ , \\
10^{999}n &= O(n) \ .
\end{aligned}
$$

■

1.6.2 Complexity of Algorithms

Now suppose that $F(n)$ represents the actual operations count for some algorithm applied to some class of problems of size n. We say that the algorithm, applied to input data of size n, has worst case asymptotic upper bound complexity $O(g(n))$ if its operations count, $F(n)$, satisfies the above definition. Thus, for all sets of inputs of size $n = 1, 2, \ldots$, the overall operations count is bounded (to within a constant) from above by some function $c_O g(n)$, for all n greater than the constant n_O. Note that this does not imply that the algorithm will actually take as much as $c_O g(n)$ operations for any given n. All we know is that it won't take more operations.

However, if, as in Figure 1.17, we can also find constants c_Ω and n_Ω such that $F(n) \geq c_\Omega g(n)$ for all n greater than the constant n_O, then we can say that $F(n)$ is $\Omega(g(n))$ as well. In this case, $F(n)$ is (asymptotically) bounded above *and* below by functions proportional to $g(n)$. As a result the function $g(n)$ is characteristic of the complexity of the algorithm.

For example, it can shown that when **procedure** LONGEST_PATH is applied to graphs $G = (V, E)$, its operations count is both $O(g(m))$ and $\Omega(g(m))$, where $g(m) = m = |E|$. With this type of characterization, we can easily determine the complexity of many simple algorithms. However, we emphasize that for some algorithms, this can be prohibitively difficult to do. For example, there are some algorithms, such as those for the so-called bipartite matching problem (See [69], Section 27.3) , for which the upper bounds have been repeatedly improved over the years, but no matching lower bounds have yet been proven.

It can similarly be shown that the the time complexity of BACK_TRACE is $O(m) = O(|E|)$, as well as $\Omega(m) = \Omega(|E|)$.

1.6.3 Practical Complexities

In VLSI applications, it is generally thought (because of the intrinsically large problem instances) that $O(n \log n)$ is the maximum tolerable algorithm complexity. In Table 1.2, we tabulate the growth of some comparable low complexity functions. We see that for graphs of size on the order of 10^3, which is modest by VLSI standards, 2^n is definitely prohibitively large, whereas n^3 is barely tractable. Similarly, for graphs of size on the order of 10^6, which occurs in practice for some VLSI problems, n^3 is definitely prohibitively large, whereas n^2 is barely tractable even for the most powerful vintage 1995 supercomputers.

Table 1.3 shows how problems of such large scale would take on the typical (for 1995) 10MIPs INTEL486 PC performing 10 million steps per second. Assuming unit coefficients and an instance of size 10^3, an $\Omega(n^3)$ algorithm would take 100 seconds, while an $\Omega(2^n)$ algorithm would take $3.2(10^{286})$ years. So, we may conclude that

log_n	n	$n \log n$	n^2	n^3	2^n
1	2	2	4	8	4
2	4	8	16	64	16
3	8	24	64	512	256
4	16	64	256	4,096	65,536
5	32	160	1,024	32,768	4,294,967,296
10	1024	10240	1048576	10^9	
20	1048576	20971520	10^{12}		

Table 1.2: Comparative growth of log, polynomial, polylog, and exponential functions.

Time for $f(n)$ instr. on a 10-MIP (10^7 instr/sec) computer					
n	$f(n) = n$	$f(n) = nlog_2 n$	$f(n) = n^2$	$f(n) = n^3$	$f(n) = 2^n$
10^3	0.1mS	1mS	100mS	100sec	$3.2(10^{286})$yr
10^6	0.1sec	2sec	27hr	3,171yr	

Table 1.3: Computing times on a 10MIP's computer, assuming unit coefficients for each complexity function.

the utility of algorithms with exponential complexity is limited to small n (typically $n \leq 40$). Similarly, we can conceive of executing instances of size 10^6 for algorithms with quadratic complexity, but instances of size 10^6 would be prohibitively large for algorithms with cubic complexity.

1.7 Brief Summary of MOS Device Behavior

We include here a few observations about MOS devices. First, we note that MOS transistors are bi-directional. That is, an n-type device is ON (or shorted) if the net voltage difference between the gate and either of the two channel nodes is logical 1 (that is, almost $V_D D$). This is because the positive gate to channel voltage attracts (negatively charged) electrons to the surface of the channel. Similarly, a p-type device is ON (or shorted) if the net voltage difference between the gate and either of the two channel nodes is negative logical 1 (that is, almost $-V_{DD}$). This is because the negative gate to channel voltage attracts (positively charged) holes to the surface of the channel.

Conversely, an n-type device is OFF (or open) if the net voltage difference between the gate and either of the two channel nodes is logical (that is, almost $-V_D D$). This is because the negative gate to channel voltage repels (negatively charged) electrons away from the surface of the channel. Similarly, a p-type device is OFF (or open) if the net voltage difference between the gate and either of the two channel nodes is positive logical 1 (that is, almost V_{DD}). This is because the positive gate to channel voltage repels (positively charged) holes away from the surface of the channel.

From a DC viewpoint, the gate to drain or gate to source is always an open circuit, although there is a small amount of capacitive coupling.

Second, we note that an MOS circuit can be regarded as a sequential circuit even in the absence of latches. For example, consider the driver circuit of Figure 1.1, which has no explicit memory storage elements in the indicated transistor model. The circuit has one output, z, and two inputs, c and d. The upper input is a control input (c) and the lower input is a data input voltage (d). If the control voltage is $c = 1$ the two outer MOS transistors have their channels shorted, so that the circuit functions as an inverter, with $z = \overline{d}$. However, if the control voltage is 0 the two outer MOS transistors have their channels shut off, so the output node is isolated from both ground and V_{DD}, leaving the output voltage in this case apparently ambiguous.

A real physical implementation of this circuit has capacitance from any circuit node to ground. The output node in particular can have a very substantial load capacitance when the driver circuit drives many other MOS transistors. This capacitance stores the voltage on the capacitor from one clock cycle to the next[7]. Thus, we can write the following equation for the output voltage:

$$z(t) = c(t)\overline{d}(t) + \overline{c}(t)z(t-1). \tag{1.6}$$

That is, if $c = 1$, $z(t) = \overline{d}(t)$, else $z(t) = z(t-1)$. The FSM (Finite State Machine) corresponding to this behavior is shown in Figure 1.18. Here the edge label 11/0

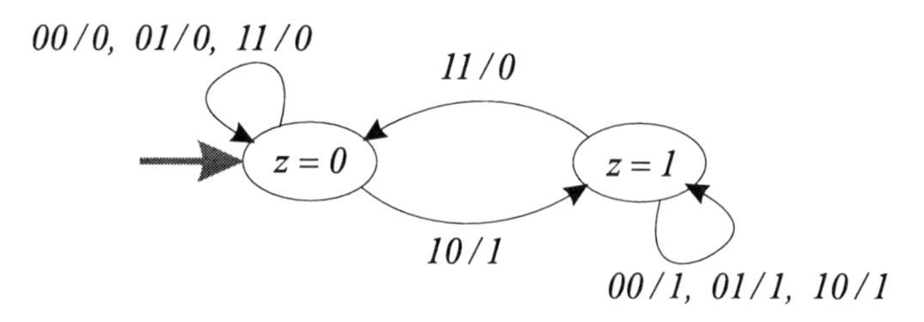

Figure 1.18: The FSM corresponding to the driver circuit of Figure 1.1 (middle) .

between the $z = 1$ state and the $z = 0$ state means " if in state $z = 1$, if $c = 1$ and $d = 1$, go to state $z = 1$, and output a 1. Note when the control input is 0, the FSM outputs the value of z which corresponds to the current state.

Third, CMOS circuits are *negative* logic. That is, a logic formula which is complemented, such as a NAND gate, for which $f = (ab)' = \overline{ab}$, is implemented naturally, whereas one which is not complemented, such as an AND gate, requires an inverter. For example, an exclusive-or function $f = x \oplus y = \overline{x}y + x\overline{y}$ would be implemented as the complement of an exclusive-nor gate, $g = (\overline{x}\overline{y} + xy)$. Then, the natural complementation of negative logic gates would provide the implicit inverter, so $f = \overline{g} = \overline{(\overline{x}\overline{y} + xy)}$. Since each negative gate requires 2 CMOS transistors per input connection, f would require 8 transistors, assuming that the signals $\overline{x}$ and $\overline{y}$ are available. If they are not, two extra inverters will be needed which will cost 2 transistors each, so the maximum cost would be 12 transistors.

Finally, note that when forming CMOS gates (without pass transistors), the the p-channel sub-network is just the graphical dual of the n-channel sub-network. That it, every series connection in the n-channel sub-network is mapped into a parallel connection in the p-channel sub-network, and conversely.

[7]Actually, even when all the MOS channels attached to this node are turned off, there is some leakage current. However, this current is usually so low that it would take seconds or even minutes for a high voltage to become low, whereas the clock cycle time is usually in the range of microseconds to nanoseconds.

1.8 Notes

At the time of writing, 1995, the IC industry was approaching $0.1\mu m$ feature sizes and 10^7 transistors on a chip. Moore's law, which predicts that processor speed would double every 18 months, was continuing to hold firm. The major VLSI CAD companies (Synopsis, Cadence, Viewlogic, Mentor Graphics, etc) continued were continuing to grow vigorously. INTEL, which has in the past been behind other large semiconductor manufacturing firms in using VLSI CAD technology, survived the infamous Pentium bug, and is now building up a world class verification group. Throughout the industry, synthesis tools were solidifying their position as a mainstream factor in the design process, and verification tools were showing signs of following the same path that synthesis tools took toward widespread adoption and designer acceptance.

1.9 Summary

In this chapter we have briefly surveyed the panoply of design methodologies which the semiconductor industry has employed in producing the powerful chips that are becoming commonplace today. We have characterized the IC design process as one of optimal tradeoff of competing design goals:

1. The area on the chip required by the circuit being designed;
2. The critical path delay of the circuit;
3. The testability of the circuit, measured in stuck-at fault coverage;
4. The power dissipated by the circuit.

Along the way we have developed basic notions of critical path delay analysis, test generation, graph models, CMOS implementation of logic functions, and complexity of algorithms. All of these notions will be recurring themes in the sequel, as we develop theoretical and intuitional background for effective use of the dominant extant VLSI CAD tools for synthesis and verification.

For those inclined to take the next step and be a VLSI CAD tool developer, this book offers many sections on advanced topics, and extensive pointers into the relevant literature. The serious student will find that mastery of the techniques covered in the "Problem" section of each chapter will put him/her into the position of being immediately, effectively, and competitively productive in a VLSI CAD or design group.

1.10 Problems

1. Consider the six-transistor cell of Figure 1.2 and the following function:

$$(a \cdot b + c)'.$$

(Here '·' means 'and,' '+' means 'or,' and ''' means 'not.')

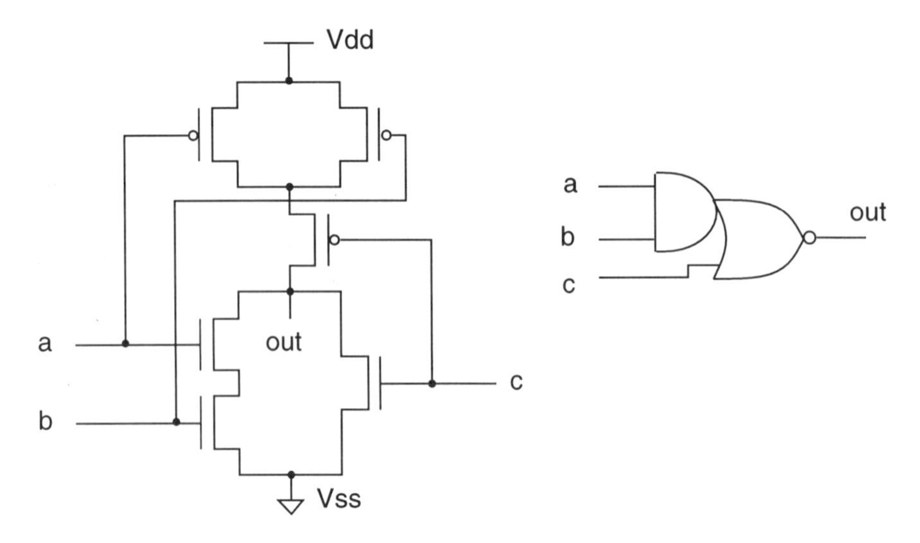

Figure 1.19: Complex CMOS gate for Problem 1.

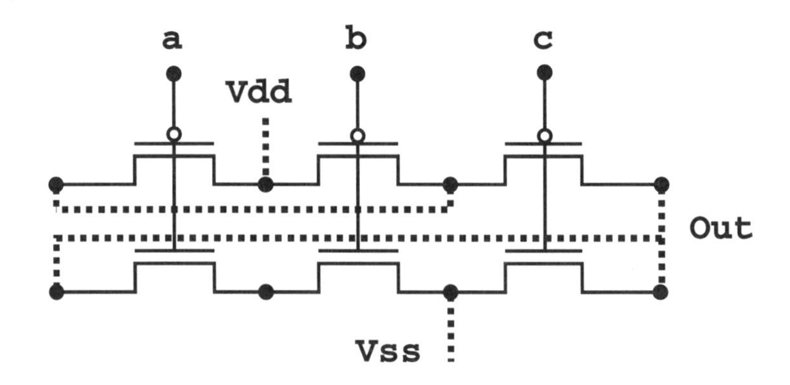

Figure 1.20: Solution of Problem 1.

Customize the six-transistor cell, so that it implements the given function. That is, you have to find the correct interconnections among the black dots in Figure 1.2.

[Hint: The transistor-level schematic and the gate-level symbol of the circuit implementing $(a \cdot b + c)'$ are shown in Figure 1.19.]

Solution. The "customization" of the cell is shown in Figure 1.20. □

2. How many transistors would be required to implement the function $(a \cdot b + c)'$ with NAND, NOR, and INVERTER gates? How many 6-transistor cells?

3. Is it possible to build a single CMOS gate like the one in Figure 1.19 that performs the function $a \cdot b + c$ (without complement)? Explain.

   Solution. It is not possible. To see why, first consider that a gate like the one in Figure 1.19 (called a fully complementary gate) is composed of two subnetworks: A pull-up subnetwork composed of p-channel transistors and a

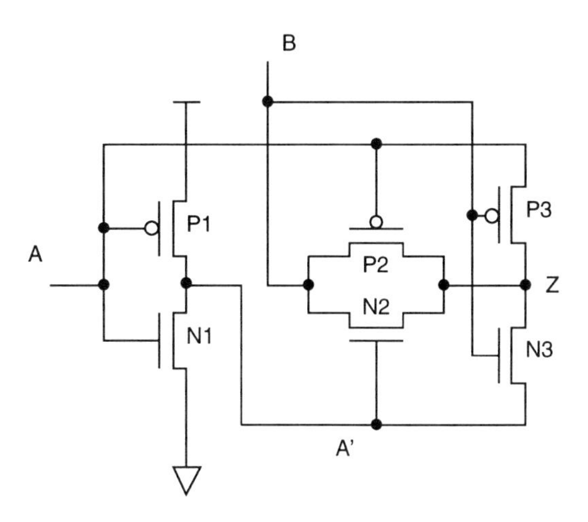

Figure 1.21: Circuit for Problem 4.

pull-down subnetwork composed of n-channel transistors. The output node is the node connecting the two subnetworks. It is necessary to have the p-channel transistors between the power supply and the output and the n-channel transistors between the output and ground, lest the transistors do not work as ideal switches controlled by the gate terminal. Indeed, MOS transistors are controlled by the gate-source voltage, V_{GS}. For a p-channel transistor to conduct, V_{GS} must be negative. If one of the two endpoints of the channel is tied to ground, the transistor stops conducting as soon as the other endpoint (that acts as source) reaches $-V_{th}$, the negative of the (negative) threshold voltage of the transistor. Therefore, the output node cannot be fully discharged through a p-channel transistor. Similarly, the output node cannot be fully charged through an n-channel transistor. To make a long story short, the p-channel transistors go on top and the n-channel transistors go at the bottom.

Having established this fact, let us now consider the effect of switching inputs from zero to one. In the target function, $a \cdot b + c$, a $0 \rightarrow 1$ transition of an input may cause no output transition or a $0 \rightarrow 1$ output transition. However, in a network of n-channel transistors, a $0 \rightarrow 1$ transition of an input increases the number of transistors that are on, and therefore may create a conducting path from the output to ground. As a consequence, if an output transition occurs, it is a $1 \rightarrow 0$ transition. This shows that a simple CMOS gate cannot implement $a \cdot b + c$. At least two gates are needed. (For instance, the gate of Figure 1.19 followed by an inverter.) □

4. Analyze the transistor-level schematic of Figure 1.21 and determine what logic function it implements.

   Solution. The function of the circuit is the exclusive OR. This can be verified by applying all four combinations of zeroes and ones to the inputs. Indeed, we have:

AB	$N1$	$P1$	A'	$N2$	$P2$	$N3$	$P3$	Z
00	OFF	ON	1	ON	OFF	OFF	OFF	0
01	OFF	ON	1	OFF	ON	OFF	OFF	1
10	ON	OFF	0	OFF	OFF	OFF	ON	1
11	ON	OFF	0	OFF	OFF	ON	OFF	0

In deciding whether a transistor is ON or OFF, we have used in this case the "correct" criterion that V_{GS} must exceed the threshold voltage in absolute value, and be of the same sign. (See the solution of Problem 3.) So, for instance, Transistor $P2$ is OFF for $AB = 00$ because all its terminals are at the same (low) voltage.

Looking at Figure 1.21, it is easy to see that $N1$ and $P1$ form an inverter with output A'. It is less obvious to realize that $N3$ and $P3$ form another inverter, which is only enabled when A is 1, and which outputs the complement of B. Finally, $N2$ and $P2$ form a *transfer gate*, that is a switch that connects B to Z when $A = 0$. The output Z can be driven by both the inverter composed of $N3$ and $P3$, and the transfer gate. However, the conditions under which the two drivers are enabled are mutually exclusive, so that no conflicts ever occur. □

5. Give the CMOS circuit for computing the function h whose complement $\overline{h}$ is given by

$$\overline{h} = a(b + c(d + ef)).$$

Your answer should be a single complex gate. Recall that the pull-up (P) subgate is just the dual of the pull-down (N) subgate.

6. Review the method given in Section 1.4.4 for deriving the set of tests for a given stuck-at fault. Apply this method to determine the corresponding set of tests for the stuck-at-faults bSA1 and bSA0 in the circuit of Figure 1.7.

Solution. Recall that for the XOR3 subcircuit, we have $T = (e^G \oplus e^F)$. For the fault bSA1, we have $e^F = 0$, so $T = e^G = (x_i\overline{y}_i + \overline{x}_i y_i)$. Thus either 01 or 10 will test for bSA1. Similarly, noting that $c_0 = 0$, for the fault bSA0, we have $e^F = \overline{a} = \overline{y}_i + \overline{x}_i$, so

$$\begin{aligned} T &= e^G \oplus e^F \\ &= (x_i\overline{y}_i + \overline{x}_i y_i) \oplus \overline{a} \\ &= (x_i\overline{y}_i + \overline{x}_i y_i) \oplus (\overline{x}_i + \overline{y}_i) \\ &= (x_i\overline{y}_i + \overline{x}_i y_i) x_i y_i + (x_i y_i + \overline{x}_i \overline{y}_i)(\overline{x}_i + \overline{y}_i) \\ &= \overline{x}_i \overline{y}_i. \end{aligned}$$

Thus only 00 will test for bSA0. □

7. Consider Procedure LEVELIZE1 of Figure 1.22 for computing the number of levels of logic of a circuit (this number is a crude estimate of the critical path

	Procedure LEVELIZE1$(V, E, INS)\{$	Ops	times/call Best	Worst
1	$n = \vert V \vert;\ m = \vert E \vert;\ q = \vert INS \vert$	c_1	1	1
2	$Q = G = V - INS;\ level = 0$	$c_2 n$	1	1
3	**for** $(g \in G)\ \ LEVELS(g) = \vert G \vert + 1$	$c_3(n-q)$	1	1
4	**for** $(g \in INS)\ \ LEVELS(g) = 0$	c_4q	1	1
5	**while** $(G \neq \emptyset)\ \{$	c_5	2	$n+1$
6	$level = level + 1$	c_6	1	n
7	**foreach**$(g \in G)\ \{$	c_7	n	$n(n+1)/2$
8	$test = 1$	c_8	n	$n(n+1)/2$
9	**foreach**$(i \in \rightarrow g)\ \{$	c_9	m	$< mn$
10	**if**$(LEVELS(i) > level - 1)\ \{$	c_{10}	m	$< mn$
11	$test = 0$; **break**	c_{11}	0	$n(n+1)/2$
	$\}$			
	$\}$			
12	**if**$(test = 1)\ \{$	c_{12}	n	$n(n+1)/2$
13	$LEVELS(g) = level;\ Q = Q - \{g\}$	c_{13}	n	n
14	**if**$(Q = \emptyset)$ **return**$(level, LEVELS)$	c_{14}	n	n
	$\}$			
	$\}$			
15	$G = Q$	c_{15}	1	n
	$\}$			
	$\}$			

Figure 1.22: Procedure LEVELIZE1.

delay). This procedure is inefficient, at least when applied to the worst case circuit graph. This is possible because for some circuit graph, some gate g might appear in the list G on the order of $|V|$ times.

(a) Apply Procedure LEVELIZE1 to the MAJORITY subcircuit of the bit serial adder circuit of Figure 1.7. Identify all critical paths, and show the slack value for every node. (See Section 1.5.4.)
(b) Consider the following 5-gate circuit (described by its blif file). Apply Procedure LEVELIZE1 to identify all critical paths, and show the slack value for every node.

```
.model bsaddr
.inputs a b
.outputs g
.names a b c
10 1
01 1
.names c b d
10 1
01 1
.names d b e
10 1
01 1
.names e b f
10 1
01 1
.names f b g
10 1
01 1
.end
```

(c) Apply Procedure LEVELIZE1 to the generalization (to n gates) of the above 5-gate circuit. Derive best case and worst case formulas for the running time of Procedure LEVELIZE1, given the data in the step table at the right of the algorithm, and show that the algorithm is $\Omega(n)$ and $O(mn)$.
(d) Present an analysis which justifies the step table entries on Lines 7-11. Then state why the bounds given in Part (c) above are tight. Use the following identity:

$$n + (n-1) + \ldots + 1 = n(n+1)/2.$$

The solution was given as part of the solution of Part (c).
(e) Would the worst case complexity be improved if only gates in the fanout of gates which have already had there level assigned were put in the set G for each pass through the **while** loop? How about if only gates in the fanout of gates at the current level, and with all fanin from gates at the current level or previous levels, were put in G?

Line	v, λ_v 6	$v\!\rightarrow$ 6	(λ_a, D_a) 10										
k			a	b	c	d	e	f	x	y	z	c-	c+
1	x/0	a,b,c,d	1/1	1/1	1/1	1/1	0/2	0/2	0/0	0/0	0/2	0/0	0/2
2	y/0	a,b,c,d	1/0	1/0	1/0	1/0							
3	c-/1	f,z						1/1			1/1		
4	a/1	e					2/1						
5	b/1	e					2/0						
6	c/1	c+											2/1
7	d/1	f						2/0					
8	e/2	z									3/0		
9	f/2	c+											3/0
10	z/3	$\emptyset$											
11	c+/3	$\emptyset$											
	final	λ:	1	1	1	1	2	2	0	0	3	0	3
	final	*slack*:	0	12	12	12	0	12	0	12	0	2	12

Table 1.4: Data trace for Procedure LONGEST_PATH, applied to Figure 1.7.

8. Apply Procedure LONGEST_PATH (Figure 1.14) to the 1-stage bit-serial adder circuit of Figure 1.7. Give a data table like Table 1.1, and then give the slack at each node. Your graph should have 11 nodes: 8 for the 8 gates (disregarding the latch), and 3 for the 3 inputs.

   To disambiguate the search order, assume that the **foreach** loop of Line 6 selects the nodes in the set $v\!\rightarrow$ in increasing lexicographical order.

   Solution. We obtained the solution of Table 1.4 from a PERL script implementing LONGEST_PATH. Consequently the symbol names are modified to fit the ASCII output. Note the critical path was (x,a,e,z). □

Chapter 2

A Quick Tour of Logic Synthesis with the Help of a Simple Example

Our purpose in this chapter is to give a quick overview of logic synthesis from the point of view of the user. We shall see what problems are faced (and solved) by a synthesis program and what general strategy may be applied to their solution. We shall use a very simple circuit as an example. No special effort has been made to make the example realistic. The sole purpose was to come up with something simple enough to be analyzed in detail with limited effort, and yet demonstrating a sufficient number of interesting problems.

In the rest of this chapter, we shall not concern ourselves with the solution of the problems, but rather with their statement. Thus we shall identify some of the particular degrees of freedom that present themselves to a designer in a practical problem. We shall then show how the designer's ingenuity exploits these degrees of freedom to create an efficient design. The sequel of this book is then devoted to showing how, and to what extent, equivalent ingenuity can be embodied in CAD tools that solve the same design problems automatically.

2.1 A Simple Case Conversion Circuit

We consider a simple circuit that will serve us as an example to illustrate the major steps of the synthesis process. The interface of the circuit is described in Figure 2.1.

A stream of alphabetic ASCII characters (a–z, A–Z) is fed to the circuit, one character for each clock cycle. The circuit performs case conversion on the incoming characters and outputs the corresponding sequence. One character is output for each clock cycle. No specification is given for the latency of the circuit. This means we can choose how many clock cycles will be required to process one character, as long as the throughput is one character per clock cycle.

The type of case conversion to be applied is specified by escape sequences[1], which

[1]An escape sequence is a sequence of two or more characters, the first of which is the escape character. Escape sequences are used to augment a character set and are typically used to carry

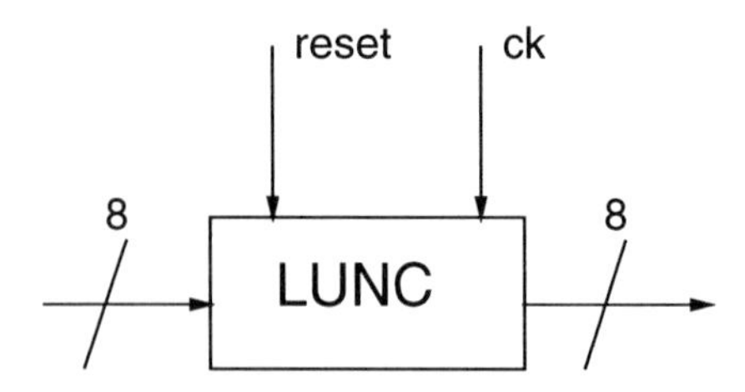

Figure 2.1: Interface of the example circuit.

are interspersed with the text characters. Our circuit recognizes four escape sequences (`^[` is our representation of the escape character):

- `^[L` Lower case;
- `^[U` Upper case;
- `^[N` No conversion;
- `^[C` Change Case.

By joining the four characters we get the name of our circuit: LUNC.

When the circuit is reset, it goes into a state where it passes the input characters unchanged. It remains in that state until it receives an escape sequence other than `^[N` .

The behavior of the circuit for escape sequences other than the four listed above and for non-alphabetic characters is don't care, represented by symbol ? (that is, the behavior is unspecified). In addition, the output of the circuit when an escape sequence is input is also ? (that is, left unspecified).

We are allowed to use the freedom resulting from the don't care specification to simplify our design as much as we can. In a more realistic situation specifications may be more complete. However, it is important to make use of such *don't care* information, whenever it is available.

As an example of possible behavior of the circuit, consider the following two streams of characters. The one on the top is the input stream and the other is the output stream. In this example we are assuming that the latency of the circuit is two clock cycles.

```
a b C d E f ^[ U a b C D...
? ? a b C d E  f ? ? A B C D...
```

The question marks indicate don't care outputs. They occur at the beginning, when the first character has not been processed yet, and in response to the escape sequence. After the escape sequence has been processed, the circuit converts all incoming characters to upper case.

---

commands. Refer to the ASCII table of Figure A.1 in the Appendix for the codes of the characters.

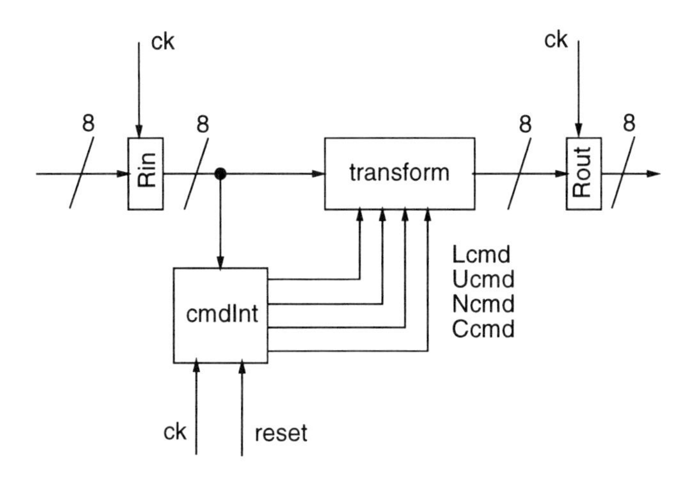

Figure 2.2: Block diagram for LUNC.

2.2 First Refinement

Our first step in the design of the LUNC circuit is to separate the data processing from the control. The result of this first step is shown in the block diagram of Figure 2.2. In this block diagram we can distinguish two main blocks: A command interpreter block (CMDINT), which parses the escape sequences and decides the state of the circuit accordingly, and a transformation block (TRANSFORM), which actually performs the case conversion. There are also an input and an output register, which determine the latency of the circuit. (We assume that there is no register inside the TRANSFORM block.)

The type of decomposition we have applied to our problem is fairly typical in the design of large digital systems. It is customary to divide the control functions from the data processing functions. It is not uncommon that different design methods are then applied to the separate parts. For instance, in a microprocessor, the design of the data path (ALUs and register files) is approached differently from the design of the instruction decoder or the cache controllers. In addition, the design of a complex circuit is carried out by a team. The block diagram is then important in defining the interfaces between the blocks designed by different people.

Defining the top-level block diagram of a circuit goes under the name of *high-level* or *architectural* design. In this course, we shall assume that this phase of the design has been carried out already—either manually or automatically—and we shall concentrate on the succeeding phases. It should be clear, however, that architectural design has a large impact on the outcome of the entire design process.

Before returning to our example, we make another general remark. In this case we have used a graphical representation of our block diagram. We could have used a textual representation as well. We are going to see examples of both in the sequel. It is typical of a real-life design system to support both ways of representation, for each has its own advantages.

Consider for instance the two major blocks of Figure 2.2. The command interpreter communicates with the transformation block by means of four signals (*Lcmd,*

Ucmd, Ncmd, Ccmd). At any time, only one of these signals is active. The active signal indicates the transformation to be applied to the character currently in the input register.

This kind of information is not easily conveyed by a drawing, but can be easily expressed by text. The previous paragraph is an example of *informal* textual description; formal languages can be used instead. These formal languages are similar to programming languages and are called *Hardware Description Languages* (HDLs).

Hardware description languages are used to describe both the structure of a circuit (what parts constitute it and how they are connected) and its behavior (how it reacts to given inputs). The syntax of HDLs is often similar to that of programming languages. For instance, VHDL,[2] one of the most widely used HDLs is derived from Ada. The semantics of the HDLs, however, differ from those of ordinary programming languages in various respects.

In this book, we shall use a fictitious language, which borrows its syntax from the 'C' language. We shall informally define its semantics as we examine the examples of its use. Our treatment of HDLs in this book is essentially restricted to these brief introductory notes. It is important to realize, though, their relevance and their relationship to automatic logic synthesis. We shall try to emphasize that relationship as we discuss the design of the transform and command interpreter blocks.

2.3 The Transform Block

In the architectural design phase we decided that the TRANSFORM block would be a combinational circuit that, given an alphabetic character, outputs either the character itself or the character obtained by changing its case. The choice is determined by the four control inputs coming from the command interpreter and by the case of the input character.

We can describe the desired function in the following piece of code, that is largely self-explanatory.

```
Procedure TRANSFORM(Rin, Lcmd, Ucmd, Ncmd, Ccmd){
    if (Lcmd)
        { mux = TOLOWER(Rin)}
        else if (Ucmd)
            {mux = TOUPPER(Rin)}
            else if (Ncmd)
                {mux = Rin }
                else if (Ccmd)
                    {mux = CHANGECASE(Rin) }
    return(mux)
}
```

Rin, Lcmd, Ucmd, Ncmd, and *Ccmd* are the inputs and *mux* is the output. All of these are either 1 or 0 on each clock cycle. In a complete description, we would

[2]VHDL stands for VHSIC Hardware Description Language; VHSIC stands for Very High Speed Integrated Circuits and is the name of a research initiative of the US Department of Defense.

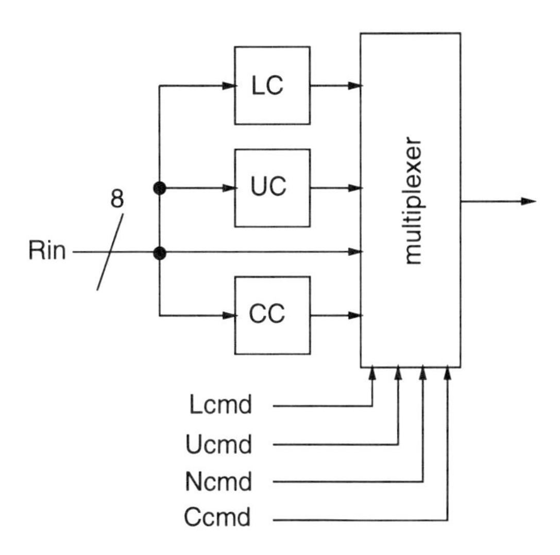

Figure 2.3: Block diagram for the transform block.

also specify how many bits each terminal has. The horizontal data path is marked as an 8-bit line, which is natural, since there are $2^8 = 256$ ASCII characters. The other lines are symbolic in principle, although in this case, as discussed below, all represent just one data bit.

Two remarks are in order here. First, a relatively high-level description like the one afforded by our simple HDL is easy to read and write. It is easier to interpret than a gate-level schematic, because it is more concise, it uses evocative names like TOLOWER, and it describes the behavior rather than the structure. For the same reasons, such a description is easier to write. In order to write it, we do not have to make up our mind on how precisely we are going to implement our circuit.

The second consideration should be suggested by the name of the circuit output. Calling the output MUX suggests that we can translate the **if-then-else** statement into a multiplexer. The translation is portrayed in Figure 2.3. In general, a synthesis program will have a scheme to translate the constructs of its input language into structure (i.e., into the interconnection of registers, multiplexers, adders, gates, and similar building blocks).

The translation scheme is not in general very sophisticated. We shall see that in our simple example, a straightforward translation of the initial description into a circuit yields an implementation that is far from optimal. We do not worry too much, though, because we rely on *optimization* techniques to improve our 'draft' circuit. This approach is followed by commercial an academic synthesis programs alike. One of its advantages is to make the final result independent—to a large extent, if not completely—from the initial description. The user of such a system can therefore save time and concentrate on clarity.

The approach we have just described relies heavily on the effectiveness and efficiency of the optimization techniques. We shall indeed concentrate on these techniques for most of this course. Returning to our LUNC circuit, we have described the TRANSFORM block in terms of simpler functions (TOLOWER, TOUPPER, CHANGECASE). We have to specify them, in order to complete our design. Here we shall only examine

Procedure CHANGECASE(Rin) {
    if (ISUC(Rin))
        { $res = Rin + 32$ }
    else
        { $res = Rin - 32$ }
    return (res)
}

Figure 2.4: Procedure CHANGECASE.

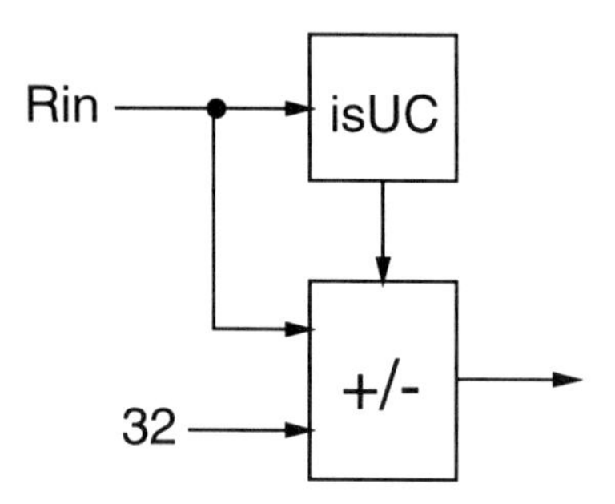

Figure 2.5: Block diagram for the CC block.

the most complex of them (CHANGECASE); the others are similar.

2.3.1 The CC Block

From Figure A.1 we see that, for each letter, the ASCII code for the lowercase character can be obtained from that of the uppercase character by adding 32 (base 10). This suggests the following definition for the CHANGECASE function. In this piece of code, ISUC is a function that says whether the input character is uppercase or lowercase. Since the output of the circuit is *don't care* when the input character is non-alphabetic, we can define ISUC by noting that bit 5 (the third most significant bit) is 0 for all uppercase letters and is 1 for all lowercase letters. Therefore, we can just define ISUC as $Rin[5]$[3]. If we notice that both addition and subtraction can be performed by a single adder/subtracter, then we can come up with the block diagram of Figure 2.5.

The TOLOWER and TOUPPER functions can also be implemented with an adder and a subtracter, respectively. Even though we have not worked out all the details, we can now get an idea of the cost—in terms of gates—of our 'draft' implementation of the TRANSFORM block. We have a total of three 8-bit adder/subtracters and an 8-bit, four-way multiplexer. It is reasonable to assume that about 200 gates are necessary for this implementation.

[3]Question: How could we hide the details of the ASCII code still further? (What if we used EBCDIC instead of ASCII?)

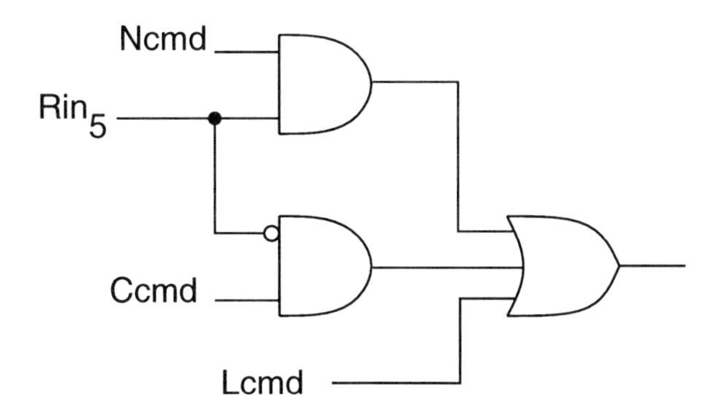

Figure 2.6: Circuit schematic for the optimized transform block.

2.3.2 An Optimized Transform Block

The discussion of the ISUC function may have already suggested to the attentive reader an alternative implementation of the TRANSFORM block. Let us consider the codes for 'a' and 'A.'

```
A = 0x41 = 01000001
a = 0x61 = 01100001
```

It is sufficient to flip bit 5 to go from lowercase to uppercase or vice versa. The same is true of all letters in ASCII. A minute's thought will show that the circuit of Figure 2.6—based on this idea—is indeed a correct implementation of the TRANSFORM block. The circuit actually produces only bit 5 of the results. All other bits of the result are identical to the corresponding input bits and therefore are not represented. Suppose $Ccmd = 1$. This implies $Ncmd = Lcmd = 0$. Then, the output is the complement of bit 5 of the input. Similarly, we can analyze the other three cases. Notice that *Ucmd* does not explicitly appear as an input to the circuit of Figure 2.6. Therefore, when $Ucmd = 1$, the output is 0.

This new implementation consists of only three gates. This is a lot less than the two hundred gates we estimated for our first 'draft.' If the optimization phase cannot pick up this slack, then the translation/optimization scheme is in trouble. Fortunately, in this case and in many others, optimization can easily get rid of the extra gates.

On the other hand, most people will agree that the purpose of the circuit of Figure 2.6, taken out of context, is not obvious. Our high-level description is more readable and eventually leads to an equally efficient implementation.

In drawing conclusions from our simple example, it is important to put things in perspective. There are cases where careful manual design is superior to the results of the best synthesis programs. The opposite also occurs, though less frequently. The importance of time-to-market should be always kept in mind when comparing manual design to automatic synthesis. One should also keep in mind that the problem of readability of a circuit description increases considerably with its size. If all design problems where of the complexity of our LUNC circuit, logic synthesis would have probably never evolved. For circuits with millions of transistors, on the other hand, the advantage afforded by logic synthesis may be decisive.

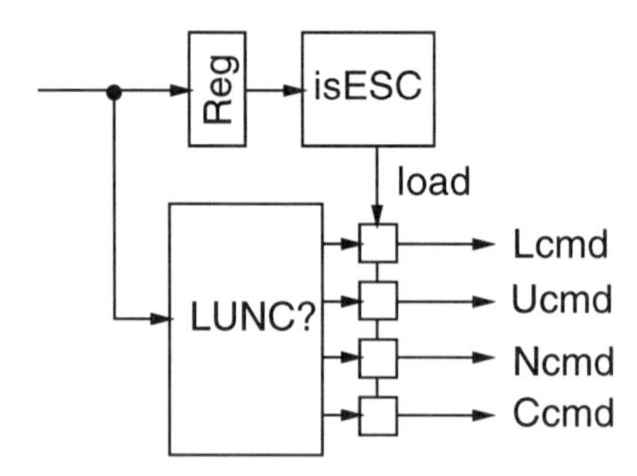

Figure 2.7: Block diagram for the command interpreter.

2.4 The Command Interpreter

Let us consider now the command interpreter. A simple block diagram for it is shown in Figure 2.7. As in the case of the *Transform* block we are initially interested in clarity and simplicity more than in efficiency.

The circuit of Figure 2.7 works by keeping a copy of the previous input character in a register. If the previous character is 'escape,' then the current character is used to determine the new state. The current input character is always decoded, but unless the previous character is 'escape,' the output of the decoder (block LUNC?) is ignored.

The function of the decoder is described by the following piece of code.

This code can be translated, for instance, in a truth table, from which a circuit can be derived. Notice that the output of the block is *don't care* for any unexpected character. This information is extremely important for the optimization of the circuit.

2.4.1 Checking for Equality

We can translate the test ISESC into a test for equality.

```
isESC(Reg) = isSame(Reg, ESC);
```

Checking two values for equality occurs frequently in digital designs. In Figure 2.9 we show the typical template used in the translation phase. Similar templates are used for tests like $x \geq y$.

Notice that this circuit simplifies considerably when one of the two operands is constant.

2.4.2 Optimizing the Command Interpreter

Also for the command interpreter we now see how things can be optimized manually. This will give a target for the optimization phase. Let us consider the codes of the four letters that may appear in an escape sequence.

```
L = 0x4C = 01001100
U = 0x55 = 01010101
```

```
Procedure SWITCH(Rin) {
    if (Rin = L) {
        Lcmd = 1
        Ucmd = Ncmd = Ccmd = 1
        break
    }
    else if (Rin = U) {
        Ucmd = 1
        Lcmd = Ncmd = Ccmd = 1
        break
    }
    else if (Rin = N) {
        Ncmd = 1
        Lcmd = Ucmd = Ccmd = 1
        break
    }
    else if (Rin = C) {
        Ccmd = 1
        Lcmd = Ucmd = Ncmd = 1
        break
    }
    else
        Lcmd = Ucmd = Ncmd = Ccmd = Don't Care
    return (res)
}
```

Figure 2.8: Procedure SWITCH

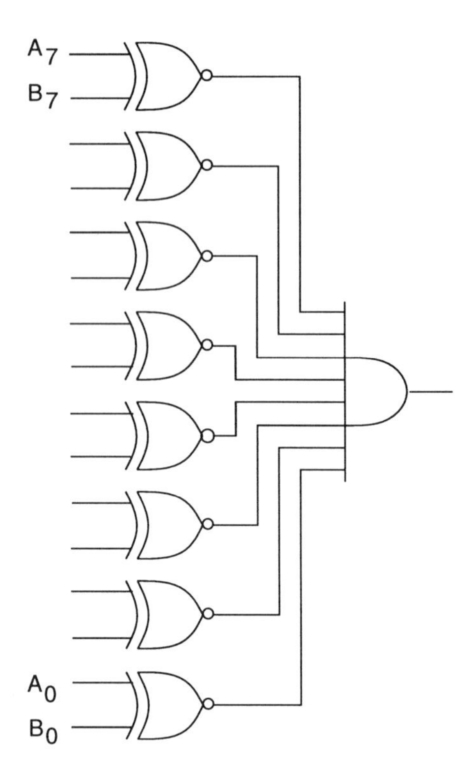

Figure 2.9: Circuit schematic for an equality checker.

```
N = 0x4E = 01001110
C = 0x43 = 01000011
```

Notice that the two least significant bits are sufficient to distinguish them. Furthermore, from the discussion of the TRANSFORM block, we know that we do not need to produce *Ucmd*. Therefore, we can implement the command decoder as shown in Figure 2.10.

Notice also that we can considerably reduce the number of flip-flops by a simple device. Instead of storing each input to test whether it is an 'escape' at the next clock cycle, we can test each input character as soon as we see it, and then store (in a single flip-flop) the result of the test. Such a transformation is called a *retiming* of

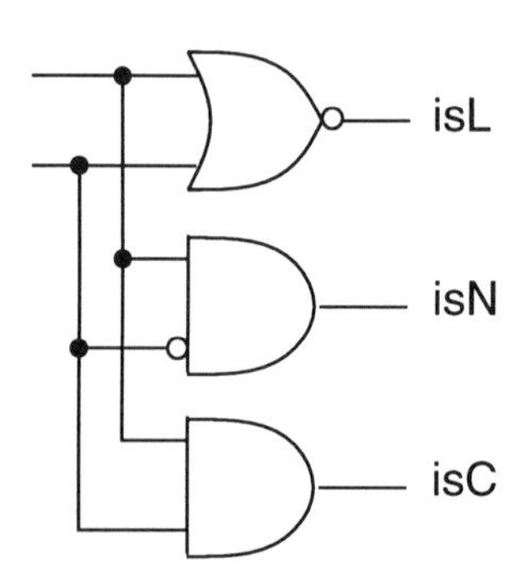

Figure 2.10: Circuit schematic for the optimized command interpreter.

the circuit. It can be formalized and automated. For lack of time, however, we shall not study it in this course and we only mention it here.

We can further reduce the number of flip-flops by noting that our command interpreter has only three possible states. Two flip-flops are sufficient to encode three states. In general, eliminating one flip-flop this way may make the combinational logic more complicated. One of the advantages of an automatic synthesis system is to make it possible for a designer to explore several possible solutions in a short time. In our case, though, a simple retiming transformation is sufficient to reduce the number of flip-flops without changing the combinational logic. It is sufficient to latch the inputs to the circuit of Figure 2.10, instead of the outputs. This transformation will make the circuit a little slower, but will also reduce power consumption.

2.5 Technology Mapping

Let us assume that our objective is to produce a netlist that may be used to fabricate a standard-cell chip. Suppose we have chosen a CMOS library. We now have to address the concerns arising from these choices.

Gates in CMOS (and in other technologies) are *negative*. This means that the basic gate is the inverter, rather than the non-inverting buffer. Consequently, NANDs and NORs are cheaper and faster than ANDs and ORs. Practical gates have limited driving capability, so that a gate typically drives four other gates or less. We say that the maximum fanout is four. This number may be further reduced if speed is a primary concern.

Likewise, a restriction is usually applied to the number of inputs to a gate. Even though it is possible in theory to build NAND and NOR gates with very large number of inputs, performance rapidly degrades as the number of inputs increases. Therefore, cell libraries do not usually provide gates with more than four or five inputs. Not all functions with those many inputs will be available either. A five-input NAND gate may be available, but a five-input EXOR may not. One must then make sure that only gates from the library are used in the circuit. All these concerns must be addressed by a synthesis program, as they are addressed by a human designer. This task is called *technology mapping*.

In the translation/optimization scheme we have examined so far, we have assumed that the result of the optimization phase is a technology-independent circuit. Our final scheme is therefore composed of three phases: Translation, optimization and technology mapping (or *techmapping* for short). The division is to some extent arbitrary. By separating optimization from technology mapping we simplify the two tasks and we can develop very powerful techniques for both. On the other hand, in the optimization phase we may have a less than perfect knowledge of the consequences of a given choice. In general the penalty for ignoring technology-specific information during optimization is higher when the target of optimization is high performance or low power. It is lower, but non-null, when area is being optimized.

As a consequence, the boundaries between the technology-independent phase and the technology-dependent phase tend to be blurred in real systems. The division is, however, useful from the didactic standpoint and we shall adopt it.

Suppose our cell library consists of inverters, two-input NAND and NOR gates,

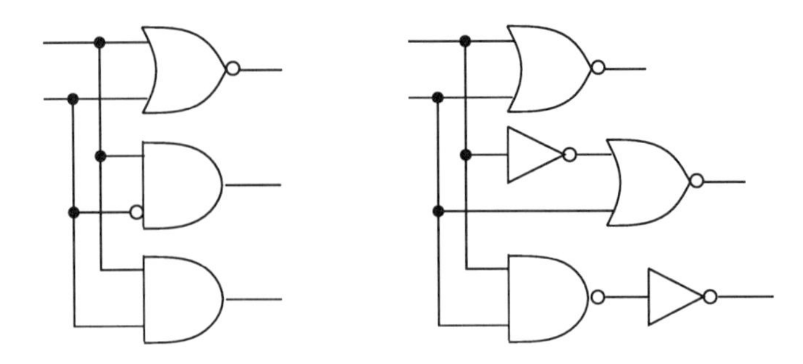

Figure 2.11: Circuit schematic for the technology-mapped decoder of the command interpreter.

and D-type flip flops. Then Figure 2.11 describes a possible mapping for the decoder of the command interpreter.

The choice we made (standard-cell chip in CMOS) is not the only one possible. On the one hand, we may be interested in full-custom design, and therefore in mapping at the transistor level, rather than at the gate level. On the other hand, we may want to implement our circuit as a Field Programmable Gate Array (FPGA). In both cases the mapping problem is different from that encountered with a fixed cell library.

2.6 Problems

1. Describe an 8-bit adder in BLIF format.

   Run SIS on your adder. Read in the circuit (with SIS command `read_blif`) and print out the statistics for it (with command `print_stats`).

   Include in your homework the description of the adder and the result of the `print_stats` command.

   Save your adder, because you will use it in other assignments as a building block. You may want to spend some time to familiarize with SIS. For instance, you may want to try the `simulate` command to verify that your description works as intended. Take a look at the man page. Obviously, at this stage, not everything will be clear: Don't worry. Notice that there is a handy UNIX `alias` command that lists all standard abbreviations. You may use `alias` to create your own abbreviations.

   Solution. Since the description of the standard "ripple-carry" adder is already in the blif documentation—albeit for four-bit numbers—we describe here another type of adder, which is faster. It is known as a carry-bypass adder. We shall have occasion to discuss it later in the course. Notice that it is composed of four blocks, each computing two output bits.

```
#------------------------- cbpadd8.blif ------------------------
# Adds two 8-bit inputs and a carry-in bit. Index 0 signals the
# least significant bit. The result is a 9-bit number. No two's
# complement overflow output is produced.
# The adder is composed of 4 modules, each computing the sum of
# two bits and based on the carry-bypass scheme.
```

```
.model cbpadd8
.inputs cin a0 a1 a2 a3 a4 a5 a6 a7 \
b0 b1 b2 b3 b4 b5 b6 b7
.outputs s0 s1 s2 s3 s4 s5 s6 s7 s8
.subckt cbp2 c0=cin a0=a0  b0=b0  a1=a1  b1=b1  s0=s0  s1=s1  c2=c2
.subckt cbp2 c0=c2  a0=a2  b0=b2  a1=a3  b1=b3  s0=s2  s1=s3  c2=c4
.subckt cbp2 c0=c4  a0=a4  b0=b4  a1=a5  b1=b5  s0=s4  s1=s5  c2=c6
.subckt cbp2 c0=c6  a0=a6  b0=b6  a1=a7  b1=b7  s0=s6  s1=s7  c2=s8
.end

# Two-bit carry bypass adder.

.model cbp2
.inputs c0 a0 b0 a1 b1
.outputs s0 s1 c2
.names a0 b0 g1
10 1
01 1
.names a0 b0 g2
11 1
.names a1 b1 g3
10 1
01 1
.names a1 b1 g4
11 1
.names c0 g1 g5
10 1
01 1
.names c0 g1 g6
11 1
.names g2 g6 g7
1- 1
-1 1
.names g3 g7 g8
10 1
01 1
.names g3 g7 g9
11 1
.names g1 g3 g10
11 1
.names g4 g9 g11
1- 1
-1 1
.names g10 g11 c0 mux
01- 1
1-1 1
.names g5 s0
1 1
.names g8 s1
1 1
.names mux c2
```

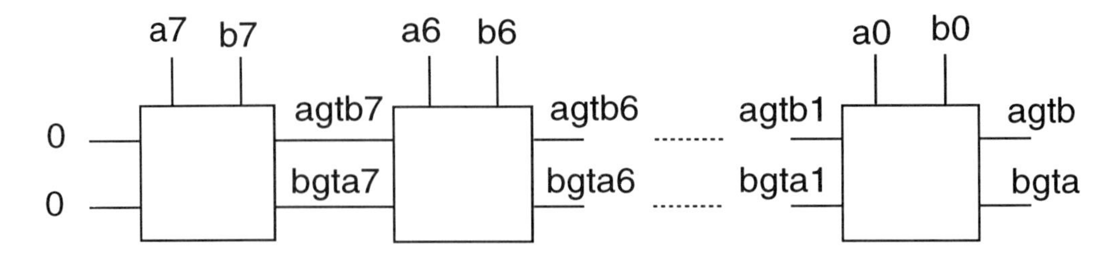

Figure 2.12: Iterative scheme for the 8-bit comparator of Problem 3.

```
1 1
.end
```

The result of running the PS command (an alias for `print_stats -f`), is the following:

```
cbpadd8          pi=17    po= 9    nodes= 51         latches= 0
lits(sop)= 139   lits(fac)= 139
```

□

2. Repeat Problem 1, this time for an 8-bit equality comparator.

3. Describe an 8-bit comparator in BLIF that takes two unsigned integers a and b as inputs and produces two outputs:

   (a) *agtb*: is 1 if and only if $a > b$;

   (b) *bgta*: is 1 if and only if $a < b$.

   Clearly, if $a = b$ both outputs are 0, and it is never the case that the two outputs are 1 simultaneously.

   Use the following "interface" for your comparator.

```
.model cmp8
.inputs a0 a1 a2 a3 a4 a5 a6 a7 b0 b1 b2 b3 b4 b5 b6 b7
.outputs agtb bgta
```

   Design the circuit by replicating a basic cell according to the scheme of Figure 2.12. Verify that your circuit works properly by using SIS to simulate the following pairs of inputs:

   (a) 0 0;

   (b) 2 3;

   (c) 3 2;

   (d) 200 100;

   (e) 5 5.

Create a script file containing the simulation commands and use the `source` command to run them.

Include in your homework your BLIF description, the simulation script, and the output of the simulation produced by SIS. (Use the `set sisout foo` command to redirect your output to file `foo`.)

This problem counts for 10.

Solution. A BLIF file for the comparator is as follows:

```
.model cmp8
.inputs a0 a1 a2 a3 a4 a5 a6 a7 b0 b1 b2 b3 b4 b5 b6 b7
.outputs agtb bgta

.names zero

.subckt comp agtb-1=zero  altb-1=zero  a=a7 b=b7 agtb=agtb7 altb=bgta7
.subckt comp agtb-1=agtb7 altb-1=bgta7 a=a6 b=b6 agtb=agtb6 altb=bgta6
.subckt comp agtb-1=agtb6 altb-1=bgta6 a=a5 b=b5 agtb=agtb5 altb=bgta5
.subckt comp agtb-1=agtb5 altb-1=bgta5 a=a4 b=b4 agtb=agtb4 altb=bgta4
.subckt comp agtb-1=agtb4 altb-1=bgta4 a=a3 b=b3 agtb=agtb3 altb=bgta3
.subckt comp agtb-1=agtb3 altb-1=bgta3 a=a2 b=b2 agtb=agtb2 altb=bgta2
.subckt comp agtb-1=agtb2 altb-1=bgta2 a=a1 b=b1 agtb=agtb1 altb=bgta1
.subckt comp agtb-1=agtb1 altb-1=bgta1 a=a0 b=b0 agtb=agtb  altb=bgta

.end

.model comp
.inputs agtb-1 altb-1 a b
.outputs agtb altb

.names agtb-1 altb-1 a b agtb
1--- 1
0010 1

.names agtb-1 altb-1 a b altb
-1-- 1
0001 1
.end
```

The script file to simulate the comparator is:

```
sim 0 0 0 0 0 0 0 0  0 0 0 0 0 0 0 0
sim 0 1 0 0 0 0 0 0  1 1 0 0 0 0 0 0
sim 1 1 0 0 0 0 0 0  0 1 0 0 0 0 0 0
sim 0 0 0 1 0 0 1 1  0 0 1 0 0 1 1 0
sim 1 0 1 0 0 0 0 0  1 0 1 0 0 0 0 0
```

When the script file is 'sourced,' the output is:

```
Network simulation:
Outputs: 0 0
```

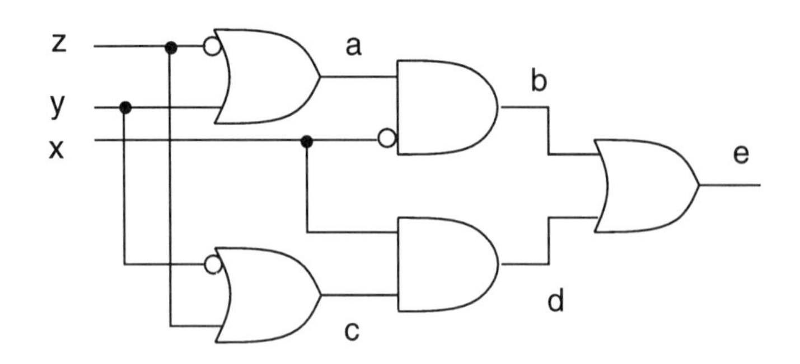

Figure 2.13: Circuit for Problem 4.

```
Next state:

Network simulation:
Outputs: 0 1
Next state:

Network simulation:
Outputs: 1 0
Next state:

Network simulation:
Outputs: 1 0
Next state:

Network simulation:
Outputs: 0 0
Next state:
```

□

4. This problem guides you through a simple example of optimization and technology mapping with SIS. Describe in BLIF the circuit of Figure 2.13. Use one *.names* directive for each gate in the drawing. Do the following.

   (a) Read your file into SIS;

   (b) read in a library for mapping with the command `rlib lib2.genlib`;

   (c) print the statistics of your circuit with `ps`;

   (d) perform techmapping with the command `map`; (ignore the warnings;)

   (e) print the statistics on your mapped network with `pg -s`;

   (f) print the equations of the network with `p`.

   (g) read in your file again;

   (h) simplify the circuit with the command ESPRESSO;

(i) repeat Steps 4c–4f.

Report in your homework the output of SIS for all steps. Include also your BLIF file. By how much does the area decrease when the circuit is optimized? (Use the areas reported by the command `pg -s`. Note that those areas only account for the *active area*; the area taken by the interconnections is not included.)
Solution. This is the `blif` file:

```
.model pb3.16
.inputs x y z
.outputs e
.names y z a
01 0
.names x a b
01 1
.names y z c
10 0
.names x c d
11 1
.names b d e
00 0
.end
```

This is the output of SIS:

```
pb3.16          pi= 3 po= 1 nodes=  5 latches= 0
lits(sop)=  10 lits(fac)=  10
inv2x             :    2 (area=928.00)
nor2              :    1 (area=1392.00)
oai21             :    2 (area=1856.00)
Total: 5 gates, 6960.00 area
     [339] = z'
     a = [339]' y'
     [338] = y'
     [355] = [338]' z' + x'
     {e} = [355]' + a' x'
pb3.16          pi= 3 po= 1 nodes=  5 latches= 0
lits(sop)=  10 lits(fac)=  10
inv2x             :    1 (area=928.00)
nand2             :    1 (area=1392.00)
oai221            :    1 (area=2784.00)
Total: 3 gates, 5104.00 area
     [400] = x'
     [415] = y' + z'
     {e} = [400]' y' + [415]' + x' z'
```

The area decreases by $6960 - 5104 = 1856$ units (in this case square microns). The percentage improvement is about 27%. Note that in both cases (with and without optimization) complex gates from the library are used to map the circuit. □

5. In this problem we use SIS to synthesize the TRANSFORM block of the LUNC circuit. (See Figure 2.3.) The purpose of this problem is twofold. On the one hand, we get to know SIS better. On the other hand, we see that the translation/optimization strategy can actually produce the results we got by manual optimization. Summarizing in a few words, the typical synthesis system translates a high-level description of the circuit into a structure composed of adders, multiplexors, comparators, etc.. This structure is then optimized. In this problem, we describe the initial structure in `blif` format and then use SIS to optimize it.

   Though the optimized transform block is small, the initial description, with several adders and multiplexors, is not so small. Therefore, we use a hierarchical description. How to write hierarchical descriptions in `blif` is described in the `blif` manual.

   To get good results from SIS, it is important that we specify whatever don't care information we have. How to specify external don't cares in `blif` is also described in the `blif` manual.

   Once we have described the TRANSFORM block, we have to optimize it with SIS. We shall use a `script`, i.e., an existing recipe that applies several commands in sequence. Scripts are run with the `source` command. Standard scripts are provided with SIS and can be used directly, without even knowing what they look like. (Though it is not advisable in general.) As all files in the default library directory, they can be referenced directly from within SIS by just giving the last component of their pathnames. One of the script is called `script.rugged`. It is a good idea to look at it and then try it out. You should also read `script.rugged.notes`. Finally, you may want to compare different scripts.

   Once we have optimized the logic, we shall perform technology mapping. We shall use the `lib2.genlib` library for that. The commands that we need are `read_library` and `map`.

   Finally, we shall run the ATPG command to generate test vectors for our optimized circuit. We shall use the -D option.

   You may find it useful to check what the `alias` command does for you. There are many handy aliases that are defined for frequently used commands in the standard configuration file.

   Now, the details of how you have to report your results.

   (a) Write separate BLIF files for each block in Figure 2.3. You will also need separate files for components like adders that you will use in separate blocks. Finally, write a master file, where you describe how the blocks are connected. Use the `.search` directive to put everything together. Remember the external don't cares!

   (b) Run SIS. Show the statistics before running `script.rugged` and afterwards. Show the equations after optimization. Draw a schematic of the circuit after optimization and compare it to what we obtained manually back at the beginning of the course.

(c) Perform technology mapping. Show the stats, the equations, and the library gates used (the last with the `print_gate` command). Draw a schematic of the mapped circuit.

(d) Run `atpg` and include the untestable faults and the tests generated in your report.

Solution. Listed below are the files comprising the description of the transform block. First we show the top-level description. We use a style that is only partially hierarchical. Indeed, the multiplexer is not describe as a nested block, but rather as a set of gates—one for each output. Notice also the description of the don't cares. It is a requirement of SIS that all don't cares be listed in the top level description—one function for each primary output.

```
#-------------------------- transform.blif ----------------------
.search toLower.blif
.search toUpper.blif
.search changecase.blif

.model transform
.inputs L U N C in0 in1 in2 in3 in4 in5 in6 in7
.outputs o0 o1 o2 o3 o4 o5 o6 o7
.subckt toLower \
        in0=in0 in1=in1 in2=in2 in3=in3 in4=in4 in5=in5 in6=in6 in7=in7 \
        o0=l0 o1=l1 o2=l2 o3=l3 o4=l4 o5=l5 o6=l6 o7=l7
.subckt toUpper \
        in0=in0 in1=in1 in2=in2 in3=in3 in4=in4 in5=in5 in6=in6 in7=in7 \
        o0=u0 o1=u1 o2=u2 o3=u3 o4=u4 o5=u5 o6=u6 o7=u7
.subckt changecase \
        in0=in0 in1=in1 in2=in2 in3=in3 in4=in4 in5=in5 in6=in6 in7=in7 \
        o0=c0 o1=c1 o2=c2 o3=c3 o4=c4 o5=c5 o6=c6 o7=c7
.names L U N C l0 u0 in0 c0 o0
10001--- 1
0100-1-- 1
0010--1- 1
0001---1 1
.names L U N C l1 u1 in1 c1 o1
10001--- 1
0100-1-- 1
0010--1- 1
0001---1 1
.names L U N C l2 u2 in2 c2 o2
10001--- 1
0100-1-- 1
0010--1- 1
0001---1 1
.names L U N C l3 u3 in3 c3 o3
10001--- 1
0100-1-- 1
0010--1- 1
0001---1 1
.names L U N C l4 u4 in4 c4 o4
10001--- 1
```

```
0100-1-- 1
0010--1- 1
0001---1 1
.names L U N C l5 u5 in5 c5 o5
10001--- 1
0100-1-- 1
0010--1- 1
0001---1 1
.names L U N C l6 u6 in6 c6 o6
10001--- 1
0100-1-- 1
0010--1- 1
0001---1 1
.names L U N C l7 u7 in7 c7 o7
10001--- 1
0100-1-- 1
0010--1- 1
0001---1 1
.exdc
.names L U N C o0
11-- 1
1-1- 1
1--1 1
-11- 1
-1-1 1
--11 1
0000 1
.names L U N C o1
11-- 1
1-1- 1
1--1 1
-11- 1
-1-1 1
--11 1
0000 1
.names L U N C o2
11-- 1
1-1- 1
1--1 1
-11- 1
-1-1 1
--11 1
0000 1
.names L U N C o3
11-- 1
1-1- 1
1--1 1
-11- 1
-1-1 1
--11 1
0000 1
.names L U N C o4
```

```
11-- 1
1-1- 1
1--1 1
-11- 1
-1-1 1
--11 1
0000 1
.names L U N C o5
11-- 1
1-1- 1
1--1 1
-11- 1
-1-1 1
--11 1
0000 1
.names L U N C o6
11-- 1
1-1- 1
1--1 1
-11- 1
-1-1 1
--11 1
0000 1
.names L U N C o7
11-- 1
1-1- 1
1--1 1
-11- 1
-1-1 1
--11 1
0000 1
.end

#--------------------------- changecase.blif -----------------------
# Changes case of an alphabetic ASCII character by adding or
# subtracting 32 (decimal). This block does not check for the
# character being non-alphabetic.

.search addsub8.blif

.model changecase
.inputs in0 in1 in2 in3 in4 in5 in6 in7
.outputs o0 o1 o2 o3 o4 o5 o6 o7
.names zero
.names one
1
.subckt addsub8 addsub=in5 \
   a0=in0 a1=in1 a2=in2 a3=in3 a4=in4 a5=in5 a6=in6 a7=in7 \
   b0=zero b1=zero b2=zero b3=zero b4=zero b5=one b6=zero b7=zero \
   s0=o0 s1=o1 s2=o2 s3=o3 s4=o4 s5=o5 s6=o6 s7=o7 s8=dummy
.end

#--------------------------- toLower.blif ----------------------
```

```
# Changes case of an uppercase alphabetic ASCII character by adding 32
# (decimal). Lowercase characters are left unchanged.
# This block does not check for the character being non-alphabetic.

.search adder8.blif

.model toLower
.inputs in0 in1 in2 in3 in4 in5 in6 in7
.outputs o0 o1 o2 o3 o4 o5 o6 o7
.names zero
.names one
1
.subckt adder8 cin=zero \
    a0=in0 a1=in1 a2=in2 a3=in3 a4=in4 a5=in5 a6=in6 a7=in7 \
    b0=zero b1=zero b2=zero b3=zero b4=zero b5=one b6=zero b7=zero \
    s0=k0 s1=k1 s2=k2 s3=k3 s4=k4 s5=k5 s6=k6 s7=k7 s8=dummy
.names in5 in0 k0 o0
11- 1
0-1 1
.names i    a0=in0 a1=in1 a2=in2 a3=in3 a4=in4 a5=in5 a6=in6 a7=in7 \
        b0=zero b1=zero b2=zero b3=zero b4=zero b5=one b6=zero b7=zero \
        n5 in1 k1 o1
11- 1
0-1 1
.names in5 in2 k2 o2
11- 1
0-1 1
.names in5 in3 k3 o3
11- 1
0-1 1
.names in5 in4 k4 o4
11- 1
0-1 1
.names in5 in5 k5 o5
11- 1
0-1 1
.names in5 in6 k6 o6
11- 1
0-1 1
.names in5 in7 k7 o7
11- 1
0-1 1
.end


#-------------------------- toUpper.blif ----------------------
# Changes case of a lowercase alphabetic ASCII character by
# subtracting 32 (decimal). Uppercase characters are left unchanged.
# This block does not check for the character being non-alphabetic.

.search addsub8.blif

.model toUpper
```

```
.inputs in0 in1 in2 in3 in4 in5 in6 in7
.outputs o0 o1 o2 o3 o4 o5 o6 o7
.names zero
.names one
1
.subckt addsub8 addsub=one \
   a0=in0 a1=in1 a2=in2 a3=in3 a4=in4 a5=in5 a6=in6 a7=in7 \
   b0=zero b1=zero b2=zero b3=zero b4=zero b5=one b6=zero b7=zero \
   s0=k0 s1=k1 s2=k2 s3=k3 s4=k4 s5=k5 s6=k6 s7=k7 s8=dummy
.names in5 in0 k0 o0
01- 1
1-1 1
.names in5 in1 k1 o1
01- 1
1-1 1
.names in5 in2 k2 o2
01- 1
1-1 1
.names in5 in3 k3 o3
01- 1
1-1 1
.names in5 in4 k4 o4
01- 1
1-1 1
.names in5 in5 k5 o5
01- 1
1-1 1
.names in5 in6 k6 o6
01- 1
1-1 1
.names in5 in7 k7 o7
01- 1
1-1 1
.end


#-------------------------- addsub8.blif ----------------------
# Adds/subtracts two 8-bit integers. Index 0 signals the least
# significant bit. Input addsub causes addition when it is 0 and
# subtraction (a-b) when it is 1. The result is a 9-bit number.
# No two´s complement overflow output is provided.

.search adder8.blif

.model addsub8
.inputs addsub a0 a1 a2 a3 a4 a5 a6 a7 b0 b1 b2 b3 b4 b5 b6 b7
.outputs s0 s1 s2 s3 s4 s5 s6 s7 s8
.names addsub b0 c0
10 1
01 1
.names addsub b1 c1
10 1
01 1
```

```
.names addsub b2 c2
10 1
01 1
.names addsub b3 c3
10 1
01 1
.names addsub b4 c4
10 1
01 1
.names addsub b5 c5
10 1
01 1
.names addsub b6 c6
10 1
01 1
.names addsub b7 c7
10 1
01 1
.subckt adder8 \
     cin=addsub a0=a0 a1=a1 a2=a2 a3=a3 a4=a4 a5=a5 a6=a6 a7=a7 \
     b0=c0 b1=c1 b2=c2 b3=c3 b4=c4 b5=c5 b6=c6 b7=c7 \
     s0=s0 s1=s1 s2=s2 s3=s3 s4=s4 s5=s5 s6=s6 s7=s7 s8=s8
.end


#--------------------------- adder8.blif -----------------------
# Adds two 8-bit inputs and a carry-in bit. Index 0 signals the least
# significant bit. The result is a 9-bit number. No two's complement
# overflow output is produced.
.model adder8
.inputs cin a0 a1 a2 a3 a4 a5 a6 a7 b0 b1 b2 b3 b4 b5 b6 b7
.outputs s0 s1 s2 s3 s4 s5 s6 s7 s8
.subckt full_adder cin=cin a=a0 b=b0 sum=s0 cout=c0
.subckt full_adder cin=c0 a=a1 b=b1 sum=s1 cout=c1
.subckt full_adder cin=c1 a=a2 b=b2 sum=s2 cout=c2
.subckt full_adder cin=c2 a=a3 b=b3 sum=s3 cout=c3
.subckt full_adder cin=c3 a=a4 b=b4 sum=s4 cout=c4
.subckt full_adder cin=c4 a=a5 b=b5 sum=s5 cout=c5
.subckt full_adder cin=c5 a=a6 b=b6 sum=s6 cout=c6
.subckt full_adder cin=c6 a=a7 b=b7 sum=s7 cout=s8
.end

.model full_adder
.inputs a b cin
.outputs sum cout
.names cin a b sum
001 1
010 1
100 1
111 1
.names cin a b cout
-11 1
1-1 1
```

```
11- 1
.end
```

If these files are read into SIS, the following initial statistics are obtained.

```
transform       pi=12   po= 8   nodes= 94       latches= 0
lits(sop)= 700  lits(fac)= 606
```

After running the script, we get the following stats.

```
transform       pi=12   po= 8   nodes=  9       latches= 0
lits(sop)=  12  lits(fac)=  12
```

These are the equations. As we can see, we obtain the same solution that was obtained manually.

```
{o0} = in0
{o1} = in1
{o2} = in2
{o3} = in3
{o4} = in4
{o5} = C in5' + L + N in5
{o6} = in6
{o7} = in7
[170] = -0-
```

These are the equations after mapping.

```
{o0} = in0
{o1} = in1
{o2} = in2
{o3} = in3
{o4} = in4
[510] = in5'
[165] = C' L' N' + C' L' in5' + L' N' [510]' + L' [510]' in5'
{o5} = [165]'
{o6} = in6
{o7} = in7
```

The library cells used in the mapped circuit are given by the `pg` command.

```
[510]       inv2x          928.00
[165]       aoi221         2784.00
{o5}        inv2x          928.00
```

Finally, running the `atpg -d` command, we get the following output.

```
38 total faults
RTG: covered 35 remaining 3
RTG: covered 1 remaining 2
36 faults covered by RTG
S_A_0: NODE: U  OUTPUT
Redundant
S_A_1: NODE: U  OUTPUT
Redundant
faults: 38      tested: 36      aborted: 0      redundant: 2
```

Notice that the redundant faults correspond to input U, that is not used in the circuit (and hence, it is not observable). The following is the list of patterns that are generated.

```
# atpg test patterns for transform
.inputs  L U N C in0 in1 in2 in3 in4 in5 in6 in7
011000000001
111001010110
010001101110
100000011000
010000111011
010110111010
101001111111
000111000101
001001011111
101001001110
```

□

6. Create a blif file and perform the SIS print_stats and sim (for the input character string of Section 2.1) commands for the overall LUNC circuit.

Part II

Two Level Logic Synthesis

A wise carpenter was fond of saying: "There's a tool for every job ...and that tool is a screwdriver". In a modern world, it can truly be said that logic is a universal tool, it being the basis of all mathematics, all science, all detective stories, most games and all puzzles. For example, the following mini-conundrum has taxed the odd adult, but was answered readily by one of our children: "If your father was my father's only son, what am I to you?".

However, logical propositions can be oximoronic and non-resolvent, so one must beware of generalizations, since it is obviously true that "All generalizations are false".

Roadmap of Part II In Part II, we treat the subfield of logic that has become associated with the design of VLSI chips and VLSI CAD tools. We break our treatment into four chapters. Chapter 3 is designed to establish a solid foundation not only for the remainder of Part II, but for the entire sequel as well. Chapters 4 and 5 are devoted to exact and heuristic methods for two level logic minimization, respectively. In Chapter 6 we look at BDDs (Binary Decision Diagrams) that has significantly impacted virtually all subareas of logic synthesis and formal verification.

Chapter 3 introduces sets, binary relations, partial orders, lattices and Boolean algebras. Along the way, we cover the theorems and logical identities that fill the back of tricks of most circuit designers. The chapter concludes with a brief treatment of the origin of don't care conditions, and their relations to intervals in the Boolean function algebras.

Chapter 4 treats three main topics: (1) the representation of logic functions; (2) the complete enumeration of the set of prime implicants of a given logic function; and (3) a branch and bound approach to finding a minimum cost subset of prime implicants that covers all the minterms in the onset of the function being minimized.

Chapter 5 looks at heuristic methods for logic minimization, paying special attention to the heuristic techniques employed in ESPRESSO [37, 239]. Here we describe the basic cycle of reduction, expansion, and redundancy removal that characterized MINI [144] and ESPRESSO. Here we also treat the unate recursive paradigm of ESPRESSO, and how this is manifested in efficient routines for tautology/equivalence checking and complementation.

Chapter 6 is included here because BDDs are engendering a whole new class of logic minimization algorithms which may someday supplant or at least become a dominant feature the methods of Chapter 4 and Chapter 5. Further, BDDs have already permeated the fields of FSM synthesis (Chapter 7.1 and Chapter 8) and formal verification (Chapter 9). We limit ourselves to a brief treatment, since these methods are still in a state of flux.

Chapter 3

Boolean Algebras

Boolean algebra is going to be our major mathematical tool in this book. Hence, we briefly recall the basics here. For more detail, you may refer to [133, 251, 235, 162, 171, 44]. Short treatments of Boolean algebras can be found in most textbooks on digital design, e.g., [187, 141]. However, since these books did not treat synthesis and verification, our treatment of Boolean algebras will be more extensive. For example we will treat "large" Boolean algebras and subalgebras, and will relate intervals and subalgebras to don't cares and sets of permissible functions, which are prominent features of synthesis algorithms.

3.1 Sets, Relations, and Functions

We summarize here some of the concepts and definitions of sets, relations, and functions, which are key to the understanding of Boolean Algebras. In turn, Boolean Algebras lie at the heart of the understanding of digital logic and the tools required for designing it.

3.1.1 Sets

A **set** is a collection of objects called *elements* or *members*. We use curly braces to indicate sets. For instance:

$$\{\clubsuit, \diamondsuit, \heartsuit, \spadesuit\}$$

or

{Bach,Beethoven,Brahms,Berlioz,Boccherini,Buxtehude,Borodin,
Bizet,Bernstein,Busoni,Berg,Bellini,Biber,Berio,Bartok,Britten}

or

$$\{(x, y) | x \in R,\ y \in R,\ x \geq y\}.$$

Here the vertical bar | can be read "such that". Also a colon : is often used for the same purpose.

The **cardinality** of a set A, written $|A|$, is the number of elements of the set. We shall only consider finite sets in the sequel. Of the three preceding examples, only the first two are finite sets. The **empty set** (written $\emptyset$) is the set with no elements. It is a subset of all sets. If a is an element of Set A, we write $a \in A$.

The elements of the sets are taken from a *universe of discourse* or **universal set**. We can define predicates and operations on sets from the same universe of discourse:

- Inclusion ($\subseteq$): Set A is included in set B ($A \subseteq B$) if and only if all elements of A are also elements of B.

- Proper inclusion ($\subset$): Set A is properly included in set B ($A \subset B$) if and only if $A \subseteq B$ and $A \neq B$.

- Complementation ($^{-}$ or $'$): The complement of A in universe U is the set of all elements of U that are not elements of A.

- Intersection ($\cap$): The intersection of A and B ($A \cap B$) is the set containing the elements that are in both A and B.

- union ($\cup$): The union of A of A and B ($A \cup B$) is the set containing the elements that are in either A or B.

We can also derive other operations from these, like the set **difference**:

$$A - B = A \cap \overline{B}.$$

Given two sets A and B, the **Cartesian product** $A \times B$ is defined by:

$$A \times B = \{(x, y) : x \in A \text{ and } y \in B\}.$$

In words, the Cartesian product of A and B is the set of all the *ordered* pairs of elements such that the first element is from A and the second element is from B. For instance, if $A = \{a, b\}$ and $B = \{1, 2\}$, then:

$$A \times B = \{(a, 1), (a, 2), (b, 1), (b, 2)\}.$$

Similarly,

$$B \times A = \{(1, a), (2, a), (1, b), (2, b)\} \neq A \times B.$$

The definition of Cartesian product is easily extended to more than two sets: The Cartesian product of n sets is the set of all ordered n-tuples taken from the n sets. The Cartesian product $A \times A$ is often indicated by A^2. Similarly, $A \times A \times A \times A$ is abbreviated A^4. The **power set** of a set A, written 2^A is defined as the set of all subsets of A:

$$2^A = \{B \subseteq A\}.$$

For instance, if $A = \{a, b\}$, then:

$$2^A = \{\emptyset, \{a\}, \{b\}, \{a, b\}\}.$$

The notation 2^A derives from the following equality:

$$|2^A| = 2^{|A|},$$

which states that there are 2^n subsets of a set with n elements. The power set of A includes the empty set (denoted $\emptyset$, and a subset of all sets), and, of course, A itself.

By definition a set is unordered and each of its elements occur exactly once. Thus $\{a, b, c\} = \{c, b, a\} = \{a, b, c, a\}$. In rare cases when one must refer to the same element more than once, the group of elements is called a **multi-set**. For multi-sets, $|\{a, b, c, a\}| = 4$, and $\{c, b, a\} \neq \{a, b, c, a\}$.

An **ordered set** is denoted with parentheses instead of braces. For ordered sets, $(a, b, c) \neq (c, b, a)$.

A simple and useful technique to employ when dealing with set operations is the **Venn Diagram**. This technique can be used as follows to show that

$$|A \cup B| = |A| + |B| - |A \cap B|.$$

The Venn Diagram for this identity is given at the left of Figure 3.1, and shows a universal set S and two subsets A and B. We see that set intersection has partitioned

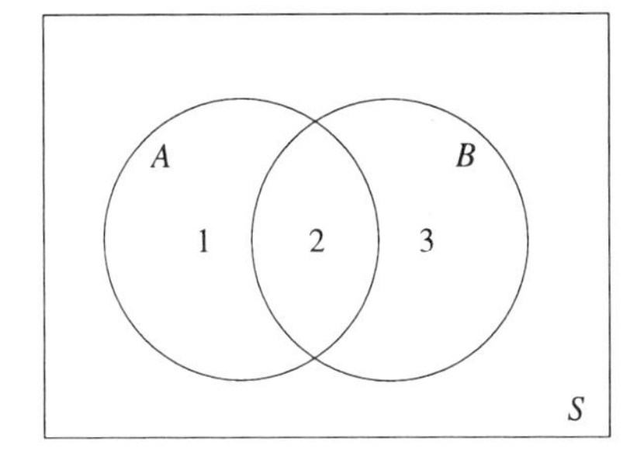

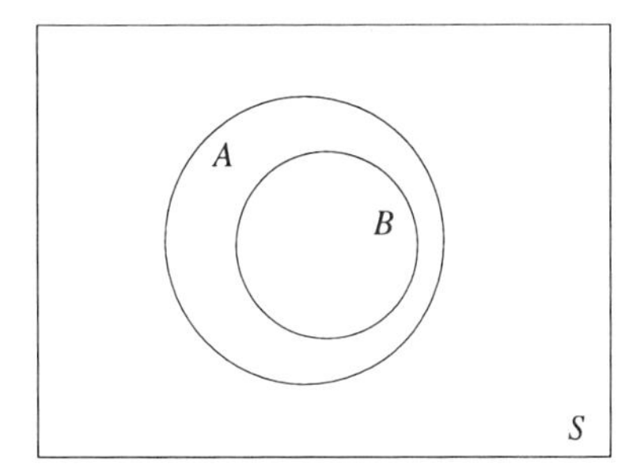

Figure 3.1: Venn Diagrams for illustrating set inclusion.

set A into two subsets: $1 = A - B$, the set of elements in A but not in B, and $2 = A \cap B$, the set of elements in A and in B. B is similarly partitioned. Thus we have:
$|A \cap B| = |\{2\}|$;
$|A| = |\{1\} \cup \{2\}| = |\{1\}| + |\{2\}|$;
$|B| = |\{2\}| + |\{3\}|$;
$|A \cup B| = |\{1\}| + |\{2\}| + |\{2\}| + |\{3\}| - |\{2\}| = |A| + |B| - |A \cap B|$.
The last identity follows from the fact that in taking the sum $|A| + |B|$ we count the elements in subset 2 twice, since these elements are in both A and B. Thus to get a correct count we must subtract $|A \cap B|$.

Similarly, The Venn Diagram for the set difference $B - A = \emptyset$, which is equivalent to $B \cap A = B$ and to $B \subseteq A$, is given at the right of Figure 3.1.

3.1.2 Relations

Given two sets A and B, a **binary relation** $\mathcal{R}$ between A and B is a subset of $A \times B$. This definition easily generalizes to more than two sets: A ternary relation over A, B, and C is a subset of $A \times B \times C$. A binary (respectively ternary) relation is sometimes called a 2-place (respectively 3-place) relation. Unless otherwise specified, a relation will be assumed to be binary. We write $x\mathcal{R}y$ if x and y are in relation $\mathcal{R}$. In formula:

$$x\mathcal{R}y \Leftrightarrow (x, y) \in \mathcal{R} \subseteq A \times B$$

For instance let

$$A = \{\text{Jane,John,Julie,Jeff}\}$$

and let B the previous set of composers. Then the relation $\mathcal{L}$ (likes) may be specified as:

$$\begin{aligned}\mathcal{L} \;=\; &\{(\text{Jane,Bach}), (\text{Jane,Biber}), (\text{Jane,Buxtehude}),\\ &(\text{John,Berg}), (\text{John,Berio}), (\text{John,Bartok}), (\text{John,Britten}), (\text{John,Bach}),\\ &(\text{Julie,Beethoven})\}.\end{aligned}$$

Such a binary relation $\mathcal{L} \subseteq J \times B$ can be visualized as a $|J| \times |B|$ sparse matrix:

$$\begin{array}{lcccccccc} & B_1 & B_2 & B_3 & B_4 & B_5 & B_6 & B_7 & B_8 \\ \\ \mathit{Jane} & \mathcal{L} & & & & & \mathcal{L} & & \mathcal{L} \\ \mathit{John} & \mathcal{L} & \mathcal{L} & & \mathcal{L} & \mathcal{L} & & \mathcal{L} & \\ \mathit{Julie} & & & \mathcal{L} & & & & & \end{array},$$

where the matrix elements indicated by $\mathcal{L}$ identify the pairs of the binary relation. Here we have encoded the composer names Bach, Bartok, Beethoven, Berg, Berio, Biber, Britten, Buxtehude with the tags $B_1, B_2, \ldots, B_8$.

The elements of a binary relation are ordered pairs. Hence we speak of a relation *from domain A into range B*. A relation is always invertible. We have:

$$\mathcal{R}^{-1} = \{(x, y) : (y, x) \in \mathcal{R}\}.$$

The inverse of $\mathcal{L}$, written $\mathcal{L}^{-1}$, is:

$$\begin{aligned}\mathcal{L}^{-1} \;=\; &\{(\text{Bach,Jane}), (\text{Biber,Jane}), (\text{Buxtehude,Jane}),\\ &(\text{Berg,John}), (\text{Berio,John}), (\text{Bartok,John}), (\text{Britten,John}), (\text{Bach,John}),\\ &(\text{Beethoven,Julie})\}.\end{aligned}$$

Jane $\mathcal{L}$ Bach means "Jane likes Bach," while Bach $\mathcal{L}^{-1}$ Jane means "Bach is liked by Jane." Hence, $\mathcal{L}$ and $\mathcal{L}^{-1}$ convey the same information—just organized differently.

3.1.3 Reflexive Binary Relations

In this section we consider binary relations $\mathcal{R} \subseteq V^2$, for which the domain and range are the same. In this case a directed graph $G = (V, E)$ is also a visually effective representation of such a relation, since each pair in the relation $\mathcal{R}$ can be viewed as an edge of G. In fact, $E = \mathcal{R}$. When a binary relation is represented by a matrix, its inverse is represented by the transpose, and when it is represented by a graph, its inverse is represented by reversing the direction of the edges. For example,the binary relation $\leq$ on the set $\{1, 2, 3, 4, 5\}$ has the matrix and graph representations of Figure 3.2.

Figure 3.2: Matrix and graph representations of a binary relation.

Reflexivity. Relation $\mathcal{R} \subseteq A^2$ is **reflexive** if and only if, for every $x \in A$, $x\mathcal{R}x$. Note the binary relation of Figure 3.2 is reflexive. In the matrix representation, this property is indicated by the fact that the relation symbol is present in all diagonal entries. In the graph representation the property is indicated by the presence of self-loops on every node of the graph. Another example of a reflexive relation over a given set of people is the subset of all possible pairs of people who have eyes of the same color. An example of a relation (over the set of natural numbers) that is not reflexive, is the subset of numbers which are, pairwise, relatively prime (that is, they have no common integer divisor other than 1).

Symmetry A relation $\mathcal{R} \subseteq A^2$ is **symmetric** if and only if, for every pair $(x, y) \in \mathcal{R}$, the pair (y, x) is also in $\mathcal{R}$. Having eyes of the same color is a symmetric relation. We can state this in propositional logic (to be described shortly) as

$$(x, y) \in \mathcal{R} \Rightarrow (y, x) \in \mathcal{R} \equiv \neg((x, y) \in \mathcal{R}) + ((y, x) \in \mathcal{R}).$$

Logically speaking the symmetry property implies that for every pair $(x, y) \in R$ there must be a matching pair (y, x). Note the symmetry property is disproved if any pair exists which is not matched in this sense. The binary relation of Figure 3.2 is not symmetric. This shown in the matrix representation by the fact the matrix is triangular, and in the graph representation by the fact there exists, for example, a pair $(1, 2)$ in the relation that is not matched by the existence of a pair $(2, 1)$.

Antisymmetry A relation $\mathcal{R} \subseteq A^2$ is **antisymmetric** if and only if the presence of both (x, y) and (y, x) in $\mathcal{R}$ implies that $x = y$. The binary relation of Figure 3.2 is antisymmetric. This shown in the matrix representation by the fact the matrix is triangular, and in the graph representation by the fact there exists no pair (i, j) in the relation that is not matched by the existence of a pair $(j, 1)$, except for the case $i = j,\ i = 1, 2, 3, 4$.

Note that a binary relation cannot be both symmetric and antisymmetric unless it consists only of a set of pairs $\{(i, i)$, which would be represented by either a diagonal matrix or a graph with only self-loops for edges.

Transitivity. A relation $\mathcal{R} \subseteq A^2$ is **transitive** if and only if the presence of (x, y) and (y, z) in $\mathcal{R}$ implies the presence of (x, z). Having eyes of the same color and being less than or equal are transitive relations. We can state this in propositional logic as

$$((x, y) \in \mathcal{R}) \wedge ((y, z) \in \mathcal{R}) \Rightarrow (x, z) \in \mathcal{R} \equiv \neg((x, y) \in \mathcal{R}) + (\neg((y, z) \in \mathcal{R}) + (x, z) \in \mathcal{R}).$$

Logically speaking the transitivity property implies that for every "2-path" of pairs $(x, y) \in \mathcal{R}$ and $(y, z) \in \mathcal{R}$ there must be a "triangulating" pair (x, z). The binary relation of Figure 3.2 is transitive. This shown in the matrix representation by the fact each pair of element pairs $(x, y) \in \mathcal{R}$ and $(y, z) \in \mathcal{R}$ are matched by a "triangulating" pair (x, z). In the graph representation, transitivity is demonstrated by the fact for every 2-path of *consecutive* edges, there is also a corresponding triangulating edge.

Equivalence Relations A relation that is reflexive, symmetric, and transitive is an **equivalence** relation. Having eyes of the same color is an equivalence relation. If we denote this as a 2-place relation $\mathcal{C}$ on $B \times B$ (B is the aforementioned set of composers), and we assume that the eye colors were as follows:

Blue: {Bach,Beethoven,Brahms,Berlioz,Boccherini,Buxtehude,Borodin}
Brown: {Bizet,Bernstein,Busoni,Berg,Bellini,Biber,Berio}
Green: {Bartok,Britten}

The relation $\mathcal{C}$ on $B \times B$ is thus

Blue: {(Bach,Bach),(Bach,Beethoven),(Bach,Brahms),. . .,
(Beethoven,Brahms),(Beethoven,Berlioz),. . .,
. . ., (Buxtehude,Borodin),(Borodin,Borodin)}
Brown: {(Bizet,Bizet),(Bizet,Bernstein),. . .,(Biber,Berio),(Berio,Berio)}
Green: {(Bartok,Bartok),(Bartok,Britten),(Britten,Britten)}

In the matrix representation, an equivalence relation is characterized by a block diagonal structure. For the above example the matrix would be

Bach	*Bl*	*Bl*	*Bl*	*Bl*	*Bl*	*Bl*									
Beethoven	*Bl*	*Bl*	*Bl*	*Bl*	*Bl*	*Bl*									
Brahms	*Bl*	*Bl*	*Bl*	*Bl*	*Bl*	*Bl*									
Berlioz	*Bl*	*Bl*	*Bl*	*Bl*	*Bl*	*Bl*									
Boccherini	*Bl*	*Bl*	*Bl*	*Bl*	*Bl*	*Bl*									
Buxtehude	*Bl*	*Bl*	*Bl*	*Bl*	*Bl*	*Bl*									
Borodin	*Bl*	*Bl*	*Bl*	*Bl*	*Bl*	*Bl*									
Bizet							*Br*	*Br*	*Br*	*Br*	*Br*	*Br*	*Br*		
Bernstein							*Br*	*Br*	*Br*	*Br*	*Br*	*Br*	*Br*		
Busoni							*Br*	*Br*	*Br*	*Br*	*Br*	*Br*	*Br*		
Berg							*Br*	*Br*	*Br*	*Br*	*Br*	*Br*	*Br*		
Bellini							*Br*	*Br*	*Br*	*Br*	*Br*	*Br*	*Br*		
Biber							*Br*	*Br*	*Br*	*Br*	*Br*	*Br*	*Br*		
Berio							*Br*	*Br*	*Br*	*Br*	*Br*	*Br*	*Br*		
Bartok														*Gr*	*Gr*
Britten														*Gr*	*Gr*

The graph representation of this matrix would be a graph with 3 disjoint completely connected subgraphs.

Note that an equivalence relation characterizes a **partition**.

Definition 3.1.1 *Given any set B a partition of B is a set of subsets $B^i \subseteq B$ with two properties:*

1. $B^i \cap B^j = \emptyset, \quad \forall i \neq j;$

2. $\bigcup_i B^i = B.$

Here we have introduced for the first time further (equally pervasive) notation: $\forall$, which is the dual of $\exists$, and is to be read as "for all". Note that the equivalence relation $\mathcal{C}$ given above characterizes, or induces, the partition $\mathcal{P}^B = \{B^{blue}, B^{brown}, B^{Green}\}$.

$$
\begin{array}{lcl}
B^{blue} & = & \{\text{Bach,Beethoven,Brahms,Berlioz,Boccherini,Buxtehude,Borodin}\} \\
B^{brown} & = & \{\text{Bizet,Bernstein,Busoni,Berg,Bellini,Biber,Berio}\} \\
B^{green} & = & \{\text{Bartok,Britten}\}
\end{array}
$$

Another example of an equivalence relation, this time on the set $I = \{0, 1, 2, \ldots$ of natural numbers is the parity of the integer. This relation induces the partition $\mathcal{P} = \{I^0, I^1\}$, where

$$
\begin{array}{lcll}
I^0 & = & \{0, 2, 4, \ldots\} & \text{(which have even parity),} \\
I^1 & = & \{1, 3, 5, \ldots\} & \text{(which have odd parity).}
\end{array}
$$

An important concept in FSM synthesis is the idea of a **refinement** of a given partition.

Definition 3.1.2 *Given two partitions P^1 and P^2 of a set S, P^1 is a* **refinement** *of P^2 if each* **block** *B_i^1 of P^1 is a subset of some block of π_2.*

For an equivalence relation ρ over the set A, the *equivalence class* of $a \in A$, denoted $[a]$, is the set $\{x \in A : a\rho x\}$. For the equivalence relation

$$
\rho = \{(1,1),(2,2),(1,2),(2,1),(1,3),(3,1),\\
(3,2),(2,3),(3,3),(4,4),(5,5),(4,5),(5,4)\},
$$

we thus have $[3] = \{1, 2, 3\}$ and $[4] = \{4, 5\}$

As defined above, a partition a set S is a collection of nonempty disjoint subsets of S whose union equals S. We now show that for ρ an arbitrary equivalence relation on S, the distinct equivalence classes of members of S form a partition of S.

By definition, each element belongs to one equivalence class, so that the union of the classes is S. Next suppose two distinct classes have some element s in common. Let t (u) be an element in the first (second) equivalence class, but not in the second (first). Then we have that $t\rho s$ and $s\rho u$. By transitivity, $t\rho u$, but this contradicts the assumption that t and u belong to different equivalence classes, since all equivalence relations are transitive by definition.

Partial Orders Relations that are reflexive, antisymmetric, and transitive are called **partial orders.** Such relations are fundamental to switching theory, and will shortly be considered in detail in Section 3.2.1.

An example of a partial order is the relation ρ defined over the set I of the positive integers by: $x\rho y$ if and only if x divides y. For instance, $3\rho 21$. The relation ρ is :

1. reflexive (x divides x, for all $x \in I$);
2. not symmetric (e.g., 3 divides 9, but 9 does not divide 3);
3. antisymmetric (x divides y and y divides x only if $x = y$);

4. transitive (x divides y and y divides z imply that x divides z; indeed, $y = kx$ and $z = ly$ imply $z = klx$, so x divides z also);
5. a partial order (because it is reflexive, antisymmetric, and transitive) ;
6. not an equivalence relation, since it is not symmetric;
7. not a function, since 3 divides both 6 and 9, which violates the definition of Section 3.1.4.

Now consider DAGs (Directed Acyclic Graphs). Since a partial order is a reflexive, antisymmetric, and transitive binary relation, and a DAG has no cycles, its edge relation is automatically antisymmetric. Consequently, the transitive closure of the edge relation E of a DAG, augmented by self-loops on each vertex, is a **partial order** on the vertex set V. Examples of DAGs include PERT charts (Performance Evaluation Review Technique — basically task completion schedules) as well as the Hasse Diagrams defined below for lattices.

Compatibility Relations A relation which is reflexive and symmetric, but not transitive, is a **compatibility relation**. Compatibility relations play a crucial role in the minimization of incompletely defined finite state machines (Cf., Section 8.1 below). For instance, a compatibility relation on a set $A = \{a, b, c\}$ is

$$\mathcal{R} = \{(a,a), (b,b), (c,c), (a,b), (b,a), (b,c), (c,b)\}.$$

Note that $\mathcal{R}$ is reflexive since $x\mathcal{R}x, \quad \forall x \in A$. Here we introduce the notation "$\forall$" to stand for "for all". Also, $\mathcal{R}$ is symmetric, since each related pair $(a,b), (b,c), \ldots$ is matched with a corresponding opposite pair $(b,a), (c,b), \ldots$. The relation is not transitive, however, since pairs (a,c) and (c,a) are missing.

3.1.4 Functions

A **function** f from A to B, written $f : A \rightarrow B$, is a rule that associates exactly one element (not two, not zero elements) of B to each element of A. A function is thus a relation, with the restriction imposed that each element of A appears in exactly one pair belonging to the relation. In other words, a relation from A to B is a function if it is **right-unique** and if every element of A appears in one pair of the relation. A is called the **domain** of the function. B is called the **co-domain**. Often the co-domain B is called the **range** of f.

If $y = f(x) : A \longmapsto B$, we say that y is the **image** of x (under f). Given a domain subset $C \subseteq A$, the set

$$\textsc{Img}(f, C) \quad = \quad \{y \in B | \exists x \in C \ni y = f(x)\}$$

is called the **image** of C under the mapping f. Here we have introduced for the first time some new notation which is pervasive throughout the literature on graphs: $\exists$, which is to be read as "there exists, and $\ni$, which is to be read as "such that", or "has the property that". Using this notation, we can compactly write otherwise lengthy sentences, in this case we interpret the formula $\{y \in B | \exists x \in C \ni y = f(x)\}$

from	0	1	3	4	5	6	7	8	9
to	9	8	7	6	5	6	7	8	9

Table 3.1: Mapping of a simple function f.

as "the set of elements y also in B for which there exists an element x of the set C such that $y = f(x)$".

Symmetrically, the set

$$\text{PRE}(f, C) \quad = \quad \{x \in A | \exists y \in C \ni y = f(x)\}.$$

is called the **preimage** of C under f.

In the remainder of this chapter, we shall focus only on the image or preimage of single point, that is, the image of a set of cardinality one. However in Chapter 7, we shall find extensive use of computing images and preimages of functions which are the next state functions of finite state machines. In essence, these computations enable us to test if a given machine reaches a certain state, has a certain temporal property, or is equivalent to another machine.

If $C = A$, that is, if the constraint set is the entire domain, then we refer to $\text{Img}(f, C)$ as the *image* of f. A simple function $f : D \mapsto D$, where $D = \{0, 1, 2, 3, 4, 5, 6, 7, 8, 9\}$, is illustrated in Table 3.1. Here the range R is the same as the domain D, that is, $R = D$. For this function, the definition gives $\text{Img}(f, \{7, 8\}) = \{7, 8\}$, but $\text{Pre}(f, \{7, 8\}) = \{1, 3, 7, 8\}$. Note also that the preimage of the whole range is necessarily the whole domain, that is, $\text{Pre}(f, R) = D$. However, the image $\text{Img}(f, D) = \{5, 6, 7, 8, 9\} \subset R$ is not the entire range for this simple function. Sometimes the image of the whole domain is called the "image of the function".

A function f is **one-to-one**, (or injective), if $x \neq y$ implies $f(x) \neq f(y)$. Every element of the domain of a one-to-one function has a distinct image. The function of Table 3.1 is not one-to-one, because for $x = 1 \neq y = 8$, $f(x) = 8 = f(y)$.

A function $f : A \rightarrow B$ is **onto** (or surjective) if, for every $y \in B$, there exists an element $x \in A$, such that $f(x) = y$. The image of the entire domain coincides with the co-domain for a function that is onto. The function of Table 3.1 is not onto, because there is no $x \in D$ such that $\text{Img}(f, \{x\}) = y, \ y = 1, 2, 3, 4$.

A function that is both one-to-one and onto is called bijective, and is invertible. This means that its inverse relation is also a function. Clearly, not all functions have an inverse function. The function of Table 3.1 is not onto, and so is not invertible.

A geometric view of image and pre-image is given in Figure 3.3.

3.2 Partial Orders

In this section we begin by introducing the idea of an **algebraic system**. Our focus will be on a specific algebraic system, called a partial order. In the ensuing sections, we then discuss particular types of partial orders called lattices and Boolean algebra. Since the behavior of logic circuits is based on the properties of Boolean algebras, the

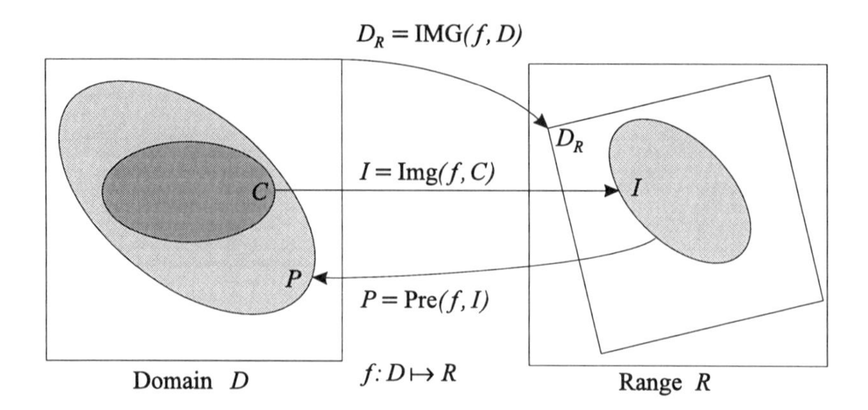

Figure 3.3: Illustration of image and preimage.

discussion will show how digital logic circuits fall into the Framework of the algebraic system known as a partial order.

An *n-ary operation* on a set A is a function $f : A^n \rightarrow A$. An *algebraic system* (or just **algebra**) is composed of a set, called the **carrier** of the algebra, a set of operations, and, optionally, a set of relations over A^n. i

For instance,

$$([0, 1], \cdot)$$

is an algebraic system where the carrier is the set of real numbers between 0 and 1, and the only operation defined is the multiplication. No relation is defined[1].

Modern algebra (sometimes also called abstract algebra) studies the properties of certain algebraic systems. In the following we examine three such systems: Partially ordered sets, lattices, and Boolean algebras. A lattice is a partially ordered set and a Boolean algebra is a lattice. Therefore, all the properties that we prove for partially ordered sets are also valid for Boolean algebras, and thus carry over to logic formulas and digital logic circuits.

3.2.1 Partially Ordered Sets

A relation $\leq \subseteq A^2$ (we shall refer to such a relation as a binary relation on A in the sequel) was defined above to be a *partial order* of A if it has the following properties:

- Reflexive ($x \leq x$);
- Antisymmetric ($x \leq y$ and $y \leq x$ imply $x = y$);
- Transitive ($x \leq y$ and $y \leq z$ imply $x \leq z$).

A set P over which a partial order is defined is called a *partially ordered set* or **poset**. The real numbers form a partially ordered set under the ordinary "less than or equal" relation. In addition to the obvious example of numerical order, inclusion between two sets and implication between two logic functions are examples of such a relation.

[1]This algebraic system is an example of **monoid** with 1 as identity [69].

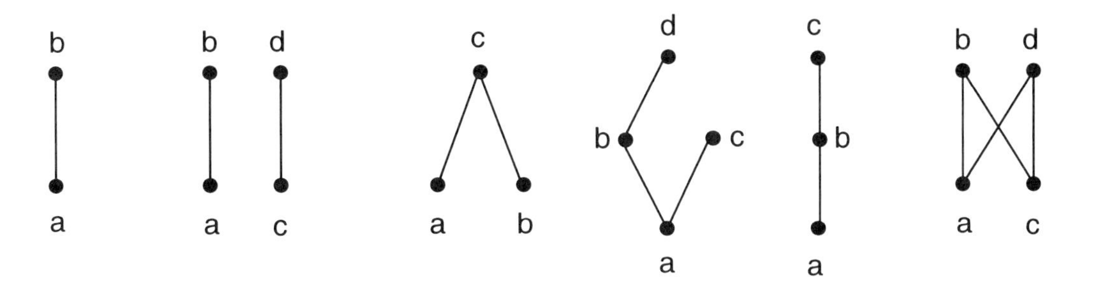

Figure 3.4: Examples of posets.

As another example, consider the points of the Cartesian plane R^2. Define an ordering relation as follows: $(x_1, y_1) \leq (x_2, y_2)$ if and only if $x_1 \leq x_2$ and $y_1 \leq y_2$. Then the algebraic system $(R^2, \leq)$ is a partial order. From this example we see clearly that $a \not\leq b$ does not imply $b \leq a$.

3.2.2 Hasse Diagrams

Figure 3.4 shows the **Hasse diagrams** of several posets. A Hasse diagram is a directed graph in which edge direction is indicated by vertical position rather than by arrowheads, In a Hasse diagram, two vertices a and b connected by a line are ordered (that is related: $a \leq b$)and the higher one (b) is greater than the other. One can think of the lines of a Hasse diagram as of arrows pointing upward. Initially, it may be convenient to actually draw arrows instead of lines. In time, however, one becomes familiar with the diagrams and the arrowheads can be dropped.

Not all lines are shown in the diagrams. For example, edges implied by the assumed transitivity property of the partial order are omitted to reduce clutter. Thus if $x \leq y$ and $y \leq z$, then $x \leq z$. The line from x to z, however, is not shown in the diagram, because it can be inferred from the lines joining x to y and y to z.

Similarly, the self-loops characteristic of graphs representing reflexive relations are also omitted. For instance, the fourth diagram from the left in Figure 3.4 corresponds to the partial order

$$\{(a,a),(a,b),(a,c),(a,d),(b,b),(b,d),(c,c),(d,d)\}.$$

Note that the majority of the edges $\{(a,a),(a,d),(b,b),(c,c),(d,d)\}$ in the graph representation of this poset are not shown in the corresponding Hasse diagram In a Hasse diagram, two elements joined by an ascending path (say from x to y) are said to be comparable (that is, related: $x \leq y$). If there is no such path the elements are not comparable ($x \not\leq y$). If one thinks of a line in a Hasse diagram as an arrow pointing upwards, then an ascending path is a directed path, i.e., a path that follows the direction of the arrows.

3.2.3 The Meet and Join Operations

We finally encounter two operations on posets, called meet and join, which are the poset analog of the operations AND and OR of digital logic.

Definition 3.2.1 *Meet and Lower Bounds An element m of a poset P is a* **lower bound** *of elements a and b of P, if $m \leq a$ and $m \leq b$. An element m of a poset P is the* **greatest lower bound** *or meet of elements a and b of P, if m is a lower bound of a and b and, for any m' such that $m' \leq a$, $m' \leq b$, $m' \leq m$.*

The **upper bound** *and the* **least upper bound** *or* **join** *are defined similarly, with $\leq$ replaced by $\geq$.*

The meet of a and b is denoted $a \cdot b$, and the join by $a + b$.

For example, in the fourth poset from the left in Figure 3.4, $a = b \cdot c$ — that is a is a lower bound of b and c. Further, it is the meet of b and c since it is the greatest (and only) such lower bound. If there were a fourth edge to a from a new element e in this Hasse Diagram, then e would also be a lower bound of b and c, but not the greatest such bound, since in this case $e \leq a$. Similarly the join of a and b is $b = a + b$. However, the join of b and c is not defined, since b and c do not have an upper bound in this poset.

The meet and join of two elements, if they exist, are unique. This follows directly from their definitions. For example, in the rightmost poset in Figure 3.4, a and c are both lower bounds of b and d, but neither is least. Thus neither of these satisfy the minimality condition in the definition of meet, so b and d have no meet in this poset.

For instance, in the fourth diagram from the left in Figure 3.4, the meet of b and c is a. The join of b and c, on the other hand, does not exist, because b and c have no upper bound. The join of a and b is b; d is an upper bound of a and b, but it is not the least. The join of two elements may not exist, even though they have upper bounds. This is the case of a and c in the sixth diagram from the left in Figure 3.4: Both b and d are upper bounds, but neither is least.

An immediate consequence of the dual definition of meet and join are the following dual properties.

Theorem 3.2.1 *In any poset, if x and y have a greatest lower bound,*

$$x \geq (x \cdot y).$$

Dually, if x and y have a least upper bound,

$$x \leq (x + y).$$

Proof. Since x and y have a greatest lower bound, we have by the definition of meet that both $x \geq (x \cdot y)$ and $y \geq (x \cdot y)$, which proves the first assertion. Dually, if x and y have a least upper bound, we have both $x \leq (x + y)$ and $y \leq (x + y)$, which proves the second assertion. □

Note the dual form of the two assertions — the second is obtained from the first be replacing $\cdot$ by $+$ and $\geq$ by $\leq$.

With this result in hand we may easily prove a second general property of posets.

Theorem 3.2.2 *In any poset,*

$$x \leq y \Leftrightarrow x \cdot y = x \quad \textit{and} \quad x + y = y.$$

Proof. Since $x \leq y$, x and y have both a lower bound x and an upper bound y. We show by contradiction that x is also the meet of x and y. Suppose not. Then by the definition of meet, there exists a lower bound m, different from x, of x and y such that $x \leq m$. However, since m was a lower bound of x and y we must have $m \leq x$ as well. Since a poset is antisymmetric, it must then be true that $m = x$. This contradicts the assumption that m was different from x. Thus x is the meet of x and y. A similar argument shows that y is the join of x and y. □

Note in this theorem we use the double implication sign to denote an "if and only if" condition.

3.2.4 Totally Ordered Sets, Well-Ordered Sets, and Induction

If all pairs of elements of a poset are comparable, then the set is *totally* ordered. A total order is thus a special case of partial order. The real numbers give an example of a totally ordered set.

If every non-empty subset of a totally ordered set has a smallest element, then the set is **well-ordered**. The natural numbers are well-ordered, whereas the rational and the real numbers are not well-ordered. In both cases, the subset of the numbers strictly greater than 1 has no smallest element.

The notion of well-ordering is the basis for the principle of **mathematical induction**. Let $\mathbf{N}$ be the set of the natural numbers (non-negative integers). Suppose we are given, for each natural number n, a proposition $P(n)$. For instance,

$$\sum_{i=0}^{n} i = \frac{(n+1)n}{2}.$$

We are interested in proving whether, for all n, $P(n)$ is true. The principle of induction gives us a very general tool to solve this type of problems, based on the following theorem.

Theorem 3.2.3 *Given, for all $n \in \mathbf{N}$, propositions $P(n)$, if:*

1. *$P(0)$ is true;*
2. *for all $n > 0$, if $P(n-1)$ is true, then $P(n)$ is true,*

then, for all $n \in \mathbf{N}$, $P(n)$ is true.

Proof. Let $F \subset N$ be the set of natural numbers for which $P(n)$ is false. If we can prove that F is empty, then we show that $P(n)$ is true for all n. To prove that F is empty, we assume the contrary—that F is non-empty—and we derive a contradiction from our assumption. Since F is a non-empty subset of a well-ordered set, it must have a smallest element. Let s be such a smallest element. Since $P(0)$ is true, s must be greater than 0; hence, $s-1$ is a natural number (≥ 0) and is not in F. Therefore, $P(s-1)$ is true, and by the second hypothesis, $P(s)$ is also true. This, however, contradicts the choice of s as the smallest element of F. Therefore, F must be empty, and $P(n)$ holds for all n. □

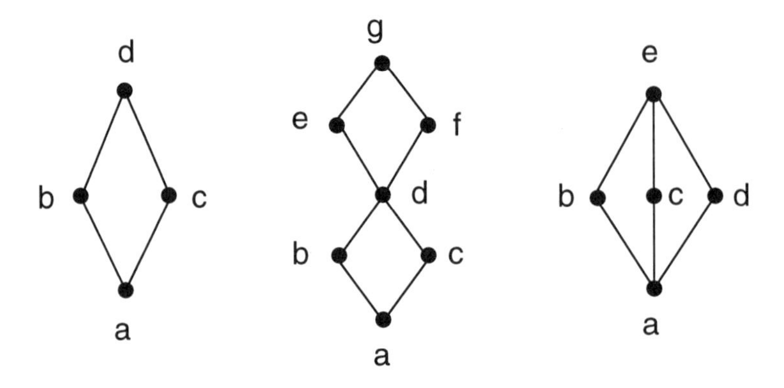

Figure 3.5: Examples of lattices.

In our example, we clearly have $\sum_{i=0}^{0} i = \frac{(0+1)0}{2} = 0$. Furthermore, for $n > 0$,

$$\begin{aligned}\sum_{i=0}^{n} i &= \sum_{i=0}^{n-1} i + n \\ &= \frac{n(n-1)}{2} + n \\ &= \frac{(n+1)n}{2}\end{aligned}$$

Therefore, our $P(n)$ is true for all $n \geq 0$.

3.2.5 Lattices

A *lattice* is a poset, $(A^2, \leq)$, in which any two elements have both meet and join. Consequently, all finite lattices have a greatest element, denoted $\mathbf{1}$, and a least element ($\mathbf{0}$), where $\mathbf{1} \in A$ and $\mathbf{0} \in A$.

All lattices enjoy the following properties:

$(P1)$ Idempotent	$x + x = x$	$x \cdot x = x$
$(P2)$ Commutative	$x + y = y + x$	$x \cdot y = y \cdot x$
$(P3)$ Associative	$x + (y + z) = (x + y) + z$	$x \cdot (y \cdot z) = (x \cdot y) \cdot z$
$(P4)$ Absorptive	$x \cdot (x + y) = x$	$x + (x \cdot y) = x$

It is actually possible to show that any non-empty algebraic system that enjoys these properties is a lattice. The Hasse diagrams of several lattices are shown in Figure 3.5. One can easily verify that the partial order we have defined on the points of the Cartesian plane in Section 3.2.1 is also a lattice.

As an exercise, we prove here one of the two forms of $P4$.

Theorem 3.2.4 *In any lattice,*

$$x \cdot (x + y) = x.$$

Proof. By definition, a lattice is a poset, in which all pairs of elements have both meet and join. Thus, by Theorem 3.2.1, $x \leq (x+y)$. Then, letting $z = (x+y)$, we have $x \leq z$, and by Theorem 3.2.2, $x \cdot z = x$. □

Another simple result we want to show is the following.

Theorem 3.2.5 *In any lattice,*

$$x \cdot (y+z) \geq (x \cdot y) + (x \cdot z).$$

Proof. From the definition of join, our result is established if we can prove that

$$\begin{aligned} (x \cdot y) &\leq x \cdot (y+z) \\ (x \cdot z) &\leq x \cdot (y+z) \end{aligned}$$

We now concentrate on the first inequality (the proof for the second is similar). We apply this time the definition of meet and say that our assertion is valid if

$$\begin{aligned} (x \cdot y) &\leq x \\ (x \cdot y) &\leq y \leq (y+z) \end{aligned}$$

These propositions are true by Theorem 3.2.1, hence our proof is complete. □

What we have proven is one of the two *distributive inequalities*. The other is

$$x + (y \cdot z) \leq (x+y) \cdot (x+z)$$

and can be proved similarly. However, we can spare some time if we resort to the *principle of duality*.

Duality Every identity is transformed into another identity by interchanging:

1. $+$ and $\cdot$;

2. $\leq$ and $\geq$;

3. $\mathbf{0}$ and $\mathbf{1}$.

This important principle follows directly from the duality of meet and join. (One definition is obtained from the other by interchanging $\leq$ and $\geq$.)

Complementation If $x + y = \mathbf{1}$ and $x \cdot y = \mathbf{0}$, then x is the *complement* of y (indicated by $\overline{y}$ or y') and vice versa. A lattice is *complemented* if all elements have a complement. In Figure 3.5, the lattice on the left is complemented: $a' = d,\ \ b' = c,\ \ c' = b,\ \ d' = a$. The lattice in the middle is not complemented, since there is no element x in this lattice such that $x + b = \mathbf{1}$, and $x \cdot b = \mathbf{0}$.

Distributivity A lattice is distributive if the two distributive properties hold by equality, i.e.,

$$x \cdot (y + z) = (x \cdot y) + (x \cdot z)$$

$$x + (y \cdot z) = (x + y) \cdot (x + z).$$

In Figure 3.5, the two leftmost lattices are distributive. However, the lattice on the right is complemented, but not distributive, since

$$b \cdot (c + d) = b \cdot e = b \neq a = b \cdot c + b \cdot d.$$

3.2.6 Definition of Boolean Algebras

A complemented, distributive lattice is a *Boolean lattice* or *Boolean algebra*. A Boolean algebra has the following properties:

$(P1)$ Idempotent	$x + x = x$	$x \cdot x = x$
$(P2)$ Commutative	$x + y = y + x$	$x \cdot y = y \cdot x$
$(P3)$ Associative	$x + (y + z) = (x + y) + z$	$x \cdot (y \cdot z) = (x \cdot y) \cdot z$
$(P4)$ Absorptive	$x \cdot (x + y) = x$	$x + (x \cdot y) = x$
$(P5)$ Distributive	$x + (y \cdot z) = (x + y) \cdot (x + z)$	$x \cdot (y + z) = (x \cdot y) + (x \cdot z)$
$(P6)$ Existence of the complement.		

It is also possible to prove that an algebraic system

$$(B, +, \cdot)$$

with those properties is a Boolean algebra. There are also other possibilities. One may define a Boolean algebra as an algebraic system $(B, +, \cdot)$ which satisfies the following postulates (Huntington):

$(P1')$ Commutative	$x + y = y + x$	$x \cdot y = y \cdot x$
$(P2')$ Distributive	$x + (y \cdot z) = (x + y) \cdot (x + z)$	$x \cdot (y + z) = (x \cdot y) + (x \cdot z)$
$(P3')$ Identities	$x + \mathbf{0} = x$	$x \cdot \mathbf{1} = x$
$(P4')$ Existence of the complement.		

The various definitions are equivalent. One can also define a *Boolean ring*.[2]

3.2.7 Examples and Properties of Boolean Algebras

Besides the well-known *two-valued* Boolean algebra or **switching algebra** (often referred to as *the* Boolean algebra), there are other Boolean algebras, both finite and infinite. As an example of an unfamiliar Boolean algebra, let us consider the following. Let n be the product of distinct relatively prime numbers, let D_n be the set of all divisors of n, and let meet and join stand for greatest common divisor and least common multiple, respectively. Then

$$(D_n, +, \cdot)$$

[2]The operations are in this case the AND and the XOR. This is the original formulation of Boole.

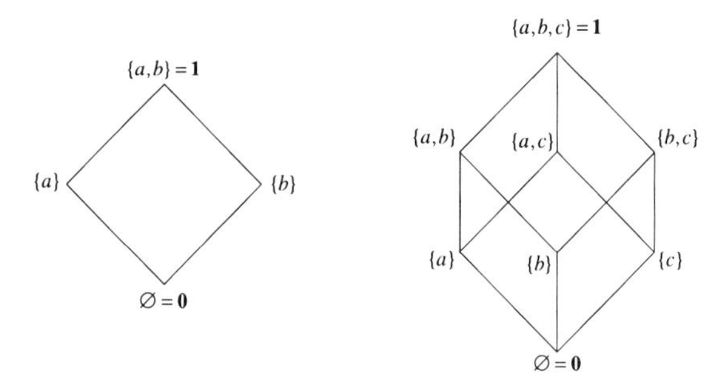

Figure 3.6: The Boolean algebra defined over the power sets of $\{a, b\}$ and $\{a, b, c\}$.

is a Boolean algebra, where

$$\mathbf{0} = 1 \quad \text{and} \quad \mathbf{1} = n.$$

In general, all finite Boolean algebras are isomorphic to the Boolean algebra defined over the power set of some finite set. The algebra defined over the power set of $\{a, b\}$ is shown in Figure 3.6. Such an algebra is called a class algebra. The carrier of the algebra is in this case $B = \{\emptyset, \{a\}, \{b\}, \{a, b\}\}$. The meet of two elements is their (set) intersection and their join is their union; complementation in the Boolean algebra coincides with complementation in the set-theoretical sense, considering $\{a, b\} = \mathbf{1}$ as the universal set in the complementation. Every Boolean algebra with four elements is isomorphic to, i.e., has the same Hasse diagram as, the Boolean algebra of Figure 3.6. This important result is known as *Stone's Representation Theorem.* We shall henceforth assume that all the algebras we consider are finite so that Stone's theorem will always apply.

In essence, the only relevant difference among the various Boolean algebras is the cardinality of the carrier. Furthermore, Stone's theorem implies that the cardinality of the carrier of a Boolean algebra must be a power of 2 and that we can use the concepts of set algebra to understand (and prove) identities in Boolean algebras.

Stone's representation theorem has the following simple application. Suppose we want to tell if a given Hasse diagram represents a Boolean algebra. We can count the number of nodes; if it is not a power of two, then we immediately conclude that the partial order is not a Boolean algebra.

All Boolean algebras have many properties in common. Here we prove some of them:

Theorem 3.2.6 *Complementation in a Boolean algebra is unique.*

Proof. Suppose both x' and y are complements of x. We show that $y = x'$. If y is a complement of x, then $x + y = \mathbf{1}$ and $xy = \mathbf{0}$. Hence:

$$y = y(x + x') = yx + yx' = yx' = x'y + x'x = x'(y + x) = x'.$$

□

Theorem 3.2.7 (Involution) *In a Boolean algebra:*

$$(x')' = x$$

Proof. By definition of complement, $xx' = \mathbf{0}$ and $x + x' = \mathbf{1}$. By commutativity, $x'x = \mathbf{0}$ and $x' + x = \mathbf{1}$. This means that x is the complement of x'. □

Theorem 3.2.8 *In a Boolean algebra:*

$$\begin{aligned} x + x'y &= x + y \\ x(x' + y) &= xy \end{aligned}$$

Proof. We can write: $x + y = x + (x + x')y = x + xy + x'y = x + x'y$. Alternatively, we can use distributivity: $x + x'y = (x + x')(x + y) = x + y$. □

Theorem 3.2.9 *In a Boolean algebra:*

$$\begin{aligned} x \leq y &\Leftrightarrow xy' = \mathbf{0} \\ &\Leftrightarrow x' + y = \mathbf{1} \end{aligned}$$

Proof. We only prove $x \leq y \Leftrightarrow xy' = \mathbf{0}$. The proof of the other result can be obtained by duality. We have:

$$x \leq y \Leftrightarrow xy' \leq yy' \Leftrightarrow xy' \leq \mathbf{0} \Leftrightarrow xy' = \mathbf{0}.$$

The property that $x \leq y \Leftrightarrow xz \leq yz$ is called the *isotone* property. To prove the isotone property, we recall that since a Boolean algebra is a poset, by Theorem 3.2.1, $x \leq y$ is equivalent to $x = xy$. From $x = xy$ we get $xz = xyz$. Now, with simple manipulation, we get $xz = xyzz = (xz)(yz)$, whence $xz \leq yz$. □

Note that by interpreting x and y as sets, and $\leq$ as the set containment operator, these inequalities may be verified with a Venn Diagram (See the discussion of Figure 3.1 on Page 79).

Theorem 3.2.10 (DeMorgan's Laws) *In a Boolean algebra:*

$$\begin{aligned} (x + y)' &= x'y' \\ (xy)' &= x' + y' \end{aligned}$$

Proof. We prove $(x + y)' = x'y'$ by proving $(x + y)' \leq x'y'$ and $x'y' \leq (x + y)'$. These in turn can be proved by applying Theorem 3.2.9 and Involution. Indeed,

$$\begin{aligned} x'y' \leq (x + y)' &\Leftrightarrow x'y'(x + y) = \mathbf{0}, \\ (x + y)' \leq x'y' &\Leftrightarrow x + y + x'y' = \mathbf{1}, \end{aligned}$$

which are easily seen to be true. □

Theorem 3.2.11 (Consensus) *In a Boolean algebra:*

$$\begin{aligned} xy + x'z + yz &= xy + x'z \\ (x + y)(x' + z)(y + z) &= (x + y)(x' + z) \end{aligned}$$

Proof. In this case we may again break down the proof of the equality into the proofs of two inequalities. Then we use duality to prove one of the two forms of the theorem. We see that $xy + x'z \leq xy + x'z + yz$ is verified using Theorem 3.2.1, which holds for any poset. The reverse inequality is verified if $yz \leq xy + x'z$. (This is actually an alternative way of formulating the theorem.) We apply again Theorem 3.2.9:

$$yz \leq xy + x'z \Leftrightarrow yz(xy + x'z)' = \mathbf{0} \Leftrightarrow yz(x' + y')(x + z') = \mathbf{0},$$

which is seen to be true. □

The consensus theorem can be used to simplify expressions. Let us see a few examples.

1. $abc + a'bd + bcd = abc + a'bd$;
2. $abc'd + c'd'e + abc'e = abc'd + c'd'e$;
3. $abe + bce + bde + ac'd' = abe + be(c + d) + a(c + d)' = bce + bde + ac'd'$;
4. $ab'c + bc'd + ad$ cannot be simplified;
5. $abc'd + abc + adc' = abc'd + abc + adc' + abd = abc + adc' + abd$.

In the simplification, we try to map the given expression into the form of Theorem 3.2.11. In the first two expressions we use this device to directly remove the redundant consensus terms. In the first, the mapping is $x = a$, $y = bc$, and $z = bd$, so the consensus term is $yz = bcbd = bcd$. In the second, after factoring out the common literal c', the mapping is $x = d$, $y = ab$, and $z = e$, so the redundant consensus term is $yz = abe$. In the third example, we use the same technique, after factoring out be from the second and third terms, and then applying DeMorgan's Laws to rewrite $c'd'$ as $(c + d)'$. As discussed in detail in Chapter 4, these manipulations are typical of the redundancy removal phase of logic minimization algorithms.

In the last example, we use the consensus theorem to *add a redundant consensus term*, and then apply the absorption property to the consensus term and one of the original terms to obtain the simpler expression. This type of two-step simplification procedure, in which the first step results in a more complicated procedure, also plays a key role in logic minimization (see Chapters 4 and 5).

In fact, instead of trying to map a given expression into the form of the consensus theorem, one might be tempted to use a so-called "Karnaugh Map" to identify the possible consensus terms for addition or deletion. This is indeed a valid approach, and will be introduced shortly, after first defining Boolean Functions.

3.3 Boolean Functions

An important example of a Boolean algebra that is distinct from the familiar switching algebra is given by the set of n-variable Boolean functions. As we shall shortly see, the Boolean functions of n variables form a Boolean algebra isomorphic to the power set of a set with $2^n \log_2 |B|$ elements. Before introducing this algebra, we must first define what we mean by a Boolean formula.

x_1	0	0	0	0	a	a	a	a	b	b	b	b	1	1	1	1
x_2	0	a	b	1	0	a	b	1	0	a	b	1	0	a	b	1
$F = x_1 + x_2$	0	a	b	1	a	a	1	1	b	1	b	1	1	1	1	1
$G = x_1 + x_1'x_2$	0	a	b	1	a	a	1	1	b	1	b	1	1	1	1	1

Table 3.2: Mapping of a two Boolean formulae representing the same Boolean function f.

3.3.1 Boolean Formulae

A function, as we have seen, is a mapping from a domain to a co-domain. We often use formulae to describe functions, but we have to keep in mind that the two things are distinct. In particular, we shall see that there are infinitely more formulae than functions (since there exist an infinite number of formulae which represent the same function).

Given a Boolean algebra B, we define **Boolean formulae** on the n variables $x_1, \ldots, x_n$. We shall assume that $x_i \in B, \quad i = 1, \ldots, n$. That is, each of the n variables is a Boolean variable which can take as its value any of the $|B|$ elements of B. Thus the set of Boolean formulae is defined recursively as follows.

Definition 3.3.1 *1. The elements of B are Boolean formulae.*

2. The symbols $x_1, \ldots, x_n$ are Boolean formulae.

3. If g and h are Boolean formulae, then so are

(a) $(g) + (h)$

(b) $(g) \cdot (h)$

(c) $(g)'$

4. A string is a Boolean formula if and only if it can be derived by applying the previous rules finitely many times.

We normally drop most parentheses, by assuming that "·" takes precedence over "+". We also drop "·" when no ambiguity arises. The following are Boolean formulae over $B = \{0, a, b, 1\}$ and the variables x_1, x_2:

$$\begin{aligned} F &= (((x_1) \cdot (a)) + (b)) \cdot (x_2) \\ &= (x_1 a + b)x_2 \end{aligned}$$

The second is obtained from the first by dropping the needless parentheses and the understood dots. Note that for $x_1 = a$, and $x_2 = b$, the formula F evaluates to $(aa + b)b = (a + b)b = b$.

Formulae are to be evaluated by a truth table with $|B|^n$ entries. For example, with $n = 2$ and $B = \{0, a, b, 1\}$, and for the two distinct formulae $F = x_1 + x_2$ and $G = x_1 + x_1'x_2$, we have the following $|B|^2 = 16$ evaluations. Note the two formulae have the same evaluation at every point in the domain B^n.

3.3.2 Boolean Functions

Whereas Boolean formulae are strings, Boolean functions are defined by the evaluation of Boolean formulae. Thus the definition of **Boolean functions** of n variables is also recursive. The functions defined by the following recursion have domain B^n and range B, that is $f(x) : B^n \longmapsto B$. The Boolean functions $f(x) : \{0,1\}^n \longmapsto \{0,1\}$ defined over the switching algebra $B = \{0,1\}$ will be called **switching functions functions**.

To distinguish formulae from functions, we shall, with some noted exceptions, denote functions with lower case symbols and formulae with upper case. When the context is clear, it shall be understood that a lower case English letter without subscript is a vector. That is, $x = \vec{x} = (x_1, \ldots, x_n)$.

Definition 3.3.2 *1. For any element $b \in B$, the* constant function, *defined by*

$$f(x_1, \ldots, x_n) = b \qquad \forall (x_1, \ldots, x_n) \in B^n,$$

is an n-variable Boolean function.

2. *For any x_i in $\{x_1, \ldots, x_n\}$, the* projection function, *defined by*

$$f(x_1, \ldots, x_n) = x_i \qquad \forall (x_1, \ldots, x_n) \in B^n,$$

is an n-variable Boolean function.

3. *If g and h are n-variable Boolean functions, then the functions $g+h$, $g \cdot h$, and g', defined by*

$$\begin{aligned} (g+h)(x_1, \ldots, x_n) &= g(x_1, \ldots, x_n) + h(x_1, \ldots, x_n) \\ (gh)(x_1, \ldots, x_n) &= g(x_1, \ldots, x_n) \cdot h(x_1, \ldots, x_n) \\ (g')(x_1, \ldots, x_n) &= (g(x_1, \ldots, x_n))', \end{aligned}$$

for all $(x_1, \ldots, x_n) \in B^n$, are also n-variable Boolean functions.

4. *Only the functions that can be derived by finitely many applications of the above rules are Boolean functions.*

As noted above in the discussion of Table 3.2, Notice that $x_1 + x_1'x_2$ and $x_1 + x_2$ are two distinct Boolean formulae, but they represent the same Boolean function. We shall show below that while not every function $f(x) : B^n \longmapsto B$ is a Boolean function, every function derived by evaluation of a Boolean formula *is* a Boolean function.

One can easily see that there are infinitely many formulae on a given set of n variables which represent the same function. For example, any literal of any term of any formula can be repeated an arbitrary number of times without changing the underlying function. However, there are only $|B|^{2^n}$ distinct Boolean functions, as we shall shortly see.

Notwithstanding this important difference, we shall not distinguish between functions and formulae, unless required by the context.

3.3.3 Boole's Expansion Theorem

In this section we discuss a fundamental result which is mostly known as Shannon Expansion, but it is actually due to Boole [27], as pointed out in [44]. We shall use the notation

$$\begin{aligned} f_{x_1'} &= f_{|x_1=0} &= f(0, x_2, \ldots, x_n), \\ f_{x_1} &= f_{|x_1=1} &= f(1, x_2, \ldots, x_n). \end{aligned}$$

to denote the function f restricted to the subdomain in which x_1 takes the value $0 = \mathbf{0}$ (respectively $1 = \mathbf{1}$). The functions $f_{x_1'}(x_2, \ldots, x_n)$ and $f_{x_1}(x_2, \ldots, x_n)$ are called the positive and negative **cofactors** of f with respect to x.

Theorem 3.3.1 *If $f : B^n \to B$ is a Boolean function, then*

$$\begin{aligned} f(x_1, x_2, \ldots, x_n) &= x_1' \cdot f(0, x_2, \ldots, x_n) + x_1 \cdot f(1, x_2, \ldots, x_n) \\ &= [x_1' + f(1, x_2, \ldots, x_n)] \cdot [x_1 + f(0, x_2, \ldots, x_n)], \end{aligned}$$

for all $(x_1, \ldots, x_n) \in B^n$.

Proof. The proof of the first identity is by induction. We give only a brief outline of this inductive proof. The statement is proved directly for constant and projection functions: That constitutes the base. The inductive step consists of considering separately, each the three mechanisms of Step 3 in the recursive Definition 3.3.2 to generate functions from other functions. That is, assuming the result is true for g and h, we prove it then to be also true for $g + h$, $g \cdot h$, and h'.

The second identity can be derived from the first by duality. Alternatively, it can be shown equivalent to the first by applying the distributive properties and consensus. □

Note we use square brackets instead of parentheses to emphasize that we are using the dual form of the expansion theorem. As an example, consider the function $f = x_1 x_2 x_3' + x_2(x_1' + x_3)$. Applying the two forms of the expansion theorem, we get:

$$\begin{aligned} f &= x_1'(x_2) + x_1(x_2 x_3' + x_2 x_3) &= (x_1' + x_1)x2 &= x_2, \\ f &= [x_1' + (x_2 x_3' + x_2 x_3)][x_1 + (x_2)] &= (x_1' + x_2)(x_1 + x_2) &= x_2. \end{aligned}$$

In these identities we have set $f(0, x_2, \ldots, x_n) = x_2(1+x_3) = (x_2)$, and $f(1, x_2, \ldots, x_n) = (x_2 x_3' + x_2 x_3) = x_2$. It is now easy to see that $f = x_2$.

Examples We can expand the functions

1. $wx'z' + xyz + w'y'z' + x'y' + w'xy$;

2. $y(xz + y'z') + xz' + wy$.

with respect to y, using *both* forms of the expansion theorem, as follows.

1. $wx'z' + xyz + w'y'z' + x'y' + w'xy = y'(wx'z' + w'z' + x') + y(wx'z' + xz + w'x)$ and $wx'z' + xyz + w'y'z' + x'y' + w'xy = [y' + wx'z' + xz + w'x][y + wx'z' + w'z' + x']$;

2. $y(xz + y'z') + xz' + wy = y'(xz') + y(xz + xz' + w)$ and $y(xz + y'z') + xz' + wy = [y' + xz + xz' + w][y + xz']$.

The expansion theorem can often be used to prove or disprove interesting identities. For example, we can show that for every Boolean function f:

$$f(x+y)\cdot f(x'+y) = f(1)\cdot f(y).$$

Proof. We expand the left-hand side with respect to x.

$$\begin{aligned} f(x+y)\cdot f(x'+y) &= x'(f(y)\cdot f(1)) + x(f(1)\cdot f(y)) \\ &= (x'+x)(f(1)\cdot f(y)) \\ &= f(1)\cdot f(y). \end{aligned}$$

□

3.3.4 The Minterm Canonical Form

Each Boolean function can be represented by infinite formulae. The problem of determining whether two formulae represent the same function is central to the minimization of Boolean functions.[3] We can impose restrictions on the form of the formulae, so that there is only one formula for each function. Such a form is called *canonical*. We shall spend more time on canonical forms in the future. For the time being, we introduce the *minterm canonical form*.

If we recursively apply Boole's expansion to a function, we eventually get

$$\begin{aligned} f(x_1,\ldots,x_{n-1},x_n) = &\quad f(0,\ldots,0,0)\, x_1'\ldots x_{n-1}'x_n' \\ &+ \ f(0,\ldots,0,1)\, x_1'\ldots x_{n-1}'x_n \\ &\ \vdots \\ &+ \ f(1,\ldots,1,1)\, x_1\ldots x_{n-1}x_n. \end{aligned}$$

The values

$$f(0,\ldots,0,0),\ \ f(0,\ldots,0,1),\ \ \ldots,\ \ f(1,\ldots,1,1)$$

are elements of B called the *discriminants* of the function f; the elementary products

$$x_1'\ldots,x_{n-1}'x_n',\quad x_1'\ldots x_{n-1}'x_n,\quad \ldots,\quad x_1\ldots x_{n-1}x_n$$

are called the *minterms*.

The *maxterm canonical form* is defined similarly.

$$\begin{aligned} f(x_1,\ldots,x_{n-1},x_n) = &\quad [f(0,\ldots,0,0)+x_1+\ldots+x_{n-1}+x_n] \\ &\cdot\ [f(0,\ldots,0,1)+x_1+\cdots+x_{n-1}+x_n'] \\ &\ \vdots \\ &\cdot\ [f(1,\ldots,1,1)+x_1'+\cdots+x_{n-1}'+x_n']. \end{aligned}$$

the elementary sums

$$x_1+\cdots+x_{n-1}+x_n,\quad x_1+\cdots+x_{n-1}+x_n',\quad x_1'+\cdots+x_{n-1}'+x_n'$$

[3] We should actually speak of minimization of Boolean formulae.

are called the *maxterms*. As an example, consider the function $f = x_1x_2 + ax_2'$ over $B = \{0, a, b, 1\}$. The minterm and maxterm canonical forms are:

$$
\begin{aligned}
f &= a \cdot x_1'x_2' + 0 \cdot x_1'x_2 + a \cdot x_1x_2' + 1 \cdot x_1x_2 \\
f &= [a + x_1 + x_2][0 + x_1 + x_2'][a + x_1' + x_2][1 + x_1' + x_2']
\end{aligned}
$$

Examples We can find the minterm and maxterm canonical forms for the functions

1. $f(x, y) = xy + y'a$;
2. $f(x, y, z) = x + y' + z$;
3. $f(x, y) = ax + bx'y$.

as follows (Assume $B = \{0, a, b, 1\}$.).

1. $f(x, y) = ax'y' + axy' + xy = (a + x + y)(a + x' + y)(x + y')$;
2. $f(x, y, z) = x'y'z' + x'y'z + x'yz + xy'z' + xy'z + xyz' + xyz = (x + y' + z)$;
3. $f(x, y) = bx'y + axy' + axy = (x + y)(b + x + y')(a + x' + y)(a + x' + y')$.

As a consequence of Theorem 3.3.1, each Boolean function has a unique minterm and maxterm canonical form. Therefore, a Boolean function is entirely characterized by its discriminants. This statement may seem inconspicuous, when referred to the Boolean algebra $B = \{0, 1\}$, which is also called the *switching algebra*. However, it is not so obvious for larger Boolean algebras, and this remarkable result is one of the the cornerstones of the theory of Boolean functions.

For example, as in Table 3.2, a 2-variable function defined over a 4-valued Boolean algebra has 16 different points in its domain. However, the image of all the 16 points is fixed once the images of the four points

$$(0, 0), (0, 1), (1, 0), (1, 1)$$

are given. As a consequence, not all possible functions $f : B^n \to B$ are Boolean functions. Furthermore, since there are exactly 2^n discriminants, each with the possibility of $|B$ values, the number of distinct Boolean functions of n variables is precisely $|B|^{2^n}$. The remaining $|B|^{|B|^n} - |B|^{2^n}$ functions are not Boolean (they cannot be represented by formulae derived according to our definition).

This counting method can be used to determine the number of Boolean functions of four variables, defined over the class algebra generated by the set $\{a, b\}$? (The carrier of the algebra is the power set of $\{a, b\}$.) The power set of $\{a, b\}$ is $\{\emptyset, \{a\}, \{b\}, \{a, b\}\}$. Hence, $|B| = 4$. The number of Boolean functions of four variables is thus:

$$|B|^{2^n} = 4^{2^4} = 2^{32} = 4294967296.$$

It should be clear that even though both the carrier and the number of variables may be relatively small, there are can be a very large number of possible Boolean functions. As we shall see at the end of this chapter, the task of designing a logic function can be viewed as the task of picking a "best" function for a given incomplete specification.

3.3.5 Pseudo-Boolean Functions

We give the following example of a 1-variable Boolean function from B to B which is not Boolean, where $B = \{0, a, b, 1\}$.

0	a
a	a
b	b
1	1

The minterm canonical form for the function is $ax' + x$, which can be simplified to $a + x$. For $x = b$, however the minterm canonical form evaluates to $a + b = 1$. Since $1 \neq b$, the given truth table is not consistent with the minterm canonical form. Hence, the function is not Boolean.

One may ask what fate is reserved to the many mappings from B^n to B that are *not* Boolean functions. Such functions are members of the class of functions known as **Pseudo-Boolean Functions**[4].

It turns out that there are *multi-valued logics* that can deal with all these mappings. The algebraic structure of these logics is somewhat weaker than the one of Boolean algebras, yet sufficient to deal with, say, heuristic minimization. (See Section 5.)

3.3.6 The Boolean Algebra of n-variable Boolean Functions

Returning to Boolean functions, we mentioned at the end of Section 3.2.7 that the Boolean functions over n variables form a Boolean algebra. The meet and join are the union $(+)$ and intersection $(\cdot)$ of functions introduced in the definition of Boolean functions. The **0** and **1** of the algebra are the constant 0 and 1 functions, respectively. This is illustrated in Figure 3.7. A function $f(x_1, \ldots, x_n)$ is greater than or equal to another function $g(x_1, \ldots, x_n)$ if and only if, for every assignment of values to the variables, $a_1, \ldots, a_n$, $f(a_1, \ldots, a_n) \geq g(a_1, \ldots, a_n)$. In Figure 3.7 we can see that, as we proceed from the smallest (0) to the largest (1) function, the number of minterms for which the functions are 1 increases. Since Boolean functions form Boolean algebras, all the theorems we have proved for Boolean algebras can be used to manipulate expressions involving Boolean functions.

Further, the theorems and properties that hold for all lattices, or for all posets, are also applicable to Boolean functions, since a Boolean algebra is also a special form of lattice and poset. For example, the absorptive property (P4) of Section 3.2.5 on Page 90 is applicable, as are the identities of Theorem 3.2.2 of Page 88. As we shall see, logic minimization algorithms are just methods of systematically applying these results.

3.3.7 Atoms of a Boolean Algebra

A Boolean algebra is characterized by its set of **atoms**.

[4] A function $f : B^n \longmapsto R$ is Pseudo-Boolean if and only if B is a Boolean algebra.

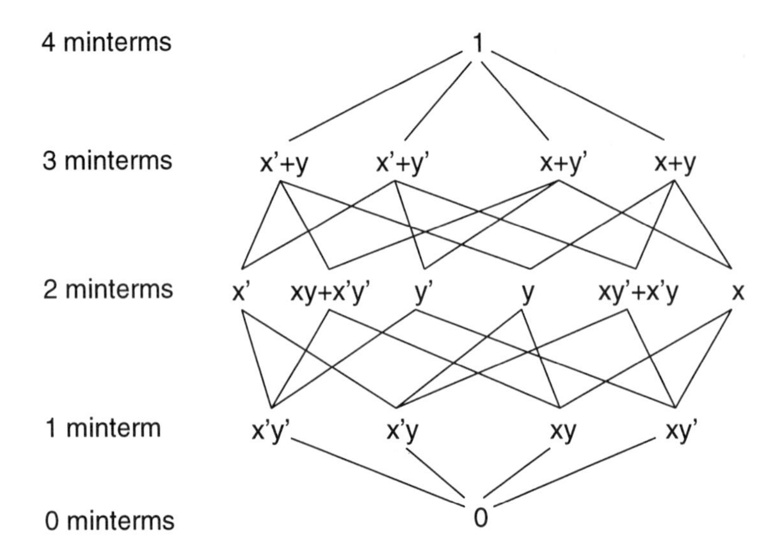

Figure 3.7: The Boolean algebra of the Boolean functions of two variables over $B = \{0, 1\}$.

Definition 3.3.3 *The atoms of a Boolean algebra are its minimal nonzero elements. They are of special significance, since in a Boolean algebra with carrier B and atoms $a \in A \subseteq B$,*

$$|B| = 2^{|A|}, \quad B = 2^A.$$

Since Stone's theorem has established that the only significant difference between Boolean algebras is their cardinality, we can also say that a Boolean algebra is uniquely characterized by its set of atomic elements A.

Since atoms are minimal, we have the following identities:

$$\begin{array}{lll} (i) & a \cdot b = \mathbf{0}, & \forall a, b \in A, \\ (ii) & a + b \notin A, & \forall a, b \in A, \\ (iii) & \sum_{a \in A} a = \mathbf{1}. & \end{array}$$

Atoms are closely related to minterms. For the Boolean function algebra $F_n(\{0, 1\})$ of n-variable switching functions, the atoms are just the minterms. For a larger base algebra, say $B = \{\mathbf{0}, a', a, \mathbf{1}\}$, the atoms of $F_n(B)$ are just the meet of the minterms (atoms) of $F_n(\{0, 1\})$ with the atoms of the base algebra B. For example, consider this base algebra for the case $n = 2$. In this case there are 4 minterms and the 8 atoms of $F_2(B)$ are

$$x_1'x_2'a', \; x_1'x_2'a, \; x_1'x_2a', \; x_1'x_2a, \; x_1x_2'a', \; x_1x_2'a, \; x_1x_2a', \; x_1x_2a,$$

Note that the atomic variable a of the base algebra B behaves just like a third switching variable.

Any element b of a Boolean algebra B with n atoms, is uniquely characterized as a set atoms, that is, as a subset of A. Thus one may employ the binomial theorem to categorize the elements of B. If $|A(B)| = n$, then the Hasse Diagram for the lattice of

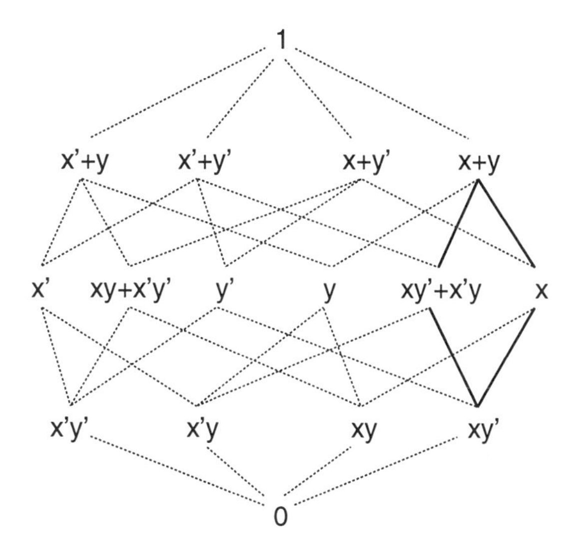

Figure 3.8: The interval $[xy', x + y]$ (represented by solid lines).

B has n levels. On the 0th level, there is only the zero element **0**. On Level 1, there are $C_{n,1} = n$ elements, each containing 1 element. Similarly, on Level l there are

$$C_{n,l} = \frac{n!}{l!(n-l)!}, \quad l = 0, \ldots, n.$$

For $n = 4$, this gives 1 element on Level 0, $\frac{4!}{1!3!} = 4$ on Level 1, $\frac{4!}{2!2!} = 6$ on Level 2, $\frac{4!}{3!1!} = 4$ on Level 3, and 1 element on Level 4, for a total of $1+4+6+4+1 = 16 = 2^4$ elements altogether.

3.4 Don't Care Conditions as Boolean Function Algebra Intervals

An **interval** $[L, U]$ in a Boolean algebra B is the subset of B defined by

$$[L, U] = \{x \in B : L \leq x \leq U\}.$$

When B is the Boolean algebra of n-variable switching functions (defined in Section 3.3.2), we can thus define an interval in the Boolean algebra $F_n(\{0,1\})$ of n-variable switching functions. For example, consider $[xy', x + y]$. This interval is highlighted in Figure 3.8. An interval of a Boolean algebra is also a Boolean algebra and can thus be understood in terms of its **atoms**, as discussed in Section 3.3.7.

Intervals in a Boolean function algebra associated with logic circuits are related to don't care conditions. These conditions are of two types, called **satisfiability don't cares** and **observability don't cares** . These don't care conditions arise due to the "filtering" action that is imposed by the environment of a given subcircuit, when it is embedded in a larger circuit. For example, a 16-bit multiplier circuit may have on the order of 6000 gates. Inside this large circuit, there are many subcircuits, for example 2-bit full adders, which find themselves embedded in a sea of gates. The portion of the circuit lying between these subcircuits and the primary inputs constitutes a "digital filter", which prevents certain local (to the subcircuit) input combinations from occurring.

For a given subcircuit g, we refer to such combinations as the **Satisfiability Don't Cares** of g, and we denote their union as D_g^{Sat}. As repeatedly demonstrated in the sequel, logic synthesis programs like SIS exploit such Don't Care conditions to synthesize efficient representations of Boolean functions.

Similarly, the portion of a circuit lying between a subcircuit g and the primary outputs constitutes a "filter", which prevents the external environment from distinguishing between certain local (to the subcircuit) input combinations. We call such combinations **Observability Don't Cares**, and we denote their union as D_g^{Obs}.

3.4.1 Satisfiability Don't Care Conditions

The origins of Satisfiability don't care conditions are illustrated in Figure 3.9. In this figure, the overall circuit is comprised of a single gate subcircuit g, which is driven by a subcircuit f, and which drives a subcircuit h, Subcircuit h is also driven by f, a phenomenon known as "reconvergent fanout". This phenomenon is ubiquitous since it usually denotes the the "re-use" of logic for area efficiency purposes. It is easily seen that

$$\begin{aligned} w &= g(x, y) \\ &= xy' + x'y \\ x &= u' \\ y &= (v' + v)' \\ &= \mathbf{0} \end{aligned}$$

Since $y = \mathbf{0}$, it is clear that the input pairs (x, y) in the set $\{(0,1),(1,1)\}$ will never occur to subcircuit g. Hence the local minterms

$$D_g^{Sat} = \{x'y, xy\}$$

can be regarded as "satisfiability don't cares". That is, we may replace g with any function in the interval

$$\mathcal{I}^{Sat} = [g_L^{Sat}, g_U^{Sat}], \text{ where}$$

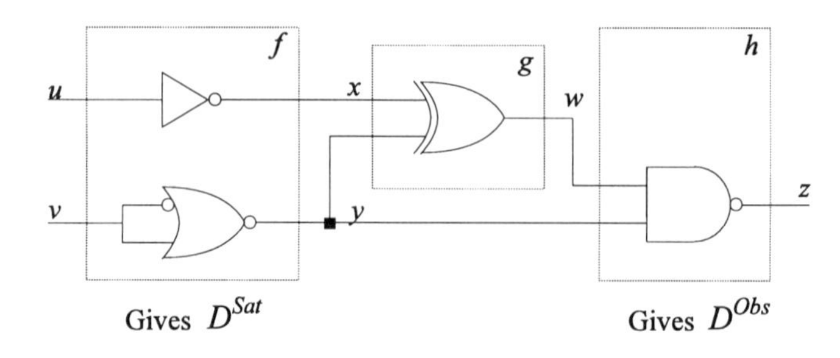

Figure 3.9: A simple example relating intervals in a Boolean function algebra to satisfiability and observability don't care conditions.

$$\begin{aligned} g_L^{Sat} &= g - D^{Sat} \\ &= \{xy', x'y\} - \{x'y, xy\} \\ &= \{xy'\}, \quad \text{and} \\ g_U^{Sat} &= g \cup D^{Sat} \\ &= \{xy', x'y\} \cup \{x'y, xy\} \\ &= x + y. \end{aligned}$$

Here we take advantage of the fact that join in a Boolean function algebra is set union. We thus see that $\mathcal{I}^{Sat}$ is just the bold interval in Figure 3.9. Note that minterm xy is in the upper bound, but, since $xy \not\geq xy' = g_L^{Sat}$, this minterm is not in the interval.

3.4.2 Observability Don't Care Conditions

The Observability don't care interval may be similarly derived in terms of D^{Obs}. In Figure 3.9, we see that $y = 0 \Rightarrow z = 1$. Hence variable w (and thus gate g) is don't care for input combinations with $y = 0$. Hence

$$D_g^{Obs} = \{xy', x'y'\}$$

can be regarded as "observability don't cares". That is, we may replace g with any function in the interval

$$\begin{aligned} \mathcal{I}^{Obs} &= [g_L^{Obs}, g_U^{Obs}], \quad \text{where} \\ g_L^{Obs} &= g - D^{Obs} = \{xy', x'y\} - \{x'y', xy'\} \\ &= \{x'y\}, \quad \text{and} \\ g_U^{Obs} &= g \cup D^{Obs} = \{xy', x'y\} \cup \{x'y', xy'\} \\ &= x' + y'. \end{aligned}$$

Note that if we take $D = D^{Sat} + D^{Obs}$, we get

$$D = D^{Sat} + D^{Obs} = \{xy', x'y'\} \cup \{x'y, xy\} = \mathbf{1},$$

which implies the overall don't care interval has a lower bound $g_L = g - \mathbf{1} = \mathbf{0}$, and upper bound $g_U = g + \mathbf{1} = \mathbf{1}$.

This is corroborated by noting that

$$\begin{aligned} w &= g(x, y) = xy' + x'y \\ x &= u' \\ y &= (v' + v)' = (\mathbf{1})' = \mathbf{0} \\ z &= (wy)' = \mathbf{1} \end{aligned}$$

While the existence and importance of don't care conditions is easily demonstrated, it may nevertheless be surprising that don't care conditions are not the exception but the rule. In fact, in many cases the difficulty is not in identifying a large set of useful don't care conditions. Instead, the difficulty is that there are so many distinct don't care conditions, it is practically impossible to enumerate and store them all for future use. The art of don't care exploitation lies in identifying a subset of all such conditions which is "sufficient" for quasi-optimal design.

3.4.3 Deriving Don't Cares From and Interval Specification

If the interval of permissible functions[164] is specified *a priori*, the corresponding don't care set is easily derived. If the interval of a Boolean function algebra is given as $[L, U]$, we see that every element (function) of the interval must contain the atoms, and no element of the interval contains an atom not in U. Thus the "discretionary" atoms are just $D = U - L$. If all the discretionary atoms are included, then element U is obtained, and if none are included on obtains the element U. Any other subset of D yields some interior element of the interval.

3.5 Incomplete Specification of Boolean Functions

We have seen that it is important to be able to capture degrees of freedom in the specification of Boolean functions. For a given Boolean algebra B, the set of n-variable incompletely specified Boolean functions is just the set of distinct intervals in the Boolean function algebra, denoted

$$\mathcal{I} = \{[a, b] \in F_n^2 B : a \leq b\}.$$

An incompletely specified Boolean function can then be defined as an element of $\mathcal{I}$.

However, before considering this general case, we first describe the simpler and more familiar case of incompletely specified switching functions ($F_n(B)$, where $B = \{0, 1\}$). We then generalize our treatment to Boolean function algebras with larger carriers, as in Figure 3.6.

3.5.1 Incompletely Specified Switching Functions

For the special case of $B = \{0, 1\}$, the base algebra has only three intervals:

$$[0, 0], \quad [0, 1], \quad [1, 1].$$

These intervals are normally indicated by

$$0, \quad d, \quad 1$$

or

$$0, \quad --, \quad 1.$$

(d and $--$ are called **don't care** values). For every particular function within a given interval in the switching algebra $F_n(\{0, 1\})$ we obtain a (completely specified) switching function. Alternatively, a given set of **permissible** switching functions form an interval in their algebra[5].

For example, the incompletely specified switching function

[5]This approach is taken, for instance, in [204].

00	0
01	–
11	–
10	1

corresponds to the interval

$$[xy', x + y]$$

of Figure 3.8. Besides the tabular representation and the interval notation, it is possible to specify an incompletely specified switching function *ff* as a triple of three completely specified switching functions [37]

$$f, d, r,$$

where f is 1 for all values for which *ff* is 1, d is 1 for all values for which *ff* is –, and r is 1 for all values for which *ff* is 0. These functions are called the **ON-set**, **don't-care**, and **OFF-set**, respectively. Clearly the three functions must have null intersection, and once two of them are specified, the third is implicitly given. Thus f, d, and r form a partition of the domain of *ff*.

If an interval $[L, U]$ is given in the Boolean algebra of switching functions, the corresponding triple (f, d, r) can be obtained as follows. The care ON-set is $f = L$, and the care OFF-set is $r = \overline{U}$. Thus don't care set is $d = U - L = \overline{f + r} = \overline{L + \overline{U}} = U\overline{L}$. Alternatively, if the triplet (f, d, r) is given, then one may compute the interval as follows. $L = f$, $U = f + d$.

The specification by means of the triplet f, d, and r is convenient from the point of view of the logic minimization algorithms. However, it does not provide much insight into the interval structure of the solutions to minimization problems, which can be useful in more advanced applications. We shall also need to generalize these concepts when we deal with functions from B^n to B^m, or multiple-output functions (See Section 4.10, Page 160).

3.5.2 Incompletely Specified Boolean Functions

For larger algebras, the interval-based approach has one additional advantage: By specifying the set of possible functions as an interval of a Boolean algebra, we are guaranteed that all elements will be Boolean functions. This problem does not arise in the case of switching functions, since all possible mappings are Boolean functions.

Of course, representation as a triplet does not work for incompletely specified n-variable Boolean function algebras $F_n(B)$ built on larger base algebras B. This is because each discriminant can be any member of the base algebra B, not just $\mathbf{0}$ and $\mathbf{1}$. For example, if B is the base (or carrier) algebra $\mathbf{0}, a', a, \mathbf{1}$, one would need an analogous quintuple: $f^{\mathbf{0}}$ to represent the set of minterms with discriminants equal to $\mathbf{0}$, $f^{a'}$ to represent the set of minterms with discriminants equal to a', f^a to represent the set of minterms with discriminants equal to a, $f^{\mathbf{1}}$ to represent the set of minterms with discriminants equal to $\mathbf{1}$, and f^{DC} to represent the set of minterms with discriminants whose value is not specified, and can be chosen to optimize the representation.

The interval notation, on the other hand, works equally well for arbitrarily large base algebras.

3.6 Notes

Detailed treatments of Boolean algebras can be found in [133, 251, 134, 235, 171, 44, 163]. In particular, Kurshan's book [163] is especially relevant to verification, given his treatment of subalgebras and the "Lifting" lemma[6]. Brown's book is especially relevant to the synthesis context, and makes an excellent reference book. The theory of Don't Cares in logic synthesis had it's beginning in the early publications by Muroga [164, 148]. Use of this theory became widespread due to the work by Brayton, Hachtel, and Sangiovanni-Vincentelli and their students, especially Rudell. Key early publications were the papers by Bartlett, et al [16, 15] appeared, followed by Muroga's transduction paper [204].

3.7 Summary

If you gave this chapter a careful reading, and went through a selection of the solved problems of Section 3.8, you should be comfortable with:

1. Sets and Venn diagrams, binary relations and functions, partitions and equivalence classes;

2. reflexivity, symmetry, antisymmetry, and transitivity of binary relations;

3. Three important kinds of reflexive binary relations: equivalence classes, partial orders, and compatibility relations;

4. The hierarachical relationship between partial orders, lattices, and Boolean algebras;

Equivalence relations are one of the cornerstones in the theoretical foundation of FSM synthesis and formal verification. They will be a principal focus of the treatment of state minimization of completely specified FSMs in Chapter 7.1. Compatibility relations play a similar role in the discussion of state minimization for incompletely specified machines, the subject of Section 8.1 of Chapter 8.

Since Boolean algebras are based on lattices and partial orders, the theory of partial orders permeates the entire book.

We have treated "large" Boolean algebras and subalgebras, and have shown how logic synthesis "lives" in the world of the "large" Boolean algebra, $F_n(\{0,1\})$, of n-variable switching functions. We have related intervals and subalgebras of $F_n(\{0,1\})$ to don't cares and sets of permissible functions, which play prominent in logic synthesis algorithms.

3.8 Problems

1. For an arbitrary set S, which of the following statements are true?

$$\begin{aligned} \emptyset &\in 2^S \\ \emptyset &\subseteq 2^S \end{aligned}$$

[6] This topic is treated in the Chapter on Finite Automata — see Section 9.5.1.

Solution. Both statements are true. Indeed, the empty set is a subset of all sets. Hence it is a subset of 2^S and S. The latter implies that $\emptyset \in 2^S$. □

2. Using the fact that $\emptyset \subseteq T$, for any set T, show that there is only one empty set.

   Solution. Suppose there are two distinct empty sets, ϕ and ψ. Then it must be $\phi \subseteq \psi$ and $\psi \subseteq \phi$. But this is equivalent to $\psi = \phi$, contradicting the assumptions that the two sets were distinct. □

3. How many relations are there from an m-element set to an n-element set?

   Solution. Suppose the two sets are P and Q. Since a relation from P to Q is a subset of the Cartesian product $P \times Q$, we have to count the number of subsets of $P \times Q$. But $P \times Q$ has $m \cdot n$ elements; hence, $2^{P \times Q}$ has $2^{m \cdot n}$ elements. □

4. Decide, for each of the following sets of ordered pairs, whether the set is a function.

   (a) $\{(x, y) : x \text{ and } y \text{ are people and } x \text{ is the mother of } y\}$
   (b) $\{(x, y) : x \text{ and } y \text{ are people and } y \text{ is the mother of } x\}$
   (c) $\{(x, y) : x \text{ and } y \text{ are real numbers and } x^2 + y^2 = 1\}$

   Solution. Relation (a) is not a function, because a mother may have more than one daughter. Relation (b) is a function, because a daughter has exactly one mother. Relation (c) is not a function: It describes a circle of unit radius. □

5. Consider the following binary relation on $A = \{a, b, c\}$ defined by

$$\mathcal{R} = \{(a, a), (b, b), (a, c), (c, a), (c, b)\}.$$

   Is $\mathcal{R}$:

   (a) reflexive?
   (b) symmetric?
   (c) antisymmetric?
   (d) transitive?
   (e) a partial order?
   (f) an equivalence relation?
   (g) a function?

   Solution. Relation $\mathcal{R}$ is:

   (a) Non reflexive ((c, c) is missing);

(b) non symmetric (there is (c, b), but (b, c) is missing);

(c) non transitive (there are (a, c) and (c, b), but (a, b) is missing);

(d) non antisymmetric (there are both (a, c) and (c, a));

(e) not a partial order;

(f) not an equivalence relation;

(g) not a function (for instance, a appears as first element in two pairs).

□

6. For each case below, give a binary relation ρ on $S = \{a, b, c, d\}$ satisfying the given conditions.

   (a) ρ is reflexive and symmetric but not transitive;

   (b) ρ is reflexive and transitive but not symmetric;

   (c) ρ is not reflexive or symmetric but is transitive;

   (d) ρ is reflexive but neither symmetric nor transitive;

   (e) ρ is neither symmetric nor antisymmetric.

7. Let $\mathcal{R}^{-1} = \{(x, y) : y\mathcal{R}x\}$. Show that $\mathcal{R}^{-1}$ is a partial order if and only if $\mathcal{R}$ is a partial order, by verifying that $\mathcal{R}^{-1}$ is reflexive, antisymmetric, and transitive if and only if $\mathcal{R}$ is so.

   Solution. To prove that $\mathcal{R}^{-1}$ has the antisymmetric property if and only if $\mathcal{R}$ has it, one proceeds as follows:

$$x\mathcal{R}^{-1}y \wedge y\mathcal{R}^{-1}x \Rightarrow x = y$$

   is equivalent (by definition) to

$$y\mathcal{R}x \wedge x\mathcal{R}y \Rightarrow x = y,$$

   which can be rewritten as

$$x\mathcal{R}y \wedge y\mathcal{R}x \Rightarrow x = y.$$

   The other properties can be proved similarly. □

8. Find $\mathcal{R}^{-1}$ for:

$$\mathcal{R} = \{(a, a), (b, b), (c, c), (d, d), (a, b), (a, c), (b, c), (d, c)\}.$$

   Is $\mathcal{R}^{-1}$ a partial order?

   Solution.

$$R^{-1} = \{(a, a), (b, b), (c, c), (d, d), (b, a), (c, a), (c, b), (c, d)\}.$$

   R^{-1} is a partial order. (See Problems 6 and 7.) □

9. For the following equivalence relation over $A = \{a, b, c\}$:

$$R = \{(a, a), (b, b), (c, c), (a, b), (b, a)\},$$

what is the equivalence class of a? (See Section 3.1.3.)

10. (a) Enumerate the total number of possible partitions of a 3-element set.

    (b) Enumerate the total number of possible partitions of a 4-element set.

11. As defined in Definition 3.1.2, given two partitions π_1 and π_2 of a set S, π_1 is a *refinement* of π_2 if each block of π_1 is a subset of a block of π_2. Show that refinement is an antisymmetric relation on the set of all partitions of S.

    Solution. If π_1 is a refinement of π_2, then for each block B_i^1 of π_1 there is a block B_j^2 of π_2 such that $B_i^1 \subseteq B_j^2$. If also π_2 is a refinement of π_1, then B_j^2 must be contained in a block B_k^1 of π_1. However, $B_i^1 \subseteq B_j^2 \subseteq B_k^1$ implies $B_i^1 \subseteq B_k^1$. Since the blocks of a partition are disjoint, this implies $i = k$. Hence $B_i^1 \subseteq B_j^2 \subseteq B_i^1$ or equivalently $B_i^1 = B_j^2$. Hence each block in one partition has an identical block in the other partition and $\pi_1 = \pi_2$. □

12. Consider the relation ρ defined over the set of the positive *integers* by: $x \rho y$ if and only if $x \cdot y = 60$. For instance, $3\rho 20$. Is ρ:

    (a) reflexive?

    (b) symmetric?

    (c) antisymmetric?

    (d) transitive?

    (e) a partial order?

    (f) an equivalence relation?

    (g) a function?

13. An example of equivalence relation on the set $I = \{0, 1, 2, \cdots$ of the natural numbers is equality modulo 3. Give the partition induced by this equivalence relation.

    Solution. $I^0 = \{0, 3, 6, 9, 12...\}, I^1 = \{1, 4, 7, 10, 13...\}, I^2 = \{2, 5, 8, 11, 14...\}$ □

14. Write, in the form of set of pairs, the partial order described by the Hasse diagram of Figure 3.10.

    Solution. The set of pairs is:

$$\{(a, a), (b, b), (c, c), (d, d), (e, e), (a, d), (b, d), (b, e), (c, e)\}.$$

□

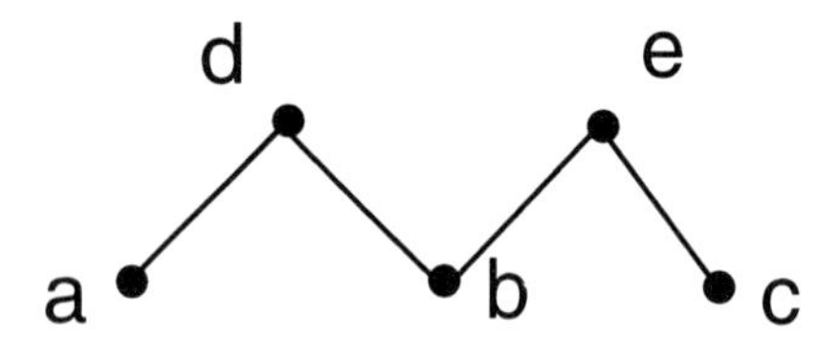

Figure 3.10: Hasse Diagram for Problem 14.

15. Consider the relation on the positive integers "x is less than or equal to y." What is the greatest lower bound of x and y? What is the least upper bound? Is the relation a lattice? Is it finite? Specifically, what is the greatest lower bound (or meet) of 64 and 56? Explain.

    Solution. The greatest lower bound is the smaller of x and y. The least upper bound is the larger of the two. If $x = y$, then both x and y are both greatest lower bound and east upper bound. The relation is a lattice, because, given two positive integers, we can always find which one is greater. It is an infinite lattice, because the positive integers form an infinite set. Finally, the meet of 64 and 56 is 56. □

16. Consider the partial order

$$\mathcal{R} = \begin{array}{c}\{(a,a),(a,b),(a,c),(a,d),(a,e),(b,b),\\(b,e),(c,c),(c,d),(c,e),(d,d),(d,e),(e,e)\}.\end{array}$$

    (a) Give the Hasse diagram for this partial order.
    (b) Is it a lattice?
    (c) Is it a Boolean algebra?
    (d) Find $\mathcal{R}^{-1}$.
    (e) Draw the Hasse diagram for $\mathcal{R}^{-1}$.
    (f) Is $\mathcal{R}^{-1}$ a lattice?
    (g) Is $\mathcal{R}^{-1}$ a Boolean algebra?

    Explain.

17. Which of the Hasse diagrams of Figure 3.11 represents a lattice? Explain.

    Solution. The diagram on the left does not correspond to a lattice. Indeed, a and b have no meet. The diagram on the right, on the other hand, corresponds to a lattice, because meet and join are defined for all pairs. □

18. Consider the lattice of Figure 3.12.

    (a) Is it complemented?
    (b) Is it distributive?

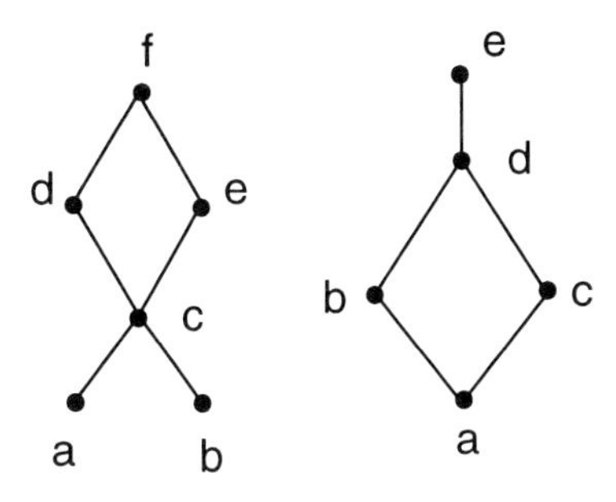

Figure 3.11: Hasse Diagram for Problem 18.

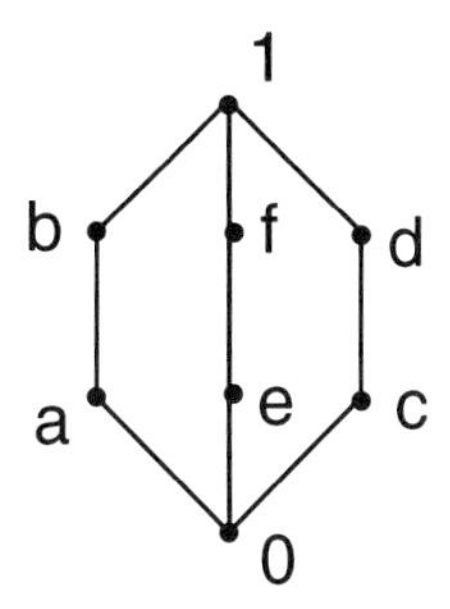

Figure 3.12: Lattice for Problem 18.

(c) Is it a Boolean algebra?

Explain.

19. Prove or disprove the following identities for Boolean algebras, *without* using truth tables. For those equalities that are not identities, give a counterexample. For those that are identities, say what theorems can be applied to prove the identities.

(a) $x' \oplus (x+y) = x + y'$

(b) $(x+y)(x+yz) = x + yz$

(c) $ab + b'cd' + acd' = ab + b'cd'$

(d) $xy + x(y+z) = x(y+z)$

(e) $ab + b'cd' + acd' = ab + cd'$

Solution.

(a) $x' \oplus (x+y) = x'x'y' + x(x+y) = x'y' + x = x + y'$, by applying the definition of exclusive or, idempotency, absorption, and $a + a'b = a + b$.

(b) Absoprtion, since $(x+yz) \leq (x+y)$.

(c) Consensus.

(d) Distributivity and idempotency.

(e) Not an identity. The right side contains the term bcd', while the left side does not.

□

20. Consider the points of the discrete Cartesian plane I^2, where $I = \{0, 1, 2, 3, 4\}$. Define an ordering relation as follows: $(i_1, j_1) \leq (i_2, j_2)$ if and only if $i_1 \leq i_2$ and $j_1 \leq j_2$. Then $(R^2, \leq)$ is a partial order.

    (a) Draw the Hasse diagram for this partial order.

    (b) Is it a lattice?

    (c) Is it a Boolean algebra?

21. Prove, without resorting to duality, the second of the two distributive inequalities for lattices:

$$x + (y \cdot z) \leq (x + y) \cdot (x + z)$$

22. Draw the Hasse diagram for the following partial order:

$$R = \{(a, a), (b, b), (c, c), (d, d), (a, b), (a, c), (b, c), (d, c)\}.$$

23. Show that any subset of a poset is itself a poset, relative to the same inclusion relation. (Use the definition of poset.)

    Solution. Using the same inclusion (or ordering) relation, means that we keep all the pairs of the original relation such that both elements are from the subset being considered. We prove that the subset satisfies the transitive property: The other two properties can be proved similarly.

    If $S \subseteq P$ does not have the transitive property, then there are two pairs (x, y) and (y, z) in the restriction of the relation to S and there is no (x, z). This, however, is impossible since the presence of (x, y) and (y, z) in P implies that (x, z) is also present, and the construction of the reduced relation would include (x, z), since both x and z belong to S. □

24. Let $S = \{1, 2, \ldots, 10\}$. Let ρ be a binary relation on S defined $a\rho b$ if $2a > b$. Is (S, ρ) a poset? Why or why not? What about (X, ρ) where $X = \{1, 2\}$?

25. Draw the Hasse diagram for each of these two partially ordered sets.

    (a) $S = \{1, 2, 3, 5, 6, 10, 15, 30\}$
    $x\rho y \leftrightarrow x$ divides y.

    (b) $S = 2^{\{1,2,3\}}$
    $A\rho B \leftrightarrow A \subseteq B$.

    What do you notice by comparing the structure of these two graphs?

    Solution.

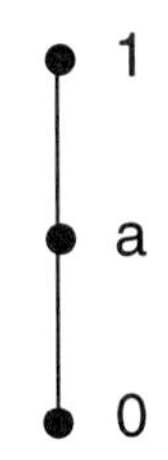

Figure 3.13: Partially ordered set (poset) for Problem 26.

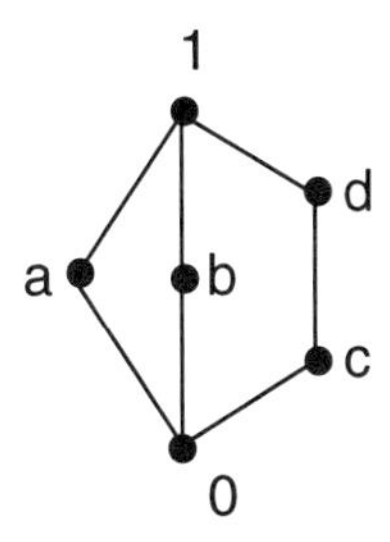

Figure 3.14: Partially ordered set (poset) for Problem 27.

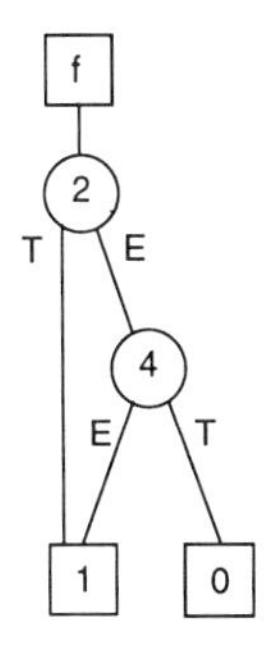

The two graphs are isomorphic, i.e., they have the same structure. This is an example of application of Stone's representation theorem. □

26. For the poset of Figure 3.13, write the ordering relation in the form of set of ordered pairs.

27. Is the poset of Figure 3.14 a lattice? Is it distributive? Is it complemented? Is it a Boolean algebra?

Solution. The poset is a lattice, since all pairs of elements have both meet and join. For instance, the meet of a and c is 0 and their join is 1. This shows that c is a complement of a and vice versa. The lattice is indeed complemented (every element has at least one complement), but it is not distributive. For instance,

$$d(c+b) > dc + db.$$

Therefore the lattice is not Boolean. This conclusion is in agreement with Stone's representation theorem that says that there are no Boolean algebras with six elements. Also, the complement of some elements is not unique (b, c, and d are all complements of a). □

28. Consider the lattice of the pairs of real numbers with the ordering relation:

$$(x_1, y_1) \leq (x_2, y_2) \text{ if and only if } x_1 \leq x_2 \text{ and } y_1 \leq y_2.$$

Find:

(a) The meet of (4,14) and (3,6);

(b) The join of (4,14) and (3,6).

29. Has the lattice of Problem 28 a greatest element? Explain.

30. Is the lattice of Problem 28 distributive? Explain.
Solution. We shall prove that

$$(x_1, y_1) \cdot [(x_2, y_2) + (x_3, y_3)] = (x_1, y_1) \cdot (x_2, y_2) + (x_1, y_1) \cdot (x_3, y_3);$$

The validity of the other distributive law will then follow by duality. We have:

$$\begin{aligned}
&(x_1, y_1) \cdot [(x_2, y_2) + (x_3, y_3)] = \\
&\quad (\min(x_1, \max(x_2, x_3)), \min(y_1, \max(y_2, y_3))) \\
&(x_1, y_1) \cdot (x_2, y_2) + (x_1, y_1) \cdot (x_3, y_3) = \\
&\quad (\max(\min(x_1, x_2), \min(x_1, x_3)), \max(\min(y_1, y_2), \min(y_1, y_3)))
\end{aligned}$$

We now prove that:

$$\min(x_1, \max(x_2, x_3)) = \max(\min(x_1, x_2), \min(x_1, x_3)). \qquad (3.1)$$

The proof for the y's is similar. We proceed by a case analysis. To reduce the number of possible cases we need to consider, we observe that Equation 3.1 is symmetric in x_2 and x_3. That is, if we exchange the two variables and make use of the commutativity of min and max, we obtain Equation 3.1 again. Because of symmetry, it is sufficient to consider only the case in which $x_2 \leq x_3$. Therefore, there are only three cases that must be examined:

case	lhs	rhs
$x_1 \leq x_2 \leq x_3$	x_1	x_1
$x_2 \leq x_1 \leq x_3$	x_1	x_1
$x_2 \leq x_3 \leq x_1$	x_3	x_3

The 'lhs' column gives the value to which the left-hand side of Equation 3.1 evaluates; the 'rhs' column gives the value to which the right-hand side of Equation 3.1 evaluates. These two columns are identical; hence our proof is complete. □

31. Which of the Hasse diagrams in Figure 3.15 represents a lattice? Explain.

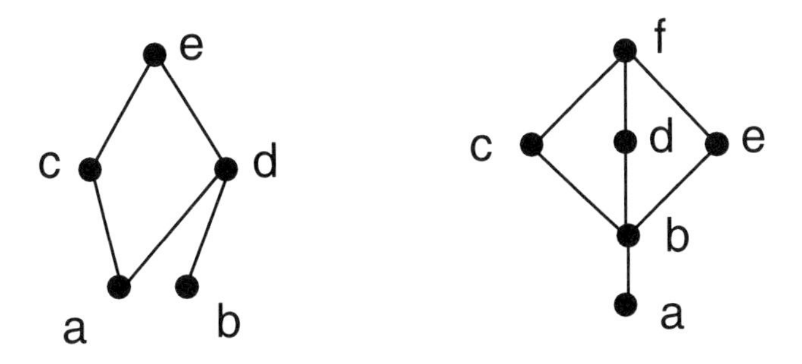

Figure 3.15: Hasse Diagrams for Problem 31.

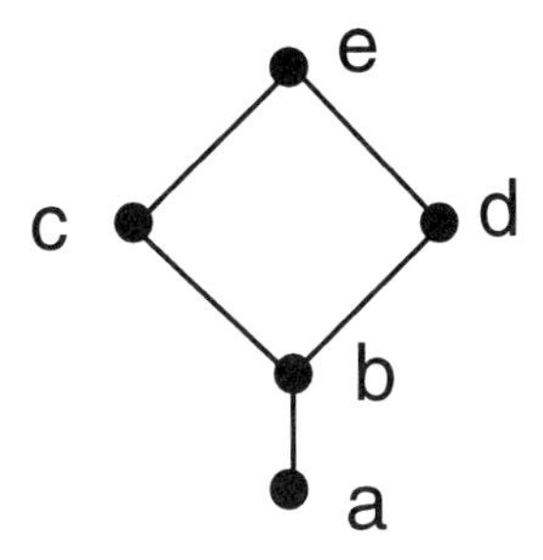

Figure 3.16: Hasse Diagram for Problem 32.

32. Is the lattice of Figure 3.16 complemented? Distributive? A Boolean algebra? Explain.

    Solution. From Stone's representation theorem, we immediately conclude that the lattice is not a Boolean algebra, because it has five elements. It is also easy to see that b, c, and d have no complements; hence the lattice is not complemented. The distributive property requires more work. As usual we choose one of the two forms and rely on duality for the other. Specifically, we consider

$$x(y+z) = xy + xz. \tag{3.2}$$

First of all, we note that for $x = a = 0$ or $x = e = 1$ the property is verified, no matter what y and z are. This is true in all finite lattices, as one can easily verify by substitution in Equation 3.2. Similarly, distributivity holds whenever $y = 0$, or $y = 1$, or $z = 0$, or $z = 1$. The cases that remain to be examined are therefore those in which x, y, and z are all different from 0 and 1. Furthermore, we can reduce the amount of work needed by noting the symmetry of y and z in Equation 3.2 and of c and d in our lattice. Distributivity also holds in all lattices when two variables have the same value; for instance, $x(x+y) = x = xx+xy$ by absorption and idempotency. All things considered, the cases we must examine are the following:

$$\begin{aligned} c(b+d) &= b = cb + cd \\ b(c+d) &= b = bc + bd \end{aligned}$$

Therefore, the lattice is distributive. □

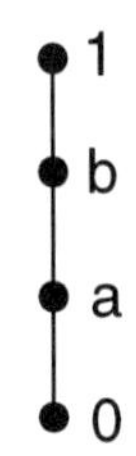

Figure 3.17: Lattice for Problem 33.

33. Is the lattice of Figure 3.17 a Boolean algebra? Explain.

34. With the help of the following lattice, prove that:

$$x + x'y = x + y$$

is not an identity in all complemented lattices (i.e., find a counterexample).

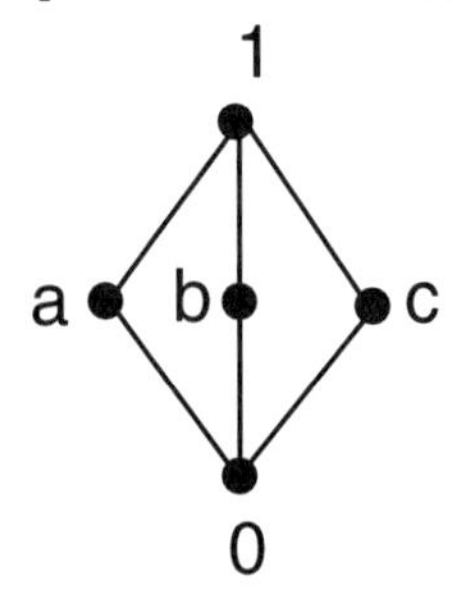

35. Prove (or disprove) the following identities for Boolean algebras, *without* using truth tables. For those equalities that are not identities, give a counterexample. For those that are identities, say what theorems can be applied to prove the identities.

> Examples:
> (a) $x = y$. Not an identity: Take $x = 1$ and $y = 0$.
>
> (b) $(x + y + z)' = x'(y + z)'$. Identity: Apply DeMorgan's.

(a) $ab + cd = (ab + c)(ab + d)$

(b) $ab \Rightarrow c = a' + b' + c$

(c) $x + y' + x'yz' = x + y' + z'$

(d) $abc'd + acd = abd + acd$

(e) $xy' \Rightarrow z = (x' + z)(y + z)$

(f) $(ab + cd)' = a'b' + c'd'$

(g) $ab + a'b'c = (a + c)(b + c)$

(h) $ab + cd = (a + b)(c + d)$

(i) $xyz + xyw + yw'z = xyw + yw'z$

(j) $(a+b')(b+c')(c+a') = (a'+b)(b'+c)(c'+a)$

36. Is the statement

$$x+y=1 \quad \text{if and only if} \quad x=1 \;\text{ or }\; y=1$$

valid in all Boolean algebras? Comment.

Solution. The statement holds for the switching algebra, but not for larger Boolean algebras. Consider $B = \{0, a, a', 1\}$. We have $a \neq 1$ and $a' \neq 1$, yet $a + a' = 1$.

This illustrates the case that there are propositions that are only valid in switching algebra, but not in all Boolean algebras. Remember, however, that all *identities* that are valid in the switching algebra are valid in all Boolean algebras. □

37. Is the following expression an identity in a Boolean algebra?

$$w'x + w'y'z + xyz \leq w'y' + xy$$

Solution. We shall present two different ways to solve the problem. First we apply the theorem $x \leq y \Leftrightarrow xy' = 0$.

$$\begin{aligned}(w'x + w'y'z + xyz)(w'y' + xy)' &= \\ = \;& (w'x + w'y'z + xyz)(w+y)(x'+y') \\ = \;& (w'x + w'y'z + xyz)(wy' + x'y) = 0.\end{aligned}$$

The second approach is based on applying the consensus theorem to the right-hand side of the inequality:

$$w'x + w'y'z + xyz \leq w'y' + xy + w'x.$$

We can now observe that all terms on the left are contained in one term of the right. In both cases we conclude that the expression is an identity. □

38. Simplify, when possible, the following formulae by applying the consensus theorem.

(a) $wxy + wx'z + wyz$;

(b) $wxy' + x'yz + wz$;

(c) $vw'y + vyz + wyz$;

(d) $xyz + wx' + wy' + wz$.

Solution.

(a) $wxy + wx'z + wyz = wxy + wx'z$;

(b) $wxy' + x'yz + wz$ cannot be simplified by application of the consensus theorem;

(c) $vw'y + vyz + wyz = vw'y + wyz$;

(d) $xyz + wx' + wy' + wz = xyz + w(x' + y') + wz = xyz + wx' + wy'$.

□

39. Let $a \Rightarrow b$ stand for $a' + b$. (The symbol $\Rightarrow$ means "implies.") Is implication associative in a Boolean algebra? (In other words, is $(a \Rightarrow b) \Rightarrow c = a \Rightarrow (b \Rightarrow c)$?) Explain.

Solution. Implication is not associative. Indeed,

$$(a' + b)' + c = ab' + c \neq a' + b' + c.$$

□

40. Let $a \oplus b$ stand for $a' \cdot b + a \cdot b'$. (This is the exclusive or.) Is $\oplus$ associative? Does $\cdot$ distribute over $\oplus$? (I.e., $a \cdot (b \oplus c) = a \cdot b \oplus a \cdot c$?) Explain.

41. Prove that 0 is the identity for $\oplus$ (i.e., $x \oplus 0 = x$).

42. Let y be an additive inverse of x if and only if $x \oplus y = 0$. Prove or disprove that in a Boolean algebra the only additive inverse of x is x itself.

    [Hint: The proof that x is an additive inverse is immediate. To prove that x is the unique additive inverse, let y be an additive inverse and, by applying the definition of additive inverse and the properties of Boolean algebras, show that it must be $x = y$.]

43. Let y be a multiplicative inverse of x if and only if $x \cdot y = 1$. (Note that 1 is the multiplicative identity, i.e., $x \cdot 1 = x$.) Does every element of a Boolean algebra have a multiplicative inverse? Explain.

44. One can easily see that

$$a + b = c + b \not\Rightarrow a = c, \tag{3.3}$$

    and similarly that

$$a \cdot b = c \cdot b \not\Rightarrow a = c. \tag{3.4}$$

    In the first case, $b = 1$ makes $a + b = c + b$, even though $a \neq c$. In the second case, a counterexample is obtained with $b = 0$.

    Prove, *without using truth tables*, that

$$a + b = c + b \text{ and } a \cdot b = c \cdot b \Leftrightarrow a = c,$$

    that is, if both (3.3) and (3.4) hold, then $a = c$.

    [Hint: $x = y \Leftrightarrow x \oplus y = 0$.]

45. Simplify, when possible, the following formulae by applying the consensus theorem (and possibly absorption).

(a) $abc'd + b'c'e' + ac'de'$;

(b) $wxy' + x'yz + wz$;

(c) $abe + a'cd' + bce + bde$;

(d) $bc'd + abc'd + abc$.

46. Enumerate the functions in the interval $[0, x' + y']$ in Figure 3.8.
Solution. There are eight functions in the interval:

$$0, x'y', xy', x'y, x', y', x'y + xy', x' + y'.$$

□

47. Given the function interval $I = [0, x+y+z']$ in the switching algebra $B = \{0, 1\}$, find the three completely specified functions f, r, and d that describe I.
Solution.

$$\begin{aligned} f &= 0 \\ r &= x'y'z \\ d &= x + y + z' \end{aligned}$$

□

48. Given $f = xy' + z$ and $d = xyz'$, find r such that f, r, and d describe an incompletely specified function; and compute $[l, u]$, the interval corresponding to that function.

49. Consider the Boolean algebra $F_2(B)$ of Boolean functions of 2 variables x and y over the Boolean algebra $B = 0, a', a, 1$, and consider the Boolean function $f \in F_2(B)$ defined by the minterm canonical form

$$f = x'y'a + x'y + xy'a' + xy\mathbf{0} = x'y'a + x'y + xy'a'.$$

(a) Give the complete truth table (Hint: there are 16 pairs in the Cartesian product $B \times B$).

(b) How many functions are there in the interval $I = [x'y'a, x' + y']$ of this Boolean function algebra?
Solution. For part (a) we give the truth table:

x	0	0	0	0	a'	a'	a'	a'	a	a	a	a	1	1	1	1
y	0	a'	a	1	0	a'	a	1	0	a'	a	1	0	a'	a	1
f	a	1	a	1	1	a	1	a	0	a'	0	a'	a'	0	a'	0

The truth table is obtained efficiently by noting that $f(0, y) = a + y$, $f(a', y) = a + y'$, $f(a, y) = a'y$, $f(1, y) = a'y'$.

For part (b) we wish to count the functions g such that $x'y'a \leq g \leq x' + y'$. Since any function g can be written in the minterm canonical form

$$g = x'y'g_{x'y'} + x'yg_{x'y} + xy'g_{xy'} + xyg_{xy},$$

and $x' + y' = x'y' + x'y + xy'$, we must have that $g_{xy} = 0$. This leaves just $|B| = 4$ ways to choose each of $g_{x'y'}$, $g_{x'y}$, and $g_{xy'}$. Thus there are $4^3 = 64$ functions less than the upper bound of the specified interval.

Similarly, every function in the interval must contain $x'y'a$, so the functions $\mathbf{0}$ and $x'y'a'$ are excluded. Looking again at the minterm canonical form, this leaves just 2 ways to select $g_{x'y'}$, $|B| = 4$ ways to select $g_{xy'}$ and $g_{x'y}$, and 1 way to select $g_{x'y'}$, so there are altogether $2 \times 4 \times 4 \times 1 = 32$ functions in the interval.

Another way to solve the problem is to count levels in the lattice. There are $|B|^{2^n} = 4^4 = 2^8 = 256$ element functions in the Boolean function algebra, so the lattice has 9 levels altogether. The upper bound $x' + y'$ is on level 7, since it has 6 atoms: $(x' + y') = x'y'a + x'y'a' + x'ya + x'ya' + xy'a + xy'a'$. Similarly $\mathbf{0}$ is the unique element on level 1, and the atom (minimal non-zero element) $x'y'a'$ must be on level 2. Since every interval of a Boolean algebra is also a Boolean algebra, this leaves $2^{7-2} = 32$ functions in the interval. □

50. (a) How many functions from B^3 to B are there, if $B = \{0, 1, a, a'\}$?

(b) What fraction of them are Boolean?

Solution. There are $|B|^{|B|^n}$ functions or 4^{4^3} from B^3 to B. Of them, $|B|^{2^n}$ or 4^{2^3} are Boolean. Therefore the fraction of functions that are Boolean is

$$\frac{4^{2^3}}{4^{4^3}} = \frac{1}{4^{56}} = 1.92 \cdot 10^{-34}.$$

□

51. Expand the following functions with respect to x.

(a) $wxy + wxz + w'x' + x'yz$;

(b) $w(x + y') + x(wx' + yz)$;

(c) $awx + awz + azx + w'y$.

Solution.

(a) $wxy + wxz + w'x' + x'yz = x'(w' + yz) + x(wy + wz)$;

(b) $w(x + y') + x(wx' + yz) = x'(wy') + x(w + wx' + yz)$;

(c) $awx + awz + azx + w'y = x'(awz + w'y) + x(aw + awz + az + w'y)$.

□

52. Using the expansion theorem, prove that for a 1-variable Boolean function

$$f(x+y)+f(xy)=f(x)+f(y)$$

Solution. By expanding the left-hand side around x, we get:

$$\begin{aligned} & f(x+y)+f(xy) \\ &= x[f(1)+f(y)]+x'[f(y)+f(0)] \\ &= [xf(1)+x'f(0)]+(x+x')f(y) \\ &= f(x)+f(y). \end{aligned}$$

It is also possible to expand with respect to $x+y$ and xy:

$$\begin{aligned} & f(x+y)+f(xy) \\ &= (x+y)f(1)+(x+y)'f(0)+(xy)f(1)+(xy)'f(0) \\ &= (x+y)f(1)+(xy)'f(0) \\ &= xf(1)+x'f(0)+yf(1)+y'f(0) \\ &= f(x)+f(y). \end{aligned}$$

This second approach illustrates the following general fact:

$$f(g(x))=g(x)'f(0)+g(x)f(1).$$

□

53. Let g and h be single-variable Boolean functions. For each of the following cases, express $f(0)$ and $f(1)$ as simplified formulae involving $g(0)$, $g(1)$, $h(0)$, and $h(1)$.

    (a) $f(x)=g(h(x))$

    (b) $f(x)=g(g'(x))$

54. Using the expansion theorem, prove or disprove that for a Boolean function f:

$$f(f(x))=f(0)f(1)+x[f(0)+f(1)].$$

55. Using the expansion theorem, prove that:

$$f(x'+y')+f(xy)=f(0)+f(1).$$

Solution.

$$\begin{aligned} f(x'+y')+f(xy) &= (x'+y')'f(0)+(x'+y')f(1)+(xy)'f(0)+(xy)f(1) \\ &= (xy+(xy)')f(0)+((xy)'+xy)f(1) \\ &= f(0)+f(1). \end{aligned}$$

□

56. Consider the Boolean algebra whose carrier is 2^A, with $A = \{a, b, c\}$. Write the minterm and maxterm canonical forms for the following function of two variables:

$$\begin{aligned} f(0,0) &= \{a,b\} \\ f(0,1) &= \emptyset \\ f(1,0) &= A \\ f(1,1) &= \{c\} \end{aligned}$$

Solution. The minterm canonical form is:

$$\{a,b\}x'y' + xy' + \{c\}xy,$$

while the maxterm canonical form is:

$$(\{a,b\} + x + y)(x + y')(\{c\} + x' + y'),$$

after performing trivial simplifications. □

57. Find the minterm and maxterm canonical forms for the following functions by repeated application of the expansion theorem. (Assume $B = \{0, a, b, 1\}$.)

(a) $f(x, y) = x(y + x') + x'y'$;

(b) $f(x, y) = ax + by$;

(c) $f(x, y, z) = axy + a'(y + z') + xz$.

58. Draw the lattice of the Boolean functions of one variable (x) over the Boolean algebra $B = \{0, a, b, 1\}$.

[Hint: Count the number of functions in the lattice and use Stone's representation theorem to see that your solution should be isomorphic to the lattice of Figure 3.7.]

Solution. The lattice has $4^{2^1} = 16$ points. Hence it is isomorphic to the the lattice of Figure 3.7. It is given in Figure 3.18. □

59. Let $B = \{0, a, b, 1\}$. Is the following mapping from B to B a Boolean function? Explain.

$$\begin{array}{c|c} 0 & b \\ a & 1 \\ b & 0 \\ 1 & 1 \end{array}$$

Solution. If the mapping corresponds to a Boolean function, then it must agree with the minterm canonical form: $bx' + x = b + x$. However, for $x = b$, the minterm canonical form gives a value of b instead of 0. Therefore, the mapping is not Boolean. □

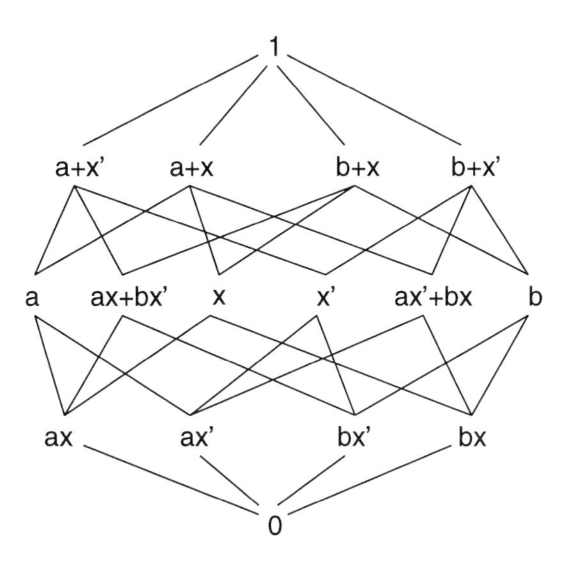

Figure 3.18: Lattice of the Boolean functions of one variable over the Boolean algebra $B = \{0, a, b, 1\}$. (Problem 58.)

60. Which of the following functions from B to B, ($B = \{0, 1, a, a'\}$) are Boolean?

x	f
0	a
a	0
a'	1
1	a'

x	g
0	a
a	a
a'	1
1	a'

61. Is $[x + y', xy]$ a non-empty interval?

    Solution. No, because $x + y' > xy$. □

62. Let $B = \{0, 1, a, a'\}$. How many Boolean functions $f : B^2 \to B$ are there that satisfy the condition

$$xy \leq f(x, y) \leq x' + y$$

for all $(x, y) \in B^2$? Do not determine the functions explicitly.

63. For $B = \{0, 1\}$, list all the functions of x, y, and z contained in the interval $[0, xy]$.

    Solution. There are four functions: $0, xyz', xyz, xy$. □

64. For the incompletely specified switching function represented by the following truth table

xy	ff
00	0
01	1
11	–
10	–

(a) find the corresponding interval of switching functions;

(b) find the corresponding triplet f, d, and r.

65. For the incompletely specified switching function represented by the following triplet

$$f = xy \qquad d = xy' \qquad r = x'$$

(a) find the corresponding interval of switching functions;

(b) find the corresponding truth table.

Solution. The lower bound is f and the upper bound is $f + d$ (or r'). Hence the interval is:

$$[xy, x].$$

The truth table is:

x	y	f
0	0	0
0	1	0
1	1	1
1	0	–

□

66. For the incompletely specified switching function represented by the following interval

$$[x'y, x' + y]$$

(a) find the corresponding truth table;

(b) find the corresponding triplet f, d, and r.

67. Given the function interval $I = [0, x + yz]$ on the switching algebra $B = \{0, 1\}$, find the three completely specified functions f, r, and d that describe I.

68. Given $f = xy'z$ and $d = x'yz$, find r such that f, r, and d describe an incompletely specified function; and compute $[l, u]$, the interval corresponding to that function.

Chapter 4

Synthesis of Two-Level Circuits

As we saw in Chapter 2, there are several points to be considered in formulating a synthesis problem:

- **Specifications.** Formal specifications are required as the starting point for synthesis. *Behavior* as well as *constraints* should be expressed. In the behavior specification, we should try to capture all possible degrees of freedom, to provide the algorithms with the largest search space that is possible.

- **Constraints.** Constraints may concern speed, testability, types of available packages, power dissipation, reliability, etc. Whenever possible, we try to incorporate the constraints in the algorithmic formulation of the problem.

- **Cost Function.** Should take into account as many factors as possible: cost of fabricating the chip, cost of testing it, cost of the package, etc. Each component of the cost depends on many factors, that are often difficult to estimate from the specifications. We shall see that we often resort to fairly crude approximations of these factors.

4.1 Design Optimality

As in the design of most complex systems, circuit designers usually have to tradeoff one design objective for another. Synthesis tools and designers try to make this tradeoff optimally, as discussed briefly in Section 1.4.1. We treat this subject at greater length here.

For example, often a designer tries to find the fastest possible circuit equivalent to a given previously designed circuit. It is often the case that overall power dissipation is strongly correlated to delay, so in seeking a faster circuit he is willing to incur a power dissipation penalty. Low power may itself be a paramount design consideration, as it is in portable computers and telecommunication devices.

However, to meet this objective, he may or may not have settle for a larger circuit. This is a typical design tradeoff, but how does he know he cannot improve both area and speed? The answer to this question depends on the optimality of the existing design. A typical design process is illustrated in Figure 4.1. In the figure, the original infeasible design (Design 0)is marked by an ellipse, the feasible design (Design 1) is

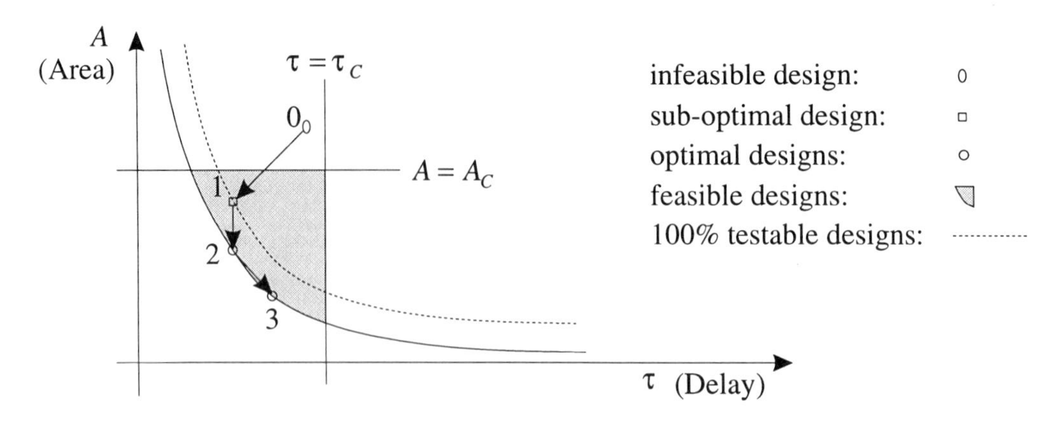

Figure 4.1: Tradeoff of area for speed for optimal designs.

represented by an small open square, and the optimal designs (Designs 2 and 3) are marked by circles. Typically the design starts out as a sub-optimal "first cut" design, which is furthest from the optimal tradeoff curve (solid line) shown in the figure. If the initial design is sub-optimal, it will be possible to redesign the circuit so as to decrease both delay and area, as in the design move from Design 0 to Design 1. As the design is improved, successive redesign iterations leave the design closer to the tradeoff curve, until a design is reached for which any attempt to increase speed (that is, decrease delay) will result in increased area. The locus of such points is called the **Pareto Critical Set** [208], and is represented by the solid curve.

At this point, the designer must establish his or her own priority, or design policy. If as is often the case, speed is the paramount consideration, the designer will continually iterate the design until a satisfactory speed objective is reached. In this process, the design moves along the critical set to the right.

Often however, the circuit eventually meets an area upper bound constraint A_U. For example, if the design becomes to large, it won't fit on a single chip. In this case, the design finalizes on the design point at the lower right. Sometimes, the designer finds that subsequent decreases in delay are not worth the area penalty they incur, so an intermediate point like the design point on the central "knee" of the curve is chosen.

It is sometimes the case that low area is the paramount design objective, so the uppermost (optimal) design point is to be chosen. Also it may be the case that delay and power are not really well correlated, so the optimal tradeoff curve becomes a 3-dimensional surface.

In any event, the purpose of logic synthesis tools is to aid the designer in reaching the optimal tradeoff curve. It is to be emphasized, however, that only the designer can set the priority for trading off area for delay and/or power dissipation.

Another complication is the issue of testability. The process of determining the testability of a design and finding all the appropriate tests (Cf., Chapter 12) for an adequate level of product assurance may be so expensive, so that design with 100% testability is sought. The family of such designs has its own tradeoff curve, generally higher that the optimal tradeoff curve, as illustrated by the dashed line in Figure 4.1.

Designs on this curve are likely to be redundant. For example, a carry-look-ahead adder has extra circuitry to get a fast output carry even though the carry propagation circuitry of a simpler ripple-carry-adder is still necessary for logic functionality. "Synthesis for Testability" tools exist whose main objective is to get onto (and stay on) this second tradeoff curve — for example, such tools might stop Design 1, the interior feasible point of the design curve, often generating all the required tests as a byproduct of this optimization.

Synthesis algorithms deal with a model of the problem, rather than the problem itself. If we want to use synthesis, we have to perform a modeling step, that converts the most important features of the design problem into features of the (mathematical) model.

We begin by looking at a restricted style of design for combinational circuits—two-level implementations—and its associated model. In spite of the restriction of two-level logic, we shall be able to derive techniques that are of very general applicability in synthesis.

4.2 Two-Level Logic

Two levels of logic are the minimum required to implement an arbitrary Boolean function. Here we assume that the primitives are AND and OR gates. AND gates are used at the first level and OR gates are used in the second level. Inverters may be present at some inputs of the gates of the first level, but we shall not count them as an additional level. Other choices are possible. In particular we could reverse the role of AND and OR gates, we could use all NAND gates or all NOR gates, or we could employ XOR gates at the second level. Other choices are possible as well.

There are two main reasons why we may want to implement a circuit in two levels, rather than multiple levels:

- Speed;
- Simplicity.

The delay of a network depends on several factors. The numbers of logic stages a signal must go through is among the important ones. So, two-level implementations tend to be fast. Notice, however, that reducing the number of levels may increase the fanin and fanout counts of gates. This may adversely impact speed.

Simplicity comes in two flavors with two-level networks. Two-level networks are easier to design and analyze, because the solution space is greatly restricted, and are easier to implement, because there are simple implementations schemes.

Historically, two-level circuits have been popular in the fifties, because they were the only circuits for which effective systematic design procedures were known. The algorithms for the optimum implementation of two-level functions were developed in the early fifties [221, 222, 188].

About twenty years later, the interest in the field was re-kindled by the advent of programmable logic: PLAs and PALs [101]. PLAs and PALs offered flexibility as well as the ability to automate a large part of the design, or even to customize the function on the field (especially PALs). Figure 4.2 shows the organization of an NMOS NAND-NAND PLA, that was popular in the seventies and early eighties.

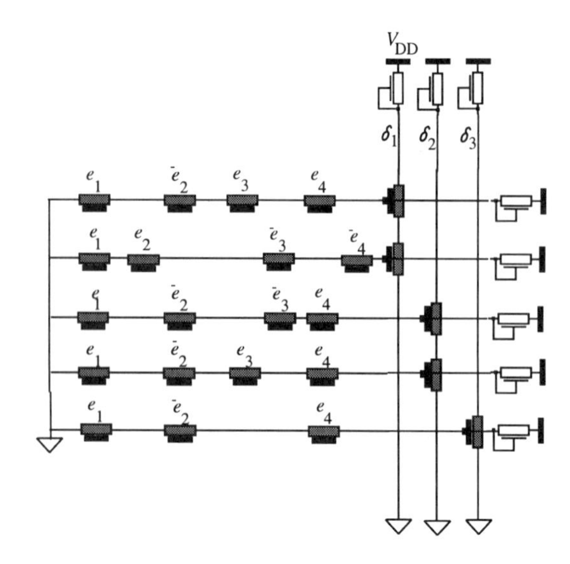

Figure 4.2: NMOS NAND-NAND PLA.

This particular architecture was fast and compact, though limited in the maximum number of inputs.

4.2.1 Cost Functions for Two-Level Implementations

Back in the fifties and sixties, it was customary to evaluate an implementation according to the number of diodes required to fabricate it. Later, people began using the number of gates and the number of gate inputs as criteria. This was a good reflection of the cost of the circuit in technologies like TTL.

In PLAs the area is primarily related to the number of product terms, which is in one-to-one correspondence with the number of rows of the array. So, in the seventies, the number of product term became a popular way to measure the cost of an implementation. Notice, however, that speed, testability, and folding (a layout optimization technique sometimes used with PLAs), all benefit from a sparser array. Hence, it is advantageous to use the number of gate inputs as a secondary criterion.

The advent of CMOS[1] and semi-custom design methodologies has marked a decline in the popularity of PLAs and PALs. (The former replaced by standard cells, the latter by more sophisticated forms of programmable logic like FPGAs.) When implementing a circuit with standard cells, it is customary to use multi-level implementations. The cost of a multi-level implementation is not directly related to the cost of an equivalent two-level circuit, but the role of the two-level techniques is still important, as we shall see. The most widely used model for the optimization of multi-level logic is actually a network whose nodes represent functions. These functions are often represented as two-level circuits. More on this in Chapter 10.

When minimizing a piece of logic for a subsequent multi-level implementation, the cost function tries to guide the optimization process towards a function that can

[1]CMOS PLAs must be dynamic in order not to draw static current; however, semi-custom design styles favor static circuitry.

be easily factored. It turns out that minimizing for number of gates and gate inputs normally provides a good starting point, especially for the testability properties of the resulting circuit.

4.2.2 Minimality and Testability

The previous overview shows that, with the partial exception of the diode count, the number of gates and the number of gate inputs has enjoyed a fairly constant success as a measure of the cost of a two-level implementation. One additional reason that we should mention is related to the concepts of testability and irredundancy.

Since a proof of correctness cannot be obtained from a black-box experiment, the testing of digital circuits is normally accomplished by checking each part for a predefined list of possible defects. Test generation is the process of finding the input sequences that cause the defect to manifest itself at the output of the circuit, in the form of errors.

As with synthesis, a good deal of modeling is required, so that the process may be carried out successfully for non-trivial devices. In particular, one idealizes to some extent the defects that may actually occur in a circuit. The most popular *fault model* is the so-called *stuck-at* (stuck-at-0, stuck-at-1) fault model. A stuck-at fault occurs when a connection (either a gate output or a gate input) is permanently stuck at one of the two logic levels. A multiple stuck-at fault is the simultaneous presence of several single stuck-at faults. From this point on we shall consider single faults unless otherwise specified. The tasks of identifying and generating tests for stuck-at faults will be discussed in detail in Chapter 12 — here we limit our discussion to a brief treatment of the connection between logic optimization and testability.

Notice that a stuck-at fault may be seen as transforming a circuit into another circuit of lower cost, according to *our* cost function. If there is no test for a given fault, then the fault is *untestable*. The connection affected by the fault is *redundant*, since the circuit can be simplified by removing the the connection itself. This link between redundant connections and untestable faults is the reason why we sometimes refer to redundant (or irredundant) faults.[2]

It is desirable to have 100% testable circuits (albeit for a restricted fault model): Hence it is good for the cost function to reflect the testability of the circuit. Specifically, if a circuit has untestable faults, then there is a cheaper (according to the cost function) implementation, which is more testable. Hence, a minimizer that can find at least a local minimum, will produce a 100% testable circuit.

For example, suppose we are given a 3-Level circuit consisting of gates g_1, g_2 and g_3, where gate g_i is characterized be the equation $y_i = F_i(x, y)$, $i = 1, 2, 3$. Suppose the primary outputs of the circuit are y_1 and y_2, and the circuit connectivity is implied by the specification

$$\begin{aligned} F_1 &= x_1' x_2' + y_3, \\ F_2 &= x_1 x_2' + x_1' x_2, \\ F_3 &= x_1 x_2 y_2' + x_1' x_2'. \end{aligned}$$

Here x_1 and x_2 are the primary inputs, and x_1' means the complement of the first

[2] We just note in passing that things are more complex in sequential circuits.

primary input. To test for the fault "input y_2 of gate g_3 stuck-at 1", we would (in principle — in practice there are much clever techniques) simply compare to this circuit to a faulty circuit. This faulty circuit is identical to the original except that the connection from gate g_2 to gate g_3 is replaced by a connection to 1 (as if this particular wire had been shorted to the power supply voltage $V_D D$). In the comparison we try to find a particular primary input combination, called a **test** or **test pattern** for which the two circuits have at least one output with different logic values.

Now consider the environment of gate g_3 in the example circuit. In the input space of variables x_1, x_2, and y_2, one can see that since g_2 is an exclusive OR gate, y_2 can't be positive while x_1 and x_2 are both positive or both negative. Consequently, the input minterms $x_1x_2y_2$ and $x_1'x_2'y_2$ are in the satisfiability don't care set $D^{Sat} = x_1x_2y_2 + x_1'x_2'y_2$ for gate g_3 — that is, these input combinations *never* occur. This has been discussed in Section 3.4. Thus it is easily seen that this fault is untestable, and therefore the literal y_2' in the logic function of gate g_3 is redundant.

The point here is that the process of identifying a redundant literal for optimization purposes is formally identical to that of testing for an input stuck-at fault. In fact we are assured that under certain assumptions about circuit cost (as discussed above), if we synthesize an area-optimal circuit, we shall be guaranteed that it is 100% testable for stuck-at faults.

In the above example it follows that $y_3 \equiv y_2'$, and therefore the following specification, with only two gates and 5 literals, is equivalent to the original, which had 3 gates and 12 literals.

$$\begin{aligned} F_1 &= y_2', \\ F_2 &= x_1x_2' + x_1'x_2, \end{aligned}$$

4.3 Sums of Products and Products of Sums

We now begin examining the minimization of two-level Boolean *formulae*. This is what normally people call "minimizing Boolean functions" and we shall also occasionally say so, since no ambiguity will arise. At this point it should be clear that our objective is to find the simplest **two-level formula** that represents a given function. The formula is related to a circuit that *implements* the given function. Simplicity is measured, as we discussed in Section 4.2.1, with respect to the number of gates and gate inputs of the circuit.

Our first step is to define formally what we mean by two-level formulae. Formulae consist of *constants*, *variables*, parentheses, and operators, combined according to the recursive definition we have seen. A **letter** is a constant or a variable. A **literal** is a letter or its complement. For instance, for $B = \{0, 1\}$ and variables $x_1, x_2, 0, 1, x_1, x_2$ are letters and $0, 1, x_1, x_1', x_2, x_2'$ are literals. For simplicity, we give the following definitions for the switching algebra only; this is the case we are most interested in.

A *product term* (or *product*, or simply *term*) is a formula of one of the following forms:

- 1;
- a non-constant literal;
- a conjunction of non-constant literals where no letter appears more than once.

A *sum term* (or *sum*, or **alterm**, or **clause**) is a formula of one of the following forms:

- 0;
- a non-constant literal;
- a disjunction of non-constant literals where no letter appears more than once.

For example, x_1x_2' is a product term, $x_1 + x_2$ is a sum term and x_1' is both. On the other hand, x_1x_1' and x_1x_1 are neither product terms nor sum terms. A **sum of products formula** is one of the following:

- 0;
- a product term;
- a disjunction of product terms.

Likewise, a **product of sums formula** is one of the following:

- 1;
- a sum term;
- a conjunction of sum terms.

For instance,

$$f = x_1x_2' + x_2'x_3 + x_1x_3' \tag{4.1}$$

is a sum of product formula for f. The product of sums dual to (4.1) is $(x_1 + x_2')(x_2' + x_3)(x_1 + x_3')$. Sum of products is abbreviated SOP or $\Sigma\Pi$ and is also called **disjunctive normal form** (DNF). Product of sums is abbreviated POS or $\Pi\Sigma$ and is also called **conjunctive normal form** (CNF).

The *cost* of a SOP formula is determined by the number of product terms and the number of literals. The cost of a POS formula is determined by the number of sum terms and the number of literals. If two SOP formulae have the same number of terms, then the one with fewer literals is cheaper. Likewise for POS formulae. It is also meaningful to compare the cost of a POS formula to the cost of a SOP formula. Indeed, every time we use a technology where the cost of a POS implementation is comparable to the cost of a SOP implementation of the same (abstract) cost, we should derive the best possible POS and the best possible SOP for the function and compare them. The cost of the SOP formula (4.1) is 3 terms and 6 literals. One can verify that the same function can be represented by the POS formula $(x_1 + x_3)(x_2' + x_3')$, whose cost is 2 terms and 4 literals.

A *two-level* formula is either a SOP or a POS. The two forms are one the dual of the other. This is very important, since it allows us to describe all our theorems and algorithms for SOP formulae, without loss of generality. A computer program does not need to know whether a formula is a SOP or a POS in order to find the cheapest equivalent formula of the same kind.

4.4 Implicants and Prime Implicants

An **implicant** of a function f is a product term p that is included in the function f ($p \leq f$). For instance, both xy' and xyz are implicants of $xy' + yz$.

A **prime implicant** of f is an implicant of f that is not included in any other implicant of f. One can easily see that if p is not prime, then it is possible to obtain another implicant of f by removing one of the literals from p. With reference to the previous example, xy' is prime, whereas xyz is not. Indeed, it is possible to remove y from the latter to get xz, which is a (prime) implicant of the given function. It is also possible to remove x to get yz.

If a prime implicant is an implicant which includes a minterm that is not included in any other prime implicant, then that prime implicant is **essential**. In the previous example, both xy' and yz are essential primes, whereas xz is not.

4.4.1 Quine's Prime Implicant Theorem

The key result for the minimization of two-level formulae is due to Quine [221].

Theorem 4.4.1 *A minimal SOP must always consist of a sum of prime implicants if any definition of cost is used in which the addition of a single literal to any formula increases the cost of the formula.*

The proof of this theorem is fairly simple. One assumes that a minimum-cost formula exists, that contains a non-prime implicant. One then shows that another formula can be obtained by replacing the non-prime implicant by a prime implicant that contains it. The cost does not increase and the formula is equivalent to the original one.

As an example, consider $f = xy' + y$. We know that x is a prime implicant of f and it includes xy'. We can then rewrite f as $x + y$, thereby saving one literal.

The consequence of the Prime Implicant Theorem is that we can focus on only those formulae that are composed of prime implicants. If we want to guarantee the optimality of the solution, we need to choose from all primes. Therefore, as the next step we analyze how to derive all the prime implicants of a given function. Later, we shall see how to select a subset of minimum cost from all the prime implicants.

Efficiency in deriving all primes is important, if we want to handle functions with more than a few inputs. The number of the prime implicants is indeed smaller in general than the number of implicants for a given functions, but still grows exponentially with the number of inputs in the worst case.

4.5 Iterated Consensus

Two common methods to generate prime implicants are based on applying the **consensus theorem**. Brown [44] notes that Blake [25] called the consensus of two terms their *syllogistic result.* To understand why, we take a short digression that will be useful in the future.

4.5.1 Consensus and Implications: A Digression

In logic, $x \Rightarrow y$ (read x implies y) is a proposition that is true if y is true whenever x is true. If x is false, then the proposition is true, regardless of the value of y. Therefore

$$x \Rightarrow y = x' + y \tag{4.2}$$

as one may find out by examining all possible cases or just from the previous discussion.

Let us consider the famous **syllogism** "Socrates is a man; all men are mortal; hence Socrates is mortal." Skipping a few formal steps, we can write it as

$$(s \Rightarrow h) \wedge (h \Rightarrow m) \Rightarrow (s \Rightarrow m),$$

where s is the truth value of the proposition "to be Socrates;" similarly for h and m. If we now rewrite it using (4.2), we get

$$(s' + h)(h' + m) \Rightarrow (s' + m).$$

But now we see that the implied term—the conclusion of our syllogism—is actually the consensus term of the two premises.

The important thing to keep in mind from this example is Equation (4.2) that we shall use liberally in the mathematical formulation of problems.

4.5.2 The Tabular Method of Computing the Prime Implicants

We are given an initial SOP formula and we want to find another SOP formula that is the sum of all prime implicants of the function represented by the initial formula.

One way to achieve our goal is to first express the function f in **minterm canonical form**. We then consider all pairs of *adjacent* terms, i.e., the pairs of terms to which consensus can be applied. The consensus terms are clearly implicants of f, though not necessarily prime. All terms that were used to form these new terms are included in the new terms, and hence they are not prime. We mark them as such.

We now take the new terms and repeat the process. We only consider pairs of terms that differ in exactly one letter, which must appear complemented in one term and uncomplemented in the other.

The process is repeated until no more consensus terms can be found. All terms that are absorbed (or contained) by the new terms are marked. Finally, the terms that are not marked constitute all the prime implicants of f.

Calculations by hand are better carried out with the help of a table like the one in Figure 4.3. To compute the complete sum for $f = x'y' + wxy + x'yz' + wy'z$, we initially compute its minterm canonical form:

$$f = w'x'y'z' + w'x'y'z + w'x'yz' + wx'y'z' + wx'y'z + wx'yz' + wxyz' + wxy'z + wxyz.$$

The minterms appearing in the canonical form are entered in the leftmost column. Notice the grouping of the terms that minimizes the number of comparisons. Each group of minterms separated by a horizontal line is composed of minterms with the same number of uncomplemented literals. Hence, the first group consists of the only minterm with no uncomplemented literals. In general, some groups may be empty.

$w'x'y'z'$ √	$w'x'y'$ √ $w'x'z'$ √ $x'y'z'$ √	$x'y'$ $x'z'$
$w'x'y'z$ √ $w'x'yz'$ √ $wx'y'z'$ √	$x'y'z$ √ $x'yz'$ √ $wx'y'$ √ $wx'z'$ √	
$wx'y'z$ √ $wx'yz'$ √	$wy'z$ wyz'	
$wxyz'$ √ $wxy'z$ √	wxy wxz	
$wxyz$ √		

Figure 4.3: Tabular Method Applied to $f = x'y' + wxy + x'yz' + wy'z$.

The separation into groups allows one to compare a minterm of a group only to minterms of the immediately successive group. Indeed, these are the only minterms that may be adjacent to it. (We do not need to consider the minterms in the immediately preceding group, because this would only cause us to repeat each comparison.) In our example, for instance, we compare $wxyz'$, from the fourth group, only to $wxyz$. Their consensus term is wxy. Both $wxyz'$ and $wxyz$ are marked: They are not prime, because there exists another implicant (wxy) that contains them.

The results of merging pairs of adjacent minterms are implicants of f with one fewer literal than the minterms; they are entered in the second column. These terms are also divided according to the number of uncomplemented literals and compared to the terms of the next group. The process is then repeated, until no new terms are formed. In our example, there are six terms that are not marked at the end of the process. (They were not used to form any new term.) They are the prime implicants of f:

$$wy'z, wyz', wxy, wxz, x'y', x'z'.$$

Starting from the third column, it is possible to form an implicant in more than one way. For instance, $x'y'$ can be obtained by merging $w'x'y'$ and $wx'y'$; or by merging $x'y'z'$ and $x'y'z$.

If the function is incompletely specified, then we shall mark appropriately the terms that are **don't care**, and drop those terms that are generated only with don't care minterms. Specifically. a product term p is a prime implicant of an incompletely specified function $\mathit{ff} = (f, d, r)$ if $p \leq f + d$, $p \cdot f \neq 0$, and p is not contained in any other implicant of ff. In words, a prime implicant of an incompletely specified function is a prime of $f + d$ that covers at least one element of f.

All prime computation procedures can be extended to handle incompletely specified functions. We examine in detail the extension of Quine's tabular method. Consider the following example:

$$f = yz' + xy'z$$

$x'yz'$ √ $x'y'z$ √ d	$x'y$ yz' $x'z$ d $y'z$
$x'yz$ √ d xyz' √ $xy'z$ √	

Figure 4.4: Tabular Method Applied to an Incompletely Specified Function.

$$d = x'z$$

In applying Quine's method, we have to keep track of what implicants have been formed by merging **don't care** terms only.

Initially, we mark with a 'd' all the minterms of d. When we merge two implicants, we mark the result with a 'd' only if both the terms that are merged are marked with a 'd.' The result of the procedure is shown in Figure 4.4.

In this example we see that there are three prime implicants. The term $x'z$ is entirely contained in d; hence, it is not a prime implicant.

The tabular method for the generation of prime implicants is due to Quine. Good accounts can be found in [186, 187].

4.5.3 Iterated Consensus in General

The tabular method is based on the application of the theorem

$$Xy + Xy' = X \tag{4.3}$$

This theorem, called **distance-1 merging**, can be seen as a specialized form of consensus, since X is the consensus term of Xy and Xy' and contains both. Because it only uses (4.3), the tabular method is simple and can avoid many comparisons. However, it requires the minterm canonical form to start with. We want to avoid expanding the function into minterms for efficiency. Therefore we look for a different approach, based on the general form of the consensus theorem. We define a **complete sum** as a SOP formula composed of all the prime implicants of the function it represents. We can restate the problem of finding all the prime implicants for f as the problem of finding a complete sum for f. Fortunately, the following result can be proven.

Theorem 4.5.1 *A SOP formula is a complete sum if and only if:*

1. *No term includes any other term.*
2. *The consensus of any two terms of the formula either does not exist or is contained in some term of the formula.*

We shall not prove this result (see [186, p. 168] or [44, Appendix A] for that), but rather suggest why the theorem works.

Suppose a SOP F representing f is given that is not a complete sum, because there is a prime implicant of f that does not appear in F. This implicant must be covered by two or more of the implicants in F. Suppose for simplicity they are two, p_1 and p_2. If we add the consensus term of these two implicants, we add one term that spans the border of p_1 and p_2 and therefore may cover the missing prime implicant. (It will actually cover it, if p_1 and p_2 are prime.)

The theorem and the discussion following it suggest a simple procedure to generate all the primes of a function, called *iterated consensus*. One starts from an arbitrary SOP formula and adds the consensus terms of all pairs of terms that are not contained in some other term. The new terms are compared to the existing terms and among themselves to see if new consensus terms can be generated. All terms that are contained is some other term are removed. When no further changes are possible, a complete sum is generated.

As in the tabular method, a clever organization of comparisons can substantially reduce the amount of work. In particular it is convenient to compare every term only to the terms that precede it in the formula. This prevents duplicate operations and also takes care naturally of the addition of new terms.

As an example, consider the computation of the complete sum starting from the following SOP formula:

$$x_1x_2 + x_2'x_3 + x_2x_3x_4.$$

We begin by comparing the second term to the first. We append the consensus term (x_1x_3) to the formula, obtaining:

$$x_1x_2 + x_2'x_3 + x_2x_3x_4 + x_1x_3.$$

We then compare the third term $(x_2x_3x_4)$ to the first and second terms. The latter gives a consensus term (x_3x_4) that we append to the formula:

$$x_1x_2 + x_2'x_3 + x_2x_3x_4 + x_1x_3 + x_3x_4.$$

Nothing happens when we compare the fourth term (x_1x_3) to those that precede it. Finally, when we compare the last term to $x_2x_3x_4$, we remove the latter, because it is included in the former. This terminates our computation. The resulting complete sum is:

$$x_1x_2 + x_2'x_3 + x_1x_3 + x_3x_4.$$

4.6 Recursive Computation of Prime Implicants

Another property of complete sums is given by the following theorem.

Theorem 4.6.1 *The SOP obtained from two complete sums F_1 and F_2 by the following procedure is a complete sum for $F_1 \cdot F_2$.*

1. *Multiply out F_1 and F_2 using the idempotent and distributive properties and $x \cdot x' = 0$.*

2. *Eliminate all terms that are contained in some other term.*

For the proof we refer to [44, Appendix A]. The result generalizes to the product of n complete sums. Since a sum term is a simple case of complete sum, if we start from a POS formula and apply the procedure of Theorem 4.6.1, we get a complete sum.

As an example, let us consider the following POS formula:

$$(x_1 + x_2)(x_2' + x_3)(x_3 + x_4).$$

After multiplying out the first two sum terms, we get:

$$(x_1 x_2' + x_1 x_3 + x_2 x_3)(x_3 + x_4).$$

No term of the first sum is contained in any other term, so we continue:

$$x_1 x_2' x_3 + x_1 x_2' x_4 + x_1 x_3 + x_1 x_3 x_4 + x_2 x_3 + x_2 x_3 x_4.$$

There are now some cases of containment. Once all the contained terms are eliminated we get the complete sum:

$$x_1 x_2' x_4 + x_1 x_3 + x_2 x_3.$$

If we are not given a POS formula, we can use the second form of Boole's expansion theorem to write f as the product of two other functions:

$$f(x_1, x_2, \ldots, x_n) = [x_1' + f(1, x_2, \ldots, x_n)] \cdot [x_1 + f(0, x_2, \ldots, x_n)].$$

This leads naturally to a recursive method where we apply the same procedure to each of $f(1, x_2, \ldots, x_n)$ and $f(0, x_2, \ldots, x_n)$ until they simplify to the point that we can derive all the prime implicants by inspection.

The simplification is guaranteed to take place, since at every step of the recursion the number of variables decreases. In the end, we shall find formulae of only one term, for which it is trivial to compute the complete sum (it is the term itself). When we return from the recursion, we combine the results of the two sub-problems using Theorem 4.6.1. We can indicate the complete sum of f by $CS(f)$ and write:

$$CS(f) = ABS([x_1 + CS(f(0, x_2, \ldots, x_n))] \cdot [x_1' + CS(f(1, x_2, \ldots, x_n))]),$$

where $ABS(f)$ returns the formula obtained by removing absorbed terms from f. For instance, $ABS(x + y + xz) = x + y$.

We shall see many recursive algorithms like this in the sequel. This one in particular is the algorithm used in ESPRESSO [37] to compute all the primes of a function, when the exact minimum-cost solution is requested. It is a good thing to spend some time to familiarize with the operation of this algorithm, especially by visualizing the **recursion tree**.

As an example, we now compute the complete sum for

$$f = v'xyz + v'w'x + v'x'z' + v'wxz + w'yz' + vw'z + vwx'z$$

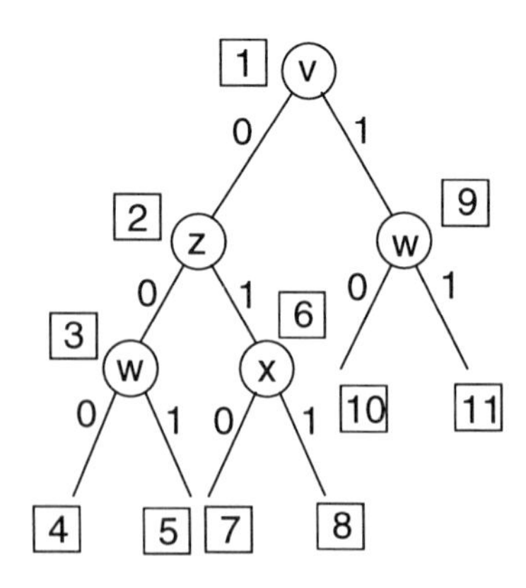

Figure 4.5: Example of Recursion Tree for the Computation of Prime Implicants.

$\boxed{1}$ $f(v,w,x,y,z) = v'xyz + v'w'x + v'x'z' + v'wxz + w'yz' + vw'z + vwx'z$

$\boxed{2}$ $f(0,w,x,y,z) = xyz + w'x + x'z' + wxz + w'yz'$

$\boxed{3}$ $f(0,w,x,y,0) = w'x + x' + w'y$

$\boxed{4}$ $f(0,0,x,y,0) = x + x' + y = 1$

$\boxed{5}$ $f(0,1,x,y,0) = x'$

$\boxed{3}$ $CS(f(0,w,x,y,0)) = ABS((w+1)(w'+x')) = w' + x'$

$\boxed{6}$ $f(0,w,x,y,1) = xy + w'x + wx$

$\boxed{7}$ $f(0,w,0,y,1) = 0$

$\boxed{8}$ $f(0,w,1,y,1) = y + w' + w = 1$

$\boxed{6}$ $CS(f(0,w,x,y,1)) = ABS((x+0)(x'+1)) = x$

$\boxed{2}$ $CS(f(0,w,x,y,z)) = ABS((z+w'+x')(z'+x))$
$= w'x + w'z' + x'z' + xz$

$\boxed{9}$ $f(1,w,x,y,z) = w'yz' + w'z + wx'z$

$\boxed{10}$ $f(1,0,x,y,z) = yz' + z = y + z$

$\boxed{11}$ $f(1,1,x,y,z) = x'z$

$\boxed{9}$ $CS(f(1,w,x,y,z)) = ABS((w+y+z)(w'+x'z))$
$= ABS(wx'z + w'y + x'yz + w'z + x'z)$
$= w'y + w'z + x'z$

$\boxed{1}$ $CS(f(v,w,x,y,z)) = ABS((v + w'x + w'z' + x'z' + xz)(v' + w'y + w'z + x'z))$
$= vw'y + vw'z + vx'z + v'w'x + w'xy + w'xz +$
$v'w'z' + w'yz' + v'x'z' + v'xz$

The recursion tree is given in Figure 4.5. The line numbers in the example correspond to the node numbers in the tree. When we apply Boole's expansion, it is important to choose a good *splitting variable.* The objective is to minimize the amount of computation, for instance, by minimizing the number of nodes in the tree. One way to heuristically achieve that is to minimize the number of terms that appear in both $f(1, x_2, \ldots, x_n)$ and $f(0, x_2, \ldots, x_n)$. This suggests the choice of the variable that appears in the largest number of terms. Indeed those terms where the selected variable does not appear, will become part of both sub-problems. By minimizing the size of the sub-problems, we increase the chance of early termination of the recursion.

Returning to our example, we chose v as initial splitting variable, because it was one of the best (the other being z) according to that criterion. We shall return to this subject in more depth when we deal with heuristic minimization in Chapter 5.

For an incompletely specified function, we need to expand the procedure we have seen by actually computing the prime implicants of $f + d$. By so doing, we may erroneously include in the list of prime implicants some terms that only cover don't care minterms. However, all the true prime implicants will be included. We shall see in Problem 13 that possibly including some terms that do not cover any 'care' minterm is not a mistake. In other words, it does not affect the correctness of the procedure.

4.7 Selecting a Subset of Primes

Recall that Quine proved that a minimum cost SOP formula can be obtained by considering prime implicants only. Since we know how to generate all the primes of a function, we now turn our attention to the selection of a subset of implicants of minimum cost. The approach to minimizing a SOP or POS formula based on computing all primes and then selecting some of them to form a cover goes under the name of the **Quine-McCluskey** procedure. As an example we consider the following formula:

$$f(x, y, z) = yz + x'y + y'z' + xyz + x'z'$$

The complete sum for f is

$$x'y + x'z' + y'z' + yz.$$

The condition that any subset of primes must satisfy to represent a valid formula of the function is that each minterm for which the function is 1 (each minterm of the function for short) be included in at least one implicant of the subset.

A subset of implicants that satisfies this requirement is called a *SOP cover* of the function, or simply a cover, when the context prevents ambiguity. We can build a **constraint matrix** that describes the conditions or constraints that a cover must satisfy. Each column of the constraint matrix corresponds to a prime implicant and each row corresponds to a minterm. Let A be the constraint matrix and let a_{ij} be the element in row i and column j. Then, $a_{ij} = 1$ if the j-th prime covers the i-th minterm. Otherwise, $a_{ij} = 0$. In our example, let p_1 stand for $x'y$, p_2 stand for $x'z'$, p_3 for $y'z'$, and p_4 for yz. A is given by:

$$\begin{array}{c} \\ x'y'z' \\ x'yz' \\ x'yz \\ xyz \\ xy'z' \end{array} \begin{array}{c} \begin{array}{cccc} p_1 & p_2 & p_3 & p_4 \end{array} \\ \left[\begin{array}{cccc} 0 & 1 & 1 & 0 \\ 1 & 1 & 0 & 0 \\ 1 & 0 & 0 & 1 \\ 0 & 0 & 0 & 1 \\ 0 & 0 & 1 & 0 \end{array}\right] \end{array} \qquad (4.4)$$

Given the constraint matrix, our problem is to find a subset of columns of minimum cost that *covers* all the rows. In other words, for every row there must be at least one selected column with a 1 in that row.

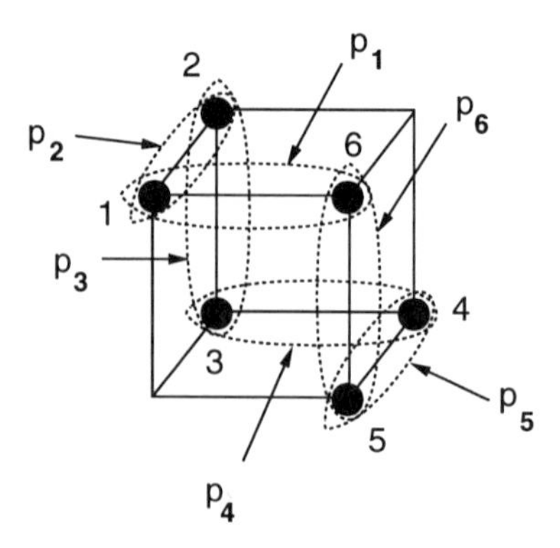

Figure 4.6: A Function with a Cyclic Core.

Before we proceed with the formal statement of our problem, we note that in our example, columns p_3 and p_4 must be part of every solution, because the last two rows are *singletons.* If a row is a singleton, there is only one column that may cover it and that column must be selected.[3] When we select some columns, we simplify the constraint matrix accordingly, by eliminating the selected columns and the rows covered by them:

$$\begin{array}{cc} & \begin{array}{cc} p_1 & p_2 \end{array} \\ x'yz' & \left[\begin{array}{cc} 1 & 1 \end{array}\right] \end{array}$$

Once this is done, we can easily see that a complete solution may be obtained by adding either p_1 or p_2 to p_3 and p_4. In the first case we obtain $x'y + y'z' + yz$; in the second, we obtain $x'z' + y'z' + yz$. These are two sums of products of minimum cost for f.

Unfortunately, we cannot always find the solution directly, as in the example of Figure 4.6, whose constraint matrix is:

$$\begin{array}{c} \\ 1 \\ 2 \\ 3 \\ 4 \\ 5 \\ 6 \end{array} \begin{array}{c} \begin{array}{cccccc} p_1 & p_2 & p_3 & p_4 & p_5 & p_6 \end{array} \\ \left[\begin{array}{cccccc} 1 & 1 & & & & \\ & 1 & 1 & & & \\ & & 1 & 1 & & \\ & & & 1 & 1 & \\ & & & & 1 & 1 \\ 1 & & & & & 1 \end{array}\right] \end{array}$$

This function is said to have a *cyclic core* (which in this case is the function itself). We shall return to the definition of cyclic core later; for the time being, it is sufficient to say that a function has a cyclic core if we cannot identify columns that must be part of the solution or that can be eliminated.

In our example, each row is covered by exactly two columns and each column covers exactly two rows. There is no apparent reason to prefer one column over another. For this matrix we must proceed by choosing one column arbitrarily and

[3] Columns p_3 and p_4 correspond—as we have seen—to essential primes, that is, primes that cover a minterm not covered by any other prime.

finding the best solution subject to the assumption that the column is selected. We must then assume that the column is not in the solution and find another solution. We then compare the two solutions obtained and keep the best. For instance, we may select p_1 and find a solution including it. Then we may find a solution not including p_1. Finally, we choose the better of the two solutions.

We have seen, in the two previous examples, two important mechanisms in action: Reduction of the constraint matrix and branching in the case of cyclic cores. We shall develop these mechanisms in detail in the sequel. Before that, however, we want to mention another possible formulation of the problem that is important to us.

One may readily see that the first row of the constraint matrix in our first example can be written as the switching function

$$(p_2 + p_3).$$

This function evaluates to one if either $p_2 = 1$ or $p_3 = 1$ or both. If we interpret $p_i = 1$ as "column p_i is selected," we see that the switching function is 1 when the first row of the matrix is covered and vice versa. We can proceed similarly for the other rows. The expressions thus obtained are switching functions that must all be 1 for a solution to be valid. Hence, their product must be 1. We can therefore write the following equation as an equivalent to the constraint matrix:

$$(p_2 + p_3)(p_1 + p_2)(p_1 + p_4)p_3p_4 = 1. \tag{4.5}$$

This equation is called the *constraint equation* of the covering problem. The covering problem can be formulated in this setting as the problem of finding an assignment of zeroes and ones to the variables that is a solution to the constraint equation and that is of minimum cost.

The equation can be simplified, for instance, to

$$(p_1 + p_2)p_3p_4 = 1,$$

using the absorption property. This simplification has a similar effect to detecting the essential columns and illustrates the general fact that we shall find a Boolean operation on the constraint equation corresponding to each operation on the constraint matrix.

We conclude this introduction by noting that all variables in the constraint equation appear uncomplemented. This is not a coincidence, but rather a direct consequence of the way the equation is built. A formula where no letter appears with both phases is said **unate**. A non-unate formula is called **binate**. Because of the form of the constraint equation that we get, the covering problem we are dealing with is called sometimes **unate covering**. The problem that is obtained by relaxing the assumption that the constraint equation is unate is called **binate covering covering**.

We shall return to unate functions and formulae and to binate covering in the future. Now, however, we concentrate on the efficient solution of the unate covering problem.

4.8 The Unate Covering Problem

Here we define UCP, the **Unate Covering Problem**, giving both a constraint matrix and a constraint equation form of the problem. Our statement of the problem will be general, in the sense that the problem will not require the context of minterms covered by prime implicants, as in the previous section.

We then describe some fundamental methods that can be used to organize and greatly simplify the solution of UCP. We first detail the reduction techniques that can be applied to simplify the constraint matrix and/or the constraint equation. We then describe how the solution of UCP can be viewed as the process of enumerating all the solutions to the constraint equation and picking out one of the solutions of minimum cost. We conclude this section with a method for determining an **a priori** lower bound on the cost of the solution. In the succeeding section, we will organize all of these ideas into an efficient algorithm for solving UCP in the general case.

Definition 4.8.1 (UCP, constraint equation form) *Let* $J = \{1, \ldots, n\}$, *and let* $p = (p_1, \ldots, p_n)$ *be a vector of* n *Boolean variables. Let* $\sigma_i = \sum_{j \in J_i} p_j$, $i = 1, \ldots, m$, *where* $J_i \subseteq J$ *is a set of* m *simple sums of single, positive literals of the variables* p_j.

Then UCP is the problem of finding a minimum cardinality subset $S \subseteq J$, *setting* $p_j = 1$, $\forall p_j \in S$, *guarantees that*

$$\prod_i \sigma_i(p) = 1.$$

That is, for any other subset S' *having the above properties, we have* $|S| \leq |S'|$. *Note each of the sums* $\sigma_i(p) : \{0,1\}^n \longmapsto \{0,1\}$ *is a Boolean function.*

In the constraint equation of Equation 4.5 we have $\sigma_1 = (p_2 + p_3)$, $\sigma_2 = (p_1 + p_2)$, $\sigma_3 = (p_1 + p_4)$, $\sigma_4 = p_3$, and $\sigma_5 = p_4$. For this problem, we showed above that $\prod_i \sigma_i(p) = (p_1 + p_2)p_3p_4$.

Definition 4.8.2 (UCP, constraint matrix form) *Let* M *be a matrix of* m *rows and* n *columns, for which* M_{ij} *is either* 0 *or* 1. *Then UCP is the problem of finding a minimum cardinality column subset* S, *such that for all* S' *such that*

$$\exists_{j \in S'} M_{ij} = 1, \ \forall i \in \{1, \ldots, n\} \Rightarrow |S| \leq |S'|.$$

That is, the columns in the set S *"cover"* M *in the sense, that every row of* M *contains a* 1*-entry in at least one of the columns of* S, *and there is no smaller set* S' *which also covers* M.

For the constraint matrix of Equation 4.4 we have two possible solutions to the matrix form of UCP: $S^1 = \{1, 3, 4\}$, and $S^2 = \{2, 3, 4\}$.

The two forms of stating UCP are totally equivalent, and applicable in a context which is much broader than the minimization of logic functions. We demonstrate this with the following example. After the example, we shall take a closer look at how the efficiency of the solution is affected by the key mechanisms of reduction, enumeration, and lower bounding.

> **Example:** Recent studies have indicated that a good diet should contain adequate amounts of proteins (P), vitamins (V), fats (F), and cookies (C). An astronaut has to choose from a menu of five different preparations with the following nutritional information labels.

- Preparation 1 contains: V and P;
- Preparation 2 contains: V and F;
- Preparation 3 contains: P and F;
- Preparation 4 contains: V;
- Preparation 5 contains: C.

(Preparation 5 is actually a ration of peanut butter cookies; the rest is yucky stuff.) Can the astronaut have a balanced diet with only two preparations?

In the constraint equation form of this instance of UCP, $n = 5$, and the 5 preparations are represented by the variables p_i, $i = 1, \ldots, 5$. Similarly, there are 4 constraint sums σ_i, corresponding to the requirement that each nutrient P, V, F, and C be included in the astronauts' diet. Thus, $\sigma_1 = (p_1 + p_3)$ represents the protein requirement, $\sigma_2 = (p_1 + p_2 + p_4)$, represents the vitamin requirement, $\sigma_3 = (p_2 + p_3)$, represents the fats requirement, and $\sigma_4 = (p_5)$, represents the cookie requirement. Thus the constraint equation is

$$\begin{aligned}
\prod_i \sigma_i(p) &= (p_1 + p_3)(p_1 + p_2 + p_4)(p_2 + p_3)(p_5) \\
&= (p_1 + p_3(p_2 + p_4))(p_2 + p3)p_5 \\
&= (p_1)(p_2 + p3)p_5 + p_3(p_2 + p_4)(p_2 + p3)p_5 \qquad (4.6) \\
&= (p_1)(p_2 + p3)p_5 + p_3(p_2 + p_4)p_5 \\
&= 1.
\end{aligned}$$

Thus we see that there are exactly 4 solutions of size 3, but none of size 2. Further, every solution to the constraint equation requires the assignment $p_5 = 1$. We thus say that p_5 is "essential" in the sense that it is the only preparation providing the essential nutrient cookies (we define this formally below).

The corresponding covering matrix is

$$\begin{array}{c} \\ P \\ V \\ F \\ C \end{array} \begin{array}{c} \begin{array}{ccccc} 1 & 2 & 3 & 4 & 5 \end{array} \\ \left[\begin{array}{ccccc} 1 & 0 & 1 & 0 & 0 \\ 1 & 1 & 0 & 1 & 0 \\ 0 & 1 & 1 & 0 & 0 \\ 0 & 0 & 0 & 0 & 1 \end{array} \right] \end{array}$$

In this representation, the essential nature of p_5 is manifested as a singleton row, meaning that only one column, p_5 can cover the cookie row C. Consequently, all minimum covers contain column 5 (p_5). Also, Preparation 4 is dominated by preparation 2. The remaining matrix is cyclic (see Figure 4.6), and, as we shall prove in Theorem 4.8.1 on Page 150, will therefore require at least two columns. Thus there are 3 solutions, which are $p_5(p_1p_2 + p_2p_3 + p_3p_1)$. Note that the solution $p_3p_4p_5$ is missing (because column 4 was eliminated by dominance). However, we are guaranteed to retain an optimum solution (of size 3). Hence, it is impossible to have a balanced diet with only two preparations. ∎

4.8.1 Reduction Techniques

We consider three reduction techniques (to be defined below):

1. elimination of rows covered by "essential columns";
2. elimination of rows through "row dominance";
3. elimination of columns through "column dominance".

For each of them, we present two forms: One applicable to a constraint matrix and one applicable to a constraint equation. We also show that the two forms are equivalent.

In general, we want to iterate the three forms of reduction in a fixed order, re-iterating as long as the matrix keeps simplifying. One common order is the one that we follow in our presentation: first, check for essential columns; second, check for row dominance; third, check for column dominance. One type of reduction often leads to another, but eventually the matrix reduces to a case in which no further reduction is possible. The iteration stops when such a case is reached.

4.8.2 Essential Columns or Variables

If a row of the constraint matrix is a singleton, the corresponding column must be part of the solution. The essential columns and all rows covered by them are removed from the constraint matrix. We saw an example of this process in Section 4.7.

The analogous process for a constraint equation consists of identifying the terms of the POS formula that consist of one literal only. The corresponding variables must be set to 1 and the equation simplified accordingly. For the following constraint equation,

$$(p_2 + p_3)(p_1 + p_2)(p_1 + p_4)p_3p_4 = 1,$$

we have to set $p_3 = p_4 = 1$. The resulting simplified equation is:

$$(p_1 + p_2) = 1.$$

In the following two sections, we show how the subproblems based on the positive and negative cofactors of the constraint POS may have essential variables, even in cases for which the original POS for F had none.

4.8.3 Row or Constraint Dominance

If a row r_i of the constraint matrix has all the ones of another row r_j, then r_i is covered whenever r_j is covered. Therefore, we do not need r_i in our matrix, because the constraint it represents is superfluous. We say that r_i dominates r_j. All dominating rows can be eliminated from the constraint matrix. For the following matrix:

$$\begin{array}{c} \\ 1 \\ 2 \\ 3 \\ 4 \\ 5 \\ 6 \end{array} \begin{array}{c} \begin{array}{cccccc} 1 & 2 & 3 & 4 & 5 & 6 \end{array} \\ \left[\begin{array}{cccccc} 1 & 1 & & & & 1 \\ 1 & 1 & & & & \\ & 1 & 1 & & & \\ & 1 & 1 & 1 & & \\ & & & 1 & 1 & \\ & & & 1 & 1 & 1 \end{array} \right] \end{array}$$

the first, fourth, and sixth rows dominate the second, third, and fifth rows, respectively and can be eliminated. The reduced matrix is:

$$\begin{array}{c} \\ 2 \\ 3 \\ 5 \end{array} \begin{array}{c} \begin{array}{ccccc} 1 & 2 & 3 & 4 & 5 \end{array} \\ \left[\begin{array}{ccccc} 1 & 1 & & & \\ & 1 & 1 & & \\ & & & 1 & 1 \end{array} \right] \end{array}$$

The corresponding reduction for the constraint equation is based on the application of the absorption property, i.e., $x(x+y) = x$. Let us consider the equation corresponding to the previous example:

$$(p_1 + p_2 + p_6)(p_1 + p_2)(p_2 + p_3)(p_2 + p_3 + p_4)(p_4 + p_5)(p_4 + p_5 + p_6) = 1.$$

By absorption, we can replace, for instance, $(p_1 + p_2 + p_6)(p_1 + p_2)$ by $(p_1 + p_2)$. The reduced equation is:

$$(p_1 + p_2)(p_2 + p_3)(p_4 + p_5) = 1.$$

We see that this equation corresponds to the reduced matrix of the previous example.

4.8.4 Column or Variable Dominance

Before we discuss this technique, we must briefly digress to consider the cost of a column or a variable. In general, each prime corresponds to one AND gate in a SOP circuit. If the number of gates is the only concern, it is correct to assign the same cost to all columns or variables. However, if the number of literals is more important, then a prime with, say, five literals, should be considered more expensive than a prime with three literals. This can be accommodated by assigning different costs to the columns or variables. The total cost of a solution is then the sum of the costs of the selected columns or, in other words, the cost of the variables set to 1.

Suppose now that a column p_i has all the ones of another column p_j. Suppose further that the cost of p_i is not greater than the cost of p_j. Then, we can say that p_i is not inferior to p_j, in that it covers all the rows that p_j covers, at a cost that is not larger. This means that we can discard p_j from the matrix, without giving up the possibility of finding an optimum solution.

As an example, let us consider the matrix obtained in the previous example by row dominance. Suppose all columns have the same cost. We see that the second column dominates the first and the third. According to our definition, the fourth column dominates the fifth and vice versa. In this case we can choose arbitrarily which column to retain. Say we choose the fourth. The result of the reduction is the following matrix:

$$\begin{array}{c} \\ 2 \\ 3 \\ 5 \end{array} \begin{array}{c} \begin{array}{cc} 2 & 4 \end{array} \\ \left[\begin{array}{cc} 1 & \\ 1 & \\ & 1 \end{array} \right] \end{array}$$

Here we see how row and column dominance reductions can engender new essential columns/variables. Note when we employ column dominance, we choose to ignore some valid solutions to the constraint equation, some of which may be optimum.

However, we may do this with the assurance that we always retain at least one optimum solution.

If we are given the constraint equation, variable dominance can be checked as follows. We say that p_i dominates p_j if the cost of p_i does not exceed the cost of p_j and p_i appears in every term where p_j appears. This criterion is a simple translation of the one employed for the constraint matrix. A more general formulation is possible, but we shall not pursue it for the time being. Returning to our example, consider the constraint equation

$$(p_1 + p_2)(p_2 + p_3)(p_4 + p_5) = 1.$$

We see that p_2 appears in all terms where p_1 appears. Hence, p_2 dominates p_1. Similarly, p_3 is dominated by p_2 and p_4 and p_5 dominate each other. (We choose p_4 arbitrarily.) The reduction consists of taking the negative cofactor of the left-hand side of the equation with respect to all dominated variables (the order is unimportant). The reduced equation is thus:

$$p_2 p_2 p_4 = 1,$$

which becomes $p_2 p_4 = 1$ by idempotency.

4.8.5 Systematically Exploring the Search Space

When the constraint matrix cannot be further reduced, we have two cases: If the matrix has no rows left, then we say that we have reached a terminal case and we have solved the problem; otherwise the problem is cyclic. If we are working with the constraint equation, the terminal case occurs when the constraint equation simplifies to $1 = 1$.

If the reduced constraint equation, written $F = 1$, has no essential variables, then, since each p_i is a binary variable, we can seek the minimum cover by a divide and conquer strategy: first, consider all solutions for which $p_i = 1$, and then consider all solutions for which $p_i = 0$. The constraint equation which corresponds to the former case is obtained by setting $p_i = 1$, and is written

$$F_{p_i} = 1.$$

Similarly, in the latter case we set $p_i = 0$ and obtain $F_{p_i'} = 1$. The two formulae F_{p_i} and $F_{p_i'}$ are the *positive* and *negative cofactors* of F with respect to p_i. Cofactors were defined in Section 3.3.3 (Page 98).

It needs to be emphasized that a zero in the constraint matrix does not correspond to a negated literal in the constraint equation. This is because we are restricting attention in this chapter to UCP, the Unate Covering Problem, in which every variable appears in just one phase. Consequently, when we take a negative cofactor of the constraint equation with respect to p_i, the corresponding action on the constraint matrix is to simply delete the column. The rows of the constraint matrix which have 0s in column p_i are *not deleted*, because the 0s mean that variable p_i did not appear in the corresponding sum of the POS form of the constraint equation.

If we want to solve large enumeration problems, we have to avoid enumerating all solutions explicitly, since their number is exponential in the variables of the problem. What we want is an **implicit enumeration** of the solutions, where many (as many as possible) solutions are not explicitly considered. The reduction techniques of the previous section are an important part of the implicit enumeration scheme for the covering problem; they allow us to determine that some variables (prime implicants in the logic minimization context) are either essential or can be left out of the solution without impairing our ability to find an optimum solution.

We have seen in Section 4.7 that when the problem is cyclic we need to select a column (or a variable) and solve two reduced subproblems: One subproblem is obtained by accepting the selecting column; the other subproblem is obtained by rejecting it. The reduced subproblems may in turn be cyclic and therefore require the selection of new splitting variables. This gives rise to a recursive process of selecting variables and tentatively solving subproblems in order to find which one yields the best solution. This process can be seen as the exploration of the *search space* of the problem. In a cyclic covering problem, we must, in principle, enumerate all possible solutions, in order to find one of minimum cost. Therefore we say that the covering problem is an enumeration problem.

In the nutrient problem discussed above on Page 144 we enumerated all the solutions to the constraint equation by transforming from POS form to SOP form. This can be called "explicit enumeration" since all solutions were examined. All possible solutions can also be obtained by recursively cofactoring. However when we use the reduction techniques of the previous section, we eliminate some solutions, including optimum solutions, from consideration. This is called "implicit enumeration", because we are enumerating some, but not all, of the possible solutions.

We now consider how to systematically explore the search space in the case of cyclic problems, so as to minimize the number of solutions that we explicitly enumerate. The implicit enumeration scheme we adopt is called **Branch and Bound**, and is a general scheme for solving enumeration problems. Branch and Bound applies to search problems where we are interested in finding the minimum or maximum cost of a feasible solution. There are many problems for which clever specialized search strategies can be applied. For instance, we do not need to enumerate all possible paths between two cities to find the shortest one. Hence we do not apply a branch-and-bound technique to solving the shortest path problem. However, no such clever search strategy is known for the covering problem (and it is unlikely to exist). Hence, we have to enumerate and we are interested in minimizing the work. In this context, Branch and Bound can be (extremely) useful.

4.8.6 Computation of the Lower Bound

We now address the problem of computing a lower bound approximation to the cost of covering a constraint matrix. The method easily translates into corresponding problem of approximating the cost of satisfying a constraint equation.

In a given covering matrix M, suppose that two rows r_i and r_j have nonzero entries in sets $R_i = \{k \mid M_{ik} \neq 0\}$ and $R_j = \{k \mid M_{jk} \neq 0\}$. If $R_i \cap R_j = \emptyset$, that is, if r_i and r_j have no nonzero columns in common, we say these two rows are independent (that is, column-disjoint). It is apparent that we need two different columns to cover these two rows. Generalizing this argument, if a matrix has n rows that are similarly disjoint (pairwise), we need at least n columns to cover the whole matrix. In this case, the n rows are said to form an **independent set** of rows. For each row of the independent set, the cost of covering it is at least the cost of the cheapest column that covers the row. Hence, by summing the costs of the cheapest columns covering each of the rows in the independent set, we get a lower bound to the cost of covering the entire matrix.

As an example, let us consider the following matrix:

$$\begin{array}{c} \\ 1 \\ 2 \\ 3 \\ 4 \\ 5 \\ 6 \end{array} \begin{array}{c} \begin{array}{cccccc} 1 & 2 & 3 & 4 & 5 & 6 \end{array} \\ \left[\begin{array}{cccccc} 1 & & & & & 1 \\ 1 & 1 & & & & \\ & 1 & 1 & & & \\ & & 1 & 1 & & \\ & & & 1 & 1 & \\ & & & & 1 & 1 \end{array} \right] \end{array} \qquad (4.7)$$

The first, third, and fifth rows are independent. Hence, we need at least three columns to cover the matrix. There is another independent set of three rows, namely the second, fourth, and sixth rows. Notice that in this case the lower bound is exact: we can cover the matrix with exactly three columns. In general, however, there will be independent sets of different sizes and the lower bound will not necessarily be exact. It is also easy to see that there is always at least an independent set of size 1. Further, for cyclic (irreducible) matrices with unit cost columns, we always get a lower bound of size 2 or more, as shown in the following theorem.

Theorem 4.8.1 *In the unate covering problem with unit costs for the columns, the lower bound for a cyclic matrix is at least 2 (even if there are no two independent rows).*

Proof. Suppose a constraint matrix of a unate covering problem contains a full column of ones. In this case, the matrix cannot be cyclic because the column with all ones dominates all the others. The matrix can therefore be reduced to one column, which is obviously essential. Thus such a matrix is not cyclic. It then follows that if the matrix is cyclic, there can be no column that covers all the rows. Hence, at least two columns are required to cover all the rows. □

An independent set of rows is called **maximal** if it intersects (that is, has a column in common with) every other row of the covering matrix. We shall use the abbreviation MIS for a Maximal Independent Set. This means that unless some of the decisions already made in building the set are reversed, the set cannot grow larger while retaining its independence (pairwise disjointness).

A simple algorithm for quickly finding an MIS is Procedure MIS_QUICK of Figure 4.7. Here $\| M \|$ denotes the number of rows left in M after deleting the rows intersecting the chosen row.

The key feature of this algorithm is Subprocedure CHOOSE_SHORTEST_ROW. In its simplest form it just chooses the "shortest" row, that is, the row with the fewest nonzero columns, and breaking ties in ascending lexicographical order. Better heuristic performance is usually obtained by a more sophisticated heuristic, in which the weight of row i is defined in terms of the column counts of its columns. That is, let

```
MIS_QUICK(M) {
    MIS = ∅
    do {
        i = CHOOSE_SHORTEST_ROW(M)
        MIS = MIS ∪ {i}
        M = DELETE_INTERSECTING ROWS(M, i)
    while (|| M ||> 0) continue
    return (MIS);
}
```

Figure 4.7: Algorithm for computing an MIS.

$C_j = \{k \mid M_{kj} \neq 0\}$ be the set of nonzero row in column j. Then the weight of row i is $\sum_{j \in R_i} |C_j|$, and CHOOSE_SHORTEST_ROW chooses the row of minimum weight, breaking ties by choosing shortest rows in ascending lexicographical order.

We first apply CHOOSE_SHORTEST_ROW to the covering matrix of Equation 4.7. Here all rows have equal weight by either heuristic, and the MIS chosen would be $\{1, 3, 5\}$, which is optimum, since $|MIS| = 3$ is the cost of the optimum solution of the covering problem. In this case the bound is sharp.

However, now let us apply this algorithm to the following example, which discriminates between the two heuristics.

$$
\begin{array}{c}
\quad\quad 1\;\;2\;\;3\;\;4\;\;5\;\;6 \\
M = \left[\begin{array}{cccccc}
1 & 0 & 0 & 0 & 1 & 0 \\
1 & 1 & 0 & 1 & 0 & 0 \\
0 & 1 & 1 & 0 & 0 & 0 \\
0 & 0 & 0 & 1 & 1 & 1 \\
0 & 0 & 1 & 1 & 0 & 0 \\
0 & 1 & 0 & 0 & 0 & 1 \\
1 & 0 & 1 & 0 & 0 & 0
\end{array}\right]
\begin{array}{lll}
1 & w_1^1 = 2 & w_1^2 = 5 \\
2 & w_2^1 = 3 & w_2^2 = 9 \\
3 & w_3^1 = 2 & w_3^2 = 6 \\
4 & w_4^1 = 3 & w_4^2 = 7 \\
5 & w_5^1 = 2 & w_5^2 = 5 \\
6 & w_6^1 = 2 & w_6^2 = 5 \\
7 & w_7^1 = 2 & w_7^2 = 6
\end{array}
\end{array}
$$

The superscripts 1 and 2 indicates row weights computed according to the first and second heuristics. Thus for the first heuristic, MIS_QUICK would obtain $MIS = \{1\}$ on the first pass, and $MIS = \{1, 3\}$ on the second (and last) pass.

However, for the second heuristic, MIS_QUICK would obtain $MIS = \{1\}$ on the first pass, after which the reduced matrix is

$$
\begin{array}{c}
\quad\quad 1\;\;2\;\;3\;\;4\;\;5\;\;6 \\
M = \left[\begin{array}{cccccc}
0 & 1 & 1 & 0 & 0 & 0 \\
0 & 0 & 1 & 1 & 0 & 0 \\
0 & 1 & 0 & 0 & 0 & 1
\end{array}\right]
\begin{array}{lll}
3 & w_3^1 = 2 & w_3^2 = 4 \\
5 & w_5^1 = 2 & w_5^2 = 3 \\
6 & w_6^1 = 2 & w_6^2 = 3
\end{array}
\end{array}
$$

The second heuristic would give $MIS = \{1, 5\}$ on the second pass, and $MIS = \{1, 5, 6\}$ on the third (and last) pass. The improved heuristic was able to identify a

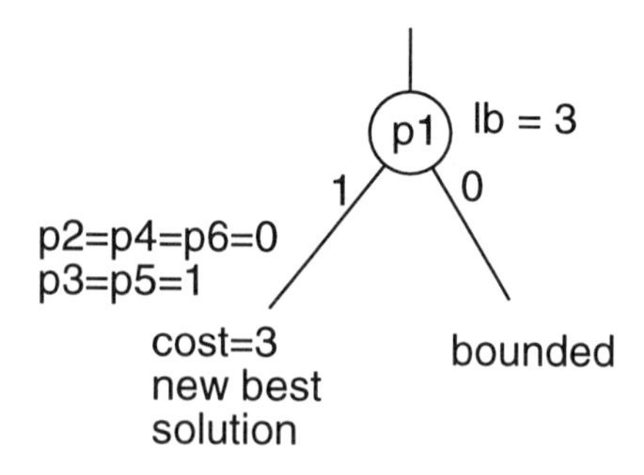

Figure 4.8: Recursion Tree for a Covering Problem.

larger MIS by paying attention to how many rows are eliminated by each choice of a row to be added to the the current MIS.

Returning to our example, we found that it was a cyclic problem. If we select Column 1 for splitting, we get the solution $\{1, 3, 5\}$. Since the cost of this solution (3) equals the lower bound, we know it is optimal and we do not need to find the best solution without Column 1. This process is illustrated in Figure 4.8, where the recursion tree is shown.

4.9 The Branch-and-Bound Algorithm

In this section we formulate an efficient procedure to solve the unate covering problem that we have informally introduced in the previous section. The procedure is essentially the one presented in [239], which in turn is based on the earlier procedure due to McCluskey [188]. We then consider cyclic problems and we present a branch-and-bound algorithm for them. Finally, we mention the connection between unate covering and integer linear programming.

The idea behind Branch and Bound is that we are only interested in finding one optimum solution (there may be many). Therefore, if we can determine that a given part of the search space does not contain any solution better than the best we have found so far, then we can avoid exploring that part of the search space altogether.

How do we come to the conclusion that there are no 'interesting' solutions in a part of the search space? In Branch and Bound, we resort to two basic ideas. The first is that the search space is organized in the form of a **search tree**(sometimes called a recursion tree). To fix ideas, we consider the case of a **binary** search tree, which is what we are going to use. Each node of the tree corresponds to a variable of the problem and the two branches out of the node correspond to the acceptance or rejection of the variable. The 'Branch' in Branch and Bound refers to the process of exploring the branches of the search tree. An example is shown in Figure 4.9. Let us examine the leftmost path in the tree. We assume that at the top node (the root) we have selected p_1 as the splitting variable. We also suppose that in the simplified subproblem we can identify that p_2 must be 1 (e.g., it is an essential variable for the subproblem) and p_6 can be set to 0 (e.g., it is a dominated variable). The resulting hypothetical simplified subproblem is still cyclic. Hence, we now select another splitting variable, p_4, whose choice allows us to set also $p_7 = 0$ (e.g., another

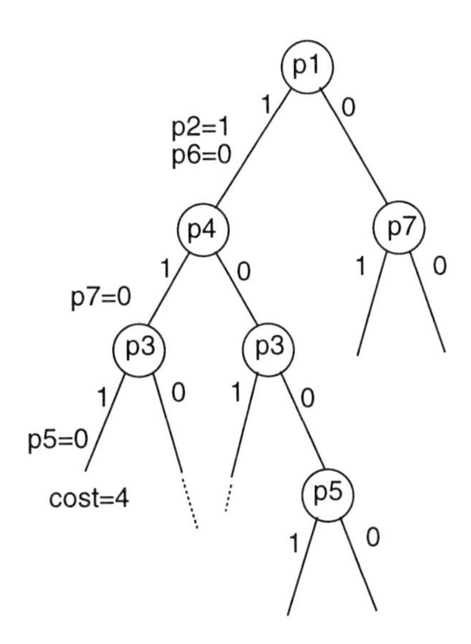

Figure 4.9: Example of Search Tree.

case of variable dominance). After choosing another splitting variable, we suppose to reach a terminal case. The cost of the solution—assuming unit cost for all variables, is given by the number of variables set to one along the path, namely 4.

At any given node of the search tree, we have selected and rejected some variables. These variables are identified by the path from the root of the tree to that node. Hence, at that node we have a partial solution. Also, we maintain an **upper bound** on the cost of the optimum solution. Initially, the upper bound is set to a suitably large number. When we find a new best solution, we set the upper bound to its cost. From that point on, we shall not be interested in solutions that are not cheaper than the new best solution.

If the cost of a partial solution exceeds or equals the value of the upper bound at a node, clearly we can abandon that node and back up. Branch and Bound goes one step further and tries to establish whether a new best solution can still be found by proceeding from the current node. This assessment is based on computing a **lower bound** on the cost of completing the current partial solution. If the cost of the current partial solution plus the lower bound on the cost of completing it exceeds the current upper bound, then the current node is abandoned. The 'Bound' in Branch and Bound takes its name from this strategy.

The way of computing the lower bound depends on the particular problem. We shall examine a lower bound that applies to the covering problem. It is obvious that a careful choice of the lower bound criterion is important. Ideally, the criterion should provide an accurate estimate of the real minimum cost incurred in completing the current solution. At the same time, the computation of the bound should be fast. Before we turn our attention to the computation of the lower bound for the unate covering problem, we now take a look at the pseudo-code for the branch-and-bound algorithm that we have delineated so far that is given in Figure 4.10.

The pseudo code assumes that we work with the constraint equation. The program is initially called with an empty current solution (*currentSol*), the initial left-hand

```
BCP(F, U, currentSol) {
1    (F, currentSol) = REDUCE(F, currentSol)
     if (terminalCase(F)) {
         if (COST(currentSol) < U) {
             U = COST(currentSol)
2            return (currentSol)
         }
3        else return ("no solution")
     }
4    L = LOWER_BOUND(F, currentSol)
     if (L ≥ U) return ("no solution")
5    x_i = CHOOSE_VAR(F)
6    S^1 = BCP(F_{x_i}, U, currentSol ∪ {s_i})
7    if (COST(S^1) = L) return (S^1)
     S^2 = BCP(F_{x_i'}, U, currentSol)
8    return BEST_SOLUTION(S^1, S^2)
}
```

Figure 4.10: Branch-and-Bound Algorithm for the Unate Covering Problem.

side of the constraint equation (F) and the upper bound (U) set to the total cost of all variables plus one. This initial value of the upper bound exceeds the cost of any possible solution and hence guarantees that the first solution found will be accepted.

The procedure REDUCE (Line 1) iteratively finds essential variables and applies row and column dominance. It also updates the current solution accordingly. If the problem is now reduced to a terminal case, the procedure checks whether the solution thus found is better than the current best. The current solution is returned only if it is the new best (Line 2).

If the problem is cyclic, the lower bound is computed by the call to LOWER_BOUND (Line 4). In this subprocedure, we find an MIS by an internal call to MIS_QUICK, and add the size of the current solution to the size of the MIS to get a lower bound. If there is still a chance of getting an optimum solution, a splitting variable is chosen (Line 5), and the first of the two subproblems are solved recursively. If the cost of this subproblem is equal to the lower bound, we immediately return (Line 7). Else, the second subproblem is solved recursively, and the best solution is returned (Line 8).

4.9.1 Choice of the Splitting Variable

The choice of the splitting variable has no effect on the correctness of the procedure, but it is important for its efficiency. A column that covers many rows is typically a better candidate than a column that covers few row. The former is more likely to be part of an optimum solution. It is convenient to find a good solution soon, so that the upper bound is close to the optimum value, and more pruning of the search tree due to bounding is possible.

A possible refinement of the above strategy consists of favoring columns that cover many short rows. (A short row is one with few ones.) This criterion is based on the assumption that shorter rows have a lower chance of being covered. It can be implemented by assigning a weight to each row inversely proportional to the row length and by summing the weights of all rows covered by a column in order to determine the value of that column. The column with the highest value is chosen. This idea is explored further in Solved Problem 18.

4.9.2 Examples of Splitting and Lower Bounding

Let us consider some examples of the application of the BCP algorithm. We begin with a simple example that illustrates the flow of execution.

$$
\begin{array}{c} \\ 1 \\ 2 \\ 3 \\ 4 \end{array}
\begin{array}{c} \begin{array}{cccccc} 1 & 2 & 3 & 4 & 5 & 6 \end{array} \\ \left[\begin{array}{cccccc} 1 & & 1 & & 1 & \\ 1 & & & 1 & & 1 \\ & 1 & 1 & & & 1 \\ & 1 & & 1 & 1 & \end{array} \right] \end{array}
$$

The given matrix has 4 rows and 6 columns and is cyclic, so BCP sets $U = 7$ and skips to Line 4 and calls LOWER_BOUND. We cannot find any set of two independent rows. Therefore, we set the initial lower bound to 2 (using Theorem 4.8.1). If we split on Column 1, we get the solution $\{1, 2\}$ in the positive half of the search space. This cost equals the lower bound. Hence, we don't need to explore the half of the search space where Column 1 is rejected. We thus return the solution $\{1, 2\}$ at Line 7.

We next consider the example of Equation 4.7. This matrix has 6 rows and 6 columns and is also cyclic, so BCP again sets $U = 7$ and skips to Line 4. There it calls LOWER_BOUND, which this time returns a lower bound of 3. BCP again chooses Column 1 as the splitting variable, and for $p_1 = 1$, the positive cofactor the matrix reduces to

$$
\begin{array}{c} \\ 3 \\ 4 \\ 5 \\ 6 \end{array}
\begin{array}{c} \begin{array}{cccccc} 1 & 2 & 3 & 4 & 5 & 6 \end{array} \\ \left[\begin{array}{cccccc} & 1 & 1 & & & \\ & & 1 & 1 & & \\ & & & 1 & 1 & \\ & & & & 1 & 1, \end{array} \right] \end{array}
$$

in which columns p_2 and p_6 are dominated. Then columns 3 and 5 become essential, leading to a terminal case with solution $S^1 = \{1, 3, 5\}$. Thus U is lowered to 3, and this solution is returned back up to the original recursive call. Then BCP compares the cost of this solution to L, S^1 is returned as the final solution. Again, we don't need to explore the half of the search space where Column 1 is rejected.

In both of these examples, lower bounding techniques have limited the implicit enumeration to the first half-space of the recursion. We now explore the application of BCP to a covering matrix with 13 rows and 13 columns, denoted by variables p_i, $i = 1, \ldots, 11$. In this richer example, all of the features of BCP are explicitly

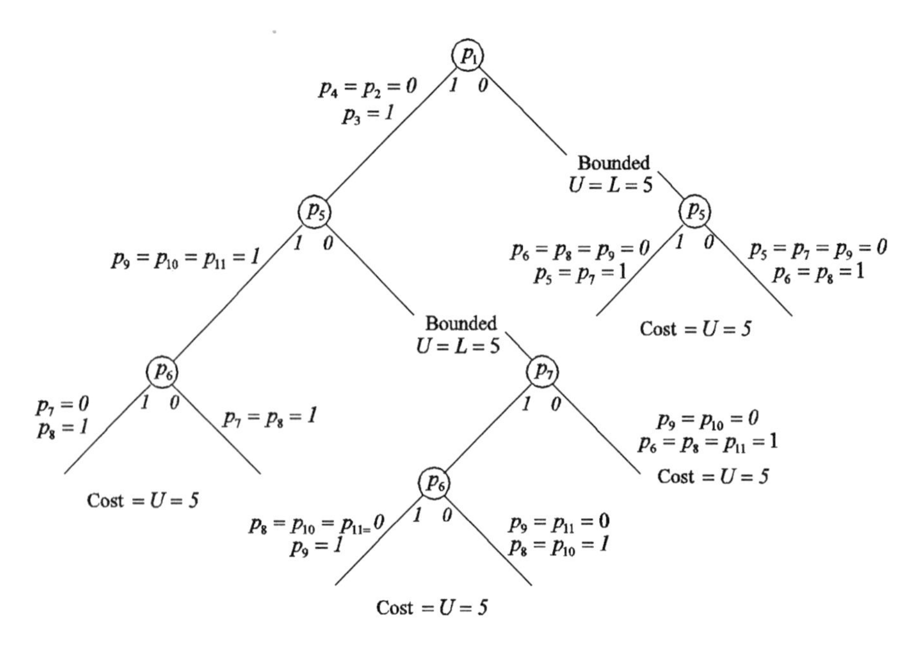

Figure 4.11: A search tree produced by Procedure BCP.

active.

$$
M = \begin{array}{c}
\begin{array}{cccc@{\quad}ccccccc} 1 & 2 & 3 & 4 & 5 & 6 & 7 & 8 & 9 & 10 & 11 \end{array} \\
\left[\begin{array}{cccc@{\quad}ccccccc}
1 & 1 & 0 & 0 & 0 & 0 & 0 & 0 & 0 & 0 & 0 \\
0 & 1 & 1 & 0 & 0 & 0 & 0 & 0 & 0 & 0 & 0 \\
0 & 0 & 1 & 1 & 0 & 0 & 0 & 0 & 0 & 0 & 0 \\
1 & 0 & 0 & 1 & 0 & 0 & 0 & 0 & 0 & 0 & 0 \\
 & & & & & & & & & & \\
0 & 0 & 0 & 0 & 1 & 1 & 0 & 0 & 0 & 1 & 0 \\
0 & 0 & 0 & 0 & 0 & 1 & 1 & 0 & 1 & 0 & 0 \\
0 & 0 & 0 & 0 & 0 & 0 & 1 & 1 & 0 & 0 & 0 \\
0 & 0 & 0 & 0 & 0 & 1 & 0 & 1 & 0 & 1 & 1 \\
0 & 0 & 0 & 0 & 1 & 0 & 0 & 0 & 1 & 1 & 1 \\
0 & 0 & 0 & 0 & 1 & 0 & 0 & 1 & 1 & 0 & 0 \\
0 & 0 & 0 & 0 & 1 & 0 & 1 & 0 & 0 & 0 & 1 \\
 & & & & & & & & & & \\
1 & 0 & 0 & 0 & 0 & 0 & 0 & 0 & 0 & 0 & 1 \\
0 & 0 & 0 & 0 & 1 & 1 & 0 & 1 & 0 & 0 & 0
\end{array}\right]
\end{array}
\begin{array}{c} \\ 1 \\ 2 \\ 3 \\ 4 \\ \\ 5 \\ 6 \\ 7 \\ 8 \\ 9 \\ 10 \\ 11 \\ \\ 12 \\ 13 \end{array}
$$

We shall that the search tree (or recursion tree) of Figure 4.11 results from the application of algorithm BCP, given appropriate heuristic choices in the lower bound computation and in the choice of the splitting variable.

BCP begins by setting $U = 12$. Because of the block structure, it is not hard to see this matrix is cyclic (no reduction), so no implied variable assignments exist. Applying MIS_QUICK and using the simpler heuristic, BCP gets MIS={1,2,5,7}, and an initial lower bound $L = 4$.

Keeping in mind the block structure, we can ignore column counts, and split on the first variable. With $p_1 = 1$ rows $1, 4, 13$ are covered, and we see that columns 2 and 4 are dominated, so we get a secondary essential column 3. The call to REDUCE leads to the following cyclic matrix.

$$M_{p_1} = \begin{array}{c} \begin{array}{ccccccc} 5 & 6 & 7 & 8 & 9 & 10 & 11 \end{array} \\ \left[\begin{array}{ccccccc} 1 & 1 & 0 & 0 & 0 & 1 & 0 \\ 0 & 1 & 1 & 0 & 1 & 0 & 0 \\ 0 & 0 & 1 & 1 & 0 & 0 & 0 \\ 0 & 1 & 0 & 1 & 0 & 1 & 1 \\ 1 & 0 & 0 & 0 & 1 & 1 & 1 \\ 1 & 0 & 0 & 1 & 1 & 0 & 0 \\ 1 & 0 & 1 & 0 & 0 & 0 & 1 \\ \\ 1 & 1 & 0 & 1 & 0 & 0 & 1 \end{array} \right] \end{array} \begin{array}{c} \\ 5 \\ 6 \\ 7 \\ 8 \\ 9 \\ 10 \\ 11 \\ \\ 13 \end{array}$$

We may then split on a longest column p_5. For the positive cofactor the partial solution is $\{p_1, p_3, p_5\}$ the reduced matrix is

$$M_{p_1 p_5} = \begin{array}{c} \begin{array}{ccc} 6 & 7 & 8 \end{array} \\ \left[\begin{array}{ccc} 1 & 1 & 0 \\ 0 & 1 & 1 \\ 1 & 0 & 1 \end{array} \right] \end{array} \begin{array}{c} \\ 6 \\ 7 \\ 8 \end{array}$$

Here we have observed that after deleting rows covered by p_5, columns 9, 10, and 11 are dominated.

This submatrix is cyclic, so we split on p_6. The assignment $p_6 = 1$ leads to a terminal case with solution $\{p_1, p_3, p_5, p_6, p_8\}$, and Cost $U = 5$. The assignment $p_6 = 0$ gives the solution $\{p_1, p_3, p_5, p_7, p_8\}$, with the same cost.

Since in this case $U > L$, we fall through at Line 7. We then obtain the solution S^2 obtained for the negative cofactor with respect to p_5. For the partial solution $\{p_1, p_3, p_5'\}$ the covering matrix is, after removing dominating row 8, as follows.

$$M_{p_1 p_3, p_5'} = \begin{array}{c} \begin{array}{cccccc} 6 & 7 & 8 & 9 & 10 & 11 \end{array} \\ \left[\begin{array}{cccccc} 1 & 0 & 0 & 0 & 1 & 0 \\ 1 & 1 & 0 & 1 & 0 & 0 \\ 0 & 1 & 1 & 0 & 0 & 0 \\ 0 & 0 & 0 & 1 & 1 & 1 \\ 0 & 0 & 1 & 1 & 0 & 0 \\ 0 & 1 & 0 & 0 & 0 & 1 \\ 1 & 0 & 1 & 0 & 0 & 0 \end{array} \right] \end{array} \begin{array}{c} \\ 5 \\ 6 \\ 7 \\ 9 \\ 10 \\ 11 \\ 13 \end{array} \quad \Rightarrow L = 2 + 3 = 5$$

Here we note that rows 5,10, and 11 are independent, so with MIS=$\{5,10,11\}$, we get a lower bound of $L = 2 + |\text{MIS}| = 2 + 3 = 5$. Thus we know there is no better solution in this subspace, and immediately return. This is indicated by the legend "Bounded $U = L = 5$". At this point BCP is through with positive cofactor with respect to p_1, and next considers the negative cofactor.

However, it is interesting to see what is saved by this lower bound operation in this case. Without the lower bounding, since there is no reduction, we would split on p_7. Thus for $p_7 = 1$, we have the following.

$$M_{p_1, p_3, p_5' p_7} = \begin{array}{c} \begin{array}{ccccc} 6 & 8 & 9 & 10 & 11 \end{array} \\ \left[\begin{array}{ccccc} 1 & 0 & 0 & 1 & 0 \\ 0 & 0 & 1 & 1 & 1 \\ 0 & 1 & 1 & 0 & 0 \\ 1 & 1 & 0 & 0 & 1 \end{array}\right] \end{array} \begin{array}{c} \\ 5 \\ 9 \\ 10 \\ 13 \end{array}$$

This matrix is again cyclic, so we split on p_6. For $p_6 = 1$, columns p_8, p_{10}, p_{11} are all dominated by column p_9, and we get the solution $\{p_1, p_3, p_7, p_6, p_9\}$ of cost 5. Similarly, for $p_6 = 0$, column p_{10} becomes essential, and after reduction p_8 does also, so we get the solution $\{p_1, p_3, p_7, p_{10}, p_8\}$ of cost 5.

Similarly for the partial solution $p_1 = p_3 = 1$, and $p_5 = p_7 = 0$, we get the following covering matrix.

$$M_{p_1 p_5' p_7'} = \begin{array}{c} \begin{array}{ccccc} 6 & 8 & 9 & 10 & 11 \end{array} \\ \left[\begin{array}{ccccc} 1 & 0 & 0 & 1 & 0 \\ 1 & 0 & 1 & 0 & 0 \\ 0 & 1 & 0 & 0 & 0 \\ 1 & 1 & 0 & 1 & 1 \\ 0 & 0 & 1 & 1 & 1 \\ 0 & 1 & 1 & 0 & 0 \\ 0 & 0 & 0 & 0 & 1 \\ 1 & 1 & 0 & 0 & 0 \end{array}\right] \end{array} \begin{array}{c} \\ 5 \\ 6 \\ 7 \\ 8 \\ 9 \\ 10 \\ 11 \\ 13 \end{array} \Rightarrow \begin{array}{c} p_9 = p_{10} = 0 \\ (p_6) = p_8 = p_{11} = 1 \end{array}$$

Here columns p_8 and p_{11} are essential, and, after deleting the rows they cover, and eliminating dominated columns, p_6 becomes (secondary) essential. Again, as expected cost = 5 for the solution $\{p_1, p_3, p_6, p_8, p_{11}\}$.

We see that a substantial amount of work is avoided by the lower bound comparison.

Now we return to the top of the recursion tree, and consider the result of excluding column p_1.

$$
M_{p_1'} = \begin{array}{c}
\begin{array}{ccccccccccc} 2 & 3 & 4 & 5 & 6 & 7 & 8 & 9 & 10 & 11 \end{array} \\
\left[\begin{array}{ccc|ccccccc}
1 & 0 & 0 & 0 & 0 & 0 & 0 & 0 & 0 & 0 \\
1 & 1 & 0 & 0 & 0 & 0 & 0 & 0 & 0 & 0 \\
0 & 1 & 1 & 0 & 0 & 0 & 0 & 0 & 0 & 0 \\
0 & 0 & 1 & 0 & 0 & 0 & 0 & 0 & 0 & 0 \\
0 & 0 & 0 & 1 & 1 & 0 & 0 & 0 & 1 & 0 \\
0 & 0 & 0 & 0 & 1 & 1 & 0 & 1 & 0 & 0 \\
0 & 0 & 0 & 0 & 0 & 1 & 1 & 0 & 0 & 0 \\
0 & 0 & 0 & 0 & 1 & 0 & 1 & 0 & 1 & 1 \\
0 & 0 & 0 & 1 & 0 & 0 & 0 & 1 & 1 & 1 \\
0 & 0 & 0 & 1 & 0 & 0 & 1 & 1 & 0 & 0 \\
0 & 0 & 0 & 1 & 0 & 1 & 0 & 0 & 0 & 1 \\
0 & 0 & 0 & 0 & 0 & 0 & 0 & 0 & 0 & 1 \\
0 & 0 & 0 & 1 & 1 & 0 & 1 & 0 & 0 & 0
\end{array}\right]
\begin{array}{c} 1 \\ 2 \\ 3 \\ 4 \\ 5 \\ 6 \\ 7 \\ 8 \\ 9 \\ 10 \\ 11 \\ 12 \\ 13 \end{array}
\end{array}
$$

Here we see that columns p_2, p_4, p_{11} have become essential, and after the rows they cover are eliminated, column p_{10} is dominated by p_5. Also row 13 is now dominated by row 5. Thus $M_{p_1'}$ is equivalent to the following reduced matrix.

$$
\begin{array}{c}
\begin{array}{ccccc} 5 & 6 & 7 & 8 & 9 \end{array} \\
\left[\begin{array}{ccccc}
1 & 1 & 0 & 0 & 0 \\
0 & 1 & 1 & 0 & 1 \\
0 & 0 & 1 & 1 & 0 \\
1 & 0 & 0 & 1 & 1
\end{array}\right]
\begin{array}{c} 5 \\ 6 \\ 7 \\ 10 \end{array}
\end{array}
$$

Here we see that for the partial solution $p_2 = p_4 = p_{11} = 1$ the reduced matrix has MIS= $\{1, 4\}$, so the lower bound is $3 + |\text{MIS}| = 3 + 2 = 5$, so again it is known that this whole half of the solution space contains no solution of cost less than 5.

However, we can again check to see what we saved by the bounding operation. If we split on p_5, we see that for $p_5 = 1$, p_7 becomes essential, so we get a cost of 5, with a solution $p_1 = p_4 = p_1 1 = p_5 = p_7 = 1$.

$$
M_{p_1' p_5} = \begin{array}{c}
\begin{array}{cccc} 6 & 7 & 8 & 9 \end{array} \\
\left[\begin{array}{cccc}
1 & 1 & 0 & 1 \\
0 & 1 & 1 & 0
\end{array}\right]
\begin{array}{c} 6 \\ 7 \end{array}
\end{array}
\Rightarrow \begin{array}{c} p_6 = p_8 = p_9 = 0 \\ p_7 = 1 \end{array}
$$

Similarly, for $p_5 = 0$, we see that p_6, and then, after row deletion and column dominance, p_8, both become essential, leading to the solution $p_1 = p_4 = p_11 = p_6 = p_8 = 1$.

$$M_{p'_{p_1}} = \begin{array}{c} \begin{array}{cccc} 6 & 7 & 8 & 9 \end{array} \\ \left[\begin{array}{cccc} 1 & 0 & 0 & 0 \\ 1 & 1 & 0 & 1 \\ 0 & 1 & 1 & 0 \\ 0 & 0 & 1 & 1 \end{array}\right] \end{array} \begin{array}{c} 5 \\ 6 \\ 7 \\ 10 \end{array} \Rightarrow \begin{array}{c} p_7 = p_9 = 0 \\ p_6 = p_8 = 1 \end{array}$$

4.9.3 The Unate Covering Problem as an Integer Linear Program

Formally, we can say that the covering problem consists of finding

$$\min \sum_{j=1}^{n} w_j p_j$$

subject to the constraint

$$A \cdot p \geq \mathbf{1},$$

where

$$p = \begin{bmatrix} p_1 \\ \vdots \\ p_n \end{bmatrix} \qquad \mathbf{1} = \begin{bmatrix} 1 \\ \vdots \\ 1 \end{bmatrix}$$

and the coefficients of the matrix and the variables can only assume the values 0 and 1. The coefficients w_j in the summation are the weights, or costs, of the columns. The product $A \cdot p$ is the standard product of real (or integer) matrices. In our first example, for instance, we can verify that the solution $p_1 = p_3 = p_4 = 1$, $p_2 = 0$ satisfies the constraint $A \cdot p \geq \mathbf{1}$:

$$\begin{bmatrix} 0 & 1 & 1 & 0 \\ 1 & 1 & 0 & 0 \\ 1 & 0 & 0 & 1 \\ 0 & 0 & 0 & 1 \\ 0 & 0 & 1 & 0 \end{bmatrix} \cdot \begin{bmatrix} 1 \\ 0 \\ 1 \\ 1 \end{bmatrix} = \begin{bmatrix} 1 \\ 1 \\ 2 \\ 1 \\ 1 \end{bmatrix}$$

This formulation of the problem exposes the fact that the covering problem is a special form of *Integer Linear Program* (ILP). We just mention here this important connection and the fact that both ILP and its special case Unate Covering are difficult to solve. The best known algorithms take time exponential in the number of the variables in the worst case.

4.10 Multiple Output Functions

It is often the case that we deal with a set of related functions over the same variables. These functions must be implemented as a *multiple-output circuit*; hence we refer to such sets of functions as **multiple-output functions**. In this section, we are concerned with the optimum **two-level functions** which implement multiple-output specifications.

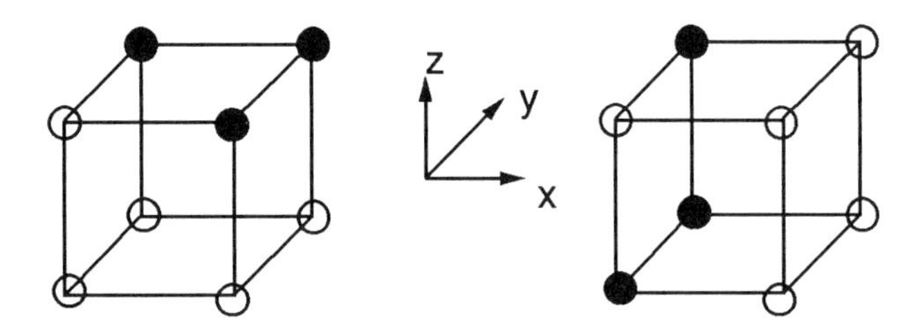

Figure 4.12: A Two-Output Function that Illustrates the Importance of Sharing Common Terms.

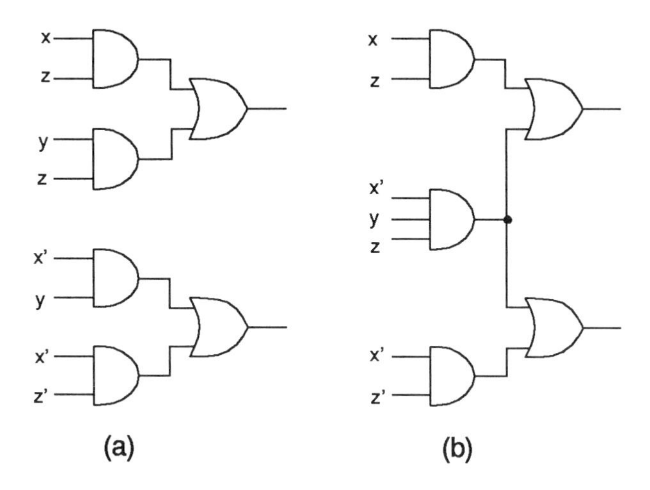

Figure 4.13: Two Implementations for the Multiple-Output Function of Figure 4.12.

One obvious way to implement such functions is to find an optimum two-level formula—either SOP or POS—for each individual output. The limitation of this approach is that it does not take into account possible terms common to two or more outputs. Consider the functions of Figure 4.12. The result of implementing the optimum cover for each output separately is represented in Part *(a)* of Figure 4.13, while a cheaper implementation is depicted in Part *(b)* of the same figure. We see that we can save one gate and one gate input (alternatively, one product term and one literal) by using $x'yz$. However, this term is not a prime of either function. Hence, we need to revise our definition of prime implicant, if we want that an optimum solution may be built out of prime implicants only.

4.10.1 Multiple-Output Primes

The revision of the definition of prime implicant for the multiple-output case is fortunately easy: Besides the prime implicants of the individual functions, we need to include also the primes of the products of the individual functions. For example, if a multiple-output function is composed of single-output functions f_1, f_2, and f_3, then we need to consider the primes of:

$$f_1, \quad f_2, \quad f_3, \quad f_1 \cdot f_2, \quad f_1 \cdot f_3, \quad f_2 \cdot f_3, \quad f_1 \cdot f_2 \cdot f_3.$$

The primes of $f_1 \cdot f_2$, for instance, are the maximal terms that may be shared by f_1 and f_2. This may appear a very bad result, since it says that for an n-output function we need to consider $2^n - 1$ functions. Fortunately, we do not need to build the product functions explicitly. Before we see how to do that in detail, we introduce a way to represent multiple-output functions that we shall use in the sequel. It is sometimes called *cubical representation* (or *cubical complex* [233]) and is best introduced with the help of an example. A cubical representation of the multiple-output function of Figure 4.12 is:

xyz	$f_1 f_2$
1–1	10
011	11
0–0	01

Cubical representations are like formulae, in that there are many cubical representations for the same function. More specifically, they correspond to two-level covers. The SOP formulae are, in the case of our example,

$$\begin{aligned} f_1 &= xz + x'yz \\ f_2 &= x'yz + x'z' \end{aligned}$$

One advantage of the cubical representation is that it emphasizes the presence of terms common to two or more outputs. Another cubical representation for the function of Figure 4.12 is:

xyz	$f_1 f_2$
1–1	10
–11	10
01–	01
0–0	01

This second representation, illustrates one important feature. It is possible for two or more *cubes* (i.e., lines of the matrix) to have overlapping input parts (in our example –11 and 01– both contain 011). In such a case, the value of the multiple-output function for the minterms in the intersection is obtained by taking the sum of all the output parts (in our example the output for 011 is $10 + 01 = 11$).

A cube is a multiple-output implicant, i.e., its input part is an implicant of all the outputs for which its output part has a 1. In other words, the output part of the cube can be regarded as a tag that indicates the outputs for which the input part is an implicant. We are now ready to see how to extend the tabular method for the determination of all prime implicants to the multiple-output case. The results of applying the extended method to the function of Figure 4.12 are illustrated in Figure 4.14.

When two adjacent implicants are merged, their output parts are intersected to form the output part of the resulting implicant. If the intersection is null, no new entry is made in the table. The other important thing to consider is the rule for marking ($\surd$) the implicants. If an implicant is used in forming a new implicant, it is marked only if the output part of the new implicant is the same as its output part. All the other details are unchanged. A good, complete account of the method can be

000|01 √	0–0|01
010|01 √	01–|01
011|11	–11|10
101|10 √	1–1|10
111|10 √	

Figure 4.14: Tabular Method Applied to the Multiple-Output Function of Figure 4.12.

found in [186, 187], that also cover the extension of the iterated consensus method. The extension of the recursive multiplication method is also possible and is left as an exercise.

4.10.2 Formulating the Covering Problem

For multiple-output functions the problem of guaranteeing the minimum number of gate inputs for the minimum number of gates is more involved than in the single-output case. A good reference for that case is again [186]. We shall limit our treatment to the simpler problem of minimizing the number of gates only, for which the solution is quite similar to the single-output case. The only thing to be noted is that minterm 011 appears twice in the covering matrix, once for each output. In our example, letting

$$\begin{aligned} p_1 &= 011|11 \\ p_2 &= 0\text{–}0|01 \\ p_3 &= 01\text{–}|01 \\ p_4 &= \text{–}11|10 \\ p_5 &= 1\text{–}1|10 \end{aligned}$$

we obtain the following covering matrix:

$$\begin{array}{c} \\ 000|01 \\ 010|01 \\ 011|01 \\ 011|10 \\ 101|10 \\ 111|10 \end{array} \begin{array}{c} \begin{array}{ccccc} p_1 & p_2 & p_3 & p_4 & p_5 \end{array} \\ \left[\begin{array}{ccccc} 0 & 1 & 0 & 0 & 0 \\ 0 & 1 & 1 & 0 & 0 \\ 1 & 0 & 1 & 0 & 0 \\ 1 & 0 & 0 & 1 & 0 \\ 0 & 0 & 0 & 0 & 1 \\ 0 & 0 & 0 & 1 & 1 \end{array}\right] \end{array}$$

One easily sees that in this case p_1, p_2, and p_5 form the unique optimum solution.

4.10.3 Incompletely Specified Multiple-Output Functions

We can extend the use of *don't cares* to multiple-output functions by letting the entries in the output parts of the cubes also take on the value '–.' If we do so, then a cube 001|01– means that for input 001, the first output must be 0 (unless otherwise specified by another cube in the same cover), the second output must be 1, and the third output may be either 0 or 1.

Prime generation and covering problem formulation extend naturally to deal with don't cares in multiple-output functions. Notice, however, that the mechanism is not general, in that it does not cover all possible situations. Consider a two-output function and try to express the requirement that, for input 000, the output should be one of 01, 10. A quick examination will show that it is not possible to express this with don't cares, at least in the unsophisticated way we have seen so far. We shall not return to the development of a more general mechanism for incompletely specified multiple-output functions until we deal with the optimization of multi-level networks in Chapter 11. In the next chapter we turn our attention to the heuristic minimization problem.

4.11 Notes

The recursive algorithm for generation of the prime implicants of a logic function derives from those found in [37] and Rudell's 1989 dissertation [236].

Many problems in the field of electronic design automation involve the selection of an optimum subset of a given set. The selection aims at minimizing a given cost function and is subject to constraints. One familiar form is the covering problem solved as part of the Quine-McCluskey procedure for Boolean function minimization [188]. Given the set of prime implicants, a subset of minimum cost is sought, such that all the vertices of the function are covered.

The problem can also be interpreted as the one of finding the minimum cost assignment that satisfies a Boolean formula in conjunctive form. In this form it is known as Petrick's method [212]. It is important to note that the Boolean formula derived from the matrix of a covering problem is (positive) unate, i.e., all variables appear in their uncomplemented forms only. For this reason we refer to this covering problem as the *unate covering problem* (UCP). In the present chapter we have introduced an algorithm, called BCP, which solves UCP, and, with minor modifications, BCP as well. The BCP form will play a key role in future chapters: for state reduction/minimization of incompletely specified FSMs (in Chapter 8) and for technology mapping (in Chapter 13).

Rudell [236] was first to introduce the name BCP, but, as he reports, BCP has been around for a while with a different name. The first problem to be cast into a binate covering problem was the state reduction of incompletely specified state machines [118]. The authors called their problem *Covering with Closure* and made the restrictive assumption that one literal at most appeared complemented in a clause of the function. This restriction was later removed, when the same authors addressed the more general problem of combined row and column reduction of flow tables [119]. Applications of binate covering include optimal design of 3-level networks, technology mapping, and, more recent, minimization of Boolean relations [35, 41]. Other problems, like the phase assignment described in [270] and optimal encoding via symbolic minimization [88], can be formulated as BCP. An iterative approach to solving BCP was given in [213], and a more recent, and surprisingly effective, contribution to the solution of BCP was given by Coudert and Madre in [78].

The reduction techniques we consider here are not the only ones that are possible. For more sophisticated techniques, like Gimpel's, we refer the interested reader to the literature [112, 229].

Significant advances in efficient techniques for solving this problem are still appearing: for example, see [78]. Also, a growth spurt in this presumably rather mature research field was stimulated by the advent of BDD-based **symbolic processing.** Symbolic processing allowed efficient representation of the huge sets of prime implicants that limited the applicability of ESPRESSO_EXACT [238]. Significant publications include [75, 175, 77, 80, 76, 81], [257], and several articles in the book by Sasao [245]. The symbolic processing is based on the BDD methods described in Chapter 6, and have much in common with the symbolic traversal methods of Section 7.10.

4.12 Summary

We have discussed design optimality and design tradeoffs, and have defined two level logic and appropriate cost functions in this context. We discussed SOP (Sums of Products) and POS (Products of Sums) as the primary two level representations. We then defined implicants and prime implicants and their role in logic minimization. Various flavors of iterated consensus were discussed, as the means of computing the complete sum (the set of all prime implicants) of a logic function.

Quine's theorem, and the Quine-McCluskey algorithm for the problem of two level logic minimization were discussed. This problem was shown to be a special case of the more general Unate Covering Problem (UCP). A branch and bound algorithm, BCP, was presented for the solution of of UCP. The key features of BCP were:

1. Reduction (essential columns, row dominance, column dominance);
2. Lower bounding (based on maximal independent row sets);
3. Recursive Boole expansion (for exploring the search space).

The surprising efficiency of BCP is due to its ability to exploit the above mechanisms to find an optimal solution while actually examining only a tiny fraction of the overall search space. It was shown how UCP could also be solved by an Integer Linear Program.

We concluded by showing how the this material, presented in the context of single output functions, could be straightforwardly extended to the multiple output case.

4.13 Problems

1. Represent the switching function

$$x'y'z + x'yz + x'z' + y'z$$

on the following three-dimensional cube, by darkening the vertices of the ON set. (The *ON-set* is the set of minterms for which the switching function is 1. Similarly, the *OFF-set* is the set of minterms for which the switching function is 0. Finally, the *don't care set* is the set of minterms for which the switching function is *don't care*.)

Solution.

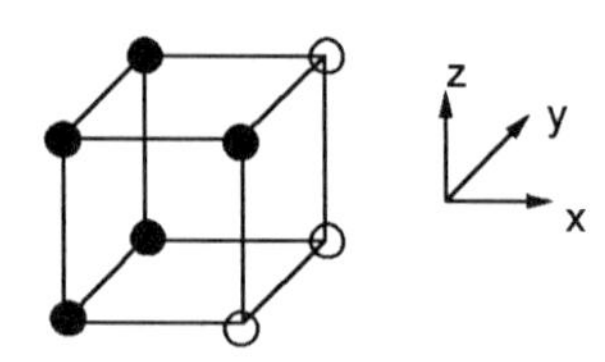

□

2. This problem is on converting formulae from SOP to POS and vice versa. One well-known method consists of 'covering the zeroes' on a Karnaugh map. Unfortunately, that method does not work well for formulae with many variables. In this problem you are to apply a method that can be the basis for an efficient program.

   The method uses the expansion theorem and the distributive property. It is illustrated in the following example. Suppose we want to convert

   $$xyz + x'y'z + xy'z'$$

   to POS form. If we apply the second form of the expansion theorem with respect to x, we get:

   $$(x + y'z)(x' + yz + y'z').$$

   We can simply put the first term in POS form by applying the distributive property:

   $$(x + y')(x + z)(x' + yz + y'z').$$

   For the second term, we now expand $yz + y'z'$ with respect to y:

   $$(x + y')(x + z)[x' + (y + z')(y' + z)].$$

   Finally, by the distributive property:

   $$(x + y')(x + z)(x' + y + z')(x' + y' + z).$$

   Notice that it is possible to convert from SOP to POS using only the distributive property or only the expansion theorem; however, by combining the two, the amount of work is greatly reduced.

   Apply the method illustrated in the previous example to convert from SOP to POS the following formula:

   $$wx + w'z + wy'z',$$

   and from POS to SOP the following formula:

   $$(x + y)(x' + z)(x + y' + z').$$

Solution.

$$\begin{aligned} wx + w'z + wy'z' &= (w+z)(w'+x+y'z') \\ &= (w+z)(w'+x+y')(w'+x+z') \end{aligned}$$

For the conversion from POS to SOP, we apply the first form of the expansion theorem with respect to x.

$$\begin{aligned} (x+y)(x'+z)(x+y'+z') &= xz + x'y(y'+z') \\ &= xz + x'yz' \end{aligned}$$

□

3. Convert to SOP form the following POS formula:

$$(x+y'+z')(x'+y')(y+z').$$

4. Find a SOP expression for the following function:

$$f(w,x,y,z) = (y+z)(w+x+z)(w'+x'+z')(w'+y+z')$$

by recursively applying the expansion theorem. Expand with respect to the most frequent variable.

Solution. We expand with respect to z, that appears in all sum terms.

$$\begin{aligned} f(w,x,y,z) &= (y+z)(w+x+z)(w'+x'+z')(w'+y+z') \\ &= z(w'+x')(w'+y) + z'y(w+x) \\ &= z(w'+x'y) + wyz' + xyz' \\ &= w'z + x'yz + wyz' + xyz' \end{aligned}$$

□

5. For the formula

$$xyz + x'y + y'z,$$

do the following:

(i) Write an equivalent POS formula;

(ii) Write a SOP formula for the complement of the function;

(iii) Compute a complete sum using the tabular method;

(iv) Compute a complete sum using iterated consensus;

(v) Compute a complete sum using the POS that you found;

(vi) Compute a complete sum using the recursive multiplication procedure.

(Make sure the complete sums obtained with the four different methods are consistent.)

Solution.

(i) Write an equivalent POS formula;

$$(y + z)(y' + xz + x')$$

$$(y + z)(y' + x' + z)$$

This can be simplified to

$$(y + z)(x' + z)$$

by consensus. Since we are going to use the POS formula later, it is convenient to spend a little extra time to simplify it.

(ii) Write a SOP formula for the complement of the function;

$$y'z' + xz'$$

by applying De Morgan to the POS form.

(iii) Compute a complete sum using the tabular method;

$x'y'z$ √ $x'yz'$ √	$x'z$ √ $y'z$ √ $x'y$	z
$x'yz$ √ $xy'z$ √	yz √ xz √	
xyz √		

The complete sum is therefore $x'y + z$.

(iv) Compute a complete sum using iterated consensus;

$$(xyz) + x'y + (y'z) + (yz) + (x'z) + z$$

The terms in parentheses are absorbed.

(v) Compute a complete sum using the POS that you found;

$$ABS((y + z)(x' + z)) = ABS(x'y + yz + x'z + z) = x'y + z$$

(vi) Compute a complete sum using the recursive multiplication procedure.

$$f(x, 0, z) = z \qquad f(x, 1, z) = xz + x' = x' + z$$

Both these formulae are complete sums; hence:

$$CS(f) = ABS((y'+x'+z)(y+z)) = ABS(y'z+x'y+x'z+yz+z) = x'y+z$$

Notice that the computations are almost identical to those of the previous case.

□

6. For the formula

$$xy + x'y'z + xy'z',$$

do the following:

(i) Write an equivalent POS formula;

(ii) Write a SOP formula for the complement of the function;

(iii) Compute a complete sum using the tabular method;

(iv) Compute a complete sum using the POS that you found;

(v) Compute a complete sum using the recursive multiplication procedure.

7. Find the complete sum for the following function:

$$f(w,x,y,z) = w'y'z + xyz + wyz' + wx'y',$$

by applying the recursive method based on the equation:

$$CS(f) = ABS([x_1 + CS(f(0,x_2,\ldots,x_n))] \cdot [x_1' + CS(f(1,x_2,\ldots,x_n))]).$$

Solution.

$$\begin{aligned}
f(w,x,y,z) &= w'y'z + xyz + wyz' + wx'y' \\
&\quad f(w,x,0,z) = w'z + wx' \\
&\quad CS(f(w,x,0,z)) = w'z + wx' + x'z \\
&\quad f(w,x,1,z) = xz + wz' \\
&\quad CS(f(w,x,1,z)) = xz + wz' + wx \\
CS(f) &= ABS((y + w'z + wx' + x'z)(y' + xz + wz' + wx)) \\
&= ABS(xyz + wyz' + wxy + w'y'z + w'xz + wx'y' + wx'z' + x'y'z) \\
&= xyz + wyz' + wxy + w'y'z + w'xz + wx'y' + wx'z' + x'y'z
\end{aligned}$$

□

8. Find the prime implicants for the following incompletely specified function:

$$\begin{aligned}
f &= wx'y' + wxz' \\
d &= xy'z + x'yz'
\end{aligned}$$

Solution.

$wx'y'z'$ √	$wy'z'$ √	wy'
$w'x'yz'$ √ d	$wx'y'$ √	wz'
	$wx'z'$ √	
	$x'yz'$ d	
$wxy'z'$ √	wxy' √	
$w'xy'z$ √ d	wxz' √	
$wx'y'z$ √	$xy'z$ d	
$wx'yz'$ √ d	$wy'z$ √	
	wyz' √	
$wxy'z$ √ d		
$wxyz'$ √		

The prime implicants are therefore wy' and wz'. □

9. Find the prime implicants for the following incompletely specified function:

$$\begin{aligned} f &= w'y'z + xyz \\ d &= wxyz' + wx'y \end{aligned}$$

10. For the following incompletely specified function:

$$\begin{aligned} f &= x_1x_2x_3' + x_1x_2'x_3 + x_1'x_2x_3 \\ d &= x_1x_2'x_3' + x_1'x_2x_3' + x_1'x_2'x_3 + x_1x_2x_3 \end{aligned}$$

Compute all prime implicants using the recursive procedure based on the expansion theorem.

Solution.

$$\begin{aligned} f+d &= x_1x_2x_3' + x_1x_2'x_3 + x_1'x_2x_3 + x_1x_2'x_3' + x_1'x_2x_3' + x_1'x_2'x_3 + x_1x_2x_3 \\ &= (x_1' + x_2x_3' + x_2'x_3 + x_2'x_3' + x_2x_3)(x_1 + x_2x_3 + x_2x_3' + x_2'x_3) \end{aligned}$$

Noting that the first factor on the right-hand side equals 1

$$\begin{aligned} &= x_1 + (x_2' + x_3 + x_3')(x_2 + x_3) \\ &= x_1 + x_2 + x_3 \end{aligned}$$

In this case, all prime implicants cover at least one minterm of f. □

11. Repeat Problem 9 using the recursive procedure based on the expansion theorem.

12. In this problem, you have to find a smallest subset of $\{x_1, x_2, x_3\}$ (a set of primes) that covers $\{x_1x_2x_3', x_1x_2'x_3, x_1'x_2x_3\}$ (a set of minterms). To do so:

(a) Let $p_1 = x_1$, $p_2 = x_2$, $p_3 = x_3$, $m_1 = x_1x_2x_3'$, $m_2 = x_1x_2'x_3$, and $m_3 = x_1'x_2x_3$. Give the corresponding constraint matrix.

(b) Solve the covering problem using BCP.

Solution.

This is the constraint matrix:

$$\begin{array}{c} \\ m_1 \\ m_2 \\ m_3 \end{array} \begin{array}{c} p_1p_2p_3 \\ \left[\begin{array}{ccc} 1 & 1 & 0 \\ 1 & 0 & 1 \\ 0 & 1 & 1 \end{array}\right] \end{array}$$

The matrix is cyclic (non reducible). The lower bound for the cost of a solution is 1. We choose p_1 arbitrarily. The residual matrix can be trivially covered by choosing either p_2 or p_3. We choose the former and set the upper bound to 2. Since the solution that we have found does not meet the lower bound, we have to examine also the solution obtained by rejecting p_1. When this is done, both p_2 and p_3 become essential and we have the solution p_2, p_3. The two solutions found have the same cost. We keep the first (p_1, p_2).

Note that this problem is the continuation of the previous one and that there is no cyclic constraint matrix for unate covering smaller than this one.

□

13. Repeat Problem 12 with the following set of primes:

$$\{x_1x_2, x_1x_3', x_2x_3\}$$

and the following set of minterms:

$$\{x_1x_2x_3', x_1x_2'x_3'\}.$$

[Hint: A column of the constraint matrix with no ones can be dropped.]

14. Solve the following covering problem:

$$\begin{array}{c} \\ 1 \\ 2 \\ 3 \\ 4 \end{array} \begin{array}{c} \begin{array}{ccccc} 1 & 2 & 3 & 4 & 5 \end{array} \\ \begin{bmatrix} 1 & 1 & 0 & 0 & 1 \\ 0 & 1 & 1 & 1 & 0 \\ 0 & 0 & 1 & 1 & 1 \\ 1 & 1 & 0 & 1 & 0 \end{bmatrix} \end{array}$$

by applying the branch-and-bound method. All columns have unit costs. When splitting, choose the longest column and, in case of tie, choose the column of lowest index. Also, when two columns dominate each other, retain the one with the lower index.

Draw the search tree, and indicate for each node the lower bound.

Solution. We set the initial upper bound to $5 + 1 = 6$. We then try to reduce the problem. Column 1 is dominated by Column 2 and Column 3 is dominated by Column 4. After the ensuing simplification, we see that Row 2 is identical to Row 4: Row 4 is dropped. The resulting matrix is:

$$\begin{array}{c} \\ 1 \\ 2 \\ 3 \end{array} \begin{array}{c} \begin{array}{ccc} 2 & 4 & 5 \end{array} \\ \begin{bmatrix} 1 & 0 & 1 \\ 1 & 1 & 0 \\ 0 & 1 & 1 \end{bmatrix} \end{array}$$

This matrix is cyclic. We then compute the lower bound, which is 1 in this case. Since $1 < 6$ we must continue. We choose to split on Column 2. We select Column 2, and then eliminate Column 5 by dominance. We then have a solution of cost 2 composed of Columns 2 and 4. We update the upper bound to 2 and then return to the original problem. (The top node in the search tree.) Now we compare the new upper bound to the lower bound. Since $1 < 2$, we need to consider also the second subproblem. Therefore, we reject Column 2.

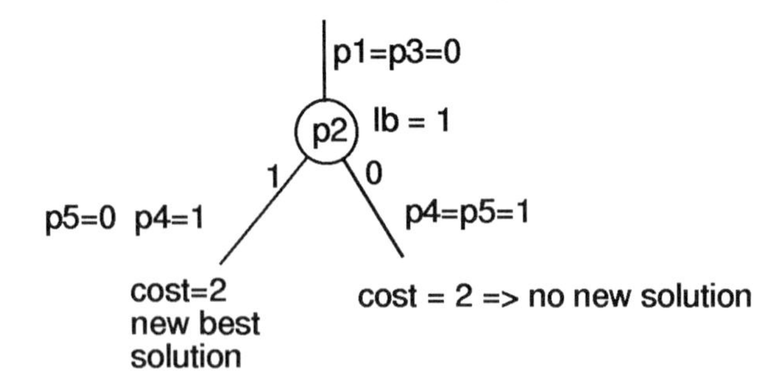

Figure 4.15: Recursion Tree for Problem 14.

This causes both Column 4 and Column 5 to become essential. Therefore d in the recursion tree of Figure 4.15. □

15. Solve the following covering problem:

$$\begin{array}{c} \\ 1 \\ 2 \\ 3 \\ 4 \\ 5 \\ 6 \\ 7 \\ 8 \\ 9 \\ 10 \end{array} \begin{array}{c} \begin{array}{ccccc} 1 & 2 & 3 & 4 & 5 \end{array} \\ \left[\begin{array}{ccccc} 1 & 1 & 1 & 0 & 0 \\ 1 & 1 & 0 & 1 & 0 \\ 1 & 1 & 0 & 0 & 1 \\ 1 & 0 & 1 & 1 & 0 \\ 1 & 0 & 1 & 0 & 1 \\ 1 & 0 & 0 & 1 & 1 \\ 0 & 1 & 1 & 1 & 0 \\ 0 & 1 & 1 & 0 & 1 \\ 0 & 1 & 0 & 1 & 1 \\ 0 & 0 & 1 & 1 & 1 \end{array}\right] \end{array}$$

by applying the branch-and-bound method. All columns have unit costs. When splitting, choose the longest column and, in case of tie, choose the column of lowest index. Also, when two columns dominate each other, retain the one with the lower index.

Draw the search tree, and indicate for each node the lower bound, computed with the **maximal independent set** method.

16. With reference to Problem 13, suppose $x_1x_2x_3' + x_1x_2'x_3'$ is the ON-set of an incompletely specified function (i.e., f) and that $x_1x_2 + x_1x_3' + x_2x_3$ is the complete sum of $f + d$. Find d.

[Hint: No prime implicant can include any point of r.]

Solution. We observe that, since f and d are disjoint, it is:

$$d = (f + d)f'.$$

To formally verify this, notice that $fd = 0$ implies $d \leq f'$ and hence $df' = d$. You can also use a Venn diagram to verify this simple fact. Let us rewrite f as

$x_1 x'_3$. (The two minterms are adjacent.) Then,

$$\begin{aligned} d &= (f+d)f' \\ &= (x_1x_2 + x_1x'_3 + x_2x_3)(x'_1 + x_3) \\ &= x'_1x_2x_3 + x_1x_2x_3 \\ &= x_2x_3 \end{aligned}$$

Notice that these f and d are those of the previous problem. □

17. For the following constraint matrix, say whether the cyclic core is non-empty.

$$\begin{array}{c} \\ 1 \\ 2 \\ 3 \\ 4 \\ 5 \end{array} \begin{array}{c} \begin{array}{cccccc} 1 & 2 & 3 & 4 & 5 & 6 \end{array} \\ \left[\begin{array}{cccccc} 0 & 1 & 0 & 1 & 1 & 0 \\ 1 & 0 & 0 & 1 & 1 & 1 \\ 1 & 0 & 1 & 0 & 0 & 0 \\ 0 & 1 & 1 & 1 & 0 & 1 \\ 0 & 1 & 0 & 0 & 1 & 1 \end{array}\right] \end{array}$$

Solution. All rows have two or more ones. There is neither row nor column dominance. Hence the cyclic core is not empty: It is the matrix itself. □

18. In Section 4.9 we have discussed a simple way of choosing the splitting variable. This problem explores a more sophisticated variant of that criterion. The idea is still to favor columns that cover many rows, but also to take into account the length of the rows that a column covers. Indeed, one may argue that a longer row (one with more ones) has a higher chance of being covered than a shorter row. Hence a column that covers shorter rows should be favored over a column that covers longer rows.

The above remark translates into the following procedure: For every row r_i of the matrix, compute:

$$z_i = \frac{1}{|r_i| - 1},$$

where $|r_i|$ is the number of ones in row r_i. For every column, compute the weight as the sum of all the z_i corresponding to rows where the column has a one. Choose a column of maximum weight.

For the matrix of Problem 17.

(a) Compute the column weights according to the criterion just presented.

(b) What columns could be selected as splitting columns?

Solution.

(a) The weights are computed as follows:

$$
\begin{aligned}
W_1 &= \tfrac{1}{3} + 1 &&= \tfrac{4}{3} \\
W_2 &= \tfrac{1}{2} + \tfrac{1}{3} + \tfrac{1}{2} &&= \tfrac{4}{3} \\
W_3 &= 1 + \tfrac{1}{3} &&= \tfrac{4}{3} \\
W_4 &= \tfrac{1}{2} + \tfrac{1}{3} + \tfrac{1}{3} &&= \tfrac{7}{6} \\
W_5 &= \tfrac{1}{2} + \tfrac{1}{3} + \tfrac{1}{2} &&= \tfrac{4}{3} \\
W_6 &= \tfrac{1}{3} + \tfrac{1}{3} + \tfrac{1}{2} &&= \tfrac{7}{6}
\end{aligned}
$$

(b) The columns that could be selected are 1,2,3,5, all having weight $\frac{4}{3}$.

□

19. Consider the following covering matrix.

$$
T = \begin{array}{c} \\ 1 \\ 2 \\ 3 \\ 4 \\ 5 \\ 6 \\ 7 \\ 8 \end{array}
\begin{array}{c}
\begin{array}{cccccccccccc} 1 & 2 & 3 & 4 & 5 & 6 & 7 & 8 & 9 & 10 & 11 & 12 \end{array} \\
\begin{bmatrix}
1 & 0 & 0 & 0 & 0 & 0 & 0 & 0 & 0 & 0 & 1 & 0 \\
1 & 1 & 0 & 0 & 1 & 0 & 0 & 0 & 0 & 0 & 0 & 0 \\
0 & 1 & 1 & 0 & 1 & 1 & 1 & 1 & 0 & 0 & 0 & 0 \\
1 & 1 & 0 & 1 & 0 & 1 & 0 & 0 & 0 & 1 & 0 & 0 \\
1 & 0 & 0 & 1 & 0 & 0 & 0 & 0 & 0 & 0 & 0 & 0 \\
0 & 0 & 1 & 0 & 0 & 0 & 1 & 0 & 1 & 0 & 0 & 1 \\
0 & 0 & 1 & 0 & 0 & 0 & 0 & 1 & 1 & 0 & 1 & 0 \\
0 & 0 & 0 & 1 & 0 & 0 & 0 & 0 & 0 & 1 & 0 & 0
\end{bmatrix}
\end{array}
$$

(a) Find a minimum cost covering set. [Hint: The solution has 3 columns.]

(b) Are there any other solutions of size 3? [Hint: Have you applied column dominance?]

20. Solve the following covering problem by the branch-and-bound algorithm. Draw the recursion tree.

$$
\begin{array}{c} \\ 1 \\ 2 \\ 3 \\ 4 \\ 5 \\ 6 \end{array}
\begin{array}{c}
\begin{array}{ccccc} 1 & 2 & 3 & 4 & 5 \end{array} \\
\begin{bmatrix}
1 & 1 & 0 & 1 & 0 \\
1 & 0 & 1 & 0 & 1 \\
1 & 0 & 0 & 1 & 0 \\
0 & 1 & 1 & 1 & 0 \\
0 & 1 & 0 & 1 & 1 \\
1 & 0 & 1 & 1 & 0
\end{bmatrix}
\end{array}
$$

Solution. The initial upper bound is 6. Column 2 is dominated by Column 4 and is therefore eliminated. Rows 3 and 6 are then eliminated because they dominate Row 1. The resulting matrix is:

$$
\begin{array}{c} \\ 1 \\ 2 \\ 4 \\ 5 \end{array}
\begin{array}{c}
\begin{array}{cccc} 1 & 3 & 4 & 5 \end{array} \\
\begin{bmatrix}
1 & 0 & 1 & 0 \\
1 & 1 & 0 & 1 \\
0 & 1 & 1 & 0 \\
0 & 0 & 1 & 1
\end{bmatrix}
\end{array}
$$

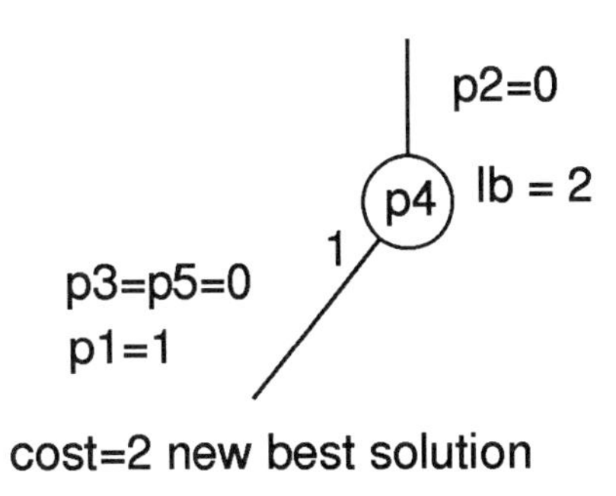

Figure 4.16: Recursion Tree for Problem 20.

This matrix is cyclic. Therefore we compute the lower bound. There are no two independent rows, but using the result of Theorem 4.8.1 of Page 150, we can set the lower bound to 2.

We now choose Column 4 as splitting variable, because it covers the most rows. The only row that remains to be covered is Row 2. Selecting Column 1 completes the cover. Since the cost equals the lower bound there is no need of exploring the part of the search tree for which Column 4 is rejected. In conclusion, $\{1, 4\}$ is an optimum solution. The recursion tree is shown in Figure 4.16.

□

21. A circuit may malfunction due to one of eight possible faults:

$$\{f_1, f_2, f_3, f_4, f_5, f_6, f_7, f_8\}.$$

A designer has come up with six tests which, when applied, detect some of the faults. In particular:

- t_1 will detect the presence of faults f_1, f_3, and f_5;
- t_2 will detect the presence of faults f_2, f_4, and f_6;
- t_3 will detect the presence of faults f_1, f_2, f_4, and f_7;
- t_4 will detect the presence of faults f_3, f_5, and f_8;
- t_5 will detect the presence of faults f_2, f_6, and f_8;
- t_6 will detect the presence of faults f_3, f_4, and f_7.

Find a minimum subset of tests that will detect any of the eight possible faults, if present. (Suppose that only one fault may occur in the circuit and that all tests have the same cost.)

[Hint: Formulate the problem as a unate covering problem.]

Solution. The tests correspond to the columns of the constraint matrix (what we have to choose), while the faults correspond to the rows (what we have to cover). The matrix is as follows.

$$\begin{array}{c} \\ 1 \\ 2 \\ 3 \\ 4 \\ 5 \\ 6 \\ 7 \\ 8 \end{array} \begin{array}{c} \begin{array}{cccccc} 1 & 2 & 3 & 4 & 5 & 6 \end{array} \\ \left[\begin{array}{cccccc} 1 & 0 & 1 & 0 & 0 & 0 \\ 0 & 1 & 1 & 0 & 1 & 0 \\ 1 & 0 & 0 & 1 & 0 & 1 \\ 0 & 1 & 1 & 0 & 0 & 1 \\ 1 & 0 & 0 & 1 & 0 & 0 \\ 0 & 1 & 0 & 0 & 1 & 0 \\ 0 & 0 & 1 & 0 & 0 & 1 \\ 0 & 0 & 0 & 1 & 1 & 0 \end{array} \right] \end{array}$$

The matrix can be entirely covered without need to branch. An optimum solution is $\{1, 3, 5\}$. □

22. Find the cheapest SOP cover for the incompletely specified function described by the following interval.

$$[w'yz + w'xz + wx'yz, yz + w'xy'z + w'xy].$$

Solution. We shall solve this problem (and the following) ignoring initially the fact that a prime may only cover don't care minterms. As discussed in Problems 10 and 13, in the constraint matrix the columns corresponding to such primes will be empty and the primes will not enter the solution.

Given this premise, we see that we can derive all primes by applying any of the three methods to the upper bound (U) of the interval, while the minterms to be covered will be provided by the lower bound (L). We decide to use in this case iterated consensus on U.

$$CS(U) = yz + (w'xy'z) + w'xy + w'xz$$

The term in parentheses is absorbed by its consensus term with the first implicant and we are left with three prime implicants:

$$p_1 = yz \qquad p_2 = w'xy \qquad p_3 = w'xz.$$

The constraint matrix is then

$$\begin{array}{c} \\ w'x'yz \\ w'xyz \\ w'xy'z \\ wx'yz \end{array} \begin{array}{c} 123 \\ \left[\begin{array}{c} 100 \\ 111 \\ 001 \\ 100 \end{array} \right] \end{array}$$

and the solution is easily seen to be $yz + w'xz$. (Two essential columns.) □

23. Find a minimum cost SOP cover for the following incompletely specified function:

$$\begin{aligned} f &= w'y'z' + xy'z + wxy \\ d &= wy'z' + w'x'y'z + w'xyz + wx'y \end{aligned}$$

Use the recursive method to compute the prime implicants. Draw the constraint matrix of the covering problem. Solve the covering problem by the branch-and-bound algorithm.

24. Find the cheapest POS cover for the incompletely specified function of Problem 22.

Solution. We indicate three ways to solve the problem (and actually solve it in the second way). The first method is to find the maxterm canonical form, use the tabular method to find the prime **implicates**, and then set up and solve the covering problem. The second method, that we shall pursue in detail, avoids the maxterm canonical form in generating the primes (but still uses it to generate the constraint matrix). This method follows closely the one we just used for Problem 22. The third method consists of solving the dual problem. This method entails essentially the same computation as the second method. We discuss it briefly at the end.

Remember that a POS formula 'covers the zeroes' of the map. This translates in the fact that the primes will be obtained from the lower bound, while the maxterms will be obtained from the upper bound. Indicating the **complete product** by CP (the product of all prime implicates), we can apply the dual of Theorem 4.6.1 and get:

$$\begin{aligned} L &= z(w'y + w'x + wx'y) \\ &= z(w'(x+y) + wx'y) \\ &= z((w'(x+y) + x')(w'(x+y) + y)(w'(x+y) + w)) \\ CP(L) &= z(w' + x')(w' + y)(x + y) \end{aligned}$$

We expand initially with respect to z because L is of the form $z \cdot L_1$ and this makes it particularly convenient to split on z. We then apply the distributive property to the terms of the sum in parentheses to get the result. We let:

$$p_1 = z \qquad p_2 = w' + x' \qquad p_3 = w' + y \qquad p_4 = x + y.$$

To form the constraint matrix, we need now to find the maxterms of the upper bound. Putting the upper bound in POS form we obtain:

$$(w' + y)(w' + z)(w + x + y' + z)(w + x + y)(w + y + z).$$

We then derive the maxterm canonical form from this formula and eventually we get the following constraint matrix:

$$\begin{array}{l} \\ w'+x'+y'+z \\ w'+x'+y+z \\ w'+x+y'+z \\ w'+x+y+z \\ w+x+y'+z \\ w+x+y+z \\ w'+x'+y+z' \\ w'+x+y+z' \\ w+x+y+z' \\ w+x'+y+z \end{array} \begin{array}{c} 1234 \\ \left[\begin{array}{c} 1100 \\ 1110 \\ 1000 \\ 1011 \\ 1000 \\ 1001 \\ 0110 \\ 0011 \\ 0001 \\ 1000 \end{array}\right] \end{array}$$

from which the minimum cover $z(x+y)(w'+x')$ is easily obtained. There is another minimum cover, namely $z(x+y)(w'+y)$.

If you still feel confused about deriving the implicates from the lower bound of the interval, consider the following. Let f^D be the dual of f. We know that $L \leq U$ is equivalent to $U^D \leq L^D$. To find the minimum POS cover for the incompletely specified function $[L, U]$, we can find the minimum SOP cover of $[U^D, L^D]$ and then take the dual of this solution. You can verify that this procedure gives exactly the dual computation with respect to our approach and, of course, the same result. For instance, the prime implicants of L^D are the duals of the prime implicates of L and the minterms of U^D are the duals of the maxterms of U.

This method can be summarized as follows: Solve the dual problem and take the dual of the solution. Though we remarked that this approach is as efficient as the one we have followed in terms of computation, it is probably more appealing, because it allows us to work with the familiar SOP form. For once, however, spending some time on the other side of the looking glass—in the dual world—may be fun and instructive. □

25. Find a minimum cost POS cover for the incompletely specified function of Problem 23. Use the recursive method to compute the prime implicates . Draw the constraint matrix of the covering problem. Solve the covering problem by the branch-and-bound algorithm.

26. For the following constraint matrix, run the program **mincov** to find an optimum solution.

$$\begin{array}{c} \\ 1 \\ 2 \\ 3 \\ 4 \\ 5 \end{array} \begin{array}{c} \begin{array}{cccccc} 1 & 2 & 3 & 4 & 5 & 6 \end{array} \\ \left[\begin{array}{cccccc} 0 & 1 & 0 & 1 & 1 & 0 \\ 1 & 0 & 0 & 1 & 1 & 1 \\ 1 & 0 & 1 & 0 & 0 & 0 \\ 0 & 1 & 1 & 1 & 0 & 1 \\ 0 & 1 & 0 & 0 & 1 & 1 \end{array}\right] \end{array}$$

Use the options `-i -v 5`. A line like:

```
ABSMIN[ 0]     5x  6 sel=  0 bnd=  7 lb=  2     0.00 sec pick=1
```

in the output means that at depth 0 (the root of the search tree), the matrix, after reduction, has 5 rows and 6 columns. No column has been selected so far, the upper bound is 7, the lower bound is 2, 0.00 seconds have been spent so far, and the variable chosen for splitting is 1. Include a printout of your input and result files.

Explain why the lower bound is 2 at the top level and why there is only one node visited at depth 1.

Solution. This is the input file:

```
1 2 +
1 4 +
1 5 +
2 1 +
2 4 +
2 5 +
2 6 +
3 1 +
3 3 +
4 2 +
4 3 +
4 4 +
4 6 +
5 2 +
5 5 +
5 6 +
```

and this is the corresponding output file:

```
# mincov -i -v5 pb1.t
# Mincov Version #1.0, Release date 10/01/92
    123456
    ------
  1:.+.++.
  2:+..+++
  3:+.+...
  4:.+++.+
  5:.+..++
Options:  Independent Set  Beta Dominance
ABSMIN[ 0]     5x  6 sel=  0 bnd=  7 lb=  2     0.00 sec pick=1
ABSMIN[ 1]     0x  0 sel=  2 bnd=  7 lb=  2     0.00 sec BEST
new 'best' solution 2 at level 1 (time is 0.00 sec)
matrix      = 5 by 6 with 16 elements (53.333\%)
cover size = 2 elements
cover cost = 2
```

```
time       = 0.01 sec
components = 0
gimpel     = 0
nodes      = 2
max_depth  = 1
**** 0 rows deleted -- 0 were not covered
Solution is  1 2
```

The lower bound at the top level (the initial node of the recursion) is 2 because there is an independent set of two rows (1 and 3 or 3 and 5). Since the cost of the solution found by selecting variable 1 equals the lower bound, we know that we cannot improve on it by rejecting variable 1. Hence, we terminate the program without creating a second node at depth 1. □

27. Use the program MINCOV to solve the covering problem of Problem 14. Use the options `-i -v 5`.

28. Find the cheapest SOP cover for the incompletely specified function described by the following interval by running the program ESPRESSO with the option `-Dexact`.

$$[w'yz + w'xz + wx'yz, yz + w'xy'z + w'xy].$$

Include a printout of your input and result files.

Solution. ESPRESSO accepts as a default a description of the ON-set and DC-set of the function to be minimized. Since we are given the function in the form of an interval, we first need to extract the don't cares. We know that the lower bound of the interval is f and the upper bound is $f + d$. Hence:

$$d = (f+d)f' = (yz + w'xy'z + w'xy)(w'yz + w'xz + wx'yz)' = wxyz + w'xyz'.$$

We can now write the input file to ESPRESSO:

```
.i 4
.o 1
.ilb w x y z
.ob f
0-11 1
01-1 1
1011 1
1111 -
0110 -
.e
```

When we feed this file to ESPRESSO, we get the following result:

```
.i 4
.o 1
```

```
.ilb w x y z
.ob f
.p 2
--11 1
01-1 1
.e
```

Which is the desired solution, corresponding to $yz + w'xz$. □

29. Use ESPRESSO to find a minimum cost SOP cover for the incompletely specified function of Problem 23.

30. Find the cheapest POS cover for the incompletely specified function of Problem 28 using the program ESPRESSO with the option `-Dexact` and any other option that may help. Include a printout of your input and result files in your homework.

    [Hint: check what the `-epos` option does and take into account that the complement of a function f can be derived from the dual of f by complementing all literals (and vice versa).]

    Solution. In this case we exploit the fact that the option `-epos` minimizes the complement of the function. If we apply De Morgan's Theorem to the minimum sum of product for the complement, we get a minimum product of sum for the original function.

    The input file is in this case the same as in the previous problem. When we run:

```
espresso -Dexact -epos pb8.pla
```

    we get:

```
.i 4
.o 1
.ilb w x y z
.ob f
#.phase 0
.p 3
---0 1
-00- 1
1-0- 1
.e
```

    The comment line `#.phase 0` reminds us that this is a cover of the complement. By applying De Morgan's Theorem, we get $z(x + y)(w' + y)$. □

31. Use ESPRESSO to find a minimum cost POS cover for the incompletely specified function of Problem 23.

32. Find a minimum cost (minimum number of cubes) cover for the following function:

xy	$f_1 f_2$
00	10
01	01
1–	11

Compute the prime implicants using the tabular method; then set up and solve the covering problem.

Solution. The computation of the prime implicants with the tabular method produces the following table:

00|10 √	–0|10
01|01 √	–1|01
10|11 √	1–|11
11|11 √	

Hence, the primes are the three terms of the second column:

$$p_1 = -0|10 \quad p_2 = -1|01 \quad p_3 = 1-|11.$$

The minterms are just those listed in the first column of the table, except that the minterms with more than a 1 in the output part must be split into as many parts as there are ones in the output part (in this case, two parts).

$$\begin{array}{c} \\ 00|10 \\ 01|01 \\ 10|10 \\ 10|01 \\ 11|10 \\ 11|01 \end{array} \begin{array}{c} p_1 p_2 p_3 \\ \begin{bmatrix} 1 & 0 & 0 \\ 0 & 1 & 0 \\ 1 & 0 & 1 \\ 0 & 0 & 1 \\ 0 & 0 & 1 \\ 0 & 1 & 1 \end{bmatrix} \end{array}$$

All three primes are essential in this case. Hence, the solution is:

xy	$f_1 f_2$
–0	10
–1	01
1–	11

□

33. Find a minimum cost SOP cover for the following multiple-output function.

xyz	f_1	f_2	f_3
110	1	1	0
011	0	1	1
100	1	1	1
001	0	1	0
000	0	1	1

Chapter 5

Heuristic Minimization of Two-Level Circuits

The main problem in exact minimization is the potentially very large number of prime implicants. For most functions with more than, say, 16 inputs, there are simply too many primes and, quite often, too many minterms. Most heuristic methods (and in particular the ones we will be concerned with) avoid computing all the primes and all the minterms in an attempt to avoid the computational cost. The usual approach is to successively modify a given initial cover of the function, until a suitable stopping criterion is met.

The price to pay is the inability to guarantee the cheapest solution, though a careful choice of the algorithms may actually provide solutions very close to the optimum (when the optimum is known) in a very reasonable time. We shall examine a highly simplified version of ESPRESSO [37, 239], arguably the most successful program for minimization of switching (and multi-valued) functions ever written. First, we shall examine a simple algorithm based on **tautology checks**; then we shall see how a more global view to the optimization problem can be taken by using a cover of the complement of the function being minimized. The two techniques are combined in ESPRESSO.

5.1 Local Search

The algorithms that start from an initial solution and try to find a better one by applying successive modifications are called *local search* algorithms. Local search algorithms are very common in optimization in general and not restricted to logic minimization. The general local search algorithm deals with a *search space* or *solution space* (the two expressions are synonyms), that is the set of all possible values that the variables of the problem may take.

For instance, in the covering problem, if we have n columns then the search space consists of 2^n points. Some of these points are real solutions and some others are not (in the covering problem, some selections of columns cover all rows and some others do not). The set of all points of the search space that are valid solutions form the so-called **feasible region**. The complement is the infeasible region. We also say a point is feasible if it corresponds to a valid solution.

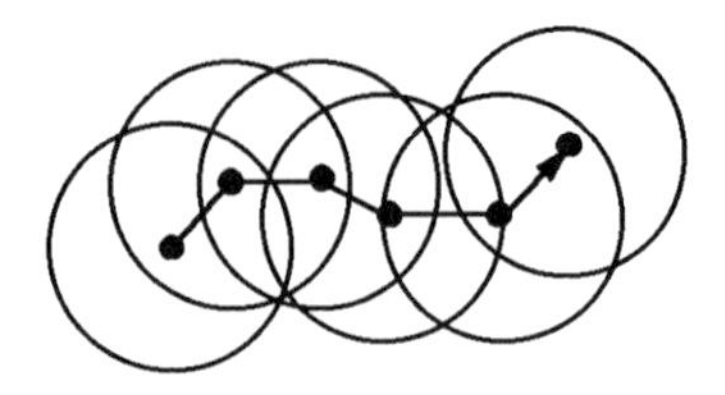

Figure 5.1: A Pictorial Representation of Local Search.

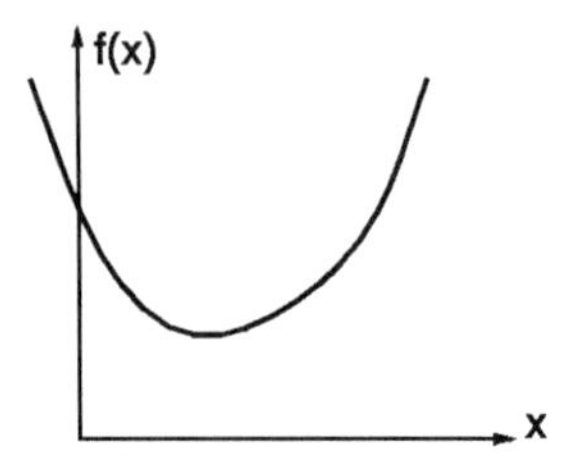

Figure 5.2: A Convex Optimization Problem.

Local search relies on the definition of a *distance* between two points in the search space. For two sets of columns in a covering problem, the distance would be the number of columns that appear in the first set, but not in the second or vice versa (i.e., the cardinality of the symmetric difference of the two sets). Different problems have generally different definitions of distance, and it is also possible to define different distances for the same problem.

Once a definition of distance is given, we can define the **neighborhood** of a point s of **radius** r in the search space as the set of points of the search space whose distance from s is less than r.

A given local search algorithm is based on a definition of distance and a radius r. Starting from the initial solution, the algorithm examines its neighborhood for a feasible point whose cost is lower than the current cost. If one is found, it is assumed as the new starting point and the process is repeated until a stopping criterion is met. See Figure 5.1 for a pictorial representation. There are clearly infinite possibility for variants of this scheme, depending on how a new feasible point is chosen when there are many in the current neighborhood, etc..

Depending on the problem and the definition of neighborhood, various degrees of optimality of the solution can be guaranteed. Suppose we are interested in the minimum of a convex function of a real variable as the one shown in Figure 5.2. Local search in this case consists of moving downward along the curve and is guaranteed to find the true minimum. However, if the function is not convex, as in Figure 5.3, then local search may get stuck in a local minimum. In the first case, we say that the solution is *minimum*, whereas in the second case we say that the solution is *minimal*.

In general, one interesting property of the solution of a local search algorithm is *local optimality*. A solution is locally optimal if its neighborhood does not contain any solution of lower cost. In order to guarantee local optimality, it is sufficient to use the following stopping criterion: Stop when there is no cheaper solution in the neighborhood of the current solution.

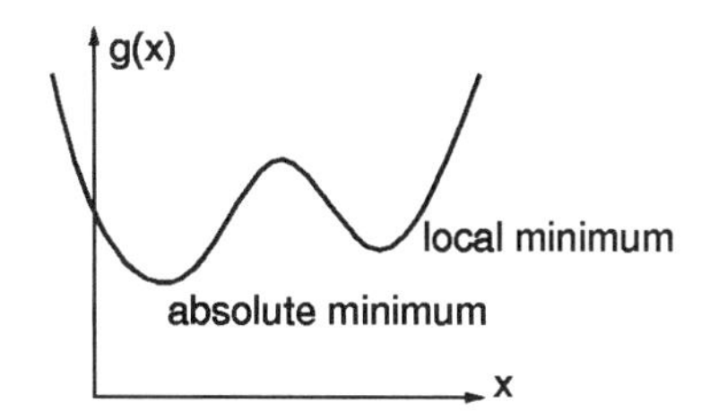

Figure 5.3: A Non-Convex Optimization Problem.

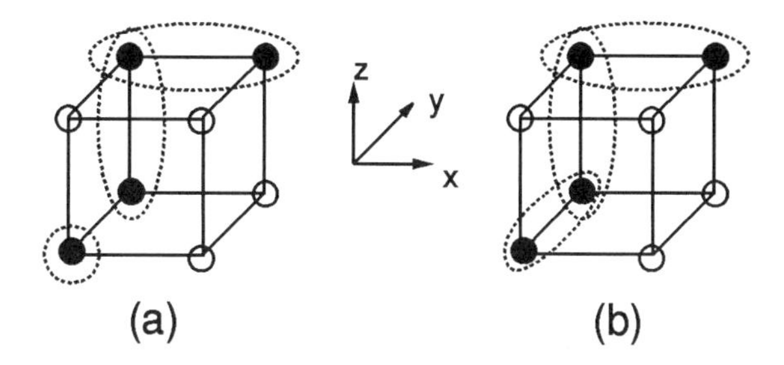

Figure 5.4: A Function with an Initial Cover *(a)* and after the Expansion of an Implicant *(b)*.

5.1.1 Local Search Applied to Logic Minimization

We have now introduced enough of the general theory of local search to start considering its application to logic minimization. Our starting point is a cover for a multiple-output function (single-output being a special case) and the cost will be the number of cubes (or number of multiple-output product-terms) with the number of literals as a tie-breaker.

The neighborhood we shall initially consider is defined as the set of covers that are obtained from the original cover by adding or removing exactly one literal to one of the cubes. A new cover thus obtained is a feasible point in the search space if it is *equivalent* to the original one, i.e., if it represents the same function.

When a cube has all zeroes in its output part, it can be dropped. Hence, our definition of neighborhood allows us to obtain solutions with fewer cubes than the initial cover. On the other hand, the solutions with more cubes that the initial cover are not part of our search space.

Before we outline the algorithm, we need to consider the effects of our *moves* in the search space. Consider the function of Part *(a)* of Figure 5.4, where the dotted lines identify the initial cover.

xyz	f
000	1
01–	1
–11	1

We see that changing 000|1 into 0–0|1 (i.e., removing an input literal) changes the cover to the one in Part (b) of Figure 5.4. From Part (b) we see that now we can

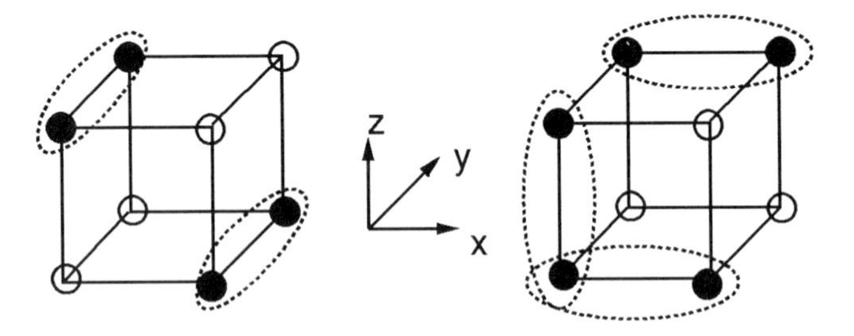

Figure 5.5: A Function and an Initial Cover Illustrating Output Expansion.

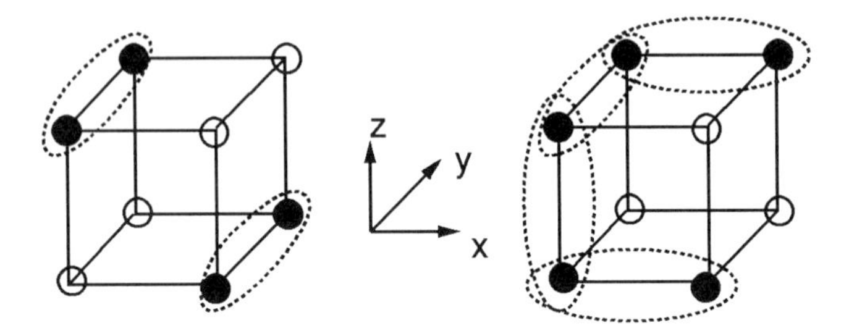

Figure 5.6: The Cover of Figure 5.5 after the Expansion of an Output Part.

remove the output literal of 01–|1, thus effectively removing the cube from the cover. In this case this yields the optimum solution.

xyz	f
0–0	1
–11	1

What we have done amounts to *expanding* the input part of a cube and *reducing* the output part of another. These transformations clearly decrease the cost of a cover. We need to consider next the effects of reducing an input part and expanding an output part.

Consider the two-output function of Figure 5.5 and the initial cover indicated by the dotted lines. The cubical representation is:

xyz	$f_1 f_2$
0–1	10
1–0	10
00–	01
–00	01
–11	01

It is readily seen that there is no way to improve the cover by expanding the input parts of the cubes or reducing their output parts. However, if we expand the output part of the first cube, we obtain the following cover, represented also in Figure 5.6.

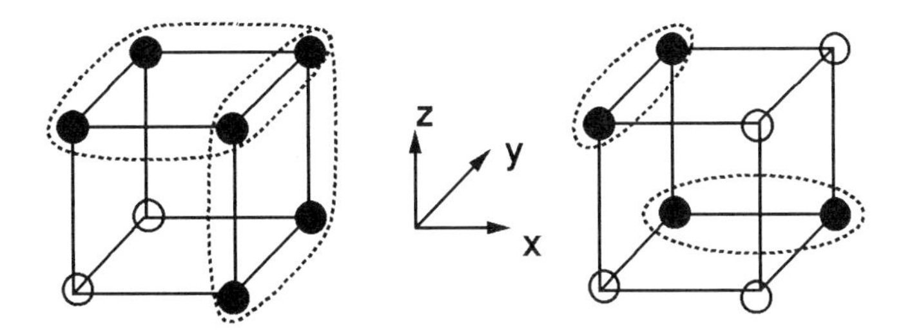

Figure 5.7: A Function and an Initial Cover Illustrating Input Reduction.

xyz	$f_1 f_2$
0–1	11
1–0	10
00–	01
–00	01
–11	01

From Figure 5.6 we see that 00–|01 is now redundant and can be eliminated, yielding:

xyz	$f_1 f_2$
0–1	11
1–0	10
–00	01
–11	01

which has one fewer cube than the initial cover. This demonstrates that sometimes we have to perform moves that do not decrease the cost (they may actually increase it) in order to escape local minima and find a better solution. The next example shows how the reduction of an input part may achieve a similar result.

Consider the two-output function of Figure 5.7 and the initial cover indicated by the dotted lines. The initial cover can be written as:

xyz	$f_1 f_2$
1——	10
——1	10
0–1	01
–10	01

Neither input expansions nor output reductions are possible. However, if we reduce the input part of the second cube to 0–1|10, we can then expand its output part and then remove the third cube.[1] The result is the following cover:

xyz	$f_1 f_2$
1——	10
0–1	11
–10	01

where we have again reduced the cost by 1 product term.

[1]Equivalently, we can expand the output part of the third cube and remove the second.

```
F = EXPAND(F,D);
F = IRREDUNDANT(F,D);
do {
        cost = |F|;
        F = REDUCE(F,D);
        F = EXPAND(F,D);
        F = IRREDUNDANT(F,D);
} while(|F| < cost);
F = MAKE_SPARSE(F,D);
```

Figure 5.8: Simple Minimization Loop.

5.1.2 A Simple Local Search Algorithm for Logic Minimization

The last two examples of the previous section suggest that a procedure based only on the simple moves that immediately decrease the cost may not be very effective. We need to consider a wider variety of moves, and combine them into sequences of moves that eventually lead to the desired reductions in cost. In other words, we need to expand the neighborhood that we are searching. We now present a simple strategy that can be found, in more elaborate forms, in many heuristic minimizers [144, 43, 37].

The idea is to iterate three phases: Initially, we expand all the input and output parts. This has the effect of increasing the overlap of the cubes. If we are lucky, one or more cubes will thus be covered completely by the union of other cubes. These redundant cubes are removed in the second phase, at the end of which we are at a local minimum (or very close to it). The third phase tries to exit this local minimum by reducing the cubes. The three phases are iterated until no further improvement is possible. The pseudo-code for this procedure is given in Figure 5.8. The inputs to the procedure are two covers for the ON-set and the DC-set of the function (F and D, respectively). The cost is measured after the cover is made irredundant by Subprocedure IRREDUNDANT. This is why the first two phases are duplicated. The loop consists of successive calls to REDUCE, EXPAND, and IRREDUNDANT. REDUCE heuristically reduces each cube of the cover. Then EXPAND heuristically expands each cube of the cover, and then IRREDUNDANT finds a minimal irredundant cover.

The loop is exited when the cost reaches a local minimum. However, at that point the output parts of the cube are still expanded. Since that may correspond to redundant connections, we perform one last pass with MAKE_SPARSE to reduce the output parts (and possibly expand some input parts that become expandable in the process). Consider, for instance, the following cover.

xyz	$f_1 f_2$
11–	10
–01	10
1–1	11
0–0	01

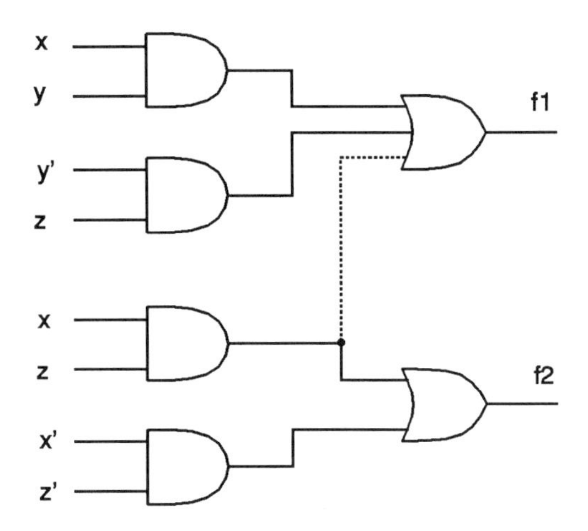

Figure 5.9: A Circuit that is Simplified by MAKE_SPARSE.

All cubes are maximally expanded in their input and output parts. The corresponding two-level circuit is shown in Figure 5.9. The dotted connection is redundant, because xz is the consensus term of xy and $y'z$; MAKE_SPARSE eliminates it.

5.2 Checking for Equivalence and Tautology

So far, we have assumed we could tell valid expansions and reductions from invalid ones. Though this may be relatively straightforward for the simple examples we have seen so far, it actually represents a non-trivial problem for larger functions. We shall devote most of the time spent on heuristic minimization to the discussion of two problems: One is the choice of the best move when many are possible. The other, that we consider now, is checking a move for validity, i.e., checking that the cover obtained by expanding and reducing some cubes is equivalent to the original one.

When a cube is expanded new minterms are added to it. We want to check that none of these minterms belong to the OFF-set of the function. If this condition is satisfied, then we are either increasing the overlap of existing terms, or covering some previously unused minterm of the DC-set. Similarly, when a cube is reduced minterms are removed from it. We should make sure that these minterms are either covered by some other cubes or belong to the DC-set.

In both cases, our equivalence check involves checking the newly added or removed minterms for containment in some other covers. Let $F = \{c_j\}$ be the current cover for the ON-set and D be the cover for the DC-set.[2] Let c_i be the cube of F that is being expanded or reduced and let $\tilde{c}_i$ be the cube that contains the newly added or removed minterms. Then the expansion or reduction is valid if and only if:

$$\tilde{c}_i \leq (F - \{c_i\}) \cup D.$$

In this formula $F - \{c_i\}$ is the cover obtained by removing c_i from F. After that, $\tilde{c}_i$,

[2] We omit the adjective current in this case, because this cover normally does not change during the computation.

$(F - \{c_i\})$, and D are regarded as sets of minterms. Interestingly enough, we can use a single algorithm to check both expansions and reductions for validity.[3]

Consider the following cover.

xyz	f
000	1
01–	1
–11	1
100	–

Suppose we want to check the validity of the expansion of the first cube from 000|1 to 0–0|1. So, in this case c_i is 000|1 and $\tilde{c}_i$ is 010|1, because the sum of 000|1 and 010|1 gives 0–0|1. We must check the containment of 010|1 in

xyz	$(F - \{c_i\}) \cup D$
01–	1
–11	1
100	1

The last term is the don't care cube, which contributes to containment as any other cube.

We know that a term may be contained in a set of terms, even though it is not contained in any single term of the set. For instance, $xz \leq xy + y'z$. Hence the containment test is not trivial. We cannot just scan the list of cubes in $(F - \{c_i\}) \cup D$ and look for one that contains $\tilde{c}_i$.[4] A possible simplistic approach is to examine all the minterms of $\tilde{c}_i$ for containment in $(F - \{c_i\}) \cup D$. This is a simple process, but there could be too many minterms to examine.

A possible improvement on the previous strategy could be as follows: We initially check for single-cube containment of $\tilde{c}_i$. If that fails, we break $\tilde{c}_i$ into halves and check the two halves separately. If both are contained we are done. If only one is contained, the other is split again, and so on. If eventually we find no minterms that are not contained, the containment test has succeeded.

Though this method is better than the previous one, we can do considerably better. However, we need some definitions and results that we now introduce.

Definition 5.2.1 *The* cofactor *(or* restriction*) of a function $f(x_1, x_2, \ldots, x_n)$ with respect to literal x_1, written f_{x_1}, is $f(1, x_2, \ldots, x_n)$. The* cofactor *(or* restriction*) of a function $f(x_1, x_2, \ldots, x_n)$ with respect to literal x_1', written $f_{x_1'}$, is $f(0, x_2, \ldots, x_n)$.*

We should be able to recognize in the cofactors as defined here the familiar expressions appearing in Boole's Expansion Theorem. Indeed we can rewrite the theorem as:

$$f = x_1 \cdot f_{x_1} + x_1' \cdot f_{x_1'}.$$

It is important to note that cofactoring is commutative, i.e.,

$$(f_{x_1})_{x_2} = (f_{x_2})_{x_1} = f_{x_1 x_2}$$

This allows us to state unambiguously the following definition.

[3] We shall see, however, that we may elect not to do so.

[4] We could actually do it if F and D were complete sums, but that would defeat the purpose of heuristic minimization.

Definition 5.2.2 *The* cofactor *(or* restriction*) of a function $f(x_1, x_2, \ldots, x_n)$ with respect to cube c is the successive cofactoring of f with respect to all the literals in c.*

Finally, it is convenient to introduce the following definition.

Definition 5.2.3 *A functions that is identically 1 is called a* tautology.

We are now ready for the punch line.

Theorem 5.2.1 *For a function f and a cube c,*

$$c \leq f \Longleftrightarrow f_c \equiv 1.$$

Intuitively, one may think of c as of a constraint applied to f. If there is a minterm for which c evaluates to 1, but f evaluates to 0, then f_c must be 0 somewhere. A real proof proceeds as follows: We observe that, for arbitrary g, $g \leq f \Rightarrow g_c \leq f_c$. We then take $g = c$ and we get:

$$c \leq f \Rightarrow c_c \leq f_c \Leftrightarrow 1 \leq f_c \Leftrightarrow f_c = 1.$$

Let us return to the example we saw earlier: Let $c = xz$ and $f = xy + y'z$. We have $f_c = y + y' = 1$, verifying that $xz \leq xy + y'z$. In view of Theorem 5.2.1, our task is to compute

$$[(F - \{c_i\}) \cup D]_{\tilde{c}_i}$$

and check whether this function is the tautology.

The advantages of this formulation are:

- cofactoring may greatly simplify the function by reducing the number of variables and cubes;
- we deal with a more uniform problem—tautology.

We have stated the definition of cofactors for single-output functions. We now examine the extension to the case of multiple outputs with the help of the following example. Let us consider the cover of Figure 5.7 and suppose we want to check the validity of reducing the second cube of the cover from —1|10 to 0–1|10, using Theorem 5.2.1. We have:

$$c_i = \text{—}1|10, \qquad \tilde{c}_i = 1\text{–}1|10,$$

and

$$F - \{c_i\} = \begin{array}{c|c} 1\text{—} & 10 \\ 0\text{–}1 & 01 \\ \text{–}10 & 01 \end{array}$$

The cofactor is taken by selecting the rows such that:

- The first and the third input columns are not 0;
- the first output column is a 1.

The general rule is:

- Eliminate the rows that disagree with the input part of $\tilde{c}_i$. A disagreement means that the row has a 1 in an input column where $\tilde{c}_i$ has a 0 or vice versa.

- Eliminate the rows that do not have at least a 1 in an output column where $\tilde{c}_i$ has a one.

After the elimination of the rows, the columns where $\tilde{c}_i$ is not '–' in the input part and is 0 in the output part are removed. This leaves us with:

$$-|1$$

which is clearly a tautology. In the case when no input columns are left, the result is a tautology if there is at least one cube left. In general, however, the result of the cofactor is not so easily understood to be (or not to be) tautologous. So we need an algorithm for **tautology checking**.

The basis for such an algorithm is, once again, the expansion theorem. In particular one can easily see that

$$f = 1 \Longleftrightarrow f_x = f_{x'} = 1.$$

In words, a function is tautologous if and only if both cofactors with respect to any variable are tautologous. It is therefore possible to recursively split the cover until a trivial case as the one in the previous example is found. The tautology check fails if any of the leaves of the recursion tree is not a tautology. Once more, however, we want to do better than that, and this requires a few more definitions and theorems that we cover next.

5.2.1 Unate Functions

Definition 5.2.4 *A function $f(x_1, x_2, \ldots, x_n)$ is* **monotonically increasing** *in variable x_1 if and only if*

$$f(0, x_2, \ldots, x_n) \leq f(1, x_2, \ldots, x_n), \qquad \forall (x_2, \ldots, x_n).$$

It is **monotonically decreasing** *in x_1 if and only if*

$$f(0, x_2, \ldots, x_n) \geq f(1, x_2, \ldots, x_n), \qquad \forall (x_2, \ldots, x_n).$$

If neither of the above is true, f is non monotonic in x_1.

For instance, $f = xy + y'z'$ is monotonic increasing in x, monotonic decreasing in z and non monotonic in y. Indeed, $f_{x'} = y'z' \leq y + z' = f_x$. Similarly we can verify that $f_z \leq f_{z'}$, while f_y and $f_{y'}$ are not comparable.

Definition 5.2.5 *A function $f(x_1, x_2, \ldots, x_n)$ is* **unate** *in variable x_1 if and only if it is either monotonic increasing or monotonic decreasing in x_1. It is* **unate** *(without further specifications) if it is unate in all of its variables.*

Our interest in unate functions is motivated by the following observation.[5] Consider a function f that is monotonic increasing in all its variables. Then f is a tautology if and only if $f(0,\ldots,0) = 1$. Clearly, if $f(0,\ldots,0) \neq 1$, f is not tautologous. On the other hand, if $f(0,\ldots,0) = 1$, then, by definition, $f(x_1,\ldots,x_n) = 1$ for every value of $(x_1,\ldots,x_n)$. If a function is monotonic decreasing in all its variables, we need to check its value for $(1,\ldots,1)$; the general case is a combination of these two.[6] In summary, checking a unate function for tautology is simple.

Deciding whether a function is unate is in general not trivial, if we want to rule out the examination of the value of the function for all possible inputs. Fortunately, there is an important special case where unateness can be determined by inspection.

Definition 5.2.6 *A cover F is* **monotonic increasing** *in variable x_1 if and only if x_1 never appears complemented in the terms of F. It is* **monotonic decreasing** *in variable x_1 if and only if x_1 never appears uncomplemented in the terms of F. If neither of the above is true, F is non monotonic in x_1.*

Returning to the previous example, the *cover* $xy + y'z'$ is monotonic increasing in x, monotonic decreasing in z and non monotonic in y. This matches what we said concerning the *function* represented by $xy + y'z'$. However, if we consider $x + x'y$, we find that the function is monotonic increasing in both x and y, while the cover is only monotonic increasing in y. Notice that $x + y$ is an alternative cover for the same function, and that $x + y$ is monotonic increasing in both x and y. This example illustrates a general point stated by the following theorem.

Theorem 5.2.2 *If function $f(x_1, x_2, \ldots, x_n)$ is unate in variable x_1, then there exists a cover of f unate in x_1. If a cover F is unate in x_1 then the function F represents is also unate in x_1.*

The second part of the theorem can be proved by observing that the product and the sum of monotonic increasing functions are monotonic increasing.[7] It follows that a monotonic increasing cover, that is the sum of products of monotonic increasing functions (the x_i) represents a monotonic increasing function. The extension to the general unate case is straightforward.

As a consequence Theorem 5.2.2, all results that we prove for unate functions can be applied to unate covers. Also, we can look for unate covers, that are easy to identify, and be sure that the corresponding functions are unate. We can then apply to them all the results we are going to derive for unate functions. By doing so we shall not identify some unate functions as such, and therefore miss some optimization opportunities. This is however a reasonable price to pay, to keep the identification of unate functions inexpensive.

The most interesting result on unate covers for us has to do with tautology checking.

---

[5] The properties of unate functions had been known long before their application to the tautology problem [223, 111]. Part of the original interest in these functions was in connection to their relation to *threshold* functions that could be realized by magnetic core devices [195].

[6] We can always prove results on unate functions by restricting our attention to monotonic increasing functions, since we can always transform a generic unate function into a monotonic increasing function by substituting each variable x_i in which the unate function is decreasing with a new variable y_i'.

[7] Just apply the definition of monotonic increasing.

Theorem 5.2.3 *A unate cover F is a tautology if and only if it contains the constant term 1.*

What we mean here is that the term appears explicitly among the terms that constitute F. For instance, $x + y$ is not tautologous, whereas $x + 1$ is. In terms of the cubical representation, a tautologous cover must contain a cube whose input part is all '–'. The proof of this result is a direct application of the fact that a monotonic increasing function can be checked for tautology by evaluating it in $(0, \ldots, 0)$. Indeed, no cube containing uncomplemented literals contains the vertex $(0, \ldots, 0)$.

Theorem 5.2.3 says that tautology checking for unate covers is trivial. The idea is therefore to adopt the usual recursive paradigm based on Boole's Expansion Theorem. At every node of the recursion tree, we test the function for unateness. This can be easily done while we look for the variable to split upon, by separately counting the number of complemented and uncomplemented occurrences of each variable. As a simple example, let us consider the following non-unate cover.

xyz	f
1–1	1
11–	1
00–	1
0–0	1

After splitting on x, we get two unate cofactors.

yz	f_x
–1	1
1–	1

yz	f'_x
0–	1
–0	1

It is sufficient to examine one of them to see that the function is not a tautology. (In practice, we would not compute the second cofactor.)

If a cover is unate only in some of the variables, it is not possible to answer the tautology question immediately. However, it is possible to simplify considerably the computation, thanks to the following observation. A cover F for a function $f(x_1, \ldots, x_n)$ that is monotonic increasing in x_1 can be written as

$$F(x_1, \ldots, x_n) = x_1 \cdot A(x_2, \ldots, x_n) + B(x_2, \ldots, x_n), \tag{5.1}$$

where A and B are SOP formulae, by regrouping terms and factoring out x_1. It is then clear that

$$F_{x_1} = A + B \qquad \text{and} \qquad F_{x'_1} = B.$$

It follows that

$$F_{x'_1} \leq F_{x_1}$$

in agreement with Definition 5.2.4, or, equivalently,

$$\begin{aligned} F_{x'_1} \equiv 1 &\implies F_{x_1} \equiv 1, && \text{whence} \\ F_{x'_1} \equiv 1 &\implies F \equiv 1, && \text{and} \\ F_{x'_1} \not\equiv 1 &\implies F \not\equiv 1, \end{aligned}$$

or, in summary,

$$F_{x_1'} \equiv 1 \Longleftrightarrow F \equiv 1.$$

Therefore, it is sufficient to test the negative cofactor for tautology to get the answer for the complete function. Similarly, if a function is negative unate in a variable, it is sufficient to test the positive cofactor for tautology. Thus, for every unate variable that we detect, we roughly cut in half the work to be done, because we only need to consider one of the two branches of the recursion. As an example, consider the following function.

$wxyz$	f
-1-1	1
0-10	1
010-	1

All minterms with $w = 1$ and $x = 0$ cause f to be 0. After simplification, no row is left; we conclude that the function is not tautologous.

5.2.2 Additional Speed-Up Techniques for Tautology Checking

The worst-case running time of the tautology algorithm we have outlined so far is exponential. Though the question of whether this worst-case behavior is optimal is still unanswered, it is extremely unlikely that a polynomial algorithm for tautology will ever be found.[8] The worst-case behavior is exponential because of the exponential number of nodes in the recursion tree. In practice, however, the behavior is not so bad, because we can heavily prune the recursion tree. Unate functions are one mechanism we use to reduce the number of nodes we need to explore. In the following we consider two additional techniques: Detection of special cases and partitioning.

The detection of special cases is a very general name for a collection of *ad hoc* techniques. Whenever we have a condition which is sufficient to answer the question and that can be tested quickly, it may be worthwhile trying it. In the case of ESPRESSO's tautology checking algorithm, the special cases that are tested include:

- A row of all '–' in the input part. If such a row is found, the function is tautologous in all the outputs that have a 1 in that row. Note that this condition is always sufficient, and is also necessary for unate functions.

- An input column of all 1s or all 0s. If such a column is found, the function is not a tautology ($B = 0$ in Equation 5.1).

- If the number of inputs is less than eight, the truth table is generated and the tautology question is answered by inspection. The rationale is that if the number of inputs is small enough, generating the truth table is faster than recurring.

- If the vertexPCcount count of the cover is insufficient, the function is not a tautology. The vertex count of the cover is computed as the sum of the vertex counts of the cubes. The vertex count of a cube is 2^d, where d is the number of

[8] In terms of complexity theory, the tautology problem is the complement of an NP-complete problem (CNF satisfiability).

dashes in the cube. The vertex count of the cover is actually an upper bound on the number of vertices in the cover. It equals the upper bound if all cubes are disjoint, otherwise, some vertices will be counted two or more times. If the vertex count of an n-variable cover is less than 2^n, then there is at least one minterm not included in the cover, and the function is not tautologous. For instance, the vertex count of the following cover is 8, which is less than $2^4 = 16$; hence the function is not a tautology.

$wxyz$	f
1-10	1
11-0	1
-00-	1

Notice the the ON-set of the function actually covers 7 vertices. (1110 is covered twice.)

Determining whether testing for a special case is worthwhile is to a large extent a matter of experimentation. The final technique we examine is partitioning. If a cover F can be written as

$$F = G + H$$

where G and H have disjoint support (i.e., no variables in common), then F is tautologous if and only if either G or H are tautologous. Let $x_1, \ldots, x_k$ be the variables on which G depends and $x_{k+1}, \ldots, x_n$ be the variables on which H depends. If $G(\tilde{x}_1, \ldots, \tilde{x}_k) = 0$ and $H(\tilde{x}_{k+1}, \ldots, \tilde{x}_n) = 0$, then $F(\tilde{x}_1, \ldots, \tilde{x}_k, \tilde{x}_{k+1}, \ldots, \tilde{x}_n) = 0$. Partitioning a search problem is normally very advantageous. The number of nodes to be visited in the worst case goes, for a uniform bipartition,[9] from 2^n to $2^{n/2+1}$ (for n even).[10]

Finding a bipartition is relatively easy. The set of columns of the first block, C_1, is initialized to empty. One then picks a cube of the cover (any cube will do). All the columns where the cube is 0 or 1 are added to C_1. Then all the cubes that intersect any of the columns in C_1 are selected (they must belong to the same block as the initial cube) and their columns added to C_1. The process is repeated until no addition to C_1 is possible. There are two possible outcomes: C_1 includes all the columns of the cover, or at least one column is left over. In the former case there is no bipartition (the second block is empty). In the latter, the leftover columns and rows identify the second block of the partition. It is possible to iterate the procedure on the second block to come up with a multi-way partition (the first block, by construction, cannot be further decomposed).

As an example of partitioning, consider the following matrix.

$$\begin{array}{c} \\ 1 \\ 2 \\ 3 \\ 4 \\ 5 \end{array} \begin{array}{c} 12345 \\ \left[\begin{array}{c} \text{1-1--} \\ \text{-1--0} \\ \text{0--0-} \\ \text{--01-} \\ \text{----1} \end{array}\right] \end{array}$$

[9] A uniform bipartition results in two blocks of the same size.

[10] Partitioning can also be applied to the covering problem.

Initially Row 1 is selected and $C_1 = \{1,3\}$. Scanning the columns in C_1, we find Rows 3 and 4. This in turn brings Column 4 into C_1. No further additions are possible. Hence there is a bipartition identified by $C_1 = \{1,3,4\}$ and $C_2 = \{2,5\}$. The two resulting subproblems are:

$$
\begin{array}{c} \\ 1 \\ 3 \\ 4 \end{array}
\begin{array}{c} 134 \\ \left[\begin{array}{c} 11- \\ 0-0 \\ -01 \end{array}\right] \end{array}
\qquad
\begin{array}{c} \\ 2 \\ 5 \end{array}
\begin{array}{c} 25 \\ \left[\begin{array}{c} 10 \\ -1 \end{array}\right] \end{array}
$$

In this case one sees immediately that neither subproblem is tautologous.

5.2.3 Examples of Tautology Checks

We conclude our examination of the tautology problem with a couple of examples. Our first example is for a single output cover.

$$
\begin{array}{c} \\ 1 \\ 2 \\ 3 \\ 4 \end{array}
\begin{array}{c} wxyz\ f \\ \left[\begin{array}{c} --1-|1 \\ 1--1|1 \\ -0-0|1 \\ -00-|1 \end{array}\right] \end{array}
$$

In this case we could proceed by producing a truth table, since the number of inputs is small. However, for humans, brute-force enumeration is seldom convenient. In this case, for instance, we are better off by noting that the cover is unate in w and x. We cofactor with respect to w' and x and we obtain:

$$
\begin{array}{c} \\ 1 \end{array}
\begin{array}{c} yz\ f \\ \left[\begin{array}{c} 1-|1 \end{array}\right] \end{array}
$$

which is clearly not a tautology. Hence, the original cover is not a tautology. As second example, we consider a multiple-output function. A multiple-output cover is tautologous if all the outputs are tautologous. Consider the following example [37, p. 71].

$$
\begin{array}{c} \\ 1 \\ 2 \\ 3 \\ 4 \\ 5 \end{array}
\begin{array}{c} wxyz\ fg \\ \left[\begin{array}{c} 1-01|01 \\ -110|10 \\ 1110|11 \\ ---0|10 \\ ---1|11 \end{array}\right] \end{array}
$$

Also in this case, the cover is unate in w and x. Cofactoring, we obtain:

$$
\begin{array}{c} \\ 4 \\ 5 \end{array}
\begin{array}{c} yz\ fg \\ \left[\begin{array}{c} -0|10 \\ -1|11 \end{array}\right] \end{array}
$$

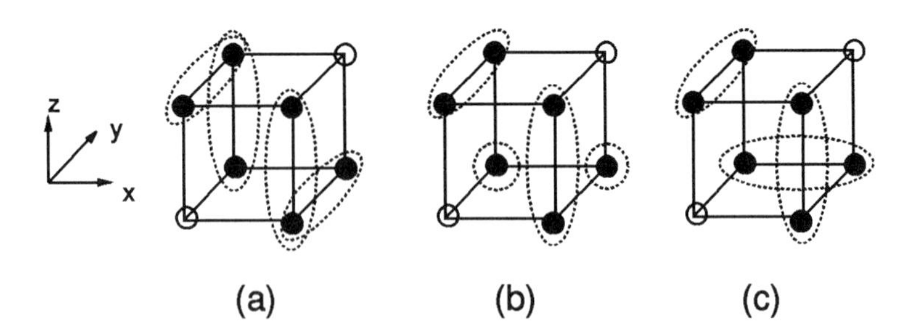

Figure 5.10: Example where the Directions in which Cubes are Expanded Matters. *(a)*: Initial Cover. *(b)*: After Reduction. *(c)*: After Expansion in the Right Direction.

We can now drop y, because the cover no longer depends on it. We now split on z, the only variable left.[11] Suppose we first analyze the positive cofactor. The only cube left is 5, with no input variables. Since both outputs have a 1 for this cube, we conclude that this half-space is tautologous. However, for $z = 0$, we are left with cube 4 only, for which g is 0. Hence, g is not a tautology and the whole cover is not a tautology. In general, multiple-output tautology requires minor extensions to single-output tautology, that are left as an exercise to the reader (or can be looked up in [37]).

5.3 Choosing the Right Direction

When we expand a cube by replacing a zero or a one with a dash in the position corresponding to variable x_i, we say that we are expanding the cube in the x_i direction. Similarly for reduction. It is often the case that a cube can be expanded in more than one direction, but not in all directions simultaneously. The choice of one direction over another may influence the quality of the solution, and therefore it is important to be able to tell what directions are likely to give the best results. We consider the example of Figure 5.10. We assume that the initial cover is the one of Part *(a)* and that the reduction step results in the cover of Part *(b)*.[12] There are two candidates for expansion and both can be expanded in two directions. For symmetry reasons, it is sufficient to consider only one of them: We choose $x'yz'$.

It is clear that we would like to expand in the x direction, since we would then cover xyz' and get the optimum solution. However, if we proceed randomly, we may as well re-expand in the z direction. There is a possibility that we return to the initial cover. The algorithm would then stop because no improvement occurred during a complete iteration.

One important distinction among various heuristic minimizers is the way they try to guide the expansion process. ESPRESSO has the most sophisticated strategy of all. It is so sophisticated, that we do not have time to examine it in detail. We rather state ESPRESSO's objectives and give a few hints on how those objectives can be pursued. Since the cost is primarily related to the number of cubes, it is reasonable

[11] In practice we would solve this problem by checking for the special cases, but here we want to illustrate the other techniques as well.

[12] Note that no expansion is possible for the initial cover and that no cubes are redundant.

to assume as an objective of the expansion process the coverage of other cubes by the expanded cube. If we know how to do that, then we can effectively solve the problem of Figure 5.10. As a secondary goal, we may want to get the largest expanded cube, thus minimizing the number of literals. In both cases, having an explicit representation of the OFF-set of the function helps. So we shall consider briefly the problem of efficient complementation of a cover. We shall then outline how we can use the complement to get the largest expanded cube, referring the interested reader to [37, 239] for the description of the complete expansion process.[13]

5.3.1 Recursive Complementation

It should come as no surprise that an efficient complementation algorithm can be based on the expansion theorem. Indeed, we have the following result.

Theorem 5.3.1 *For a Boolean function f,*

$$f' = x \cdot f'_x + x' \cdot f'_{x'}.$$

Proof. Let $g = x \cdot f'_x + x' \cdot f'_{x'}$. We prove that g is the complement of f. It is sufficient to prove that $f \cdot g = 0$ and $f + g = 1$. This in turn is easily established by expanding f with respect to x and substituting into $f \cdot g$ and $f + g$. □

We can therefore select a variable according to the usual criterion, compute recursively the complement of the two cofactors, and combine the results. When the subproblems reduce to a single cube, we can simply apply De Morgan's Laws to get the complement.

Since we want to keep the size of the cover for the complement small, we normally check for cubes that appear in both cofactors and merge them into one ($xc + x'c = c$). As an example, consider the following cover.

xyz	f
10–	1
110	1
0–1	1

We first consider the positive cofactor with respect to x.

yz	f_x
0–	1
10	1

We then split with respect to y, to get f_{xy} and $f_{xy'}$. These cofactors are simple enough that we may find their complements by inspection.

z	f_{xy}
0	1

z	f'_{xy}
1	1

z	$f_{xy'}$
–	1

z	$f'_{xy'}$

Notice that $f'_{xy'} = 0$. We can now form f'_x.

[13]There are differences between ESPRESSO, described in [37], and ESPRESSO-MV, described in [239]. However, the basic ideas are the the same.

yz	f'_x
11	1

We now compute $f_{x'}$ and its complement.

yz	$f_{x'}$
–1	1

yz	$f'_{x'}$
–0	1

Finally, we combine the two partial results into the complement of f.

xyz	f'
111	1
0–0	1

As in the tautology check, we can exploit the properties of unate functions. Specifically, we can prove that for f monotonic increasing in x,

$$f' = f'_x + x' \cdot f'_{x'}$$

and for f monotonic decreasing in x,

$$f' = x \cdot f'_x + f'_{x'}.$$

Additional details on how to speed-up the complementation of unate functions can be found in [37], from which we take the following example.

wxyz
11––
1–––
–1–1
–11–
–111

Since the function is unate, we adopt a special variable selection strategy. We notice that if we select w, a row of all dashes will appear in f_w. Since the complement of the tautology is trivial to compute, selecting w is a good thing. Since we know that

$$f'_w = 0,$$

we only need to compute $f'_{w'}$. By cofactoring f, we get

xyz
1–1
11–
111

We notice that this cover can be written as $x \cdot G$, with G given by:

yz
–1
1–
11

Applying De Morgan, $f'_{w'} = x' + G'$, or

xyz
0—
–00

which gives the following final result:

wxyz
00—
0–00

When we need to complement a multiple-output function, we can complement one output at the time.

5.3.2 Using the OFF-set in the Expansion

The complementation techniques we just saw can be used to compute a SOP cover of the OFF-set, given the SOP covers for ON-set and DC-set. We now see how the cover for the OFF-set can be used to find the maximal expansion of a cube. We know that a cube can be expanded as long as it does not intersect any cube of the OFF-set. For this to hold, the cube being expanded should conflict in at least one position with each cube in the OFF-set. We can build a matrix, called the **blocking matrix**, that has a row for each cube of the OFF-set and a column for each variable. Each row has ones in the positions where the cube to be expanded conflicts[14] with the cube of the OFF-set. A row of the blocking matrix tells us that it is O.K. to expand the cube, as long as at least one of the directions corresponding to one of its ones is not expanded. It is not difficult to see that finding a maximum expansion corresponds to finding a minimum subset of the columns of the blocking matrix that covers all the rows.

It should be noted that the covering problem we need to solve in this case has much fewer variables than the covering problem we face in the exact minimization method. It is also easy to see that a covering problem of manageable size can be used in performing the IRREDUNDANT step. One finds a minimum subset of the cubes in the current cover that covers all minterms in the ON-set.

5.4 Identifying Essential Primes

It is advantageous to identify the essential primes, because we know that they will be part of every optimal solution. Interestingly enough, after the initial expansion all cubes in the cover are primes and therefore all the essentials are present in the cover. If we can identify them, then we can put them aside and avoid silly things like reducing them in an attempt to expand them in a different direction.

If we had all primes of the function, we could easily tell whether a prime is essential, by checking whether it was covered by the union of all other primes. When we are not given all the primes, the identification of the essential primes is based on the following theorem [244].

[14] A conflict here means 0 for the cube being expanded and 1 for the cube of the OFF-set or vice versa.

Theorem 5.4.1 *Let F be a cover composed of prime implicants. Let e be one of the primes in F and let G be the cover composed of the remaining primes. Then, e is an essential prime if and only if it is not covered by the union of:*

1. *The consensus terms of e and each term of G;*

2. *The intersections of e and each term of G.*

As an example, let us consider

$$y'z' + xy' + xz$$

and test $y'z'$. This prime intersects xy' (the intersection is $xy'z'$). It also has a consensus term with xz (xy'). To see whether $y'z'$ is essential, we check $y'z' \leq xy' + xy'z'$. Since $xy'z' \leq xy'$, we only need to check whether $y'z' \leq xy'$. The answer is clearly no; hence $y'z'$ is essential. If we now consider xy', we find that we need to check whether xy' is covered by $xy'z' + xy'z$. The answer is in this case positive and xy' is not essential.

The usefulness of this approach to the identification of essential implicants appears more clearly in our next example.

$$f = x'z' + x'y + xz.$$

The cover in this case is not a complete sum. It is not therefore immediate to see whether a given prime is essential. Suppose we want to test $x'y$. Its intersection with $x'z'$ is $x'yz'$, while its consensus term with xz is yz. Hence, we test whether:

$$x'y \leq x'yz' + yz.$$

Cofactoring the right hand side of the inequality, we get $z' + z$, which is clearly a tautology. Hence, $x'y$ is not essential.

5.5 Multiple-Valued Logics

When we design by successive refinements (or top-down), we normally find beneficial to postpone decisions on details until the context is detailed enough to provide guidance for the choices to be made. A typical example could be the assignment of operation codes to machine instructions (or micro-instructions) or the assignments of codes to the states of a finite state machine (FSM). In such situations, we want to use symbolic names when we design, and possibly leave the choice of the encoding to an optimization program.

It is easy to see that the general situation is when we have two sub-circuits that communicate via wires that are not primary outputs of the circuit, as depicted in Figure 5.11. The special case of a finite state machine is obtained when the two blocks coincide. If we use symbolic names, we can write expressions involving them, like:

> If the op-code is *move* and the addressing mode is *indirect* go to state *bigmess*.

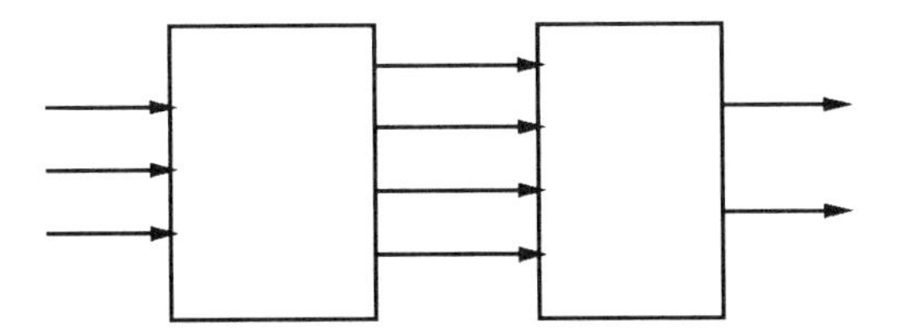

Figure 5.11: The Interconnection of Sub-Circuits Gives Rise to Encoding Problems.

We are interested in a formal way to deal with expressions of this kind. Large Boolean algebras are not general enough, because we know that the number of values in the carrier must be a power of two and furthermore only a small fraction of the possible mappings are Boolean functions. We need to relax these assumptions and this corresponds to dealing with multiple-valued logics. We shortly discuss the basic ideas of multiple-valued logics, following [244, 239]. Suppose we have a function of n multiple-valued variables that returns a binary value (i.e., 0 or 1). Suppose that p_i, for $i = 1, \ldots, n$, is the number of values the i-th variable can assume. Define the set $P_i = \{0, \ldots, p_i - 1\}$ as the set of values the i-th variable can assume. A *multiple-valued input, binary-valued output function* f is a mapping

$$f : P_1 \times \cdots \times P_n \to B,$$

where $B = \{0, 1\}$, as usual. Let X_i be a variable taking a value from the set P_i, and let S_i be a subset of P_i. Then, $X_i^{S_i}$ represent the mapping from P_i to B defined by

$$X_i^{S_i} = \begin{cases} 0 & \text{if } X_i \notin S_i \\ 1 & \text{if } X_i \in S_i \end{cases}$$

$X_i^{S_i}$ is called a *literal* of variable X_i. We build sum of products formulae out of multiple-valued literals much in the same way as we do in ordinary Boolean algebras. We can also define implicants, prime implicants, and essential prime implicants. Many of the laws of Boolean algebras are still valid in this new setting, though not all of them. In particular, it is possible to extend the expansion theorem:

$$f = X_1^{S_i} f_{X_1^{S_i}} + X_1^{S_i'} f_{X_1^{S_i'}}.$$

This forms the basis for the extension of most of the techniques we have seen so far to multiple-valued logics. It is therefore possible to write a minimizer for multiple-valued functions and ESPRESSO-MV [239] is one example.

5.6 Notes

In some sense the modern era of logic minimization began with MINI [144], a logic minimizer based on multiple-valued logics. MINI was very general and fast on some problems, but not so robust on larger or more difficult problems. MINI was followed by ESPRESSO from IBM. Rudell, who had worked on ESPRESSO, then wrote the variant ESPRESSO-MV, which allowed multivalued logics, and was much more efficient in dealing with the complement. ESPRESSO-MV has stood, since 1987, as the minimizer of choice at most industrial and university laboratories worldwide.

Although heuristic methods are still widely used today, advances in exact minimization techniques (See Section 4.11 on Page 164) have made significant inroads into application areas traditionally dominated by heuristic methods such as ESPRESSO-MV.

5.7 Summary

We began in this chapter by identifying some common characteristics of local Search algorithms (algorithms that start from an initial solution and try to find a better one by applying successive modifications). Then we studied the set of methods that combine to make the local search algorithm known as the ESPRESSO algorithm for logic minimization.

Two methods were discussed for quickly checking whether a given implicant is prime: one based on tautology/equivalence checking, and one based on intersection with the OFF-set of the function to be minimized. Fast methods for tautology checking were discussed, based on unateness or partial unateness of the cover being checked. And similarly fast methods for computing the OFF-set were discussed.

We then gave a method for computing the essential primes, and closed by showing how the whole paradigm of heuristic minimization could be extended to multiple-valued logics.

5.8 Problems

1. Find a sequence of moves that transforms:

xyz	$f_1 f_2$
—1	10
110	11
-01	01
10-	01

into:

xyz	$f_1 f_2$
—1	10
11-	10
-01	01
1-0	01

All moves must consist of expansion or reduction of a cube in one position. For instance,

$$—1|10 \rightarrow -01|10$$

is a valid reduction move, though not a useful one (you can verify that it does not preserve the equivalence of the covers). As you can see the second cubical complex (or simply cover) has a lower cost. Sometimes, as in this example, decreasing the amount of sharing may lead to a better solution.

Solution. We can verify that no cube of the original description can be expanded in the input part or reduced in the output part. Hence, we are in a local minimum. We start with a reduction:

$$10-|01 \rightarrow 100|01.$$

We can now re-expand this cube in a different direction:

$$100|01 \;\rightarrow\; 1\text{-}0|01.$$

This in turn enables the output reduction of the only shared cube:

$$110|11 \;\rightarrow\; 110|10.$$

Finally, we obtain the desired result by expanding this cube in the z direction:

$$110|10 \;\rightarrow\; 11\text{-}|10.$$

□

2. Find a sequence of moves that transforms:

xyz	$f_1 f_2$
00-	10
-10	11
1-1	10
-01	01
11-	01

into:

xyz	$f_1 f_2$
-01	11
-10	01
0-0	10
11-	11

All moves must consist of expansion or reduction of a cube in one position. Cubes with all-zero output part can be discarded.

3. Find a sequence of moves that transforms:

$wxyz$	$f_1 f_2$
0-0-	10
-10-	10
1-11	11
00--	01
0-1-	01

into:

$wxyz$	$f_1 f_2$
000-	11
-10-	10
1-11	11
0-1-	01

All attempted moves must consist of expansion or reduction of a cube in one position. For instance,

$$0\text{-}0\text{-}|10 \;\rightarrow\; 0\text{---}|10$$

is a valid expansion move, though not a useful one (you can verify that it does not preserve the equivalence of the covers and hence it may be attempted, but it must be rejected).

4. Find the cofactor of:

xyz	$f_1 f_2$
1-0	10
-10	01
0-1	11
11-	01

with respect to the cube –1–|01.

Solution.

xz	f_2
–0	1
01	1
1–	1

We can observe that this cofactor is a tautology. □

5. Find the cofactor of:

$vwxyz$	fgh
1–0–1	101
001––	001
–111–	110
01––0	111
110–0	010
1010–	101

with respect to the cube ––110|110.

6. In what variables is the following cover unate?

$wxyz$	f_1
1–0–	1
–11–	1
0––0	1
100–	1
–1–0	1

Solution. The only variable in which the cover is unate is z. All other columns contain both zeroes and ones. □

7. In what variables is the following cover unate?

xyz	f
11–	1
011	1
10–	1

8. In what variables is the function of Problem 6 unate? Note that this problem refers to the unateness of the function as opposed to the unateness of the cover.

   [Hint: Is any cube of the cover redundant?]

   Solution. The fourth cube is contained in the first and is therefore redundant. When we eliminate it, we see that the column corresponding to x has no zeroes left. Hence, the function is positive unate (monotonic increasing) in x, even though the original cover is not. One can verify that the function is not unate in the other two variables (w and y), by computing the cofactors and checking that one is not less than or equal to the other. □

9. In what variables is the function of Problem 7 unate? Note that this problem refers to the unateness of the function as opposed to the unateness of the cover.

10. Check whether the following function is the tautology.

$wxyz$	f
111–	1
1–10	1
000–	1
0—1	1
–10–	1
0–1–	1

Use the properties of unate functions whenever possible.

Solution. We split initially on w, because the function is not unate in any variable. For the positive cofactor, we have:

xyz	f_w
11–	1
–10	1
10–	1

This function is unate positive in x and unate negative in z. If we take the cofactor with respect to $x'z$ we get the 0 function. Hence f is not tautologous. □

11. Prove or disprove that the following function is the tautology.

$abcdefghij$	pq
0–1–1——	10
–11–1——	10
—11——	10
———0–	10
0–1–1——	01
–11—1—	01
—1—1—	01
———0	01
——11—	11

Show your argument.

12. Check whether the following function is a tautology:

$wxyz$	f
1100	1
1–––	1
–0–1	1
01––	1
0–1–	1

13. Check whether the following function is the tautology.

$vwxyz$	f
1–1––	1
–0––0	1
100–1	1
––11–	1
1–00–	1

Use the properties of unate functions.

Solution. We note that the function is unate positive in v and unate negative in w. It is then sufficient to check $f_{v'w}$ for tautology:

xyz	$f_{v'w}$
11–	1

This function is clearly not a tautology. Hence, the original function is not a tautology. □

14. Check that for the following cover:

xyz	$f_1 f_2$
––1	10
110	11
–01	01
100	01

the expansion:

$$100|01 \rightarrow 1\text{–}0|01$$

is valid. Do so by computing

$$(F - \{c_i\})_{\tilde{c}_i}$$

(note that D is empty in this example) and checking if this function is the tautology.

Solution. In this case, $\tilde{c}_i = 110|01$. The cofactor of $F - \{c_i\}$ is clearly tautologous, since F contains $110|11$ that covers $\tilde{c}_i$. Hence, the expansion is valid. □

15. Check whether for the following cover:

$vwxyz$	f
00—	1
—101	1
-01—	1
0-1-0	1
0—1	1
1101-	1

the expansion:

$$00\text{—}|1 \rightarrow 0\text{—}|1$$

is valid. Do so by computing

$$(F - \{c_i\})_{\tilde{c}_i}$$

(note that D is empty in this example) and checking if this function is the tautology. Use the properties of unate functions to simplify the computation.

16. This problem is on the recursive algorithm for complementation. Consider the following function:

xyz	f
1-0	1
-01	1
01-	1

We apply the algorithm by selecting a splitting variable, say x, and computing f'_x and $f'_{x'}$ as follows:

yz	f_x
-0	1
01	1

yz	f'_x
11	1

and

yz	$f_{x'}$
01	1
1-	1

yz	$f'_{x'}$
00	1

Notice that it is easy to compute the complement by simple enumeration on the cofactors f_x and $f_{x'}$. Hence, we only split once. On large examples, we may need to split several times. We finally compute $f' = x'f'_{x'} + xf'_x$.

xyz	f'
111	1
000	1

Apply this procedure to compute the complement of the following function:

$wxyz$	f
1–0–	1
–11–	1
0–1–	1
–1–0	1
––00	1

Choose y as initial splitting variable.

Solution. If we split on y, we are going to compute $f' = y' f'_{y'} + y f'_y$. We have:

wxz	$f_{y'}$
1––	1
–10	1
––0	1

The second cube is redundant and we can just complement by applying De Morgan's Laws:

wxz	$f'_{y'}$
0–1	1

Similarly:

wxz	f_y
–1–	1
0––	1
–10	1

This time, the third cube is redundant. We get:

wxz	f'_y
10–	1

Putting things together, we finally obtain:

$wxyz$	f'
101–	1
0–01	1

which is the desired solution. □

17. Compute the complement of

$$f = wx'y + w'xy + yz' + wxy' + wy'z',$$

by applying the recursive algorithm.

18. The Boolean difference (or Boolean derivative) of $f(x_1, \ldots, x_n)$ with respect to x_1 is defined as:

$$\frac{\partial f}{\partial x_1} = f(0, x_2, \ldots, x_n) \oplus f(1, x_2, \ldots, x_n).$$

Prove that if $f(x_1, \ldots, x_n)$ is monotonic increasing in x_1,

$$\frac{\partial f}{\partial x_1} = f(1, x_2, \ldots, x_n) f'(0, x_2, \ldots, x_n).$$

Solution. We have:

$$\frac{\partial f}{\partial x_1} = f(0, x_2, \ldots, x_n) f'(1, x_2, \ldots, x_n) + f'(0, x_2, \ldots, x_n) f(1, x_2, \ldots, x_n).$$

However, since f is monotonic increasing, we have:

$$f(0, x_2, \ldots, x_n) \leq f(1, x_2, \ldots, x_n),$$

which is equivalent to:

$$f(0, x_2, \ldots, x_n) f'(1, x_2, \ldots, x_n) = 0.$$

Therefore, we can drop the first term of the sum and get the desired result. □

19. Use ESPRESSO to solve Problem 17. Include your input and output files and the command line you used.
Solution. This is the input file:

```
.i 4
.o 1
.ilb w x y z
.ob f
101- 1
011- 1
--10 1
110- 1
1-00 1
.e
```

Giving the command *espresso -epos -Dexact infile*, we get:

```
.i 4
.o 1
.ilb w x y z
.ob f
#.phase 0
.p 4
```

```
0-0- 1
00-1 1
-001 1
1111 1
.e
```

which agrees with our manual solution. □

20. Is xz an essential implicant of

$$f = w'xy' + wy'z + w'yz + wyz' + xz?$$

Solution. For each implicant other than xz we check whether it has an intersection or a consensus term with xz. Notice that if there is intersection, then there cannot be consensus and vice versa.

(a) Intersection with $w'xy'$: $w'xy'z$;

(b) intersection with $wy'z$: $wxy'z$;

(c) intersection with $w'yz$: $w'xyz$;

(d) consensus with wyz': wxy.

The problem is therefore to check whether:

$$xz \leq w'xy'z + wxy'z + w'xyz + wxy.$$

We cofactor the right-hand side of the inequality with respect to the cube xz (set $x = z = 1$) and get:

$$w'y' + wy' + w'y + wy,$$

which is clearly a tautology. Hence, xz is not essential. □

21. Find all maximal expansions of $w'xyz'$ in

$$f = w'y'z' + xz + x'yz' + w'xyz',$$

by doing the following.

(a) Compute the complement of f by the recursive algorithm. (Split on z.)

(b) Build the blocking matrix.

(c) Enumerate all minimum cost solutions of the covering problem.

Which of the maximal expansions is more advantageous? Explain.

Solution.

(a) Compute the complement of f.

$$\begin{aligned} f' &= zx' + z'(w'y' + x'y + w'xy)' \\ &= x'z + z'(y(x' + w'x)' + y'w) \\ &= x'z + z'(yxw + y'w) \\ &= x'z + wxyz' + wy'z' \\ &= x'z + wxz' + wy'z'. \end{aligned}$$

(b) Build the blocking matrix.

$$\begin{array}{l} \\ x'z \\ wxz' \\ wy'z' \end{array} \quad \begin{array}{c} \begin{array}{cccc} w' & x & y & z' \end{array} \\ \left[\begin{array}{cccc} 0 & 1 & 0 & 1 \\ 1 & 0 & 0 & 0 \\ 1 & 0 & 1 & 0 \end{array}\right] \end{array} \quad \begin{array}{c} \\ 1 \\ 2 \\ 3 \end{array}$$

(c) Enumerate all minimum cost solutions. Clearly w' is essential and must be chosen. We can then cover the first row with either x or z'. Therefore we have two minimum cost solutions:

$$w'x \quad \text{and} \quad w'z'.$$

We can see that $w'z'$ is advantageous, because it will cover $w'y'z'$.

□

22. This problem is based on the LUNC circuit that we considered in Chapter 2.1. We saw that a typical synthesis system proceeds through two phases. Initially, a description of the behavior of the circuit is translated into a structure, using adders, comparators, and multiplexors to represent the various language constructs. The initial structure is normally far from optimal. Hence the second phase deals with its minimization. In our review of the LUNC example, we saw how the initial structure could be optimized by hand. In particular, we considered the command interpreter block that decodes four possible escape sequences. In this problem, we shall see how we can use ESPRESSO to optimize the combinational logic of the decoder: Specifically, the logic that decodes the character following the escape character. We shall see the important role that *don't cares* play in this process.

    The following input to ESPRESSO describes the decoder:

```
.i 8
.o 3
.ilb x7 x6 x5 x4 x3 x2 x1 x0
.ob L N C
01001100 100
01001110 010
01000011 001
.e
```

The input part of each cube corresponds to the ASCII code of one of the three characters L, N, and C. (Remember that we do not need to generate a signal for U.) If we minimize this function, we get:

```
.i 8
.o 3
.ilb x7 x6 x5 x4 x3 x2 x1 x0
.ob L N C
.p 3
01000011 001
01001100 100
01001110 010
.e
```

which is exactly the same function. (Only the order of the terms has been changed.) This result is actually expected. When we simplified the decoder manually, we actually used the assumption that only L, U, N, and C could follow an escape character. In other words, all other codes are *don't cares* for the decoder. Our current specification, on the other hand, actually says that the output for all other codes should be 000.

In this problem, you are requested to:

(a) Augment the specification of the decoder given above by adding the description of the DC-set. This can be done in (at least) two ways. (You should refer to the man page for ESPRESSO in Section 5 for the details.)

  i. The first way consists of using the default type (*fd*) and of specifying a cover for the DC-set.

  ii. The second way is probably less intuitive, but much more efficient in this case. It consists of using the type *fr*, where you specify the ON-set and the OFF-set. Everything else is considered *don't care*.

(b) Run your augmented specification with and without the `-Dexact` option. You should get two different results. (Both correct, of course.)

(c) Discuss why the two results obtained are different. Keep in mind that the exact minimization in ESPRESSO uses only the number of cubes as cost criterion.

Include your input and output files from running ESPRESSO.

Solution. We need to specify that all codes different from L, U, N, and C are don't care. The straightforward, yet laborious way to do that is as follows:

```
.i 8
.o 3
.ilb x7 x6 x5 x4 x3 x2 x1 x0
.ob L N C
01001100 100
01001110 010
```

```
01000011 001
1------- ---
00------ ---
011----- ---
010001-- ---
010010-- ---
010100-- ---
010110-- ---
010111-- ---
0100000- ---
010000-0 ---
010011-1 ---
0101011- ---
010101-0 ---
.e
```

On the other hand, we can notice that the only thing that is missing to completely specify both ON-set and OFF-set is the indication that all three outputs should be zero when the input is U. This approach is more convenient in this case because there are only four input values for which the outputs are not all *don't care*. Therefore we can write:

```
.i 8
.o 3
.ilb x7 x6 x5 x4 x3 x2 x1 x0
.ob L N C
.type fr
01001100 100
01010101 000
01001110 010
01000011 001
.e
```

This specification is equivalent to the previous one. ESPRESSO is given f and either d or r, but it internally generates the other one. When we run *espresso* on one of the two input files with the `-Dexact` option, we get:

```
.i 8
.o 3
.ilb x7 x6 x5 x4 x3 x2 x1 x0
.ob L N C
.p 3
------11 001
------10 010
------00 100
.e
```

If we do not request exact minimization, we get:

```
.i 8
.o 3
.ilb x7 x6 x5 x4 x3 x2 x1 x0
.ob L N C
.p 3
-----0-- 001
------00 100
------10 010
.e
```

which has the same number of product terms, but one fewer literal. This can be explained by noting that the cost considered by *espresso* in exact minimization is just the number of terms or cubes. Therefore, *espresso* returns the first solution with three cubes that it finds.

On the other hand, when *espresso* is run in heuristic mode, it tries to maximally expand every cube. When 01000011|001 is maximally expanded, -----0--|001 is found to be a better solution than ------11|001, because it has one fewer literal.

Note that the solution found by the heuristic minimizer is better in terms of literals. However, the minimized function depends on three variables, x_2, x_1, and x_0, instead of two. This may be sub-optimal, if long connections have to be drawn from where the inputs are available to where the function is placed in the layout of the circuit. Though in our simple circuit this is not likely to be the case, this problem gives an example of how abstracting the cost into a simple formulation—the number of cubes and the number of literals—may sometimes hide important details. It should be mentioned that it is possible to use as cost criterion the number of variables a function depends on. In that case we say that we minimize the *support* of the function. □

Chapter 6

Binary Decision Diagrams (BDDs)

In this chapter we develop theory, algorithms, and data structures for the treatment of BDDs (Binary Decision Diagrams) and their applications. Along the way we discuss the relative advantages of Canonical and Non-Canonical Representations, and introduce the reader to BDDs by way of examples.

Many synthesis, verification, and testing algorithms manipulate large switching formulae. It is therefore important to have efficient ways of representing and manipulating such formulae. In recent times Binary Decision Diagrams (BDDs) have emerged as the representation of choice for many applications. Though BDDs are relatively old [169, 5], it was only with the work of Bryant [47], which brought out their advantages as canonical representations, that they began attracting the attention of many researchers.

The impact of BDDs has been enormous. Since 1986, when BDDs were unknown in synthesis circles, BDDs have penetrated virtually every subfield in the areas of synthesis and verification. As we shall see, BDDs have two remarkable properties. First, they are canonical, so if you correctly build the BDDs for two circuits, the two circuits are equivalent if and only if the BDDs are identical. This has led to significant breakthroughs in circuit optimization, testing, and equivalence checking. Second BDDs are amazingly effective at representing combinatorially large sets. This has led to stunning breakthroughs in FSM equivalence checking (and many other forms of formal verification [71]) and in two level logic minimization [78, 77, 80, 76]. This recent research is summarized below in Section 6.4.

Clearly a whole book can and should be written about DDs[1]. However, in this book we limit ourselves to a brief, introductory treatment in support of our discussion on "Symbolic" FSM equivalence checking. In our description, we shall follow [47] and [29], though with slightly different notation.

[1]Public domain BDD packages are freely available, for example our own CUDD package — see Section 6.4.

6.1 Representing Logic Functions with BDDs

We have already discussed several ways of representing a Boolean function, for example by Boolean formulae, or by the minterm or maxterm canonical form. The latter two forms are **canonical**. A form is canonical if the representation of a function in that form is unique.[2]

A canonical form is desirable because it makes equivalence tests easy. From the definition, it follows that f_1 and f_2 have the same representation in a canonical form if and only if $f_1 = f_2$. The minterm and maxterm canonical forms, however, have a serious drawback: The representations tend to be quite large (they are exponential in the number of variables). So they are not used except for the simplest cases.

Among the non-canonical forms, the formula type known as the sum of products (SOP) and the product of sums (POS) have been widely used. These two-level representations are discussed in detail in Section 4.3. They are two-level representations, and quite useful in synthesis and verification. However, there are also some fundamental difficulties with these forms.

- The two-level representations of some functions are too large to be practical (e.g., EXCLUSIVE-OR);
- Passing from SOP to POS and vice versa is difficult. As a consequence:
  - Taking the complement is difficult;
  - Taking the AND of two sums of products (or the OR of two products of sums) is difficult.
- Since SOP and POS are not canonical forms, answering the equivalence question for two functions is difficult.
- Furthermore, deciding whether a product of sums is satisfiable (a sum of product is a tautology) is NP-complete (coNP-complete).[3]

In the next sections we discuss, and the define formally, a canonical form—the BDD—that has the advantage of being compact for many functions and definitely superior to the other known canonical forms in that respect.

6.1.1 Binary Decision Diagrams by Way of Examples

We first introduce BDDs with the help of two examples and then give a formal definition. First, a BDD is a DAG (Directed Acyclic Graph — see Section 7.9.1, Page 305), such as the DAG shown at the right in Figure 6.1. Note the BDD nodes are in one to one correspondence with the gates of the MUX circuit at the left of the figure. This shows how BDDs can be viewed as a shorthand representation for MUX circuits, just as an SOP form can be viewed as a shorthand representation for an OR of ANDS.

[2] We normally disregard reordering of terms.

[3] The best known algorithms for these problems have exponential worst-case run times.

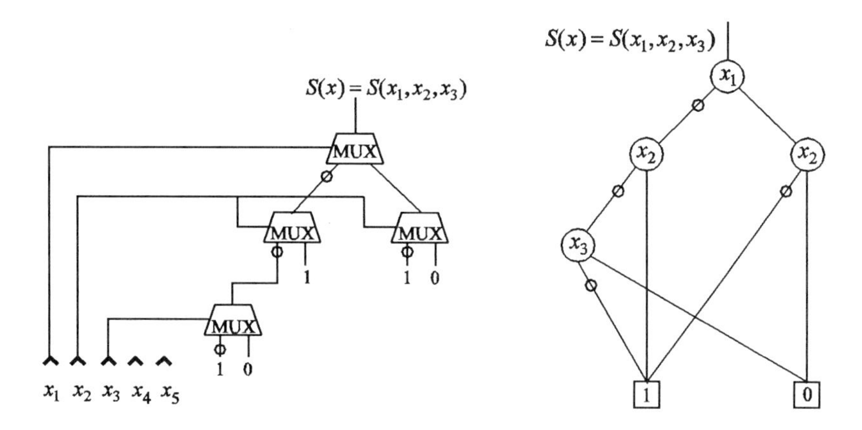

Figure 6.1: A MUX circuit and the corresponding BDD.

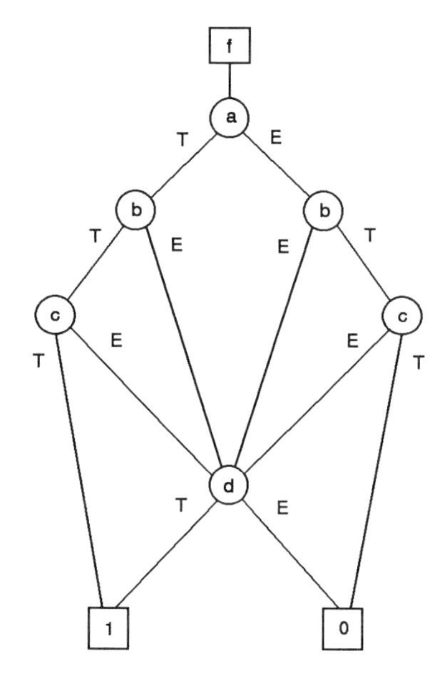

Figure 6.2: A binary decision diagram.

Second, let us consider the following function, given in SOP form:

$$f = abc + b'd + c'd.$$

A BDD for this function is given in Figure 6.2. If we want to know the value of f for a particular assignment to the variables a, b, c, and d, we just follow the corresponding path from the square box labeled f (this node is the *root* of the BDD[4]). Suppose we want to know

$$f(1, 0, 1, 0).$$

The first variable encountered from the root is a, whose value is 1. We then follow the edge labeled T (which stands for *then*). We then come across a node labeled b.

[4]In the following we will sometimes omit the root from the figures, when that does not generate confusion.

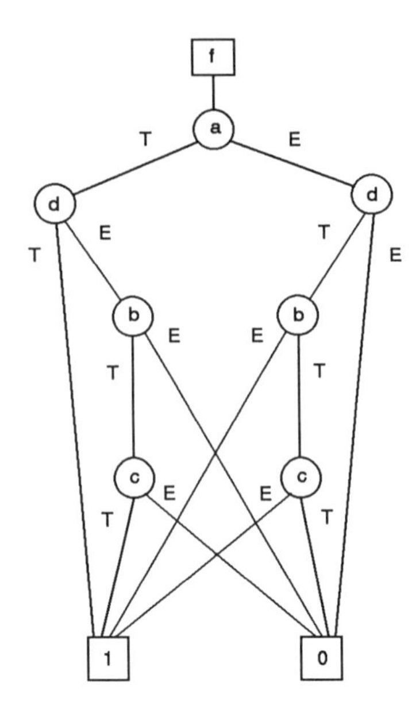

Figure 6.3: Another BDD.

Since the value of b is 0, we follow the edge labeled E (*else*). The next node is labeled d, which implies that for $a = 1$ and $b = 0$, the value of f does not depend on c. Following the E edge we finally reach the *leaf* labeled 0. This tells us the value of the function is 0, as can be easily verified from the SOP formula.

The BDD of Figure 6.2 is an *ordered* binary decision diagram, because the variables appear in the same order along all paths from the root to the leaves. The ordering in this case is

$$a \leq b \leq c \leq d.$$

The appearance and the size of the BDD depend on the variable ordering. This is illustrated in Figure 6.3, where a different BDD for f is given according to the following variable ordering:

$$a \leq d \leq b \leq c.$$

Finally, for the ordering

$$b \leq c \leq a \leq d,$$

we obtain the BDD of Figure 6.4. This is an optimal ordering, since there is exactly one node for each variable. Whenever not otherwise specified, we shall assume that our BDDs are ordered. Figure 6.5 gives the BDDs for some elementary functions. Notice the similarity of the BDDs for $f = a$ and $f = a'$. One is obtained from the other by swapping the two terminal nodes.

6.1.2 Formal Definition of BDDs

We now give a formal definition of a binary decision diagram. We shall then outline the algorithms for BDD manipulation and finally, based on the requirements of those algorithms, we shall devise the data structures and the details of the algorithms.

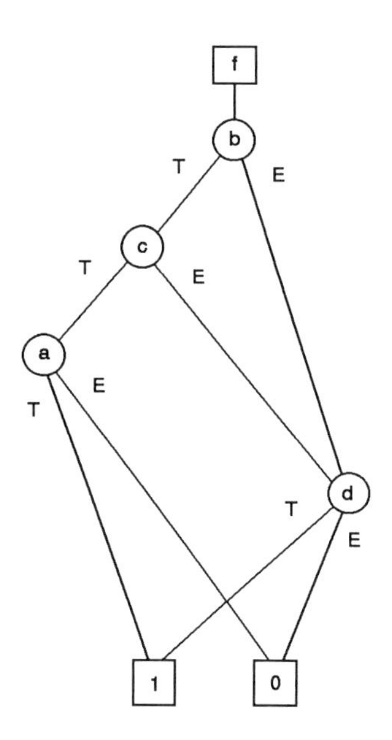

Figure 6.4: An optimal BDD.

Definition 6.1.1 *A BDD is a directed acyclic graph* $(V \cup \Phi \cup \{\mathbf{1}\}, E)$ *representing a multiple-output switching function* F. *The nodes are partitioned into three subsets.* V *is the set of the internal nodes. The outdegree of* $v \in V$ *is 2. Every node* v *has a label* $l(v) \in S_F$. *Here* $S_F = \{x_1, \ldots, x_n\}$ *denotes the* **support** *of* F, *i.e., the set of variables on which* F *actually depends. Thus,* $l(v)$ *is one of the variables variable* $\{x_i\}$. $\mathbf{1}$ *is the terminal node: Its outdegree is 0.* Φ *is the set of the function nodes: The outdegree of* $\phi \in \Phi$ *is 1 and its indegree is 0. The function nodes are in one-to-one correspondence with the components of* F. *The outgoing edges of function nodes may have the* complement *attribute. The two outgoing edges for a node* $v \in V$ *are labeled* T *and* E, *respectively. The* E *edge may have the* complement *attribute. We use* $(l(v), T, E)$ *to indicate an internal node and its two outgoing edges. The variables in* S_F *are ordered and if* v_j *is a descendant of* v_i $(v_i, v_j \in V)$, *then* $l(v_i) < l(v_j)$. *The function* F *represented by a BDD is defined as follows:*

1. *The function of the terminal node is the constant function* $\mathbf{1}$.

2. *The function of an edge is the function of the head node, unless the edge has the complement attribute, in which case the function of the edge is the complement of the function of the node.*

3. *The function of a node* $v \in V$ *is given by* $l(v)f_T + l(v)'f_E$, *where* f_T (f_E) *is the function of the* T (E) *edge.*

4. *The function of* $\phi \in \Phi$ *is the function of its outgoing edge.*

■

An edge with (without) the attribute is called a **complement (regular) edge**.

BDDs are canonical (the representation of F is unique for a given variable ordering) if:

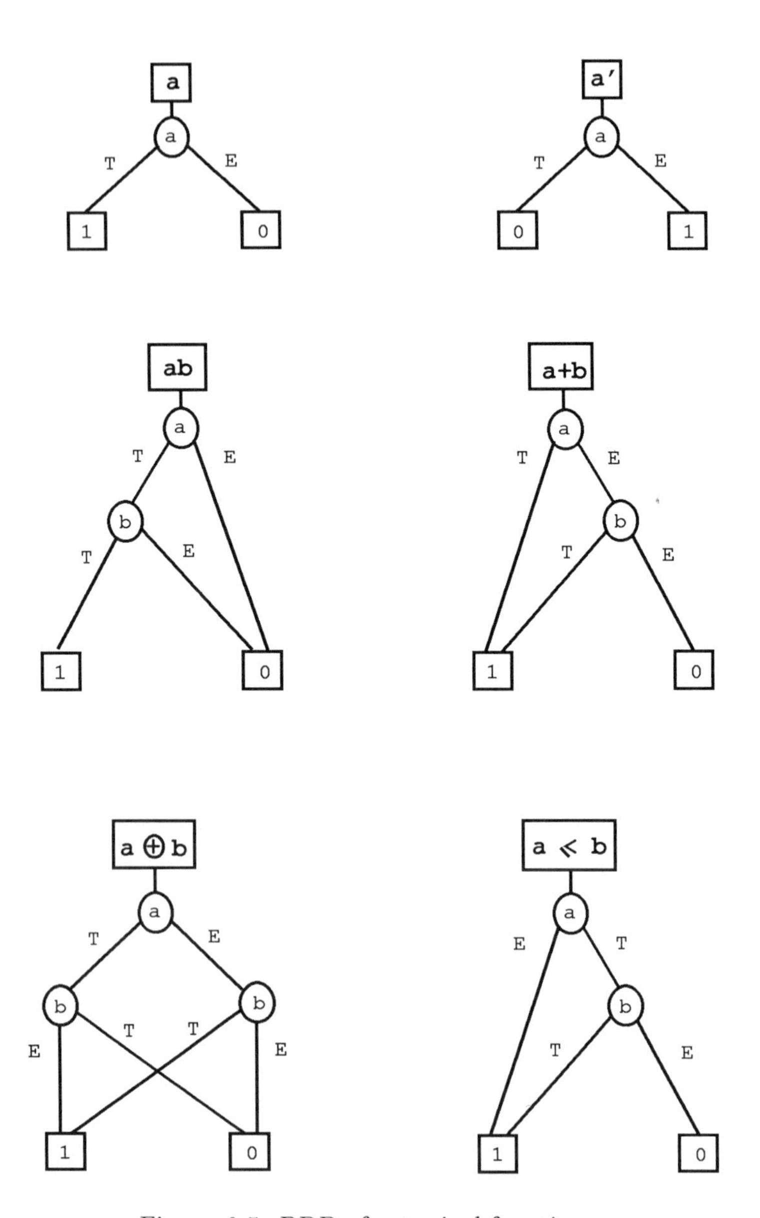

Figure 6.5: BDDs for typical functions.

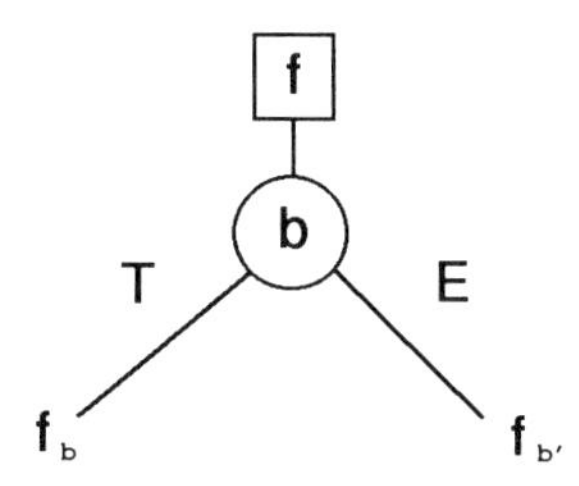

Figure 6.6: Partial BDD after expansion with respect to b.

1. All internal nodes are descendants of some node in Φ.
2. There are no isomorphic subgraphs.
3. For every node, $f_T \neq f_E$.

In the following, we only consider BDDs that conform to these requirements. Note that the restriction that the T edge may not be complemented is imposed to guarantee canonicity.

The distinction between the function of a node and the function of an edge allows us to deal with attributed edges in a natural way.

6.1.3 How to Build the BDD for f

BDDS can be built from recursive use of Boole's expansion theorem (See Section 3.3.3). We shall see that the expansion theorem plays a central role in the definition and manipulation of BDDs.

As an example, consider how the BDD is built for

$$f = abc + b'd + c'd$$

under the variable ordering:

$$b \leq c \leq d \leq a.$$

We start by computing the cofactors of f with respect to b, the first variable in the ordering. We get:

$$f_b = ac + c'd$$

and

$$f_{b'} = d + c'd.$$

We can summarize this initial result by a partial diagram as the one of Figure 6.6. It is true in general that the two children of a node represent the two cofactors of the function represented by the node with respect to the variable labeling the node. We then compute the cofactors of f_b and $f_{b'}$ with respect to c. This yields:

$$(f_b)_c = f_{bc} = a \qquad f_{b'c} = d$$

$$f_{bc'} = d \qquad f_{b'c'} = d$$

We observe that three of these four cofactors are identical. Hence we create a single node for them in the new partial BDD, shown in Figure 6.7. Recognizing that some

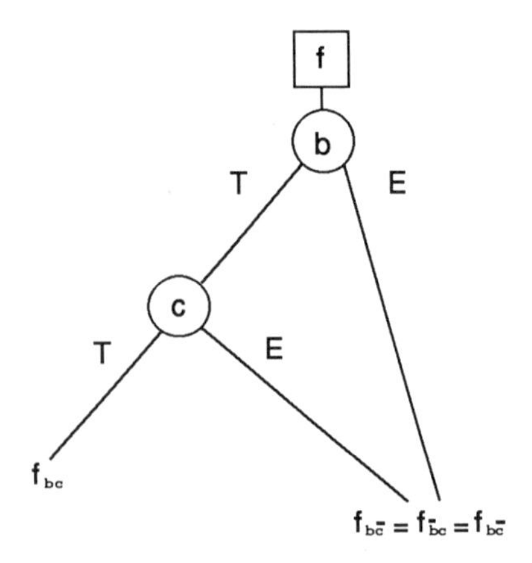

Figure 6.7: Partial BDD after expansion with respect to b and c.

cofactors are identical guarantees that the BDD will be *reduced* (intuitively, it does not contain duplicated and superfluous nodes). This is an important property, as we shall see.

Finally, noting that

$$(x_i)_{x_i} = 1 \quad \text{and} \quad (x_i)_{x_i'} = 0$$

and

$$(x_i)_{x_j} = (x_i)_{x_j'} = x_i \quad \text{for} \quad i \neq j$$

we get the BDD of Figure 6.8. Notice the similarity to the BDD of Figure 6.4. This example illustrates that the optimal variable order is not unique in general.

6.1.4 Reduced BDDs

Notice that if we had not identified the identical cofactors, we would have obtained the tree of Figure 6.9. This BDD, unlike the previous ones, is not *reduced.* There are *isomorphic* subgraphs. A non-reduced BDD can be systematically transformed into a reduced one. Consider the two subgraphs highlighted in Figure 6.10. They represent the same function and therefore they can be merged, as shown in Figure 6.11. We now notice that the node pointed by the arrow is redundant (it correspond to no decision), hence it can be removed. This is shown in Figure 6.12. By iteratively applying:

- Identification of isomorphic subgraphs;
- Removal of redundant nodes;

we obtain the initial reduced graph.

Given an ordering, the reduced graph for a function is unique. Hence, the Reduced Ordered BDD (ROBDD) is a canonical form. This is the first important property of binary decision diagrams, that is extremely useful for verification. Two functions are equivalent if and only if they have the same BDD.

Other interesting properties of BDDs are:

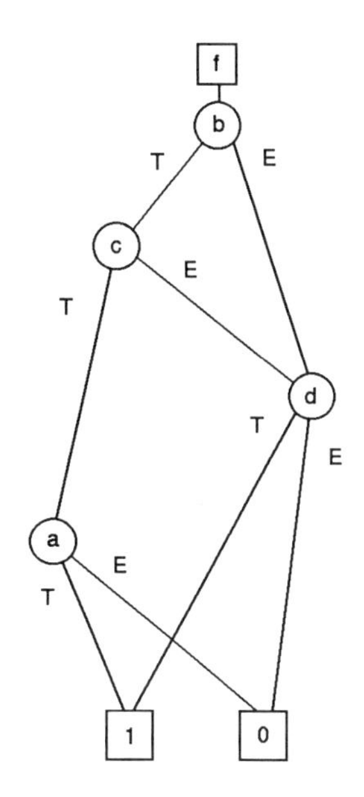

Figure 6.8: Final BDD.

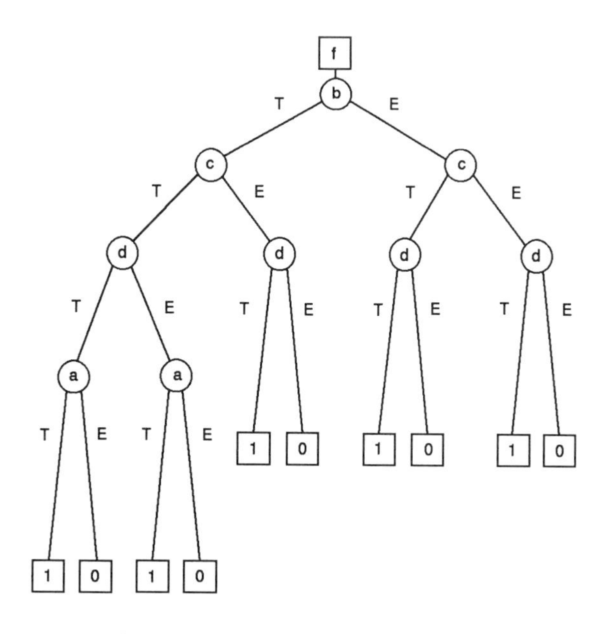

Figure 6.9: Non-reduced BDD.

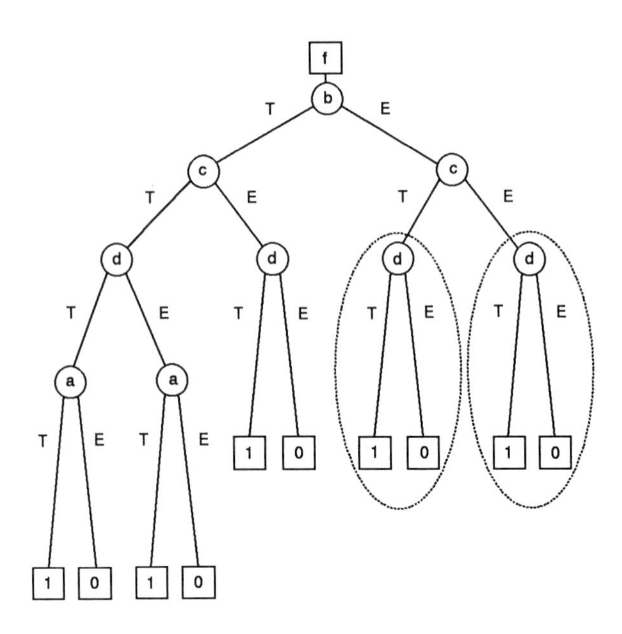

Figure 6.10: Two isomorphic subgraphs.

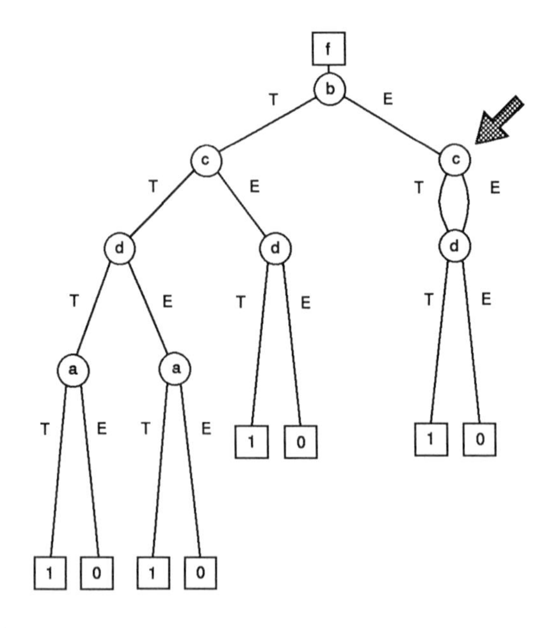

Figure 6.11: Merging two isomorphic subgraphs.

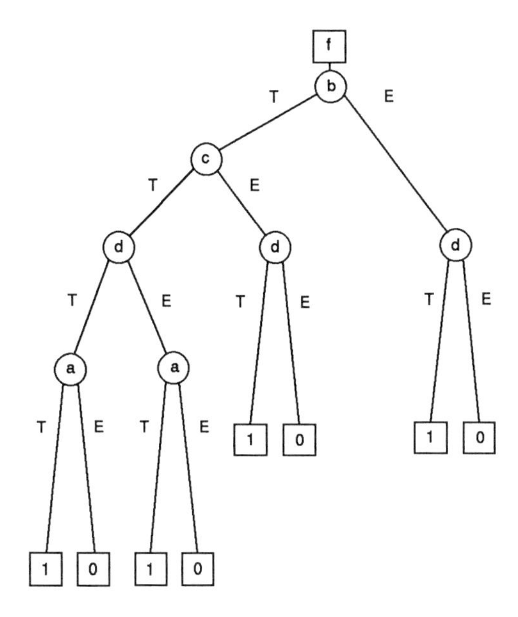

Figure 6.12: Elimination of a redundant node.

- The size of the BDD (the number of nodes) is exponential in the number of variables in the worst case (e.g., multipliers); however, BDDs are well-behaved for many functions that are not amenable to two-level representations (e.g., EXCLUSIVE-OR).

- The logical AND and OR of BDDs have the same complexity (polynomial in the size of the operands). Complementation is inexpensive.

- Both satisfiability and tautology can be solved in constant time. Indeed, f is a tautology if and only if its BDD consists of the terminal node 1.

- Covering problems can be solved in time linear in the size of the BDD representing the constraints.

On the other side:

- BDD sizes depend on the ordering. Finding a good ordering is not always simple.

- There are functions for which the SOP or POS representations are more compact than the BDDs. Unfortunately, many constraint functions of covering problems fall into this category.

- In some cases SOP/POS forms are closer to the final implementation of a circuit. For instance, if we want to implement a PLA, we need to generate at some point a SOP or POS form.

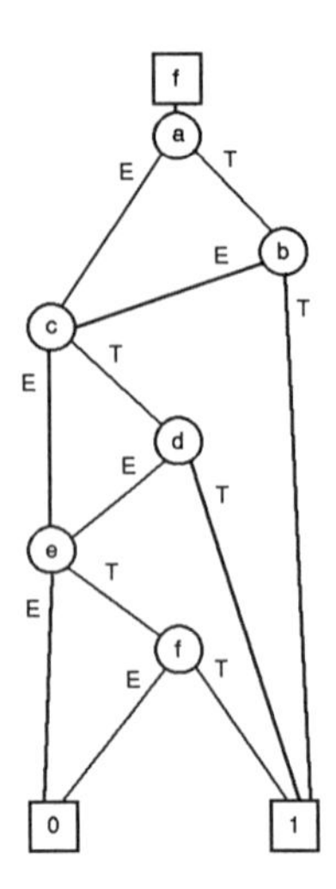

Figure 6.13: BDD illustrating the advantages of a good ordering.

6.1.5 Why Ordering is Important

Before looking in more detail at the manipulation of BDDs, let us try to better understand why ordering is important. We consider

$$f = ab + cd + ef$$

with ordering

$$a \leq b \leq c \leq d \leq e \leq f.$$

The BDD is shown in Figure 6.13. Let us now considering the ordering

$$a \leq c \leq e \leq b \leq d \leq f.$$

The resulting BDD, shown in Figure 6.14, is considerably more complex. (In that BDD, the E edges go to the left.) The reason for the big difference is the following. In the decision making process that eventually gives us the value of the function for a given assignment, we follow two opposite strategies depending on the ordering.

With the first ordering we consider one product term at the time. After the first two variables have been examined, we know whether the first product term (ab) is 0 or 1. If it is 1, we are done. If it is not, we just have to remember that it evaluated to 0. We don't need to know which variable (a or b) caused it to be 0. After the first four variables have been considered, we just need to remember whether either product term evaluated to 1, or whether both were 0. The specifics are not important to determine the value of the function. Since we have very little to remember, we get by with very few nodes.

With the second ordering, we process all product terms "in parallel." If the first three variables are all 1's, we cannot tell the value of any product term. In addition, the value of a along any given path must be remembered until b is met—three levels below. Similarly for c and e. This fact prevents the recombination of different paths.

We can imagine a bit-serial processor that examines the values of the variables one at the time. The size of the BDD is related to the amount of information that

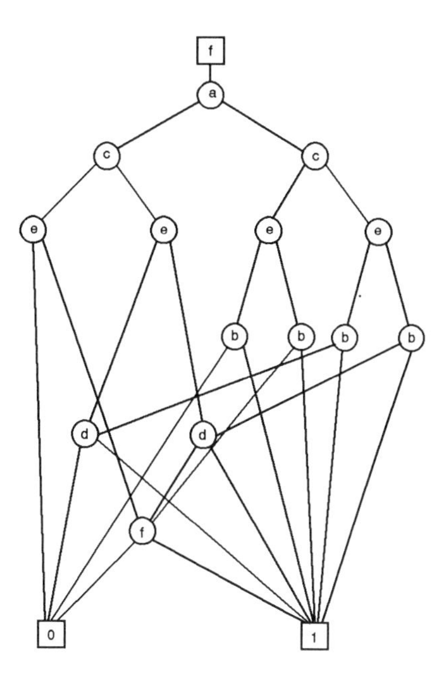

Figure 6.14: BDD illustrating the drawbacks of a bad ordering.

must be stored to compute the result. We shall return to these considerations when we consider algorithms for finding optimal—or simply good—orderings.

6.2 Design Considerations for a BDD Package

We have seen how switching functions can be represented as ordered reduced BDDs. We now consider the efficient implementation of BDDs, in terms of both memory and CPU. Before we proceed in detailing data structures and algorithms, and before we give a formal definition, we need to add a few design considerations. We shall closely follow [29].

Shared BDDs. We have seen that each node of a BDD has a function associated with it. If we have several functions, chances are that they will have subexpressions in common. For instance, if we have $f_1 = b + c$ and $f_2 = a + b + c$, we would like to represent them like in Figure 6.15. As a special case, two equivalent functions could be represented by the same BDD (not just two identical BDDs). This amounts to dealing with a single multi-rooted directed acyclic graph (DAG) with a root for each function we are explicitly interested in. All functions share the same DAG.

Unique Table. We are ultimately interested in reduced BDDs. Rather than generating non-reduced BDDs and than reducing them, we are interested in guaranteeing that at any time there are no isomorphic subgraphs and no redundant nodes in the multi-rooted DAG. This can be achieved by checking for the existence of a node representing the function we want to add, prior to the creation of a new node. A straightforward approach would consist of searching the whole DAG every time we want to insert a new node. However, that would be far too inefficient. Instead, we

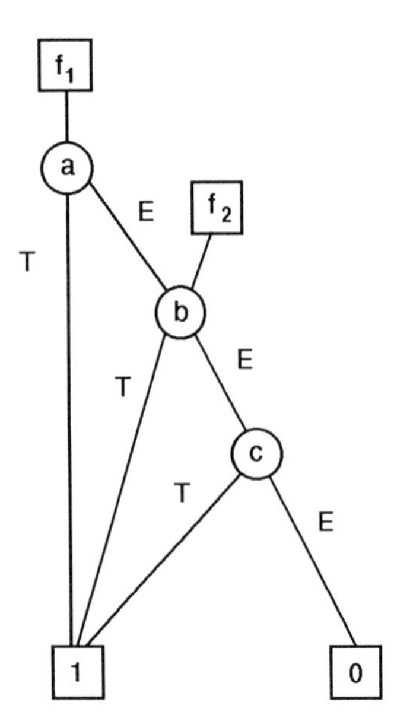

Figure 6.15: Shared BDD.

shall keep a dictionary of the functions represented in the DAG. This dictionary is called *unique table* and is best implemented as a hash table.

Strong Canonicity. Because of the unique table, two equivalent functions end up sharing exactly the same subgraph. Hence checking for equivalence just requires checking that the pointers in the DAG associated with the two functions are identical. This property is called *strong canonicity* and makes constant-time equivalence check possible—a very desirable consequence.

Attributed Edges. We have seen that the BDDs for f and f' are very similar. The only difference being the values of the leaves that are interchanged. This suggests the possibility of actually using the same subgraph to represent both f and f'. Suppose the BDD actually represents f. If we are interested in $g = f'$ it is then sufficient to remember that the function we have in the multi-rooted graph is the complement of g. This can be accomplished by attaching an attribute to the edge pointing to the top node of f. An edge with the complement attribute is called a *complement edge*. The edges without the attributes are called *regular edges*. The use of complement edges slightly complicates the manipulation of the BDDs, but has two advantages. Obviously, it decreases the memory requirements.[5] However, the most important consequence of using complement edges is the fact that complementation can be done in constant time—the BDD is already in place—and checking two functions for one being the complement of the other also takes constant time. Note that with complement edges we need only one constant function (we choose **1**) and hence only one leaf in the multi-rooted DAG. The attribute mechanism is quite general: Other attributes have been used for other purposes [201, 152].

[5]We are assuming that the overhead for storing the attributes is negligible. This will be justified in the sequel.

Computed Table. As a speed improvement device, we shall keep a table of recently computed functions. The purpose of this table is different from that of the unique table. With the unique table we answer questions like: "Does there exist a node labeled v with children g and h?" On the other hand, the computed table answers questions like: "Did we recently compute the AND of f_1 and f_2?" We can ask this question before we actually know that the AND of f_1 and f_2 is a function whose top node is labeled v and whose children are g and h. Hence we can avoid recomputing the result.

Memory Management and Dynamic Re-Ordering. In a typical application we build and then dispose of many BDDs. An efficient memory management is important. We shall adopt a strategy based on **garbage collection**, i.e., we shall not immediately free nodes that are no longer used. Instead, from time to time we shall visit our data structure to recover all the unused memory.

Also, an *a priori* ordering of the shared BDDs may continue to grow as the BDDs are manipulated in a particular application. If this growth is allowed to go unchecked, it may (and often does, for large problems) occur that we run out of memory. In many cases this can be alleviated by **dynamically re-ordering** the BDD variables. This can be quite dramatic for some circuits.

Both of these mechanisms are essential to the operation of a robust BDD package.

6.3 Algorithms

We now outline the algorithms for BDD manipulation. Then, based on the requirements of those algorithms, we shall devise appropriate data structures and give the details of the algorithms.

The usual way of generating new BDDs is to combine existing BDDs with connectives like AND, OR, EX-OR. As a starting point one generates the simple BDDs for the functions $f_i = x_i$, for all the variables in the functions of interest[6]. We are therefore interested in an algorithm that, given BDDs for f and g, will build the BDD for $f\langle op\rangle g$, where $\langle op\rangle$ is a binary connective (a switching function of two arguments). The basic idea comes—not surprisingly—from the expansion theorem, since:

$$f\langle op\rangle g = v(f_v\langle op\rangle g_v) + v'(f_{v'}\langle op\rangle g_{v'}).$$

So, if v is the top variable of f and g, we can first cofactor the two functions with respect to v and solve two simpler problems recursively, and then create a node labeled v that points to the results of the two subproblems (if such a node does not exist yet; otherwise we just return the existing node).

Finding the cofactors of f and g with respect to v is easy: If f does not depend on v, $f_v = f_{v'} = f$, that is, the cofactors are the function itself. If, on the other hand, v is the top variable of f, the two cofactors are the two children of the top node of f. Similarly for g.

[6]The function $f_i = x_i$ is called a *projection* function, and was defined in Section 3.3.2.

Table	Name	Expression	Equivalent Form
0000	0	0	0
0001	AND(F,G)	$F \cdot G$	ITE$(F,G,0)$
0010	$F > G$	$F \cdot G'$	ITE$(F,G',0)$
0011	F	F	F
0100	$F < G$	$F' \cdot G$	ITE$(F,0,G)$
0101	G	G	G
0110	XOR(F,G)	$F \bigoplus G$	ITE(F,G',G)
0111	OR(F,G)	$F+G$	ITE$(F,1,G)$
1000	NOR(F,G)	$(F+G)'$	ITE$(F,0,G')$
1001	XNOR(F,G)	$(F \bigoplus G)'$	ITE(F,G,G')
1010	NOT(G)	G'	ITE$(G,0,1)$
1011	$F \geq G$	$F+G'$	ITE$(F,1,G')$
1100	NOT(F)	F'	ITE$(F,0,1)$
1101	$F \leq G$	$F'+G$	ITE$(F,G,1)$
1110	NAND(F,G)	$(F \cdot G)'$	ITE$(F,G',1)$
1111	1	1	1

Figure 6.16: Two-argument operators expressed in terms of ITE.

The approach based on the expansion theorem[7] can be further improved by considering the *if-then-else* operator ITE, which is a ternary operator defined as follows:

$$\text{ITE}(F,G,H) = F \cdot G + F' \cdot H,$$

where F, G, H are three arbitrary switching functions. One interesting property of the ITE operator is that all two-argument operators can be expressed in terms of it, as shown in Figure 6.16. Therefore, most of the standard manipulation of BDDs can be done with ITE.

6.3.1 The ITE Algorithm

The algorithm for ITE is clearly a recursive one. It is based on the following formulation. Let v be the top variable of F, G, H (the variable with the lowest index). Then,

$$\begin{aligned}
\text{ITE}(F,G,H) &= F \cdot G + F' \cdot H \\
&= v \cdot (F \cdot G + F' \cdot H)_v + v' \cdot (F \cdot G + F' \cdot H)_{v'} \\
&= v \cdot (F_v \cdot G_v + F_v' \cdot H_v) + v' \cdot (F_{v'} \cdot G_{v'} + F_{v'}' \cdot H_{v'}) \\
&= (v, ite(F_v, G_v, H_v), \text{ITE}(F_{v'}, G_{v'}, H_{v'})).
\end{aligned}$$

The terminal cases of the recursion are:

$$\text{ITE}(1,F,G) = \text{ITE}(0,G,F) = \text{ITE}(F,1,0) = \text{ITE}(G,F,F) = F.$$

[7] The algorithm that takes f, g, and $\langle op \rangle$ as arguments and returns $f \langle op \rangle g$ is called APPLY in the literature.

Whenever we encounter one of these cases, we just return the pointer to F. Otherwise we find the top variable of the three functions and we apply the recursive formula. This is the basic idea of the algorithm. However, before we examine the pseudo-code, we need to consider some of the design issues of Section 6.2. In particular, we need to consider the unique table and the computed table.

As we said, we use a hash table for the unique table, i.e., a data structure that stores an item in a location of a table identified by its **key**. The key in our case is the triple (v, G, H) that identifies a node in the DAG. In the key, v is an integer—the index of the variable—while G and H are pointers. The **hashing function** maps the key into the location in the table. The mapping is not one-to-one, so that several keys may correspond to the same location in the table, thereby producing collisions. All colliding entries are kept in a linked list called the **collision chain**. A typical hashing function may shift v, G, and H by different numbers of bits, add the results, and finally compute the remainder of the sum on division by the size of the table.[8]

If we want to see if a given key is present in the hash table, we compute the hashing function with the key as argument. This identifies one entry in the table, i.e., a collision chain. We then examine all the elements in the collision chain until we find one that matches our entry or until all elements have been examined.[9] The advantage of a hash table is that the collision chains are short if the table is large enough and if the hashing function is good; therefore search is performed in expected constant time.

In the unique table, the pointer to the node with label v and children F and G is found in the collision list corresponding to the key (v, G, H). In the recursive algorithm, we first find G and H recursively, and then ask whether we should introduce a new node labeled v and pointing to G and H. The unique table tells us if such a node already exists. If not, a new entry is created in the table.

For the computed table we initially make the assumption that we also use a hash table. This corresponds to assuming infinite memory, which is a convenient assumption for our purpose of analyzing the qualitative operation and complexity of the algorithm[10]. The key for the computed table is the triple (F, G, H). At every level of the recursion, we check the computed table for the result. If it is already there, we just return. If not we recur and, before returning, we insert the newly computed result in the table.

We are now ready to look at the pseudo-code of the ITE algorithm, reported in Figure 6.17. For the time being, we ignore the complement edges.

Notice how the algorithm maintains the BDD reduced by checking if T equals E and by consulting the unique table. An example of application of the algorithm is in Figure 6.18. The result consists of two newly created nodes and two already existing nodes. (Those labeled C and D.) This is a fairly typical situation.

In terms of complexity, let us first examine what would happen without the computed table. Assuming the accesses to the unique table take constant time, then all operations performed by the algorithm require constant time, except for the recursive

[8]The size of the table should be a prime number for best results. Another approach uses v to select a sub-table and then G and H to select a position in the sub-table.

[9]This particular scheme is called *open hashing*. Hash tables are treated in [3, 69].

[10]This is not realistic in practice. However, management of the computed table is intricate, and beyond the scope of this text.

```
ITE(F, G, H) {
        (result, terminal_case) = TERMINAL_CASE(F, G, H)
        if (terminal_case) {
                return (result)
        }       (result, in_computed_table) = COMPUTED_TABLE_HAS_ENTRY(F, G, H)
        if (in_computed_table) {
                return (result)
        }
        v = TOP_VARIABLE(F, G, H)
        T = ITE(F_v, G_v, H_v)
        E = ITE(F_v', G_v', H_v')
        if (T = E) return (T)
        R = FIND_OR_ADD_UNIQUE_TABLE(v, T, E)
        INSERT_COMPUTED_TABLE((F, G, H), R)
        RETURN (R)
        }
}
```

Figure 6.17: Pseudo-code of the ITE algorithm.

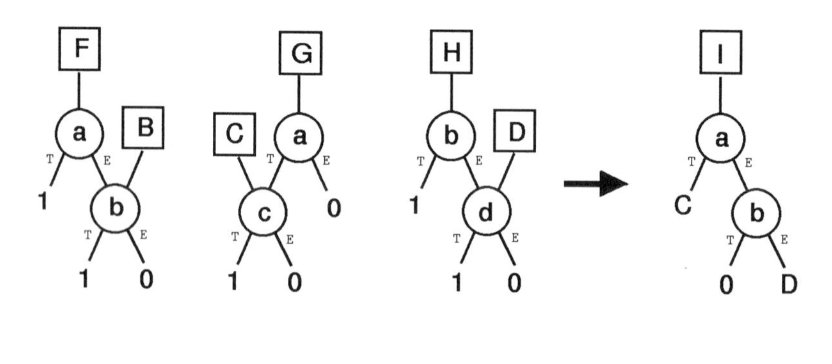

$$
\begin{aligned}
I &= \text{ITE}(F, G, H) \\
&= (a, \text{ITE}(F_a, G_a, H_a), \text{ITE}(F_{a'}, G_{a'}, H_{a'})) \\
&= (a, \text{ITE}(1, C, H), \text{ITE}(B, 0, H)) \\
&= (a, C, (b, \text{ITE}(B_b, 0_b, H_b), \text{ITE}(B_{b'}, 0_{b'}, H_{b'}))) \\
&= (a, C, (b, \text{ITE}(1, 0, 1), \text{ITE}(0, 0, D))) \\
&= (a, C, (b, 0, D))
\end{aligned}
$$

Figure 6.18: Example of application of ITE.

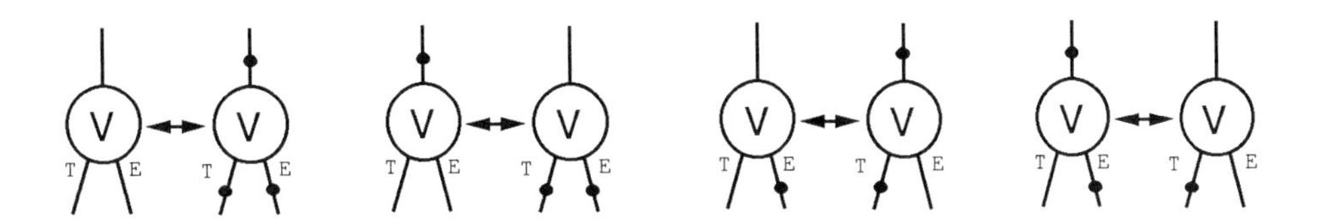

Figure 6.19: Equivalent pairs of functions.

calls. However, every call to the procedure generates two other calls unless we are in a terminal case, so that the total number of calls—and hence the execution time—is exponential in the number of variables. However, if we consider the computed table, things change dramatically. Let $|F|$ be the number of nodes in the BDD for F. The effect of the computed table is to cause ITE to be called at most once for each distinct combination of nodes in F, G, and H. Hence, ITE can be called $|F| \cdot |G| \cdot |H|$ times at most and the complexity of the procedure is $O(|F| \cdot |G| \cdot |H|)$. Note that when using ITE to compute the AND of two functions, one of the arguments to ITE is $\mathbf{0}$; in such a case the complexity is quadratic in the size of the operands. In practice performance is normally better than quadratic and typically closer to the size of the resulting BDD. In the worst case, $|F \cdot G|$ is comparable to $|F| \cdot |G|$. However, in many cases, $|F \cdot G| \ll |F| \cdot |G|$. In these cases the time taken by ITE is typically proportional to $|F \cdot G|$.

If we remove the infinite memory assumption, as we do in practice, the worst-case complexity returns to be exponential. However, the typical performance is not affected significantly by the loss of infinite memory, if the computed table is implemented carefully.

6.3.2 Complement Edges

We mentioned that edges may carry attributes that specify how to derive the function of the edge from the function of the node it points to. The most common attribute is the complement. It allows the BDDs for f and f' to be shared, thus saving space and making complementation a constant-time operation. We use a dot to indicate a complement edge.

To maintain canonicity, we must constrain where complement edges can be used. The four pairs of functions of Figure 6.19 are equivalent. Each pair consists of two functions that can be obtained from each other by application of the identity

$$f' = v \cdot f'_v + v' \cdot f'_{v'},$$

whence

$$f = (v \cdot f'_v + v' \cdot f'_{v'})'.$$

Also, the functions represented in Figure 6.19 represent all the eight possible ways of placing dots on the incoming and outgoing edges of a node. To guarantee canonicity, we impose that the *then* edge of every node be regular. Notice that this is true of exactly one function for each of the four pairs of Figure 6.19. It will be the task of the algorithms, ITE in particular, to insure that every non-canonical case be transformed into the corresponding canonical case. The overhead is indeed minimal. Since the dots may only appear on *else* edges, in the sequel we shall adopt the convention of marking an *else* edge with an empty circle if it is regular and with a dot if it is complemented. The *then* edges will be unmarked. The outgoing edges of function nodes may have the complement attribute: In that case they will have a dot. We shall also convene that dangling edges point to the constant node, unless otherwise stated.

The use of complement edges makes testing for another special case of ITE possible. Indeed,

$$\text{ITE}(F, 0, 1) = F'.$$

Though this is true in general, without complement edges we could not terminate the recursion because the BDD for F' would not necessarily exist and could not be easily found if it existed. Another advantage of complement edges, to be discussed in Section 6.3.3, is the ability to use a single entry of the computed table for both F and F'.

We end this section with one observation that is useful when drawing BDDs by hand. Since the *then* edges are never complemented, the value of a function for 1,1,...,1 will be 0 if and only if the outgoing edge of the function node is complemented. Computing the value of the function for which a BDD must be drawn when all variables equal 1 allows one to determine how to start (with a regular or complement edge) and normally prevents one from pushing too many complement dots around.

6.3.3 The Computed Table

The computed table stores results of recent computations. Under the assumption of infinite memory—all previously computed results are stored—we have seen that the computed table makes the complexity of ITE polynomial in the size of the operand BDDs. In general memory is at a premium; it is important to implement the computed table efficiently. However, discussion of this crucial issue is beyond the scope of this text.

6.3.4 Conditioning of the ITE Calls

For every triple (F, G, H) there are other triples (F_i, G_i, H_i) such that $\text{ITE}(F, G, H) = \text{ITE}(F_i, G_i, H_i)$ though at least one of the three arguments is different. For instance, if

$$F = x_1' + x_2 x_4 \quad G = (x_1 + x_2) x_4 \quad H = 0,$$

and

$$F_1 = x_2 x_4 (x_1 + x_3') \quad G_1 = 1 \quad H_1 = x_1' x_2 x_3 x_4,$$

then

$$\text{ITE}(F, G, H) = \text{ITE}(F_1, G_1, H_1) = x_2 x_4.$$

If we could identify all triples that give the same result, we could map all of them into the same entry of the computed table. In general, finding all such triples is too expensive, but there are special cases that can be identified with very little effort. For instance,

$$\text{ITE}(F, G, F) = \text{ITE}(F, G, 0) = \text{ITE}(G, F, G) = \text{ITE}(G, F, 0) = F \cdot G.$$

For these special cases, we want to transform the triple into a standard form before looking up the computed table. In this way, all the triples that map into the same standard form share the same entry. This saves both memory and time.

A mapping of a triple into another can be based on the identification of the following occurrences:

- Two arguments are the same function (e.g., $\text{ITE}(F, F, G)$);
- Two arguments are one the complement of the other (e.g., $\text{ITE}(F, G, F')$);
- One or more arguments are constants (e.g., $\text{ITE}(F, G, 0)$).

Note that these checks can be performed in constant time. A first set of transformations replaces as many arguments as possible with constants:

$$\begin{aligned} \text{ITE}(F, F, G) &\Rightarrow \text{ITE}(F, 1, G) \\ \text{ITE}(F, G, F) &\Rightarrow \text{ITE}(F, G, 0) \\ \text{ITE}(F, G, F') &\Rightarrow \text{ITE}(F, G, 1) \\ \text{ITE}(F, F', G) &\Rightarrow \text{ITE}(F, 0, G). \end{aligned}$$

A second set of transformations permutes the arguments based on the following equalities.

$$\begin{aligned} \text{ITE}(F, 1, G) &= \text{ITE}(G, 1, F) \\ \text{ITE}(F, G, 0) &= \text{ITE}(G, F, 0) \\ \text{ITE}(F, G, 1) &= \text{ITE}(G', F', 1) \\ \text{ITE}(F, 0, G) &= \text{ITE}(G', 0, F') \\ \text{ITE}(F, G, G') &= \text{ITE}(G, F, F'). \end{aligned}$$

If one of these cases occurs, we choose the triple with the first argument having the smallest index for its top variable. For instance, if the top variable of F is x_3 and the top variable of G is x_5, we prefer $(F, G, 1)$ to $(G', F', 1)$. In case of a tie, we compare the addresses of the top nodes of the first arguments—they are guaranteed to be different—and choose the triple with the lower address.

A third and final set of transformations is based the following equalities.

$$\text{ITE}(F, G, H) = \text{ITE}(F', H, G) = \overline{\text{ITE}(F, G', H')} = \overline{\text{ITE}(F', H', G')}.$$

As in the other cases, we want to choose only one of the equivalent forms and map the others onto it. We notice that there is only one form such that the first two arguments should be pointed via regular edges. For instance, if F is regular and G is complement, ITE will replace (F, G, H) by (F, G', H') and take the complement of the result before returning. If, on the other hand, F is complement and H is regular, ITE will replace (F, G, H) by (F', H, G), and will not take the complement of the result before returning.

Among the effects of these transformations is the ability to detect that $F \cdot G = (F' + G')'$, i.e., the ability to apply De Morgan's laws. This demonstrates the power and flexibility of the approach based on the ITE operator.

```
ITE_CONSTANT(F, G, H) {
    (result, terminal_case) = TERMINAL_CASE(F, G, H)
    if (terminal_case) {
        return (result)                    result is 0, 1, or non_constant
    }
    (result, in_computed_table) = COMPUTED_TABLE_HAS_ENTRY(F, G, H)
    if (in_computed_table) {
        return (result)
    }
    v = TOP_VARIABLE(F, G, H)
    T = ITE_CONSTANT(F_v, G_v, H_v)
    if (T ≠ 0 and T ≠ 1) return non_constant
    E = ITE_CONSTANT(F_v', G_v', H_v')
    if (T ≠ E) return non_constant
    INSERT_COMPUTED_TABLE({F, G, H}, T)
    return T
    }
}
```

Figure 6.20: Pseudo-code of the ITE_CONSTANT algorithm.

6.3.5 The ITE_CONSTANT Algorithm

It is often the case that one wants to check whether, for two functions F and G, $F \leq G$, or equivalently $F \Rightarrow G$, holds. This amounts to checking if $F' + G$ is identically one. We could use ITE to compute $F' + G$ and then test the result for being the tautology. However, there is a more efficient way of proceeding, which avoids building the intermediate result; it is based on a specialized version of ITE called ITE_CONSTANT, and is outlined in Figure 6.20. The check for implication $F \Rightarrow G$ can then be performed by computing ITE_CONSTANT$(F, G, 1)$. The procedure returns one of three values: 1, 0, and *non_constant*. It is based on the observation that for the resulting function to be constant, the two cofactors must be identical and constant. As soon as a violation of this condition is detected, ITE_CONSTANT returns *non_constant*. This early termination in unsuccessful cases, combined with not building intermediate results, makes this procedure much faster than the general approach. Notice that the speed-up occurs only for cases where $F \not\Rightarrow G$. Hence, the speed-up observed in practice depends on the problem at hand. A typical value could be 20.

As an example of ITE_CONSTANT, consider the following functions.

$$\begin{aligned} f &= ab + bcd' \\ g &= ab + ac'd \end{aligned}$$

We are interested in checking whether $f \Rightarrow g$. This translates into checking whether ITE_CONSTANT$(f, g, 1) = 1$. The BDD for these functions and the computation are given in Figure 6.21. Note that unlike in ITE, the fact that the two cofactors of the

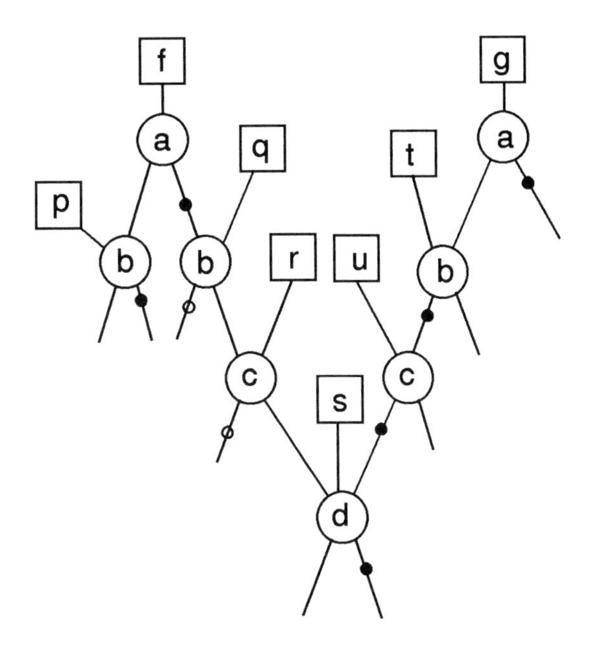

$$
\begin{aligned}
\text{ITE\_CONSTANT}(f, g, 1) &= (a, \text{ITE\_CONSTANT}(p, t, 1), \text{ITE\_CONSTANT}(q', 0, 1)) \\
&= (a, (b, \text{ITE\_CONSTANT}(1, 1, 1), \text{ITE\_CONSTANT}(0, u', 1)), \\
&\quad \text{ITE\_CONSTANT}(q', 0, 1)) \\
&= (a, (b, 1, 1), \text{ITE\_CONSTANT}(q', 0, 1)) \\
&= (a, 1, \text{ITE\_CONSTANT}(q', 0, 1)) \\
&= (a, 1, \mathit{non\_constant}) \\
&= \mathit{non\_constant}
\end{aligned}
$$

Figure 6.21: An example of computation by ITE_CONSTANT.

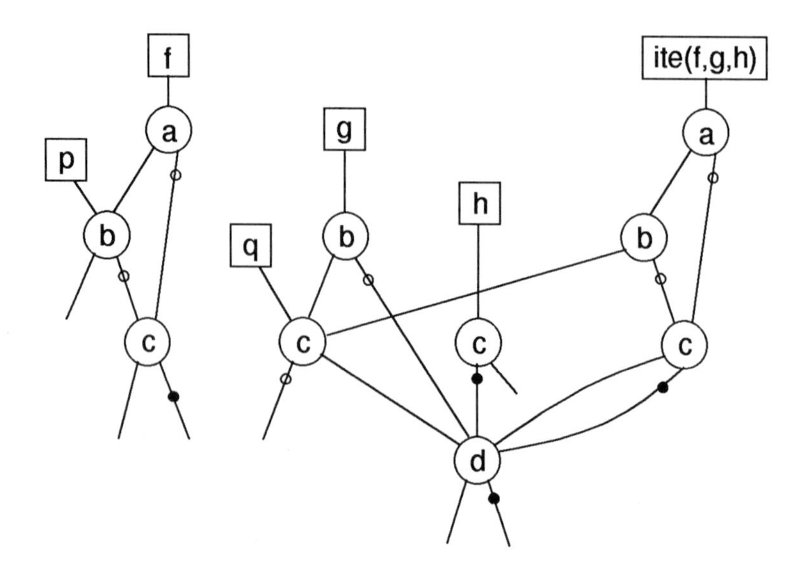

Figure 6.22: BDDs f, g, h and $\text{ITE}(f,g,h)$.

result are computed serially does matter.[11] This is taken into account in Figure 6.21 by not expanding $\text{ITE\_CONSTANT}(q',0,1)$ until the result of the other branch is known. The computation of $\text{ITE\_CONSTANT}(q',0,1)$ itself is a special case, since q' is not constant and therefore $q' \cdot 0 + q \cdot 1 = q$ is not constant either.

The correctness of the result can be verified by observing that for $a'bcd'$, $f = 1$ and $g = 0$. The procedure, however, does not even need to find such an example.

We conclude with a more meaningful example, which shows how to deal with complement edges in ITE and ITE_CONSTANT. Consider the functions

$$
\begin{aligned}
f &= ab + c, \\
g &= bc' + d, \\
h &= c + d',
\end{aligned}
$$

with the variable ordering

$$a \leq b \leq c \leq d.$$

Now suppose that we are given the shared BDD (with complement edges) for f, g, h, and need to compute the BDD for $\text{ITE}(f,g,h)$. The results, shown in Figure 6.22, are obtained as follows.

$$
\begin{aligned}
\text{ITE}(f,g,h) &= (a, \text{ITE}(p,g,h), \text{ITE}(c,g,h)) \\
&= (a, (b, \text{ITE}(1,q,h), \text{ITE}(c,d,h)), (b, \text{ITE}(c,q,h), \text{ITE}(c,d,h))) \\
&= (a, (b, q, \text{ITE}(c,d,h)), (b, \text{ITE}(c,q,h), \text{ITE}(c,d,h))) \\
&= (a, (b, q, (c,d,d')), (b, (c,d,d'), (c,d,d'))) \\
&= (a, (b, q, (c,d,d')), (c,d,d')).
\end{aligned}
$$

[11] Specifically, it matters for the number of recursive calls. It does not affect the final result.

Next, if we wanted to compute $\text{ITE\_CONSTANT}(f, g, h)$, we would proceed as follows.

$$\text{ITE\_CONSTANT}(a, \text{ITE\_CONSTANT}(p, g, h), \text{ITE\_CONSTANT}(c, g, h)) =$$
$$(a, (b, \text{ITE\_CONSTANT}(1, q, h), \text{ITE\_CONSTANT}(c, d, h)), \text{ITE\_CONSTANT}(c, g, h)) =$$
$$(a, (b, non\_constant, \text{ITE\_CONSTANT}(c, d, h)), \text{ITE\_CONSTANT}(c, g, h)).$$

and then all pending calls return *non_constant*.

6.4 Notes

There has been a surprisingly steady influx of "new" types of **DDs (Decision Diagrams)**. The work of Bryant in [47] focused on Boolean functions and decision diagrams that were based on the Boole-Shannon expansion of Section 3.3.3 and specific types of reduction to canonical form. However, some new forms of DDs focus on **Pseudo-Boolean Functions** have been introduced: [67] (called MTBDDs), [11, 12, 131, 132] (called ADDs), [165] (called EVBDDs), [46] (called BMDs).

These works have successfully extended the applicability of DDs into the realm of functions with Boolean domains but integer or real valued ranges. This has opened up new vistas for research on Markov Analysis and Probabilistic Verification, Shortest Path Analysis on graphs of vast size, "Exact" (that is free of false paths) timing analysis of combinational circuits, low power synthesis, technology mapping, etc.

Other work on extending the applicability of DDs has focused on the reduction to canonicity. Different reduction or ordering rules have been studied: [200] (Zero-Suppressed BDDs), [23] (Free BDDs), [91] (Functional DDs), [218] (Extended Decision Diagrams). Some or all of these variants can be mixed into a single "Hybrid Decision Diagram". Zhao and Clarke of CMU have developed this DD in a package for word level verification of sequential arithmetic circuits which has won internal awards at INTEL.

A serious issue with DDs is ordering. In every type of DD, the variables in the support of the function being represented must be ordered to obtain canonicity. Unfortunately, in many cases, the size of the BDD depends critically on the the specific ordering chosen. This dependence is so drastic that in many cases it is impractical to build the BDD at all. For example it is known that BDD size grows exponentially with variable count, independent of the order. This limits the use of BDDs for multipliers to 16 bits or less.

An exact algorithm for BDD variable ordering is the improved version by Ishiura *et al.* [150] of Friedman and Supowit's [104]. Various iterative techniques of are discussed in [106, 150, 237, 207, 206]. Heuristic techniques to find a good order for a circuit are discussed in [105, 183, 201, 262]. A comparison of the performance of various methods can be found in [154]. Partial BDDs have been studied in [230, 53]. Other references on variable ordering are: [269, 102, 48, 26, 90].

At the time of writing, several BDD packages of high quality were in use. Rudell's package was proprietary to Synopsys. David Long's package, developed at CMU, is employed in SMV (CMU verification package), VIS (UC Berkeley Verification Package) and SIS (UC Berkeley Synthesis Package) as well as AT&T and other industrial sites. A more recent entry into the BDD package field is the CUDD package from the University of Colorado at Boulder. This package is used in VIS and in

Motorola's VERDICT verification package as well as at several university sites. Of all of these, the CUDD package has the most efficient dynamic reordering package, and goes the farthest in supporting diverse BDD types (it currently supports BDDs, ADDs/MTBDDs, and ZDDs (zero-suppressed BDDs)).

The CUDD package is available on the web via anonymous FTP at

monk.colorado.edu

(login as anonymous, give your email address as password, and then type "cd pub").

6.5 Summary

In this chapter, we have introduced BDDs and summarized briefly the enormous impact they have had on the merging fields of synthesis and verification. We have shown how to build and manipulate them (see the solved problems for applying the usual Boolean operators to two functions represented by BDDs). We have shown how an ordered BDD is made canonical by imposing certain reduction rules.

We have studied the qualitative operation of the ITE and ITE_CONSTANT algorithms, and show how many of the BDD manipulations that might be needed in practice can be done efficiently, given efficient implementation of these algorithms.

We have avoided the intricacies of memory management, but they are all-important to the robustness of a BDD package[12].

We have given enough background so that the discussion of FSM traversal using binary decision diagrams of Section 7.9.1 can be appreciated. This subject is of pervasive importance in synthesis, testing, and verification [61, 63].

We have demonstrated by example why (and how) variable ordering is important. In Section 6.4 we discussed the state of the art in variable ordering techniques. We also discussed design considerations for BDD packages, and how to obtain efficient and recently developed BDD packages.

6.6 Problems

1. Find the reduced BDD without complement edges for the BDD of Figure 6.23. Repeat the problem, drawing this time a reduced BDD with complement edges. Finally, write a sum-of-product expression for f.

   Solution. The two BDDs are in Figure 6.24. A sum-of-product expression for f can be derived from the BDDs:

   $$f = abd' + ab'd + a'c + a'c'd.$$

   □

2. Write a sum-of-product expression for the function f represented by the BDD with complement edges of Figure 6.25.

[12] detailed notes on this subject are available from the authors

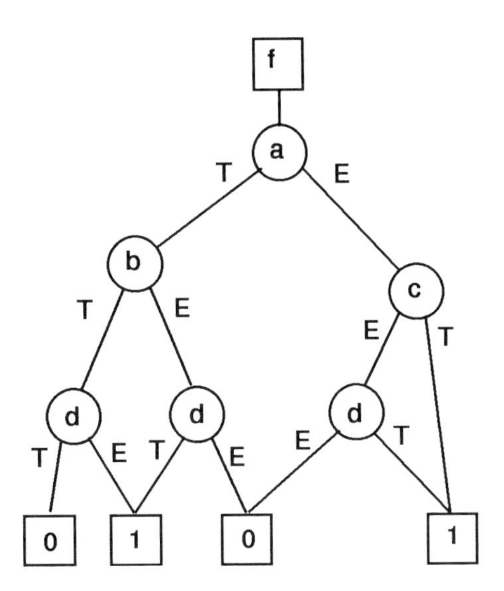

Figure 6.23: BDD for Problem 1.

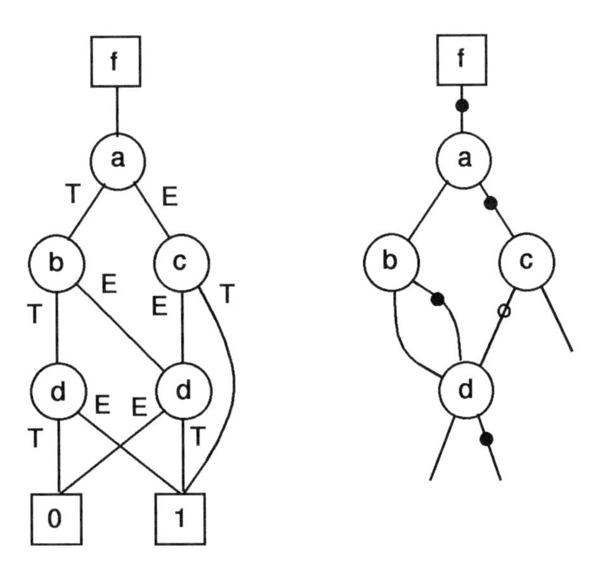

Figure 6.24: Solution for Problem 1.

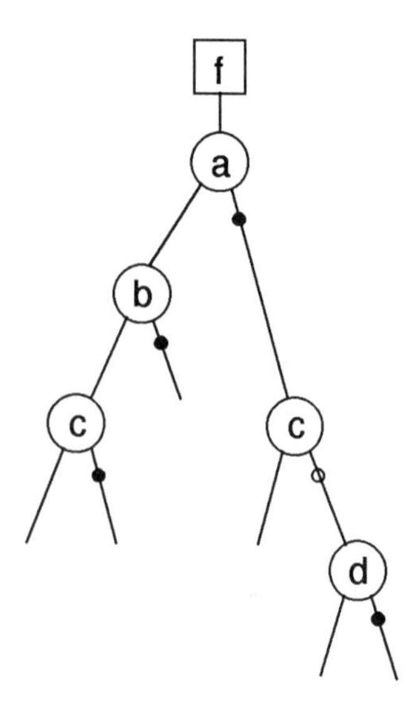

Figure 6.25: BDD for Problem 2.

Solution. One sum-of-product expression for f is:

$$abc + a'c'd'.$$

□

3. Find a good variable ordering for

$$f = (a + be)(d + c).$$

Discuss what was your reasoning. Draw the BDD for your ordering using complement edges.

Solution. Since f is the product of two functions with disjoint support, we keep the variables in the two supports separate in the order. We then notice that b and e on the one hand, and d and c on the other hand, are symmetric. Hence, their relative order is immaterial. Finally, we can apply the same argument we applied to f to $a+be$ so that a and b, e are not interleaved. These consideration leave several orders possible. One is simply obtained by listing the variables in the order in which they appear in the given expression for f:

$$a < b < e < d < c.$$

The resulting BDD is shown in Figure 6.26. The BDD is clearly optimal because each variable labels exactly one node. □

4. For the following functions:

$$\begin{aligned} f &= a + b'c \\ g &= b'c + d \\ h &= b + c' + d \end{aligned}$$

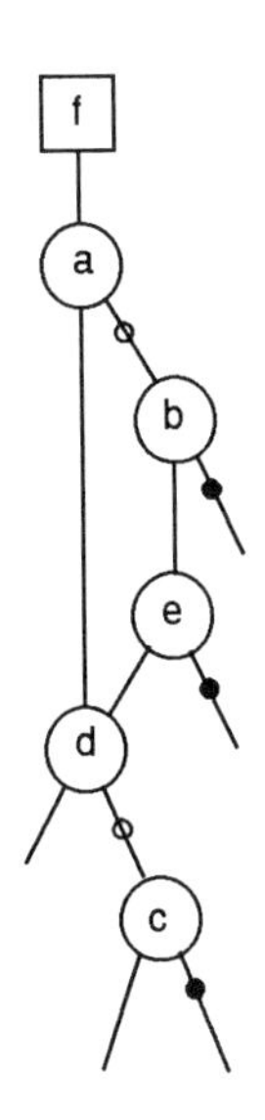

Figure 6.26: Solution for Problem 3.

(a) Draw BDDs F, G, and H with complement edges using the variable ordering

$$a \leq b \leq c \leq d;$$

(b) Compute $\text{ITE}(F, G, H)$;

(c) Draw the corresponding BDD with complement edges;

(d) Compute $\text{ITE\_CONSTANT}(F, G, H)$.

Note that you have to figure out the details of how to deal with complement edges in the ITE algorithm.

Solution. The BDD for the operands and the result is shown in Figure 6.27. The computation of $\text{ITE}(F, G, H)$ proceeds as follows.

$$\begin{aligned}
I &= (a, \text{ITE}(1, G, H), \text{ITE}(p', G, H)) \\
&= (a, G, (b, \text{ITE}(0, d, 1), \text{ITE}(c, q, r))) \\
&= (a, G, (b, 1, (c, \text{ITE}(1, 1, d), \text{ITE}(0, d, 1)))) \\
&= (a, G, (b, 1, (c, 1, 1))) \\
&= (a, G, (b, 1, 1)) \\
&= (a, G, 1)
\end{aligned}$$

The computation of $\text{ITE\_CONSTANT}(F, G, H)$ proceeds as follows.

$$\begin{aligned}
I &= (a, \text{ITE_CONSTANT}(1, G, H), \text{ITE_CONSTANT}(p', G, H)) \\
&= (a, \text{non_constant}, \text{ITE_CONSTANT}(p', G, H)) \\
&= \text{non_constant}
\end{aligned}$$

This happens because the result from the positive branch is known to be G, and G is known not to be constant. □

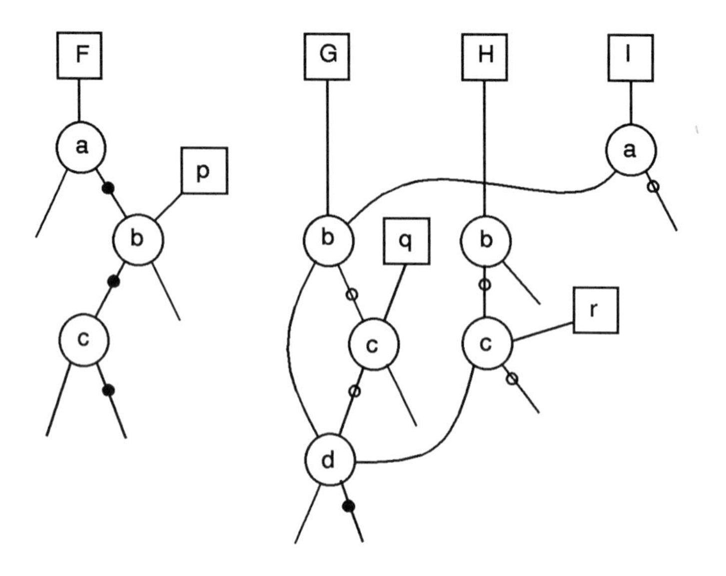

Figure 6.27: Solution for Problem 4.

5. Write pseudo code for the APPLY operation, defined here as a procedure that will take two BDDs and a Boolean connective operator as arguments. Assume the operation is specified by passing the name of a function that actually computes the desired operation. Write the function needed to compute the OR of two BDDs.

   Solution. The pseudo code for APPLY is shown in Figure 6.28. The pseudo code for the OR of two BDDs is shown in Figure 6.29. Notice that we could return 1 if $F \geq G'$. However, this test would not be possible in constant time. □

6. For the following functions:

$$\begin{aligned} f &= b + cd \\ g &= b + cd \\ h &= a'c + b \end{aligned}$$

   (a) Draw BDDs F, G, and H with complement edges using the variable ordering

$$a \leq b \leq c \leq d;$$

   (b) Compute $\text{ITE}(F, G, H)$ by putting the triple in standard form first;

   (c) Draw the resulting BDD with complement edges.

   Solution. The BDD for the operands and the result is shown in Figure 6.30. The computation of $\text{ITE}(F, G, H)$ proceeds as follows.

$$\begin{aligned} I &= \text{ITE}(F, G, H) \\ &= \text{ITE}(F, 1, H) \\ &= \text{ITE}(H, 1, F) \end{aligned}$$

```
APPLY(F,G,OP) {
        R = OP(F,G)
        if (R ≠ non_terminal) {
                return (R)
        } else if (computed-table has entry result = (F,G,OP)) {
                return (result)
        } else {
                let v be the top variable of {F,G}
                T = APPLY(F_v, G_v,OP)
                E = APPLY(F_v', G_v',OP)
                if (T = E) return (T)
                R = FIND_OR_ADD_UNIQUE_TABLE(v,T,E)
                INSERT_COMPUTED_TABLE((F,G,OP),R)
                return (R)
        }
}
```

Figure 6.28: Pseudo-code of the APPLY algorithm.

```
OR(F,G) {
        if (F = 1 or G = 1 or F = G') return (1)
        if (F = 0) return (G)
        if (G = 0) return (F)
        return non_terminal;
}
```

Figure 6.29: Pseudo-code of the OR operation.

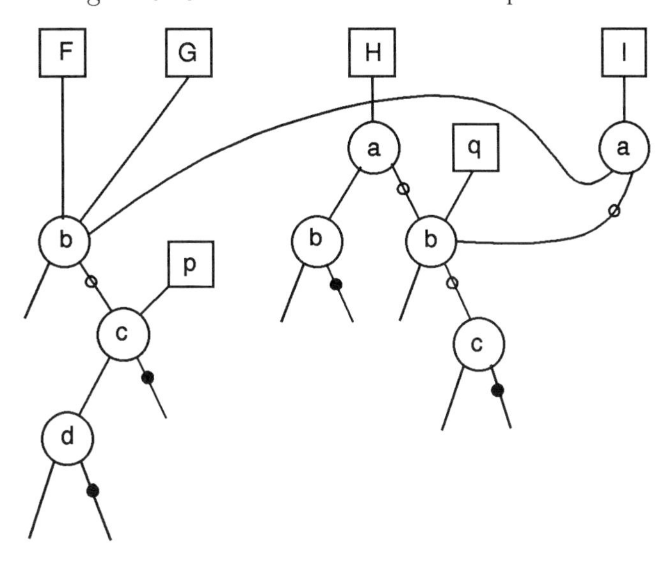

Figure 6.30: Solution for Problem 6.

$$
\begin{aligned}
&= (a, \text{ITE}(b, 1, F), \text{ITE}(q, 1, F)) \\
&= (a, (b, \text{ITE}(1, 1, 1), \text{ITE}(0, 1, p)), (b, \text{ITE}(1, 1, 1), \text{ITE}(c, 1, p))) \\
&= (a, (b, 1, p), (b, 1, (c, \text{ITE}(1, 1, d), \text{ITE}(0, 1, 0)))) \\
&= (a, F, (b, 1, (c, 1, 0))) \\
&= (a, F, (b, 1, c)) \\
&= (a, F, q)
\end{aligned}
$$

Note that in this case the sharing between the operands and the result is detected by looking at the unique table. □

Part III

Models of Sequential Systems

Although our humanity cannot be characterized in a finite dimensional space, our lives are controlled by the ubiquitous FSMs (Finite State Machines). Our computers, planes, cars, and appliances are run by them, as are traffic lights, internet traffic controllers, and efficient logic simulators.

C, PERL, and JAVA programs (in most practical cases) are finite state machines. So too is your wristwatch. It may be hard to admit, but too often we behave only in a finite and cyclic portion of our infinitely human and humanly infinite state space. As we respond finitely to finite stimuli, we are driven by genetic class inheritance and socio-biological controllers. Thus our lives tend to be processed in orderly stages, like instructions in the FSM-controlled P6 pipeline.

Roadmap of Part III This part is comprised of three chapters. First, in Chapter 7, we treat the related problems of state minimization (restricted to the completely specified case), and behavioral equivalence of FSMs. Roughly speaking, two FSM $\mathcal{M}^1$ and $\mathcal{M}^2$ are equivalent if their respective initial states s_0^1 and s_0^2 are equivalent. Optimization and synthesis of circuits are based on equivalence preserving transformations which reduce some measure of cost. Consequently, this elementary theory enables to treat virtually all synchronous sequential circuits, assuming there are no environmental don't care conditions. We include in this chapter a brief sketch of the variety of FSMs and their variants. For example, we define Moore and Mealy FSMs, as well as deterministic and nondeterministic finite automata, and state how the basic theory just established applies to all these variants.

Second, in Chapter 8, we extend our treatment to include incomplete specification and don't care conditions. Don't cares arise in two ways: first, they may come from the the external environment; second, they may arise from incomplete specification. For example, a designer might say: "... I don't care what else it does, as long as it does these specific things.". Secondly, for the sake of tractability, a large FSM is often decomposed into a set of (smaller) interacting subFSMs. This also derives in many practical cases from a "distributed control" design policy. Once decomposed, the behavior of one submachine depends on the behavior of the other submachines that it interacts with. For example, in a serial decomposition, the input sequences that come into FSM $\mathcal{M}^2$ are limited to those that may be produced as output sequences by submachine $\mathcal{M}^1$. In this case, the sequences that are not producible by $\mathcal{M}^1$ are "don't care" in the design of $\mathcal{M}^2$. In this chapter, we also treat the issue of state encoding, which has a profound effect on the optimality of a design.

Taken together, the algorithms of these two sections are capable of addressing most specific problems that arise in the automatic compilation of an FSM specification into a sequential logic circuit (we call this process FSM synthesis).

Third, in Chapter 9, we discuss formal language aspects of FSMs and their variants, mainly DFAs (Deterministic Finite Automata). These **language acceptors** complement the FSMs being designed, which can be thought of as **language generators**. This provides a means of proving that an FSMs correctly does the job it was designed for. This task is called **FSM verification**, and is currently (in the wake of the infamous Pentium Bug of 1995) emerging as co-equal in importance to synthesis and post-manufacturing test in the overall design cycle. In this chapter, we also deal with the issue of non-determinism, which is shown to be a powerful tool for simplifying the enormously complex task of verification.

Our main objective in this part is to give a solid foundation in FSM and automata theory, and at the same time provide the theoretical insight necessary for efficacious utilization of VLSI CAD tools for synthesis and verification. Thus the solved problems will again emphasize the use of supplied CAD tools for each synthesis step. In this part the primary synthesis tools will be STAMINA and MINCOV (for FSM state minimization), and JEDI for state encoding. All of these tools are available through the SIS synthesis environment.

Chapter 7

Models of Sequential Systems

7.1 Introduction to Finite State Machines

In the just concluded study of two level logic minimization we have used switching algebra as our principal formalism. In the study of sequential circuits our main formalism will be the theory of finite state machines. We shall now show how a specification of a synchronous sequential system can be converted into an optimized logic circuit consisting of logic gates and flip-flops.[1] We shall initially formulate our specification as a finite state machine. We shall then see how to minimize the number of states and assign binary codes to them. The resulting logic can then be optimized by using the subtools of SIS or other CAD tools or techniques.

Our treatment of FSMs (Finite State Machines) will be rather elementary. In particular, we shall restrict our treatment to single machines, rather than treating networks of machines. An exception to this rule is our treatment of product machines in Section 7.7.5. To some extent, this restriction is analogous to considering only two-level logic in the study of combinational circuits. There are many circuits of practical interest that cannot be conveniently modeled as a single finite state machine, because they would result in machines with too many states. For those systems, more powerful modeling techniques are needed. However, as for multi-level logic with respect to two-level logic, there are more powerful formalisms that relate to finite state machines as Boolean networks do to sums of products. Furthermore, large, complex systems can be normally decomposed into controllers and data paths. For the purposes of synthesis, one can consider each part separately (at the possible expense of losing some optimality). It is not uncommon that controllers can be modeled as single finite state machines and indeed finite state machines are used in practice to design controllers and simple sequential circuits, as in the following example.

A Simple Example of FSM Design, Using an Informal Approach Suppose that we want to design a circuit according to the following specifications. The circuit receives a stream of bits, one per clock cycle, on input x and at every clock cycle it indicates, on the three outputs z_2, z_1, z_0, the difference between the number of ones and the number of zeroes in the last three bits received. The difference is positive

[1] We shall restrict our attention to D-type flip-flops.

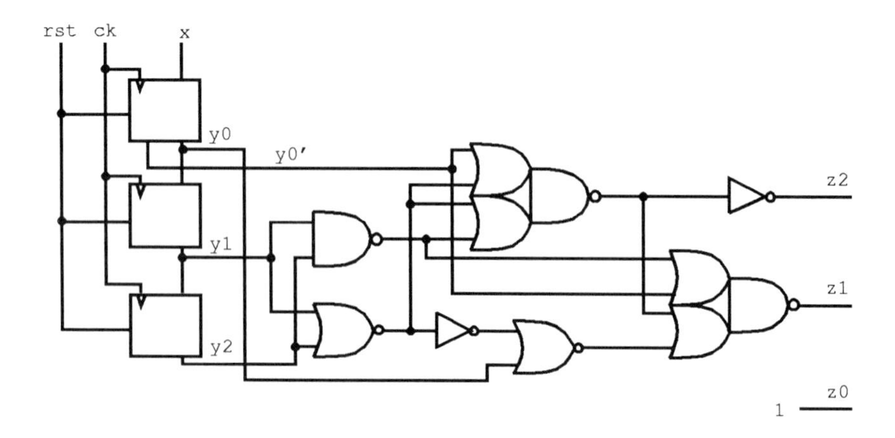

Figure 7.1: Simple Sequential Circuit.

if the number of ones exceeded the number of zeroes and it is negative otherwise. The difference is represented as a two's complement number, with z_2 the sign bit and z_0 the least significant bit. At reset, the circuit is assumed to have received an arbitrarily long string of zeroes, so that the output is -3.

Proceeding informally, we may decide to build our circuit around a 3-bit shift register. The shift register holds the last three bits received, so that a simple combinational circuit can determine the difference in the numbers of ones and zeroes and produce the correct output. Letting y_0, y_1, and y_2 be the three bits stored in the shift register, the truth table for the combinational circuit is:

$y_2 y_1 y_0$	$z_2 z_1 z_0$
0 0 0	1 0 1
0 0 1	1 1 1
0 1 1	0 0 1
0 1 0	1 1 1
1 1 0	0 0 1
1 1 1	0 1 1
1 0 1	0 0 1
1 0 0	1 1 1

and one possible resulting circuit, after optimization of the combinational logic, is shown in Figure 7.1[2].

Though in this case the informal procedure is immediate and results in a good circuit, in general it does not allow us to answer questions like: "Is it possible to design the circuit using one fewer flip-flop? Can we simplify the output logic, by making the next state logic slightly more expensive?" These two questions are related to the general issue of the exploration of the design space. We would like a systematic approach to considering different implementations. The informal approach may be effective, but it is not systematic.

Furthermore, some design problems, unlike our example, may not be amenable to the informal approach. In particular, if the initial description of the problem is

[2]Recall that when a number n is represented by the two's complement encoding z_2, z_1, z_0, we have $n = -z_2(2^2) + z_1(2^1) + z_0(2^0) = -4z_2 + 2z_1 + z_0$.

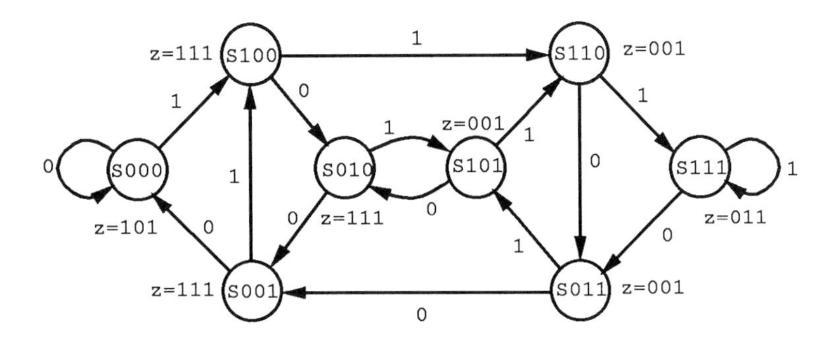

Figure 7.2: State Transition Graph for the Circuit of Figure 7.1.

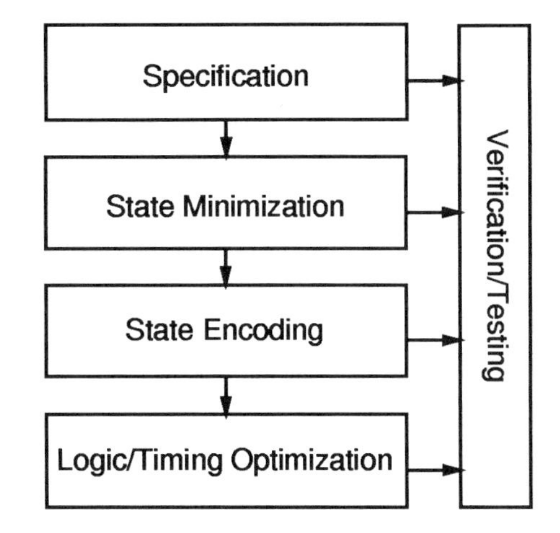

Figure 7.3: Simplified FSM Design Flow.

in a high-level format such as an HDL description, we need an approach to design that may be easily automated. Finally, if we are interested in verifying that the implementation is correct, we need a description of the intended behavior that is not ambiguous and that can be used for both synthesis and verification.

One such description of our example problem is the **state transition graph** of Figure 7.2, that we shall introduce formally in the next section. In the figure, the label $s100$ identifies the state the machine will be in if the last three bits were 100 (with the last bit on the left). Though this name suggests a possible encoding of the state, we are not committed to that particular one. The state transition graph is the same as the graph of a shift register. However, there is no implicit stipulation that a shift register will be used.

7.2 Synthesis of Finite State Machines

A simplified view of the design flow for an FSM is given in Figure 7.3. As an introductory example, consider the FSM represented by the STG on the left of Figure 7.4. We can observe that there is no way to distinguish state S_1 from state S_2. (This will be explained in detail in Section 7.4.1.) These two states are equivalent. Hence we

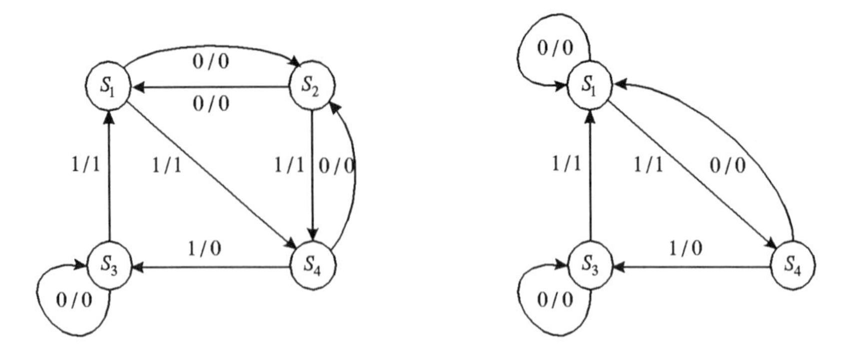

Figure 7.4: An FSM with Redundant States.

can reduce the machine accordingly, obtaining the minimized machine on the right of Figure 7.4. It is often true that a reduction in the number of states is accompanied by a simplification in the final implementation, though there are exceptions [93]. This illustrates one common theme that we already encountered in the synthesis of combinational circuits. We have a complex optimization problem, namely translating an STG into an optimal circuit. We break down the problem into sub-problems that are more manageable. In doing so, we need to formulate cost criteria for the sub-problems. This may be an arduous task. In the case of state minimization, we normally assume that we want to minimize the number of states, but what we are really interested in is the quality of the final implementation.

If the FSM is to be implemented as a PLA, then the cost is usually given by a simple function of the number of product terms and the number of bits used to encode the states.

If the encoded FSM is passed to a multilevel optimization program, then it is customary to consider the number of literals in the optimized circuit as the figure of merit. This is an approximation, since the cost of interconnections is not modeled, but it is usually a good one.

Once the number of states has been minimized, we need to assign binary codes to the states. This process is usually referred to as **state assignment** or **state encoding**. In general, we may also have to encode the inputs and/or the outputs. Such encoding problems were studied in [176, 92], and will treated in Chapter 8 below.

The high number of possible assignments make explicit enumeration of all possible solutions impossible for all but the simplest cases. In addition, it is not possible to determine a priori which number of encoding bits results in the optimum solution. Hence, practical algorithms for state assignment are heuristic. Most fall in one of these categories:

- algorithms that try to minimize the number of product terms in the two-level representation of the encoded machine;
- algorithms that try to minimize the dependence among state variables;
- algorithms that try to estimate the impact of encoding choices on the cost of a multi-level representation.

We shall review the last two categories briefly, by comparing two choices for encoding the minimized FSM on the right of Figure 7.4. First, suppose that the following assignments were made:

$$S_1 : 00 \quad S_4 : 01 \quad S_3 : 10.$$

(The two state variables are s_1 and s_2, in this order.) Then the equations for the next-state function and the output are:

$$\begin{aligned}
\delta_1(s, x) &= s_1\overline{s}_2\overline{x} + \overline{s}_1 s_2 x \\
\delta_2(s, x) &= \overline{s}_1\overline{s}_2 x \\
\lambda(s, x) &= \overline{s}_1\overline{s}_2 x + s_1\overline{s}_2 x = \overline{s}_2 x
\end{aligned}$$

Note that these SOP expressions require 11 literals, and may be derived from the minterm expansion of these three functions as follows.

s_1	s_2	x	$\delta_1(s, x)$	$\delta_2(s, x)$	$\lambda(s, x)$	*transition*
0	0	0	0	0	0	$S_1 \overset{0/0}{\rightarrow} S_1$
0	0	1	0	1	1	$S_1 \overset{1/1}{\rightarrow} S_4$
0	1	0	0	0	0	$S_2 \overset{0/0}{\rightarrow} S_1$
0	1	1	1	0	0	$S_2 \overset{1/0}{\rightarrow} S_3$
1	0	0	1	0	0	$S_3 \overset{0/0}{\rightarrow} S_3$
1	0	1	0	1	1	$S_3 \overset{1/1}{\rightarrow} S_1$
1	1	0	–	–	–	
1	1	1	–	–	–	

By collecting the terms of the truth table where functions evaluate to 1, we get SOP (Sum Of Products) expressions for $\delta_1(s, x)$, $\delta_2(s, x)$ and $\lambda(s, x)$. These may be optimized to produce the three functions given above. However, these representations do not account for the don't care conditions which correspond to the unused code 11. If we assume that this code never appears as the present state, we may form the don't care function $d = s_1 s_2$ and further minimize the above representation to the following.

$$\begin{aligned}
\delta_1(s, x) &= s_2 x + s_1\overline{x} \\
\delta_2(s, x) &= \overline{s}_1\overline{s}_2 x \\
\lambda(s, x) &= \overline{s}_1\overline{s}_2 x + s_1\overline{s}_2 x = \overline{s}_2 x,
\end{aligned}$$

which require only 9 literals.

Second, suppose that the following assignments were made:

$$S_1 : 01 \quad S_4 : 10 \quad S_3 : 11.$$

Note that there are more 1s in this encoding. Then the equations for the next-state function and the output are:

$$\begin{aligned}
\delta_1(s, x) &= s_1 s_2\overline{x} + \overline{s}_1 x + \overline{s}_2 x \\
\delta_2(s, x) &= s_1 + \overline{x} \\
\lambda(s, x) &= s_2 x
\end{aligned}$$

Note that these SOP expressions require 11 literals, and may be derived from the minterm expansion of these three functions as follows.

s_1	s_2	x	$\delta_1(s,x)$	$\delta_2(s,x)$	$\lambda(s,x)$	*transition*
0	0	0	–	–	–	
0	0	1	–	–	–	
0	1	0	0	1	0	$S_1 \overset{0/0}{\to} S_1$
0	1	1	1	0	0	$S_1 \overset{1/1}{\to} S_4$
1	0	0	0	1	0	$S_2 \overset{0/0}{\to} S_1$
1	0	1	1	1	0	$S_2 \overset{1/0}{\to} S_3$
1	1	0	1	1	0	$S_3 \overset{0/0}{\to} S_3$
1	1	1	0	1	1	$S_3 \overset{0/1}{\to} S_1$

By again collecting the terms of the truth table where functions evaluate to 1, we get SOP (Sum Of Products) expressions for $\delta_1(s,x)$, $\delta_2(s,x)$ and $\lambda(s,x)$ for the new encoding. These may be optimized to produce the three functions given above. However, in this optimization we did account for the don't care conditions which correspond to the unused code 00.

We see that the first encoding produces a result requiring 2 fewer literals than the result obtained with the second encoding. In this case there is very little room left for optimization. In general, one would then submit the resulting equations to, say, SIS, and optimize them using the SIS tool suite. Note SIS would likely catch the fact that $\delta_2 = \overline{s}_1\lambda$, thus producing a final result requiring only 8 literals. It frequently occurs in such circuits that the logic cones of the state transition function and the output function have considerable overlap and should be optimized jointly in the synthesis process.

After technology mapping, one would add the two flip-flops for δ_1 and δ_2 and finally submit the resulting netlist to a placement and routing tool. As illustrated in Figure 7.3, it is important to verify at each step that no errors are introduced by the synthesis procedure, as described in Section 7.8 on FSM equivalence checking. It is equally important to generate test patterns for the circuit, so that the correctness of the devices that are fabricated can be established, as described in Chapter 12.

A pictorial representation of an encoded FSM is given in Figure 7.5. As in the above discussion, each letter $a \in X$ of the input alphabet set is given a unique encoding $x(a) = (x_1, \ldots, x_n)$. The states $S_i \in S$ are similarly encoded. In the figure, the inputs are provided by the external environment, and the machine changes state to a next state $t = \delta(s,x)$. FSMs have output functions $\lambda(s,x) : I(x) \times S(s) \mapsto S(t)$, which on each clock cycle map an input x and present state $s \in S$ into a next state t. Note that such a physical implementation cannot directly represent nondeterminism. However, the encoding method of Section 9.2.3 can be used to represent the nondeterminism implicitly [66]. This technique is used extensively in Berkeley's verification tool VIS [225].

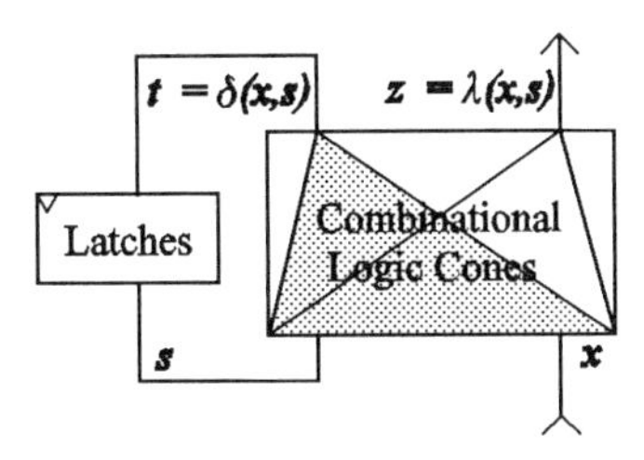

Figure 7.5: A Finite State Machine.

7.3 FSMs: Definitions, Notation, and Examples

In mathematical terms, a *(completely specified, deterministic) Finite State Machine* (FSM) of Mealy type is a 6-tuple

$$\langle I, S, \delta, S_0, \ \ O, \lambda \rangle,$$

where:

- I is the input alphabet, i.e., a finite, non-empty set of input values;
- S is the (finite, non-empty) set of states;
- $\delta : S \times I \rightarrow S$ is the next-state function;
- $S^0 \subseteq S$ is the the set of initial (reset) states.
- O is the output alphabet;
- $\lambda : S \times I \rightarrow O$ is the output function;

For a Moore type machine, $\lambda : S \rightarrow O$, i.e., the outputs do not depend on the present inputs.[3] In some contexts, like state minimization, we shall disregard the initial states. We shall mostly deal with Mealy machines, because they are more general. In practice, Moore machines are encountered when the outputs need to be latched: The algorithms we shall examine will work equally well for both types of machines.

7.3.1 Examples

An example of Mealy-type FSM is as follows:

- $I = \{0, 1\}$;
- $S = \{S_1, S_2, S_3\}$;
- $O = \{0, 1\}$;
- $\delta(S_1, 0) = S_1$, $\delta(S_1, 1) = S_2$, $\delta(S_2, 0) = S_1$,
  $\delta(S_2, 1) = S_3$, $\delta(S_3, 0) = S_3$, $\delta(S_3, 1) = S_1$;

[3]The two types of machine take their names from two pioneers of the field [197, 202].

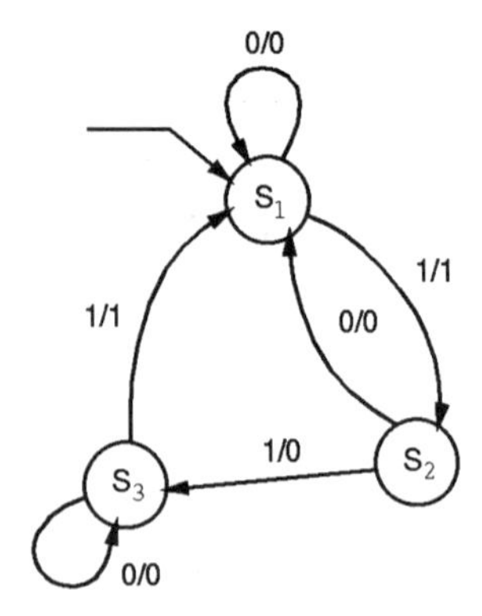

Figure 7.6: Example of State Transition Graph.

Flow Table

	$x = 0$	$x = 1$
s_1	$s_1, 0$	$s_2, 1$
s_2	$s_1, 0$	$s_3, 0$
s_3	$s_3, 0$	$s_1, 1$

Cube Table

I	PS	NS	O
0	s_1	s_1	0
1	s_1	s_2	1
0	s_2	s_1	0
1	s_2	s_3	0
0	s_3	s_3	0
1	s_3	s_1	1

Figure 7.7: Tabular Representations of FSMs.

- $\lambda(S_1, 0) = 0$, $\lambda(S_1, 1) = 1$, $\lambda(S_2, 0) = 0$, $\lambda(S_2, 1) = 0$, $\lambda(S_3, 0) = 0$, $\lambda(S_3, 1) = 1$;
- $S^0 = \{S_1\}$.

This way of specifying an FSM is cumbersome and difficult to read, so that we usually resort to more expedient ways. Various representations of FSMs are in use. Figure 7.6 shows the State Transition Graph (STG)[4] for our example FSM. Every node in the graph corresponds to a state; every arc corresponds to a transition. The label on the arc indicates the input that enables the transition and the output produced when the transition takes place. In Moore-type FSMs, the output values are normally associated with the states. An example was given in Figure 7.2. Other representations include the flow table and the cube table described in Figure 7.7. The cube table representation is very close to the input format of the CAD programs we shall use to manipulate FSMs.

The alphabets I and O and the set of states S are just sets of symbols. In particular, the symbols can be strings of zeroes and ones, in which case we have a straightforward way to implement the finite state machine as a circuit. Initially,

[4]Such graphs should not be confused with the **signal transition graphs** used in the synthesis of asynchronous circuits, which are also referred to as STGs.

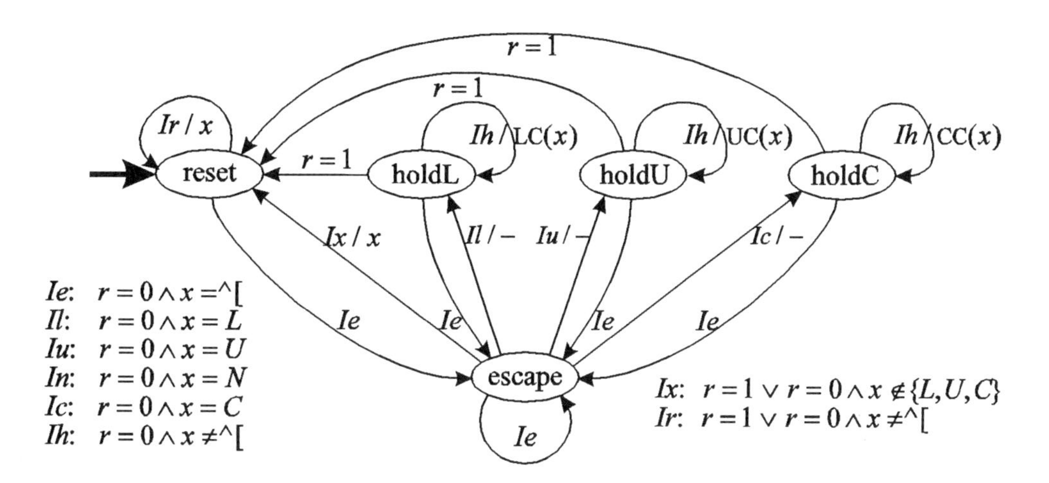

Figure 7.8: STG of the symbolic LUNC FSM.

though, the symbols may just be mnemonics, in which case the synthesis process will involve the assignment of binary codes to them.

Another example of an FSM is

$$\mathcal{M}_{\text{LUNC}} = \langle X, S, \delta, S_0, \ \ O, \lambda \rangle,$$

which is a higher level representation of the LUNC example of Section 2.1. $\mathcal{M}_{\text{LUNC}}$ is illustrated in Figure 7.8. In this FSM, X is the set of the 256 ASCII codes (plus the reset signal rst), S is the set {holdL,holdU,holdN,holdC}, S_0 ={holdN}, and δ is as specified by the given STG. The output function λ is not specified explicitly, but is defined to be x, LC(x), UC(x), or CC(x), or unspecified (don't care), depending on the input.

Programs exist, some as part of the SIS package (STAMINA[228],MUSE[92], and JEDI [92, 176, 179]) which can synthesize high quality logic directly from such a symbolic (non-Boolean) specification. Of course, a low-level description, as provided for the LUNC circuit in Chapter 2, can usually lead a creative designer to a better solution (if the problem is not too large and she/he is given enough time).

7.3.2 Incomplete Specification

An FSM is **incompletely specified** if the next-state function and/or the output functions are specified only for some combinations of inputs and present states. Examples of incompletely specified machines are given in Figure 7.9. Notice that the cube table representation is in this case slightly more powerful than the STG, since it can express the fact that the output for state s_3 and input 1 is specified, whereas the next state is not.

This is not a major drawback for the STG, since one can always find an equivalent FSM where only the output function is incompletely specified, by adding a dummy state, often called a **trap state**, and denoted by $s_?$. The use of a trap state is illustrated in Figure 7.10. Unless otherwise specified, we shall assume throughout the sequel that $\delta(s, x)$ is deterministic and complete. If a partial specification of δ is

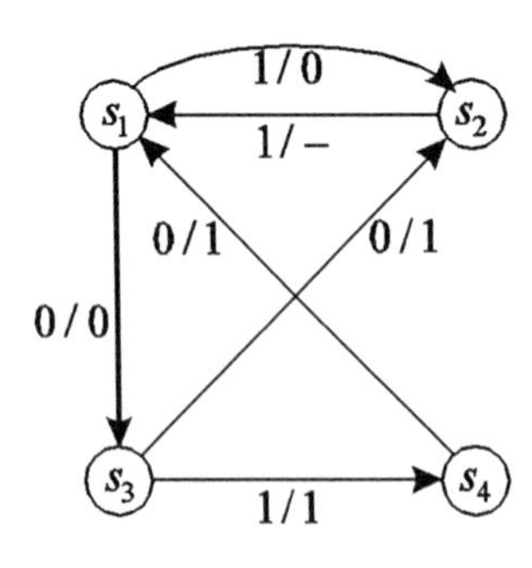

I	PS	NS	O
0	s_1	s_4	0
1	s_1	s_2	0
0	s_2	–	–
1	s_2	s_1	–
0	s_3	s_1	1
1	s_3	–	1
0	s_4	s_2	1
1	s_4	s_3	1

Figure 7.9: Example of Incompletely Specified FSM.

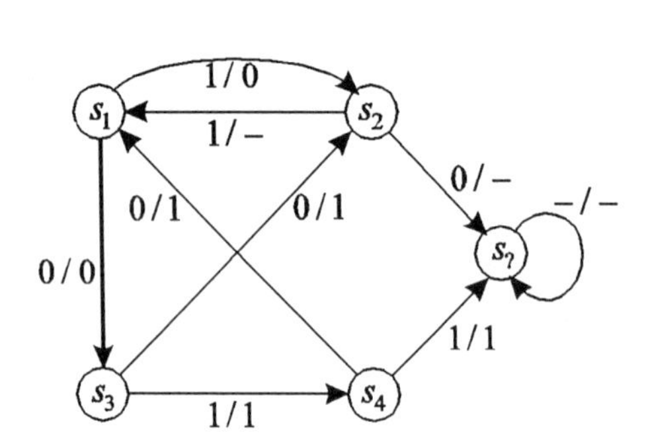

I	PS	NS	O
0	s_1	s_4	0
1	s_1	s_2	0
0	s_2	s_d	–
1	s_2	s_1	–
0	s_3	s_1	1
1	s_3	s_d	1
0	s_4	s_2	1
1	s_4	s_3	1
–	s_d	s_d	–

Figure 7.10: Machine Equivalent to the One of Figure 7.9.

given, in which transitions are not specified for some $x \in X$, then we will augment S with a trap (or "dummy") state, state $s_?$, which becomes the next state for every unspecified argument s, x of δ. Further we specify $\delta(x, s_?) = s_?, \ \forall x \in X$. Often it is unnecessary to explicitly show the trap state, so unless otherwise specified, it will be omitted in the sequel.

Finite state machines are devices that receive inputs and produce outputs. They perform an input/output transformation. There are devices, very similar to finite state machines, whose task is only to declare inputs accepted or rejected. They are called *finite automata*. Though we are primarily interested in finite state machines, we also define finite automata here for completeness. Our treatment here is brief and informal — the subject will be treated formally and in detail in Chapter 9.

A **deterministic finite automaton** is a 5-tuple $\langle I, S, \delta, s_0, \ F\rangle$, where I, S, δ, and s_0 are as before. F is a set of **final** or **accepting** states.

We distinguish deterministic automata, where the next-state function δ maps an input and a present state into a next state, from **nondeterministic** automata, where the next-state function δ maps an input and a present state into a set of next states, i.e.,

$$\delta : S \times I \rightarrow 2^S.$$

The set of next states can be interpreted as uncertainty as to what state the nondeterministic automaton will actually enter: hence the nondeterministic attribute.

It is also possible to define nondeterministic finite state machines, and we shall

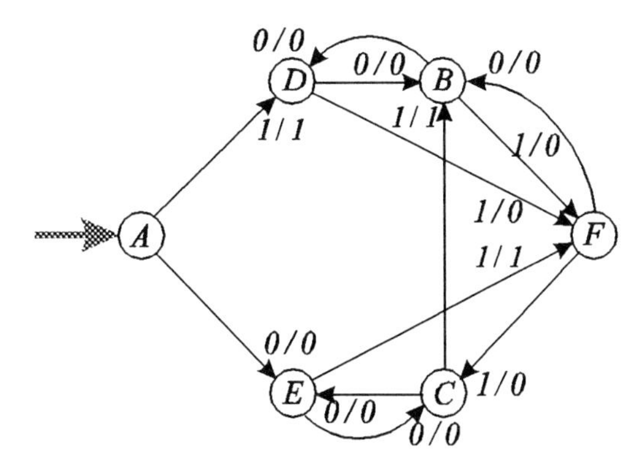

Figure 7.11: The STG of a simple FSM.

do so in Section 7.7.1. We shall see that incomplete specification can be interpreted as a restricted form of nondeterminism. At first, one may think that nondeterminism is interesting mainly from a theoretical standpoint. However, it is actually important in many practical applications of finite state machines and automata, because it provides for compact representations and may be used to express abstraction[5]. We mention here only a few of those applications: the design of efficient pattern matching algorithms, like those used in text editors; the design of compiler compilers, the verification of communicating processes.

We shall return to the subject of incomplete specification in Section 8.1, which treats the state minimization of incompletely specified machines.

7.4 FSM Minimization for Completely Specified Machines

State minimization plays a crucial role in logic synthesis. Especially in the context of a top-down design style starting from a high-level HDL specification, such as VERILOG, VHDL, or even a lower level specification such as the SIS blif-MV language. In this section we shall demonstrate why the 6-state machine of Figure 7.11 is equivalent to the 4-state machine of Figure 7.12.

Further we shall give an automatic procedure for synthesizing the smallest, in terms of the number of states, machine which is equivalent to a given machine. Often the circuit implementation of the reduced machine is much smaller and faster, and consumes less power, than the original. To our knowledge, Pixley [214] was the first to present a truly symbolic state minimization procedure, capable of handling machines with more than 10^9 states.

7.4.1 Identifying the Equivalent States of an FSM

We shall begin by defining the concept of equivalence between two states of a given FSM, and then show how this can be generalized into a procedure for state minimization. In the example of Figure 7.11 we will show that states A and C are equivalent,

[5]By abstraction we mean a process of reducing a given sequential system to a simpler system which retains essential properties of the original.

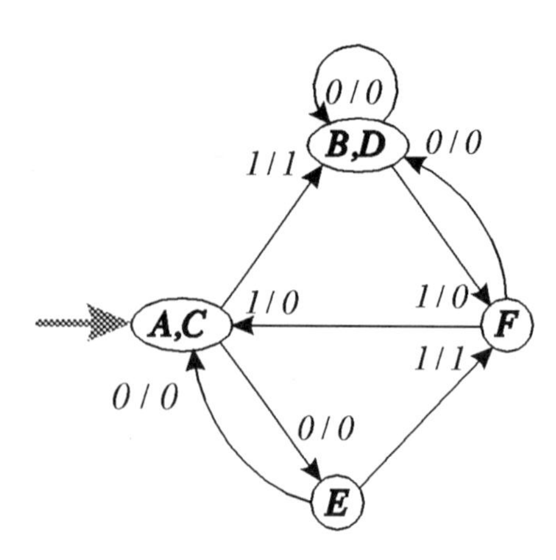

Figure 7.12: The STG of an FSM equivalent to the one of Figure 7.11.

and that states B and D are equivalent.

Definition 7.4.1 *Consider two states s and t of a given FSM, and a k-string (sequence of input symbols) $\mathbf{x} = (x_0, x_1, \ldots, x_{k-1})$. Suppose the* **string** $\mathbf{x}$ *produces one* **run** $\mathbf{s} = (s_0, s_1, \ldots, s_k)$, *with $s_0 = s$, when starting from s and another run $\mathbf{t} = (t_0, t_1, \ldots, t_k)$, with $t_0 = t$, when starting from t. Further, let $\mathbf{z}^s = (z_0^s, z_1^s, \ldots, z_{k-1}^s)$ and $\mathbf{z}^t = (z_1^t, z_2^t, \ldots, z_{k-1}^t)$ be the corresponding output strings .*

The string $\mathbf{x}$ is said to be a length-k distinguishing sequence for states s and t if and only if

$$z_{k-1}^s \neq z_{k-1}^t$$

Definition 7.4.2 *Two states s and t are k-equivalent, written $s \equiv_k t$, if and only if there does not exist a distinguishing sequence of length k or less for these states. Two states s and t are equivalent if and only if they are n-equivalent, where $n = |S|$.*

Given this simple definition of equivalence and distinguishability, we give the following theorem as the basis of state minimization.

Theorem 7.4.1 *$s \equiv_{k+1} t$ if and only if $s \equiv_k t$ and $\forall x \in X, \quad s_x \equiv_k t_x$, where s_x and t_x are the x-successors of s and t.*

Proof. (If):
Suppose $s \equiv_{k+1} t$. Then $s \equiv_k t$ by Definition 7.4.2. Further consider an arbitrary $(k+1)$-string $\mathbf{x} = x, \mathbf{x}^k$, the catenation of an arbitrary $x \in X$ followed by an arbitrary k-string $\mathbf{x}^k \in X^k$. Since $s \equiv_{k+1} t$, we also have $s \equiv_1 t$, so the prefix x does not distinguish s and t. Further, since $s \equiv_{k+1} t$, it must be true that the suffix $\mathbf{x}^k$ does not distinguish the x-successors s_x and t_x of s and t, $\forall x \in X$. This proves $\forall_{x \in X} s_x \equiv_k t_x$, thus completing proof of the If part of the theorem.

(Only If):
Assume $s \equiv_k t$ and $\forall_{x \in X} s_x \equiv_k t_x$, Since $s \equiv_k t$, we also have $s \equiv_1 t$. Thus no 1-string distinguishes between s and t. Further, since $s \equiv_k t$ and $\forall_{x \in X} s_x \equiv_k t_x$, no k-string distinguishes between the x-successors s_x and t_x of s and t. However an arbitrary $k+1$-string is composed as the catenation of a 1-string and a k-string, and we have

just shown that no such string distinguishes s and t. Hence $s \equiv_{k+1} t$, completing the proof of the Only If part and of the theorem. □

Definition 7.4.3 *We define the binary relation $\xi^k(s,t) \subset S \times S$ as*

$$\xi^k = \{(s,t) \mid s \equiv_k t\}.$$

We also denote the m_k equivalence classes of the relation ξ^k by B_i^k, $i = 1, \ldots, m_k$.

ξ^k is an equivalence relation since it is reflexive (a state is k-equivalent to itself), symmetric ($s \equiv_k t \Rightarrow t \equiv_k s$), and transitive ($s \equiv_k t, \quad t \equiv_k u \Rightarrow s \equiv_k u$). As discussed in Section 3.1.3, Page 80, an equivalence relation on a set S partitions S into the disjoint equivalence classes B_i^k. We shall denote the set of equivalence classes as the partition $P^k = \{B_i^k\}$.

In the example of Figure 7.11 we see from the above definition that states in the set $B_1^1 = \{A, C, E\}$ are 1-equivalent, and states in the set $B_2^1 = \{B, D, F\}$ are also 1-equivalent. However, since states A and B are in different 1-equivalence classes, they are 1-distinguishable — in fact, an input of 1 produces different outputs.

Similarly, we see that states $B_1^1 = \{A, C, E\}$ are also 2-equivalent, since they are 1-equivalent, and for all $x \in X$, their successors (respectively) E, E, C (under input 0) are 1-equivalent, and D, B, F (under input 1) are also 1-equivalent. This means that no input sequence of length 2 can distinguish between these states. Further, the same reasoning shows that states B and D are 2-equivalent.

However, state F in $B_3^2 = \{F\}$ is not 2-equivalent to state B, since F goes to C under input 1, whereas B goes to D under input 1 and C and D are not 1-equivalent. Thus we conclude that the 2-equivalence classes are $B_1^2 = \{A, C, E\}$, $B_2^2 = \{B, D\}$, and $B_3^2 = \{F\}$. This shows that F is not 2-equivalent to any other state.

Similarly, we can show that states $B_1^3 = \{A, C\}$ are also 3-equivalent, and that states $B_2^3\{B, D\}$ are also 3-equivalent, but that there are no other 3-equivalent pairs (thus $B_3^3 = \{E\}$ and $B_4^3 = \{F\}$). Since the same two pairs are also the only 4-equivalent pairs, the following proposition and theorem show that $\{A, C\}$ are equivalent, and $\{B, D\}$ are equivalent.

proposition 7.4.1

$$\xi^k = \xi^{k-1} \Rightarrow \xi^{k+j} = \xi^{k-1}, \quad j = 1, \cdots, (n-k).$$

The following theorem is due to Kohavi [162, Chapter 10].

Theorem 7.4.2 *Two states of a given FSM with n states are equivalent if and only if they are $(n-1)$-equivalent.*

Proof. Kohavi's proof is based on the preceding proposition. □

```
STATE_EQUIVALENCE (X, S, δ, λ) {
/* Partition-Refinement algorithm for identifying state equivalence */
        P^0 = {B^0_1}; B^0_1 = S; P^1 = P^0
1       for(x ∈ X) {
2           for(s_j ∈ S)  o^x_j = λ(s_j, x)
3           P^1_x = PARTITION(o^x, s)                  Sort and group by like outputs
4           P^1 = REFINE(P^1, P^1_x)
        }
        if (P^1 = P^0) return (P^0, 0)                 Return: all states are 1-equivalent
5       for(k = 2, ..., |S|) {                         Distinguishing path lengths
            P^k = {}                                   Initialize P^k
6           for(B_i ∈ P^{k-1}){
                P^k_i = {B_i}                          Initialize refined block partition
7               for(x ∈ X) {
8                   for(s_j ∈ B_i)  t^x_j = δ(s_j, x)
9                   b^x = BLOCK_INDEX(t^x, P^{k-1})
10                  Pb^x_i = PARTITION(b^x, s)         Sort and group by like block indices
11                  P^k_i = REFINE(P^k_i, Pb^x_i)
                }
12              P^k = P^k ∪ P^k_i
            }
13          if (P^k = P^{k-1}) return (P^k, k)
        }
}
```

Figure 7.13: Procedure for Finding Equivalent States of an FSM.

7.4.2 State Equivalence Checking: the Partition/Refinement Approach

Using these results we can formulate a simple **partition-refinement** procedure, which determines the set $\xi^n(s)$ of all equivalent state pairs of a given n-state FSM. Procedure STATE_EQUIVALENCE is given in Figure 7.13. The procedure employs a state partition P^k to represent the equivalence relation ξ^k of k-equivalent state pairs. It consists of two phases. The first determines the 1-equivalence classes by examining the outputs of the state transitions. The second phase iteratively determines the$k$-equivalence classes by examining the partition block indices of the head state of each transition. We shall describe the two phases separately, illustrating each phase on the same example.

The first phase begins by initializing P^0 and P^1 as partitions with one block, which contains all the states in S. It then proceeds to execute two nested **for**-loops (Lines 1,2). it gathers a vector o^x of the x-outputs $\lambda(s_j, x)$ of the states s_j in B_i. Then Subprocedure PARTITION uses o^x to partition the vector s into blocks with like output values, returning this partition as P^1_x.

At Line 4, the current approximation of the partition P^1 is then refined, that is, intersected, with P^1_x. We interpret this operation according to the following definition.

Definition 7.4.4 *If $P^a = \{B^a_i\}$, and $P^b = \{B^b_i\}$, then $P^a \cdot P^b$, called the meet, or intersection, of partitions P^a, and P^b, is given by*

$$P^a \cdot P^b = \cup_{ij}(B^a_i \cap B^b_j).$$

Note that according to Definition 3.1.2 of Page 83, the meet P^c of two partitions P^a, and $P^b$is a **refinement** *of both P^a and P^b.*

For example, consider the application of Procedure STATE_EQUIVALENCE to the example of Figure 7.11. The initial partition is $P^1 = P^0 = \{B^1\}$, where $B^1 = \{A, B, C, D, E, F\}$. On the $x = 0$ pass through the first **for** loop, we get output vector $o^0 = (0, 0, 0, 0, 0, 0)$ and corresponding state vector $s = (A, B, C, D, E, F)$. Since there is only 1 unique value in o^0, Subprocedure PARTITION returns $P^1_0 = \{S\}$. Thus at Line 4 REFINE intersects $\{S\}$ with itself, so at this point P^1 still is equal to $\{S\}$.

However, for the $x = 1$ pass, we get $o^1 = (1, 0, 1, 0, 1, 0)$, and the same state vector s. This time, PARTITION returns $P^1_1 = \{B^x_1, B^x_2\}$, where $B^x_1 = \{A, C, E\}$, and $B^x_2 = \{B, D, F\}$. Then at Line 4 REFINE intersects $\{S\}$ with P^1_1, which gives P^1_1, so at the end of **for**-x loop, we have $P^1 = P^1_1 = \{\{A, C, E\}, \{B, D, F\}\}$.

If Procedure STATE_EQUIVALENCE returns P^0 at the end of the first phase, it indicates that all the states of the specified machine are pairwise equivalent [6]. If this is the case, the FSM does not exhibit any sequential behavior and is in fact equivalent to a combinational circuit. An example of such an FSM is given in Figure 7.14.

We now discuss the second phase of Procedure STATE_EQUIVALENCE comprised of a set of four nested **for**-loops (Lines 5,6,7,8). This time, the FSM next state function

[6] This must be checked for and is a practical possibility. In fact, a 27 state machine, called DONFILE, was once distributed as part of a benchmark set and later proven to be equivalent to a combinational circuit.

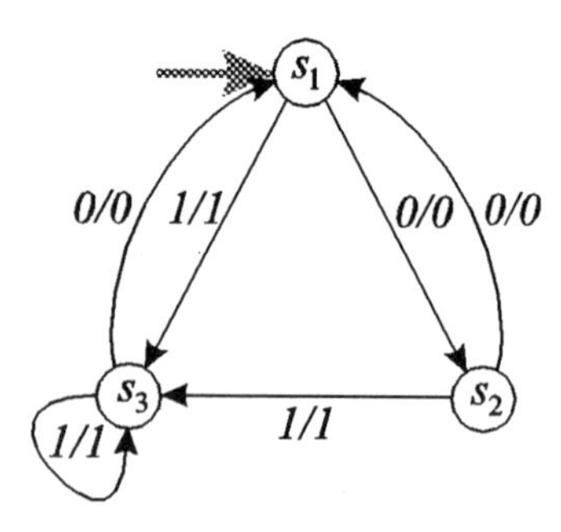

Figure 7.14: The STG of an FSM in which all state pairs are equivalent.

$\delta(s, x)$ is used to update the approximation P^k to the partition of k-equivalent pairs. On each pass through the loop over k, we initialize P^k to the empty partition. Then we loop through the blocks B_i of P^{k-1}, and compute P_i^k as a partition of B_i. This works as follows.

Inside the **for**-x loop, Procedure STATE_EQUIVALENCE gathers a vector t^x of the x-successors of the states in B_i. Then Subprocedure BLOCK_INDEX is called to compute b^x, the corresponding vector of partition block indices (in partition P^{k-1}). Then PARTITION(b^x, s) is called to partition B_i into blocks with like block indices, calling the result Pb_i^x. REFINE then intersects this with the current P_i^k, according to the above definition.

The application of the second phase of Procedure PARTITION_REFINEMENT to the example of Figure 7.11 proceeds as follows. When Line 6 is first entered, $P^1 = \{\{B, D, F\}, \{A, C, E\}\}$, and P^2 is the empty partition. Before entering the **for**-x loop, we initialize P_i^k to B_i. Then we upgrade P_i^k to a partition of B_i through refinement as follows.

On the first pass through the loop of Line 7, $x = 0$, and we get next state vector $t^0 = (D, B, B)$ and corresponding present state vector $s = (B, D, F)$. Since the block indices of t^0 are $b^0 = (1, 1, 1)$, PARTITION returns the trivial partition $Pb_1^0 = \{B_i\} = \{\{B, D, F\}\}$. However, for $x = 1$, and block $\{B, D, F\}$ we get $t^1 = (F, F, C)$, and $b^1 = (1, 1, 2)$. This time, PARTITION returns $Pb_1^1 = \{\{B, D\}, \{F\}\}$. Then REFINE intersects Pb_1^0 and Pb_1^1, thus refining the partition of the first block B^1 of partition P^1 into two blocks. After the set union of Line 12, $P^2 = \{\{B, D\}, \{F\}\}$.

Similar processing of block $B_2 = \{A, C, E\}$ of P^1 yields no refinement, which shows that these three states are 2-equivalent. The set union of Line 12 then correctly gives $P^2 = \{\{B, D\}, \{F\}, \{A, C, E\}\}$.

For $k = 3$, the first two blocks of P^2 yield no refinement. When the third block $B_3 = \{A, C, E\}$ of P^2 is processed BLOCK_INDEX first returns $t^0 = (E, E, C)$ and $b^0 = (3, 3, 3)$, which is uneventful. But for $x = 1$, $t^1 = (D, B, F)$ and $b^1 = (1, 1, 2)$, which causes PARTITION to return $Pb_3^1 = \{\{A, C\}, \{E\}\}$. This leads to $P^3 = \{\{B, D\}, \{F\}, \{A, C\}, \{E\}\}$.

Note that $A \equiv_2 E$, but $A/not \equiv_3 E$. Thus, there is a length-3 distinguishing sequence, 111, but no length-2 or length-1 distinguishing sequence. For 111, the run from A is A, D, F, C, with corresponding output sequence $1, 0, 0$. However, the run from E is E, F, C, B, with corresponding output sequence $1, 0, 1$.

Present State	Next/Output States	
	$x = 0$	$x = 1$
A	E/0	C/0
B	C/0	A/1
C	B/0	G/1
D	G/0	A/0
E	F/1	B/0
F	E/0	D/0
G	D/0	G/0

Figure 7.15: Flow Table for a Completely Specified Mealy Machine.

As per Theorem 7.4.2, and its preceding proposition, Procedure STATE_EQUIVALENCE terminates (Line 13) when it is observed that $P^k = P^{k-1}$ (or, equivalently, $\xi^k = \xi^{k-1}$), which is bound to occur for some $k \leq n = |S|$. In our example, for $k = 4$ no further refinement occurs, and STATE_EQUIVALENCE finally returns P^3. Note that an input sequence of length at least 2 is required to distinguish between non-equivalent states B and F, whereas an input sequence of length at least 3 is required to distinguish between non-equivalent states A and E.

Another example which shows more directly the partition and refinement operations is given in the flow table of Figure 7.15. While applying the first phase of STATE_EQUIVALENCE to this Mealy machine, we get $o^0 = (0,0,0,0,1,0,0)$, and $o^1 = (0,1,1,0,0,0,0)$, which PARTITION turns into $P_0^1 = \{\{A,B,C,D,F,G\},\{E\}\}$ and $P_1^1 = \{\{A,D,E,F,G\},\{BC\}\}$. Then REFINE takes the intersection of these partitions, yielding

$$P^1 = \{\{A,D,F,G\},\{B,C\},\{E\},\emptyset\} = \{\{A,D,F,G\},\{B,C\},\{E\}\}.$$

In the second phase of STATE_EQUIVALENCE for this machine, we consider the x-successors of the two non-trivial blocks of P^1 in turn. For $\{A,D,F,G\}$ we get $t^0 = (E,G,E,D)$ and $b^0 = (3,1,3,1)$ for input $x = 0$, and $t^1 = (C,A,D,G)$ and $b^1 = (2,1,1,1)$ for input $x = 1$. Thus $Pb_1^0 = \{\{A,F\},\{D,G\}\}$ and $Pb_1^1 = \{\{A\},\{D,F,G\}\}$, and the second REFINE step gives $P_1^2 = \{\{A\},\{D,G\},\{F\}\}$. The second and third blocks are not refined, so the 2-equivalent state partition is

$$P^2 = \{\{A\},\{B,C\},\{D,G\},\{E\},\{F\}\}.$$

For $k = 3$, we find that both non-trivial blocks are split apart. For $B_2 = \{B,C\}$ and $x = 1$, we get $t^1 = (A,G)$, $b^1 = (1,3)$, and $Pb_2^1 = \{\{B\},\{C\}\}$. Similarly, we get $Pb_3^1 = \{\{D\},\{G\}\}$, so

$$P^3 = \{\{A\},\{B\},\{C\},\{D\},\{E\},\{F\},\{G\}\},$$

which is the finest partition possible. This proves that every pair of states has a distinguishing input sequence of length 3 or less. If we performed another iteration, we of course would find that no refinement occurs and therefore we can stop.

The partition-refinement procedure we have presented can be implemented so that it takes time proportional to $|I||S|\log|S|$ [145]. A naive implementation based directly on our pseudo-code runs in time quadratic in the number of states.

7.4.3 Finding the Reduced Machine

The solution of the state minimization problem for completely specified machines is completed by producing an equivalent machine with a minimum number of states.

Procedure STATE_EQUIVALENCE returns the partition induced by the equivalence relation. It also returns an upper bound k on the length of distinguishing sequences required to distinguish any pair of states in the given FSM. As discussed above, each equivalence class can be combined into a composite state, called the **representative** of the class. In the above example this leads to 4 composite states and 4 representatives. The symbol associated with these 4 states is immaterial. One may select one distinct representative from each class, in which case the new set of state symbols will be a subset of the original. Alternatively, one may use the set of states in each class as the symbol of the class.

This latter choice is the one made in Figure 7.12, All the transitions in the original machine from one class to another class are represented by a single edge in the reduced machine. For example, state AC goes to E under input 0 and into BD under input 1, the first edge represents two edges in the original machine: (A, E) and (C, E). As a result of the minimization procedure, it is guaranteed that both of these edges in the original STG produced the same outputs. Similarly, the (AC, BD) edge in the reduced machine represents 4 output-consistent edges in the original machine.

7.4.4 Moore Machines and DFAs

This method can also be employed to minimize Moore machines or DFAs. For Moore machines, since the output depends only on the state (and not on the input) all edges emanating from a given state can be thought to have the same input label (in practice the label would be on the state node rather than on the edge).

For DFAs, we can minimize the corresponding FSM as constructed by the following definition, which describes the construction of Mealy and Moore Machines equivalent to a given FA.

Definition 7.4.5 *A DFA $\mathcal{A}$ is defined as $\mathcal{A} = \langle X, S, \delta, S_0, A\rangle$ where A is a set of* accepting *states (see Chapter 9 for a formal treatment). Since an FA has no output alphabet, we begin by defining the output alphabet of the corresponding FSM $\mathcal{M}$ to be $\{0, 1\}$. We can then define a Mealy Machine by defining the output function of the FSM so that all edges leading from an accepting state produce a 1 output (independent of the input), whereas all other edges produce a 0 output. Alternatively, we can even more directly define a Moore Machine by defining the output function to be a 1 for all accepting states and 0 otherwise.*

The determination of k-equivalence for $k \geq 1$ is thus similar for Mealy machines, Moore Machines, and DFAs.

For example, consider the Moore machine of the tabular specification of Figure 7.16. To apply the first phase of STATE_EQUIVALENCE to a Moore machine, we

Present State	Next States		Output
	$x = 0$	$x = 1$	
1	3	7	0
2	4	8	0
3	1	6	1
4	2	5	1
5	2	4	1
6	1	3	1
7	4	4	0
8	3	3	0

Figure 7.16: Flow Table for a Completely Specified Moore Machine.

imagine that the outputs given are the same for all $x \in X$. In this example, we just assume the outputs are the same for $x = 0$ and $x = 1$, and as a result, PARTITION returns $P_0^1 = P_1^1 = \{\{3,4,5,6\},\{1,2,7,8\}\}$.

In the second phase of STATE_EQUIVALENCE for Moore machines or DFAs, Processing is identical to what it would be in a Mealy machine. Let us consider the x-successors of $\{3,4,5,6\}$. For input $x = 0$, all states are mapped into state 1 or 2. These two states belong to the same block of P^1, and hence no refinement occurs here. A similar result obtains for input $x = 1$.

We then consider the x-successors of $\{1,2,7,8\}$. Nothing new happens for input $x = 0$. However, the 1-successors are $t^1 = (3,4,7,8)$, which leads to $b^x = (1,1,2,2)$. Consequently, Subprocedure PARTITION breaks $\{1,2,7,8\}$ into $\{1,2\}$ and $\{7,8\}$. Thus, at the end of the $k = 2$ iteration, we have the partition $P^2 = \{\{1,2\},\{3,4,5,6\},\{7,8\}\}$, thus identifying the states that are 2-equivalent. If we perform another iteration, we find that no refinement occurs and therefore we stop. The partition we have found identifies all the equivalent states.

Finding the reduced machine, given the equivalent states, is easy. We assign one state of the new machine to each equivalence class of the original machine. In our example, let us make the following assignments:

$$\{1,2\} = a \quad \{3,4,5,6\} = b \quad \{7,8\} = c.$$

In order to find the next states of a, we look at the next states of any state in $\{1,2\}$. Say we choose 1. The next state of 1 for input 0 is 3. Hence, the next state of a for input 0 is b. Similarly, since the next state of 1 for input 1 is 7, the next state of a for the same input is c. One can easily verify that the result would be identical, had we chosen State 2 as representative of its equivalence class.

Continuing in this procedure, we eventually obtain the table of Figure 7.17.

7.4.5 The Iterative Collapsing Approach

Another way to look at state equivalence checking is presented in [190]. This method combines and interleaves the techniques of Section 7.4.2 and Section 7.4.3. We shall state this method informally — the reader is referred to [190] for details.

	0	1	
a	b	c	0
b	a	b	1
c	b	b	0

Figure 7.17: Result of Reducing the FSM of Figure 7.16.

Present State	Next State/Output	
	$x = 0$	$x = 1$
A	E/0	B/0
B	B/0	A/1
C	B/0	A/1
D	A/0	A/0
E	A/1	B/0
F	E/0	A/0
G	A/0	A/0

Figure 7.18: First Collapsed Flow Table.

In this approach, one begins by performing phase 1 of STATE_EQUIVALENCE. Then, a representative (say the lexicographically smallest) of each class (block) of the partition P^1 is chosen. Then, a "collapsed" flow table is constructed, in which each occurrence of any member of a given class as a next state is replaced by its representative. Then, one checks to see if the partition P^1 is "consistent" with this modified table. This means that all members of each equivalence class must have identical entries in the new flow table. If so, we are finished. Else, we refine the partition into classes which do have identical entries, and again check for consistency.

Let's try this procedure out on the example of Figure 7.15. From our previous discussion, we know that $P^1 = \{\{A, D, F, G\}, \{B, C\}, \{E\}\}$. We then select representatives A, B, E from these 3 classes, and construct the collapsed flow table of Figure 7.18. This is not consistent, since rows A, D, F, G are not identical, although rows D and G are and B and C are. The partition is thus refined into

$$P^2 = \{\{A\}, \{B, C\}, \{D, G\}, \{E\}, \{F\}\},$$

as before. The next step is to collapse the *original* flow table by only replacing G by block 3 representative D, and C by block 2 representative B. This leads to the collapsed flow table of Figure 7.19. This flow table is still not consistent, since rows D, G are not identical, neither are B, C. At this point the partition is refined to the minimum partition, which of course is consistent, so we correctly get

$$P^3 = \{\{A\}, \{B\}, \{C\}, \{D\}, \{E\}, \{F\}, \{G\}\}.$$

An advantage of this approach is that the reduced machine is always apparent at each stage of refinement, so there is no reduction step at the end.

Present State	Next State/Output	
	$x = 0$	$x = 1$
A	E/0	B/0
B	B/0	A/1
C	B/0	D/1
D	D/0	A/0
E	F/1	B/0
F	E/0	D/0
G	D/0	D/0

Figure 7.19: Second Collapsed Flow Table.

7.4.6 Summary of State Equivalence Checking Methods

We emphasize that these procedures are applicable only to completely specified machines, and we shall shortly turn our attention to the more interesting case of incompletely specified machines. For this case we develop a more general procedure, which is capable of exploiting all the myriad don't care conditions which arise in practical cases like the LUNC FSM of Chapter 2.

We note in passing that with some effort and modificationSTATE_EQUIVALENCE can be mapped into a *symbolic* procedure [214] (Procedure EQUIV_N of Solved Problem 7 can be more straightforwardly mapped into such a procedure). Then, if BDDs were employed for efficient representation (Cf., Section 6.1) of the various characteristic functions, the modified procedure would be applicable to the large practical FSMs arising in VLSI CAD. Unfortunately, the analogous procedure for incompletely specified machines is not easily rendered into a similarly efficient symbolic procedure.

7.5 Graph Algorithms for FSM Traversal

Having established the basic concepts of state minimization and equivalence, our next goal is to describe the procedures for determining whether two given FSMs are equivalent. We shall consider only the case in which the two machines submitted for equivalence checking have specified initial states. Although the case in which initial states or state sets are not specified is of significant practical importance, the detailed algorithms for this case are much more involved, and beyond the scope of this book. Therefore we omit treatment of this case, and refer the reader to the excellent recent literature on this case: [252, 217], as well as the references cited in these papers.

There are two key steps in FSM equivalence checking: first,finding the FSM which is the product of the two comparison machines, and second, traversing (that is searching) the STG of the derived product FSM. Before doing this, we establish some background in graph search, and discuss the three main graph search procedures that have been extensively used as subprocedures in FSM synthesis algorithms.

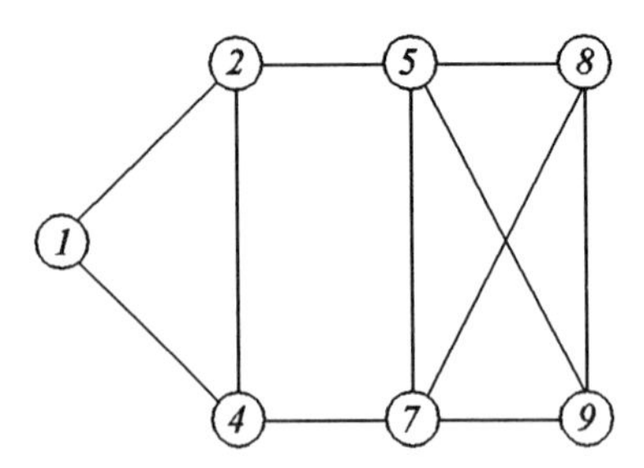

Figure 7.20: A simple undirected graph.

7.5.1 Graphs, Subgraphs, and Components

A **subgraph** of a given graph $G = (V, E)$ is another graph (V^-, E^-), where $V^- \subseteq V$, $E^- \subseteq E$. Given a graph $G = (V, E)$, we say that the subgraph $G^S = (V^S, E^S)$ that is "induced" by a given vertex subset S is the subgraph with $V^S = S$, and E^S equal to the edge subset of E whose edges have *both* vertices in S. When the context is clear, we shall simply use subgraphs to stand for induced subgraphs , and refer to them by their specified vertex set.

A subgraph is sometimes called a component. A **connected subgraph** of G is one in which for every $u \in V^-$, $v \in V^-$, there is a path from u to v or from v to u. In the context of logic circuits represented by a graph $G = (V, E)$, if v is a vertex representing an output signal of the circuit, then the subgraph (V^v, E^v) defined by vertices u such that $u \overset{*}{\rightarrow} v$ is often called the **fan-in cone** (or **logic cone**) of v.

In the circuit of Figure 1.12, the fan-in cone of gate 10 are gates in the set $\{7, 4, 3, 2, 1\}$ and the fanout cone of buffer gate 2 are the gates in the set $\{4, 5, 6, 7, 8, 9, 10, 11$

If $G = (V, E)$ is undirected, a **maximal completely connected subgraph** will be called a CCC. A CCC $G^- = (V^-, E^-)$ of G is subgraph in which $(u, v) \in E^-$, $\forall u \in V^-$, $\forall v \in V^-$. CCCs are sometimes called **cliques** or maximal cliques. Note that CCCs are maximal in the sense of the following definition.

Definition 7.5.1 *A subgraph (or subset) is* **maximal** *with respect to a given property if it is not a proper subgraph (or subset) of another subgraph (or subset) which also has the property. Minimal subgraphs (or subsets) are similarly defined.*

Illustrated in Figure 7.20 is a simple undirected graph, whose CCC's have vertex sets $\{1, 2, 4\}$, $\{2, 5\}$, $\{4, 7\}$, and $\{5, 7, 8, 9\}$. Note that the subgraph $\{5, 7, 8\}$ is completely connected, but it is not a CCC, since it is properly contained in the CCC $\{5, 7, 8, 9\}$.

If G is directed, a **strongly connected subgraph** is a maximal subgraph for which every pair of included vertices lies on a cycle. That is, for a strongly connected subgraph (V^-, E^-) of $G = (V, E)$, for every $u \in V^-$, $v \in V^-$, there is a path from u to v and from v to u, that is $u \overset{*}{\leftrightarrow} v$. It is also customary, although somewhat contradictory, to identify nodes not lying on any cycle as single node SCCs, as if each node had an implicit self-edge.

A simple digraph with 4 strong components (circled by dotted lines) is illustrated in Figure 7.21. Note that nodes 1 and 10 do not belong to any cycle and yet are identified as single node SCCs. Note at the bottom of Figure 7.21, that a DAG is formed by collapsing the SCCs of a digraph into a single node.

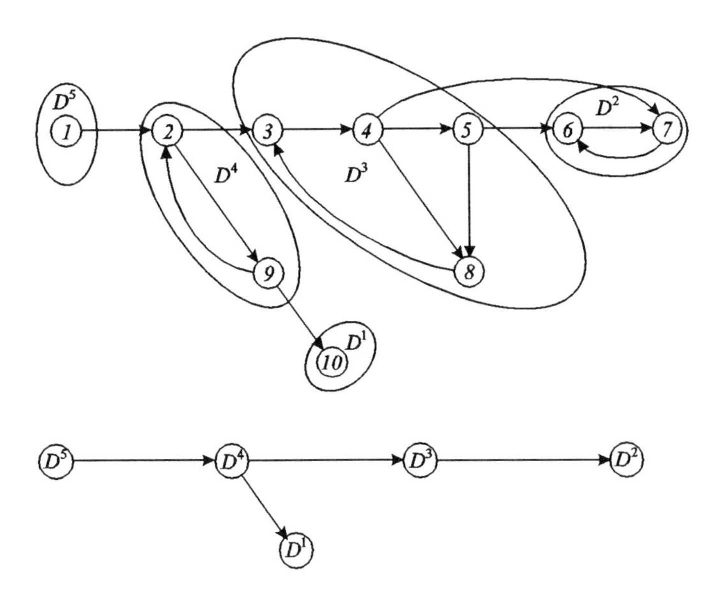

Figure 7.21: A digraph and its strong components

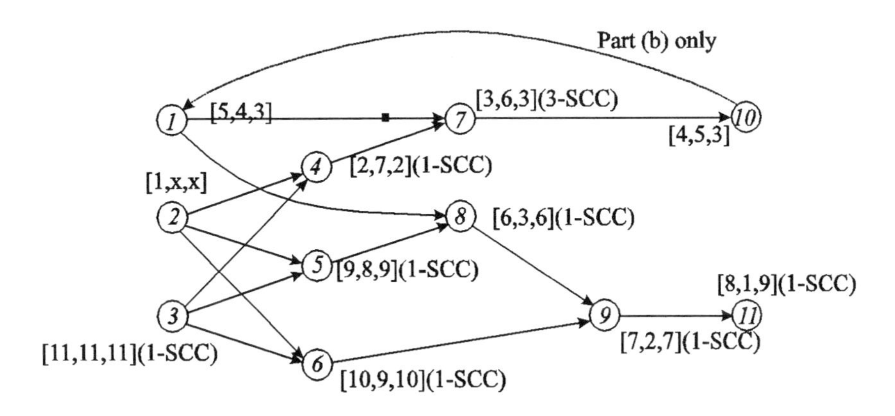

Figure 7.22: A directed graph representing the connectivity of the circuit of Figure 1.12.

Tarjan has given an algorithm DFS_SCC (discussed in Section 7.5.3) for finding the strong components of a digraph $G(V, E)$ whose complexity is **linear** in $|E|$. That is, the total number of operations performed by the algorithm is proportional to $|E|$. [258] (See Section 1.6.1 on Page 34). Tarjan's algorithm traverses the the digraph with depth first search, starting from the source node(s). In depth first search, the search must be completed for all the descendants of a node before it can be completed for the node itself. The superscript on the SCCs indicates the order in which Tarjan's DFS_SCC algorithm (starting at node 1) discovers the SCCs.

The strongly connected components of circuit graphs are of special interest. For example, if we modified the circuit of Figure 1.12 with a wire connecting output buffer 10 with input buffer 1, and replaced each gate with a circled node, we obtain the directed graph of Figure 7.22. The strong components of this digraph can be recognized visually in this simple case, in which the vertex set $\{1, 10, 7\}$ is identified as a 3-SCC, and all other vertices are identified as 1-SCCs. We shall discuss the node labels (enclosed in square brackets in the figure) in Section 7.5.3 below.

An (S, T) **cutset** C of a directed graph $G = (V, E)$ is a set of nodes (edges) of G such that *any path* from $s \in S$ to $t \in T$ passes through one of the nodes (edges) in C. If the graph is a DAG, it is often understood that S and T are the sets of source nodes and sink nodes, respectively.

7.5.2 Graph Traversal — Breadth First Search

A traversal of a graph $G = (V, E)$ is defined as any search from a given node u which systematically visits every $v \in V$ reachable from u. Node v is reachable from u if and only if $u \overset{*}{\rightarrow} v$, that is, if there exists a path in G from u to v. There are many orders in which a graph may be traversed [190]. In the so-called preorder traversal, each node is visited before any of its (transitive) successors. In so-called postorder traversal, each node is visited after all of its (transitive) successors. These two traversal methods will be discussed in Section 7.5.3.

Another traversal method, called BFS **Breadth first search**, effectively levelizes the graph with respect to a specified start set V_0, and then visits all the nodes at level k before visiting any nodes at level $k+1$. The success of BDDs in representing characteristic functions has made BFS the method of choice in exploring and analyzing the large directed graphs used to model the behavior of sequential circuits and software (that is, computer programs).

Breadth first search was first used in VLSI CAD to find minimum length interconnect routing algorithms of first generation VLSI chips. As we shall see, BFS has the property of finding a minimum length path to all vertices in a graph which are reachable from a given vertex. Thus BFS can always find a shortest route from one pin to another in the layout connection graph if one such route existed. It is still extensively used in all phases of VLSI physical design as well as in Logic Synthesis and Formal Verification.

Breadth first search relies on the computation of the "image" of a vertex subset. Images and pre-images, of a graph are defined (as follows) are defined like they were for functions: defined as follows.

$$\begin{aligned} \textsc{Img}(E(u,v), C(u)) &= \{v \mid \exists u \in C \ni (u,v) \in E(u,v)\}, \\ &= \exists_u C(u) \cap E(u,v). \end{aligned}$$

$$\begin{aligned} \textsc{Pre}(E(u,v), C(v)) &= \{v \mid \exists v \in C \ni (u,v) \in E(u,v)\}, \\ &= \exists_v E(u,v) \cap C(v). \end{aligned}$$

Here the notation $\exists_u C(u) \cap E(u,v)$ means that "the set of vertices v such that there exists a vertex u such that $u \in C$ and $(u,v) \in E$". BFS can be viewed as the process of iteratively computing the image, $I \subseteq V$, of a set $C \subseteq V$, from which the search is started. After adding I to the set R of vertices reached from the initial C, C is replaced by I, and the image computed again. This is repeated until on some iteration R is the same on two consecutive iterations.

In the graph context BFS is an efficient way of finding all vertices reachable by shortest paths from a given start vertex (or set of start vertices). Starting at vertex

$v \in V$ of a graph $G = (V, E)$, BFS proceeds as follows. First, all the vertices v in a given set V_0 are marked as "reached" (in the BFS algorithm, this property will be denoted by membership in a set *Reached*). Next, the set of all unreached successor vertices a of any $v \in V_0$ is determined. If no such vertex exists, the search terminates. Procedure BREADTH_FIRST_SEARCH of Figure 7.23 is a formal specification of BFS.

```
Line Procedure BFS(V, E, V0) {
 0      Reached = From = New^0 = V0; k = 0
        do {
           k = k + 1
 1         To = Img(E, From)
 2         New^k = To − Reached
           From = New^k
           Reached = Reached ∪ New^k    Reached is augmented
 3      } while (New^k ≠ ∅)             with vertices newly reached.
 4      return(k − 1, New^j, j = 0, ..., k − 1)
     }
```

Input: A graph $G = (V, E)$, and a starting vertex set V_0.

Output: The sequential depth of the graph (with respect to V_0) $k - 1$, and a set of sets (New^j, $j = 0, \cdots, k-1$), representing the set of vertices which occur at (minimum length) paths of length k from V_0.

Figure 7.23: Procedure for basic Breadth First Search.

Procedure BFS can be applied to the graph of Figure 7.24 as follows. Suppose $V_0 = \{1\}$. Thus *From* $= V_0$, when the loop body is entered (Line 1). The set *To* $= \{2, 3, 4\}$ of vertices reachable by paths of length 1 is computed as the image of the constraint set *From*, according to the above definition of IMG. Since none of

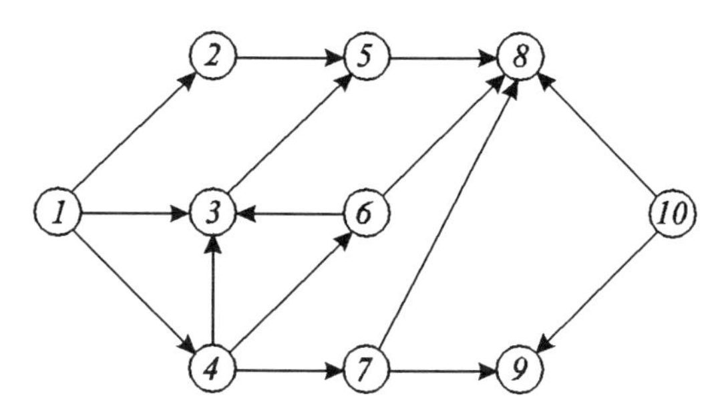

Figure 7.24: A directed acyclic graph.

these vertices were previously reached, we then set New^1 = *To* − *Reached* $= \{2, 3, 4\}$ (Line 2). Since $New^1 \neq \emptyset$ (Line 3), the loop body is executed again, this time yielding New^2 = *To* − *Reached* $= \{5, 6, 7\}$. Since $New^2 \neq \emptyset$, the loop body is executed again, this time yielding New^3 = *To* − *Reached* $= \{8, 9\}$. However, in the next loop, we get $New^4 = \emptyset$ at Line 3, so the algorithm returns the sequential depth $k - 1 = 3$ of the graph, relative to the initial vertex set $V_0 = \{1\}$. Note that the length of the shortest paths from V_0 to vertices in New^j is j.

Note by **sequential depth** of the graph (with respect to V_0) we mean the length of the *longest* of these shortest paths. The sequential depth corresponding to the particular subset V_0 (note that there are $2^{|V|}$ of these) leading to the largest sequential depth is called the **diameter** of the graph.

7.5.3 Traversal — Depth First Search

In this section we explore DFS **Depth first search** as a means of performing either preorder or postorder traversal DFS was originally developed by the French mathematician Hadamard as a formal way of traversing the elegant garden mazes of the courts of Louis XV France. Like BFS, its first use in VLSI CAD came in interconnect routing (or, especially, re-routing of hard-to complete nets) algorithms of first generation VLSI chips. Since depth first search has the property of finding a path to all vertices in a graph which are reachable from a given vertex, it could always find a route from one pin to another in the layout connection graph if one existed. DFS performs especially well in traversing the maze like configuration of interconnect layers and insulating obstacles that characterize a modern IC chip. It is still extensively used in all phases of VLSI physical design as well as in Logic Synthesis and Formal Verification.

In the graph context it is an efficient way of finding all vertices reachable by paths from a given (single) start vertex. The algorithm emphasizes locality and is typically implemented recursively. These two characteristics make it less natural to be implemented in terms of characteristic functions and BDDs. Hence it is less useful in modern synthesis and verification than BFS.

One version of DFS is given as Procedure DFS of Figure 7.25 below, which computes the preorder and postorder search indices of a given graph $G = (V, E)$. Note in the comments header that the search counters *kpre* and *kpost*, and the search index arrays *preorder* and *preorder* are global variables of the recursive algorithm, initalized to all 0s.

Recursive algorithms like this one are typically implemented with a (usually hidden, as in the present case) systems stack, S_u, which can be viewed as a LIFO (Last In First Out) queue. When Line 3 is reached, S_u contains, in the order that DFS was started, all nodes whose DFS has started but not ended. When DFS is completed for a given node, that node is popped off the stack. Thus at Line 3, $S_u = (u_1, \ldots, u_k)$, where u_1 is the node of the originating DFS, and $u_k = u$. It should be clear that $u_1 \overset{*}{\rightarrow} u$, that is, there must exist a path in G from u_1 to u. Note that if it is also true that $u \overset{*}{\rightarrow} u_1$, then u and u_1 must lie on a cycle of G.

When loop conditions are empty, the loop is to be skipped. Thus in the **foreach** loop in Procedure DFS, if node u has no (direct) successors, we will have $u\!\rightarrow\; = \emptyset$ and execution control will pass directly to Line 3.

Note that Procedure DFS does not specify what order the successors of the active node u are to be searched. Thus we shall disambiguate in the sequel by assuming that the successor lists are to be searched in descending lexicographical order.

When Procedure DFS is thus applied to the graph of Figure 7.26, the following

Global Variables:
Graph $G = (V, E)$
Integers $kpre, kpost$ Initially 0
Arrays $preorder$, $postorder$ Initially all 0s

```
    Procedure DFS(u){
1     kpre = kpre + 1;   preorder[u] = kpre
      foreach(a ∈ u →) {                          Scan successors of u.
2        if(preorder[a] = 0) DFS(a)               Recur if unsearched.
      }
3     kpost = kpost + 1,   postorder[u] = kpost
    }
```

Figure 7.25: Algorithm for Depth First Search Traversal of graph $G = (V, E)$ from start vertex u (first call).

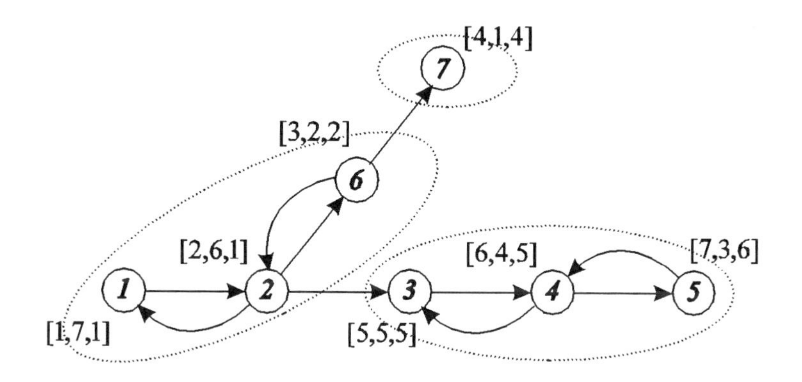

Figure 7.26: Directed acyclic graph with 5 nodes labeled by [*preorder*, *postorder*, *lowlink*].

values appear in arrays *preorder* and *preorder*:

$$\begin{aligned} v &= 1,2,3,4,5,6,7 \\ preorder[v] &= (1,2,5,6,7,3,4) \\ postorder[v] &= (7,6,5,4,3,2,1) \end{aligned}$$

The *preorder* and *preorder* indices appear as the first and second integers of the triple of node labels in Figure7.26. The third member of the triple will be discussed below in context of finding the SCCs of the given graph.

When searching vertex 2, we disambiguated by recursively searching successor vertex 6 before successor vertex 3. Thus the search of node 7 was started 4th (at this point the stack held $S_7 = (7,6,2,1)$), but was the first to finish. After the return from the recursive call for node 7, the stack held $S_6 = (6,2,1)$.

7.5.4 Finding the SCCs of a Directed Graph

In such small examples, the SCCs can easily be identified by inspection — in this case, they are $\{1,2,6\}, \{7\}, \{3,4,5\}$. However, a variant of DFS, called DFS_SCC in Figure 7.27[7], efficiently identifies the SCCs of any given graph.

As discussed in its header comments, Procedure DFS_SCC uses preorder and postorder index counters *kpre*, initialized to 0, the corresponding array *preorder*, initialized to all 0s, and an array *lowlink*, initialized to all $|V|$s. Procedure DFS_SCC further utilizes a global stack SCC, onto which each node name is pushed when (in postorder) the recursive call for node u is completed.

When a graph $G = (V, E)$ is traversed by DFS from start node v^0, DFS_SCC(u) of Figure 7.27 is called recursively for every node

$$u \in Reached = \{v \mid v^0 \xrightarrow{*} v\}.$$

In what follows we shall assume for simplicity that $V = Reached$, which means that all nodes are reachable from the start node (v^0)[8].

DFS thus creates a "spanning tree", T, in the sense that it partitions E into four disjoint subsets $E = E_T \bigcup E_L \bigcup E_X$ as follows:

$$\begin{aligned}
E_T =&\ \{(u,a) \in E \mid preorder[a] = preorder[u] + 1\} && \text{(tree edges);}\\
E_U =&\ \{(u,a) \in E \mid u \xrightarrow{T*} a,\ \ (u,a) \notin E_T\} && \text{(up-links);}\\
E_F =&\ \{(u,a) \in E \mid preorder[a] < preorder[u]\}, && \text{(fronds);,}\\
E_X =&\ \{(u,a) \in E \mid preorder[a] < preorder[u]\} && \text{(cross-links).}
\end{aligned}$$

Here the notation $u \xrightarrow{T*} a$ means there is a path in the tree from u to a. The edges in the set E_T are called tree edges, denoted $u \xrightarrow{T} a$, and represent node pairs which occurred consecutively on the system stack S_u. The subgraph $G^T = (V^T, E^T)$ induced by the nodes of E_T is a (spanning) tree with root v^0. Since this tree is a DAG, the edges of E_T are insufficient to form any cycles. The up-links are of no interest, because for any cycle containing such an edge, there is another cycle with the up link replaced by a path in the tree. Thus E_U will not be considered further.

The frond edges $(u,a) \in E_F$ are of special interest, because they run from ancestors to descendants $(a \xrightarrow{T*} u)$ in the tree, and thus always complete cycles $a \xrightarrow{T*} u \rightarrow a$. The cross-links run from one subtree to another and also may contribute significant cycles.

The cycles created through fronds and cross-links determine the SCCs of G by assigning node label $lowlink[v]$ to each node $v \in V$. Tarjan ([258]) defines $lowlink[v]$ as $preorder[u]$, where node u is "the node of lowest preorder index which is in the same SCC as v, and which is reachable from v by traversing zero or more tree edges followed by at most one frond or cross-link",

[7] Cormen, et al discuss an interesting alternative to Tarjan's algorithm, based on the spanning trees produced by two separate depth first searches: one on the original graph, and one on the graph obtained by reversing the direction of all edges. This technique has the same complexity and does not use *lowlink*.

[8] As described in [258], if some nodes are unreachable from v^0, the complete graph can still be traversed by starting other DFSs from nodes in the parts of the graph unreachable from the start node (v^0)

```
Global Variables:
    Graph     G = (V, E)
    Counter   kpre                                  Initially 0
    Array     preorder                              Initially all 0s
    Array     lowlink                               Initially all |V|s
    Stack     SCC                                   Initially () (empty)

  Procedure DFS_SCC(u){
1     kpre = kpre + 1; lowlink[u] = preorder[u] = kpre
      push(SCC, u)
      foreach(a ∈ E(u, v)_u) {                      Scan successors of u.
         if(preorder[a] = 0) {
            DFS_SCC(a)                              Tree edge — recur.
2           lowlink[u] = MIN(lowlink[u], lowlink[a])
         } elseif(preorder[a] < preorder[u]) {      Skip up-links and self-loops.
            if (a ∈ SCC){                           Frond or cross-link
3              lowlink[u] = MIN(lowlink[u], preorder[a])
            }     If, at Line 3, (u, a) is a cross-link, it's not sterile since a ∈ SCC
         }
      }
4     if (preorder[u] = lowlink[u]) SCC_POP(u)
  }
```

Note: At Line 4, *lowlink*$[u]$ is lowest preorder index of nodes in the same SCC as u and reachable from u by traversing zero or more tree edges followed by at most one frond or cross-link.

Figure 7.27: Recursive procedure for depth first search, modified to identify SCCs.

Algorithm DFS_SCC proceeds by traversing G by recursive DFS, and keeping track of the global arrays *preorder*$[v]$ and *lowlink*$[v]$. The active node u of each recursive call is pushed onto a global stack SCC, and when the cycle containment condition $preorder[u] = lowlink[u]$ is satisfied, SCC_POP(u) of Figure 7.28 is called to **pop** all the nodes $v \in SCC$ satisfying $lowlink[v] \geq lowlink[u]$ off the stack. The proof that this correctly identifies an SCC is found in [258].

```
  Procedure SCC_POP(u){
1    while(SCC ≠ ()) {                          Pop SCC off stack SCC.
        v = pop(SCC)
2       if(lowlink[v] ≥ lowlink[u]) print v
        else {
3          push (SCC, v)
           break
        }
     }
  }
```

Figure 7.28: Algorithm for popping the SCC stack in DFS_SCC.

Example:

We apply algorithm DFS_SCC to the graph of Figure 7.29, starting at vertex A, and always searching first the fanout vertex of maximal (lexicographical) index. As we traverse the graph, the edges are labeled with T, F, X, or U to identify their place in the aforementioned edge partition. The strong components, in order of detection, are (E, D), with

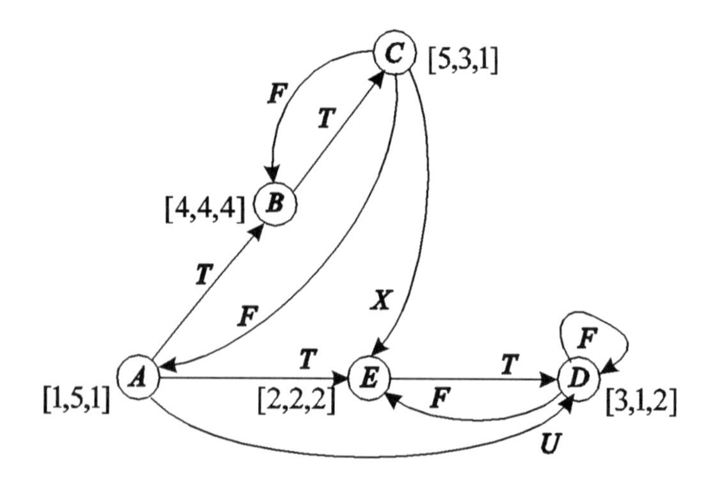

Figure 7.29: Directed acyclic graph with labeled edges and nodes labeled by [*preorder*, *postorder*, *lowlink*].

root labeled [2, 2, 2], (B, C), with root labeled [4, 4, 4], and (A), with root labeled [1, 5, 1].

In this example, the triples [*preorder*[v], *preorder*[v], *lowlink*[v]] given for each $v \in \{A, B, C, D, E\}$ lead to a graph partition with 4 tree edges, 3 fronds, 1 up-link ($(A, D) \in E_U$), and 1 cross-link ($(C, E) \in E_X$). In this case, the cross link is sterile in the sense that node E has already been popped out of the stack SCC when this cross-link is examined during DFS_SCC(C). Thus in the **foreach** loop, Line 3 will not be reached for $a = E$. ■

The cycles created through fronds and cross links determine the SCCs of G by assigning node label *lowlink*[v] to each node $v \in V$. Tarjan ([258]) defines *lowlink*[v] as *preorder*[u], where node u is "the node of lowest preorder index which is in the same SCC as v, and which is reachable from v by traversing zero or more tree edges followed by at most one frond or cross-link".

Algorithm DFS_SCC proceeds by traversing G by recursive DFS, and keeping track of the global arrays *preorder*[v] and *lowlink*[v]. The active node u of each recursive call is pushed onto a global stack SCC, and when the cycle containment condition *preorder*[u] = *lowlink*[u] is satisfied, SCC_POP(u) of Figure 7.28 is called to **pop** all the nodes $v \in SCC$ satisfying *lowlink*[v] $\geq$ *lowlink*[u] off the stack.

Now suppose that the DFS traversal which started at node v^0 has arrived at Line 4, and that the condition *preorder*[u] = *lowlink*[u] is satisfied for the first time — that is, no strong components have yet been identified. Given the active node u, we denote the q nodes of the subtree T_u of T rooted at the active node u by the ordered set

$$V_u = (v_1, v_2, \ldots, v_q).$$

These nodes have the following properties:

$$\begin{array}{lll} (a) & v_i \in SCC & 1 \leq i \leq q, \\ (b) & preorder[v_i] \geq lowlink[v_i] \geq lowlink[u] & 1 \leq i \leq q, \\ (c) & \text{DFS\_SCC}(v_i) \text{ has completed} & 1 \leq i \leq q. \end{array}$$

Note we count DFS_SCC(u) as completed, even though it has yet to finish the execution of Line 4. These nodes are of special interest and and we shall sort them in ascending order of completion — thus v_1 is the node that completes first, and $v_q = u$ is the node whose DFS_SCC finishes last of all the nodes in V_u.

It can be seen that if the set V_u were re-ordered in ascending order of their *preorder* index, the re-ordered set would be either identical to, or an initial prefix of, SCC.

We shall repeatedly refer to V_u in the proof of the following lemma, which is central to the proof that Algorithm DFS_SCC correctly finds the SCCs of the graph G.

Lemma 7.5.1 *For every v_i in the set V_u produced by* DFS_SCC, *there exists a corresponding $v_j \in V_u$ such that:*

$$\begin{array}{ll} (d) & v_i \xrightarrow{*} v_j \xrightarrow{T^*} v_i; \\ (e) & lowlink[v_j] \leq lowlink[v_i]. \end{array}$$

7.5.5 Shortest Paths

We have seen that Procedure BFS[9] implicitly identifies topologically shortest paths from a specified source vertex to all other vertices in a graph. When edges are labeled with edge lengths, the shortest path problem becomes more complicated. Shortest and longest path algorithms like Procedure SHORTEST_PATH of Figure 7.30 are often based on breadth first traversal of undirected and directed graphs. Such algorithms typically use some variant of BFS, and use a priority queue with keys determined by shortest distance from the initial vertex set.

The shortest path problem is typically defined on a weighted graph $G = (V, E, L)$, where $L_{u,w}$ is the specified length of edge $(u, w) \in E$. The problem is to find the shortest path to all vertices $v \in V$, from a specified vertex v_0. Here the length of a v_0, v path $\pi(v_0, v)$ is $l(\pi) = \sum_{(u,w)\in\pi} L_{u,w}$. If π^* is a shortest path from v_0 to v, then

$$l(\pi^*(v_0, v)) \leq l(\pi(v_0, v)) \ , \quad \forall \pi(v_0, v) \ .$$

Thus SHORTEST_PATH of Figure 7.30 accepts an edge length vector L (in one-to-one correspondence with E) and computes a vector λ (in one-to-one correspondence with V) defined by $\lambda_v = l(\pi^*(v_0, v))$. The vector λ_v is initialized to ∞ in line 1. SHORTEST_PATH, and also employs two data objects, S and *Reached*. S is a set of vertices for which at least one path from the start vertex, v_0, has been traversed in the breadth first search. Since a path from v_0 to $s \in S$ has been traversed, an upper bound, stored in the current value of λ_v, is known on the quantity $l(\pi^*(v_0, v))$. Each time a "new" path to v from v_0 has been traversed, the new upper bound is compared to the old one, and if lower, is stored in λ_v.

Reached$_v$ is initialized to $\emptyset$, and is augmented by $\{v\}$ when it can be identified that λ_v has converged to $l(\pi^*(v_0, v))$, by Lines 5 and 6. This is based on the identification, in Line 5, of vertex v as one of the currently unreached vertices which have the *smallest* current upper bound. Here

$$\lambda^* = \min_{s\in S}\{\lambda_s\} \ , \quad V^* = \{v \in V | \lambda_v = \lambda^*\}$$

That is, V^* is *set of minimizers* of λ_v.

Example:

SELECT1 makes an arbitrary selection of one member of V^* Thus if $S = (2, 3, 4, 6)$, and $\lambda_2 = 2$, $\lambda_3 = 3$, $\lambda_4 = 4$, and $\lambda_6 = 2$ in line 4, we have $V^* = \{2, 6\}$, the set of vertices v^* which have $\lambda^* = 2$ as the length of the shortest path from v_0 to v^* in the graph of Figure 7.31. In this case SELECT1 arbitrarily selects vertex $2 \in V^*$ as the next active vertex. ■

[9] Implementations of BFS often use a FIFO (FirstInFirstOut) queue. A queue is just an indexed list with two defined operations: QUEUE and DEQUEUE. The QUEUE operation adds an element at the end of the list, and the DEQUEUE operation takes the top element of the list. The queue is used to keep track of nearest neighbors of the already reached vertices. A variant of BFS uses a so-called priority queue. A priority queue is also an indexed list with two defined operations: QUEUE and . However, in a priority queue, each list element is associated with a key, and DEQUEUE takes the element with the smallest key out of the list.

```
Line Procedure SHORTEST_PATH(V, E, L, v0) {
 1      for(v ∈ V)λ_v = ∞; Reached = ∅
 2      S = {v0}; λ_v0) = 0
 3      while(S ≠ ∅) {
 4         λ* = min_{s∈S}{λ_s}            Vertex whose shortest path
           V* = {v ∈ V|λ_v = λ*}          length is known, and is
 5         v =SELECT1(V*)                 minimum, and becomes active.
 6         S = S − {v}
           Reached = Reached ∪ {v}
 7         for (a ∈ADJ(V, E, v)){
 8            if(a ∉ Reached){            Merge untouched vertices with
 9               S = S ∪ {a}              working set, and update upper bound.
10               λ_a = min(λ_a, λ_v + L_{v,a})
              }
           }
        }
11      return(λ)
    }
```

Input: A graph $G = (V, E)$, an edge length vector, L, and a start vertex v_0.

Output: Vector λ such that λ_v is the length of a shortest weighted edge length path from v_0 to $v \in V$. By convention, $\lambda_v = \infty$, for all v such that there is no path to v from v_0.

Figure 7.30: Procedure for finding shortest paths in a weighted graph.

The proof that the set V^* contains all vertices v^* whose shortest path distance, $\lambda(v^*)$, is already known, rests on a principle called **dynamic programming**, and can be shown by mathematical induction.

The implementation of SHORTEST_PATH is straightforward, with the only subtlety being that of finding a suitable member of V^* (this requires $O(log|S|)$ operations since there is an implied sort)[10].

A partial data trace of the application of SHORTEST_PATH to the graph of Figure 7.31 is shown in Table 7.1.

The first header row of the table identifies the type of data in each column and the line number of the procedure at which the data has been computed. The second header row identifies which of the 11 vertices the data applies to. There is one additional row for each pass through the **while** loop of Line 3. Note there must be fewer than $|V|$ such passes for any graph $G = (V, E)$. The first data column represents two quantities, the currently active vertex, v, and the minimum, λ^*, of the upper bounds of path lengths from v_0 to $v \in S$ (separated by a slash). The second column represents

[10] The notation $O(\log(|S|))$ means that there exists constants c_0 and n_0, such that $Count(|S|)$, the operations count of SHORTEST_PATH, expressed as a function of $|S|$, satisfies the inequality $Count(|S|) \leq c_0 \log(|S|), \quad \forall |S| > n_0$. That is, for large enough $|S|$, the cpu time complexity of SHORTEST_PATH is essentially linear in $\log(|S|)$. See Section 1.6.1 on Page 34.

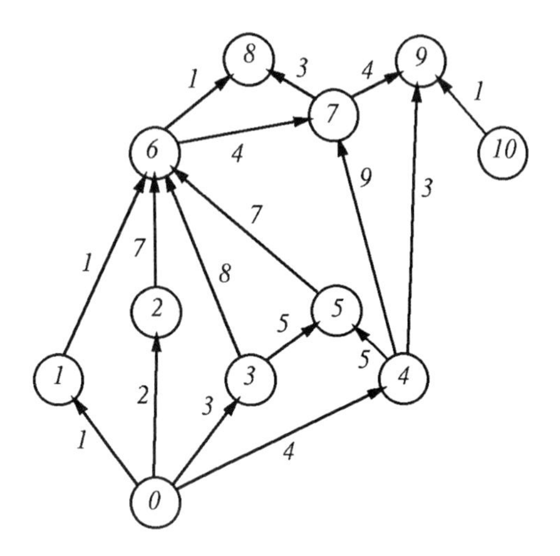

Figure 7.31: A weighted directed acyclic graph.

$v/\lambda^*/$ ADJ(v) Line 6	$(\lambda_v/(v \in \mathit{Reached}))$ Line 10										S Line 10
	1	2	3	4	5	6	7	8	9	10	
0/0/ {1,2,3,4}	1/0	2/0	3/0	4/0	∞/0	∞/0	∞/0	∞/0	∞/0	∞/0	{1,2,3,4}
1/1/{6}	1/1					2/0					{2,6,3,4}
2/2/{6}		2/1									{6,3,4}
6/2/{7,8}						2/1	6/0	3/0			{3,8,4,7}
final λ:	1	2	3	4	8	2	6	3	7	∞	

Table 7.1: Partial data trace for Procedure SHORTEST_PATH, applied to the graph of Figure 7.31($v_0 = 0$).

ADJ(v), which here denotes the successor set of the active vertex v. The next set of columns also give two data items separated by slashes. The first is λ_v, the second being 1 if $v \in$ *Reached*, else 0. The initial values of these quantities are shown in the first data row. Updates are entered in subsequent rows only if a change has occured, else (for emphasis), the data fields are left blank.

The potential efficiency of the algorithm is evidenced by the sparsity of these updates. In fact the whole algorithm has an operations count that is $(O(|E|)$ in all) (Cf., Section 1.6.1). The last set of columns gives the set S, *sorted in ascending order* with respect to the current estimate of shortest path lengths. Example 7.5.5 of V^* and λ^* came from row 2 of the table.

7.6 Models of Sequential Systems

In this section we shall briefly review some of the currently important models and algorithms that have been used in VLSI CAD tools to characterize, analyze, and synthesize sequential digital systems. Some of these models are illustrated in Figure 7.32, and are itemized as follows. The main point of this section is to show how FSMs and finite automata are built on top of an underlying structure we call an FST **Finite State Transition Structures**

- FSTs — In a given state, FSTs receive an input symbol, and make a transition to a new state;
- FAs (**Finite Automata**) are FSTs which also take notice when a favorable state (called *accepting*) is entered;
- FSMs (**Finite State Machines**) are FSTs which emit a specified output symbol when they make a transition from one state to another;
- **Regular Languages** — Languages are just sets of sequences of input or output symbols (called *strings*). Regular languages are just the kind of sets of strings that can be accepted by FAs or generated by FSMs — their regular structure is determined by the paths and cycles in the STGs of their underlying FSTs. This will be discussed in detail below.

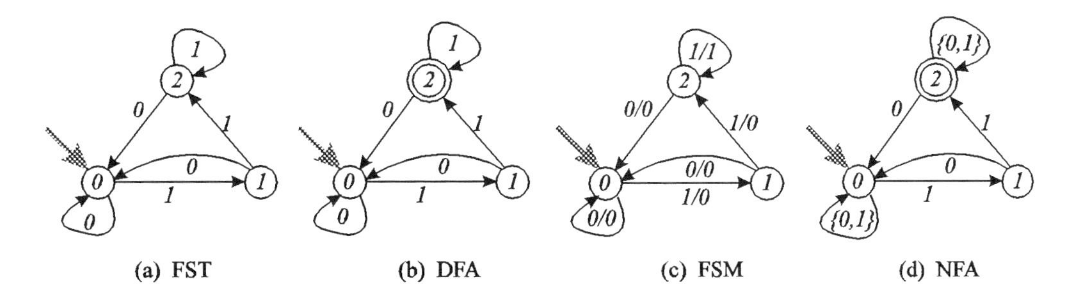

Figure 7.32: Models of finite-state transition systems.

An FST (Finite State Transition structure) (represented by the graph of Figure 7.32a) models how the underlying physical system changes its state in response to external stimulus (input symbol). An initial state (marked by the block arrow) is part of the FST specification. FSTs are closely related to Markov Chains [255, 121, 231]. FSTs may be represented either by flow tables, as in Section 7.3 or by the kind of labeled directed graph which we refer to as an STG (State Transition Graph), as in Figure 7.32.

In the STG representation, the edges of the graph are labeled with the set of input symbols that cause the indicated state transition. In Figure 7.32, an input of 1 takes the FST from state 0 to state 1, whereas an input of 0 takes it from state 0 back to state 0. When modeling an implementable physical system the edge labels can be thought of as sets of symbols, where the sets labeling two edges emanating from the same state are disjoint sets. In the deterministic cases of Figure 7.32a-c, these sets are disjoint.

A Finite Automaton (Figure 7.32b) is, in turn, built on top of an FST by adding an acceptance condition — accepting states are marked with a double circle. We shall see that the language of a given FA is the set of sequences of input symbols, called strings, which are accepted in the sense that they take the FA from its initial state, 0, to some accepting state, 2 in this case. In the figure, an input sequence of $(0, 1, 1)$ takes the FST from state 0 back to itself, through state 1, and on to accepting state 2. Similarly, an input sequence of $(0, 1, 0)$ takes the FA through state 0 to state 1 and then back to state 0. Since the string $(0, 1, 1)$ brings the FA to an accepting state, it is therefore part of the language accepted by the FA. In contrast, the string $(0, 1, 0)$ does not leave the automaton in an accepting state, and is therefore not accepted. Thus this string is not part of the language **accepted** by the FA.

An FSM is similarly built on top of an FST. A Finite State Machine (Figure 7.32c) is formed by labeling each edge of the FST's STG with an additional output symbol. Note that output symbols are those to the right of the slash which separates the two fields of the edge labels. In the figure, when an input of 1 takes the FSM from state 0 to state 1, an output of 0 is generated for this transition. The set of output strings generated by the FSM, over all possible input strings, is called the language **generated** by the FSM.

It should be kept in mind that for each given FA, an FSM can be constructed that has the same behavior[11]. FSMs and FAs are commonly used to model synchronous and asynchronous sequential circuits, communication protocols, parsers and lexical analyzers, table-lookup-based discrete event simulators[12]. Similarly, software, for example any terminating C-program, can be modeled as an FSM in many practical cases, although such models can be enormously complex.

Such a program is an example of an abstract FSM — one whose state, input, and output alphabets are sets of abstract symbols. In Section 8.3 we discuss algorithms for the **binary encoding** of such input, output, and state symbols. Encoding is the mechanism by which any abstract FST, FSM, or FA can be modeled by a physical synchronous sequential circuit. Thus all the applications mentioned in the previous paragraph can be built in hardware.

---

[11]Definition 7.4.5 of Page 272 describes the construction of Mealy and Moore Machines equivalent to the given FA.

[12]This type of simulation is employed in the most ultra-efficient commercial logic simulators and test generators.

FSMs and FAs are among the most important and fundamental mathematical models used in computer and software engineering, as well as in computer science. They are to the design of digital systems what the customary lumped resistor-capacitor-transistor circuit model is to the design of electrical circuits. Given a little background in logic circuits and graphs, they are easily understood, and powerful for modeling digital systems.

The term FSM is common in the literature, but often the more specific terms DFA **Deterministic Finite Automata** and NFA **Nondeterministic Finite Automata** are used instead of FA. Nondeterminism is one of the key concepts developed in Chapter 9. An NFA can be viewed simply as an FA with only partially specified behavior. An NFA is thus an abstract, non-physical representation. Partial specification is important because often the partial specification has a more tractable (smaller or more compact representation) and/or is significantly easier to formulate, or simply reflects current ignorance of the more complete specification.

The NFA of Figure 7.32d is derived from the DFA of Figure 7.32b by adding extra input labels to edges issuing from states 0 and 2. Thus the edge input label sets may no longer be disjoint from one another. This specification is non-physical because when in the initial state, 0, an input of 1 tells the system to transition to either state 0, state 1, or to both. In effect, we are specifying that we don't care what the system does with a 0 input. It can be shown that in this case, the added edge labels have no effect on the language accepted by the FA. That is, the FAs of Figure 7.32b and Figure 7.32d accept the same set of strings.

We shall show in Chapter 9 why the languages of the DFA of Figure 7.32b and the NFA of Figure 7.32d are identical. In each case, *only* strings of the form ...11 (that is, strings ending in 11) are accepted[13]. The language $\mathcal{L}$ of both the DFA and the NFA can be expressed as

$$\{11, 011, 111, 0011, \ldots, 1111, 0\cdots011, \ldots, 1\cdots111, \ldots\}.$$

Note that we don't succeed here in exhaustively enumerating this language, because it is an infinite set. Nevertheless every string in $\mathcal{L}$ is finite and ends in the string suffix 11.

There is a well known equivalence between DFAs, NFAs, and regular languages [66, 190, 163], which we shall discuss in detail in Chapter 9. Because of this correspondence, a system is often designed first as a simpler, but non-physical NFA, and then automatically converted to a physical, but possibly more complicated, DFA. This involves two steps. The first step is first, **determinization** (that is converting an NFA to an equivalent DFA, as discussed this in detail in Section 9.2.2). The second step is state minimization, discussed in detail in Section 7.4. Since minimization leads to a canonical representation (modulo isomorphism) the DFA corresponding to the given NFA is unique.

[13]This may seem paradoxical at first, since the string 011 also can take NFA of Figure 7.32d back to the initial state.

7.7 FSTs: Strings, Runs, Reachability and Products

Before we directly discuss the problem of equivalence checking of pairs of FSMs, it is helpful to establish context by looking more closely at the the behavior of an FSM. We begin by introducing the notion of an FST. An FST (Finite State Transition structure) (represented by the graph of Figure 7.32a) models how the underlying physical system changes its state in response to external stimulus (input symbol). An initial state (marked by the block arrow) is part of the FST specification. FSTs are closely related to Markov Chains [255, 121, 231]. It is the directed graph of the FST that is referred to as an STG (State Transition Graph). The edges of the STG are labeled with the set of input symbols that cause the indicated state transition. In Figure 7.32, an input of 1 takes the FST from state 0 to state 1, whereas an input of 0 takes it from state 0 back to state 0. When modeling an implementable physical system the edge labels can be thought of as sets of symbols, where the sets labeling two edges emanating from the same state are disjoint sets. In the simple cases of Figure 7.32a-c, these sets are disjoint.

7.7.1 Finite State Transition Structures

An FST (**Finite State Transition Structure**) is defined as a 4-tuple

$$\mathcal{T} = \langle X, S, \delta, S_0 \rangle,$$

where

X is the **input alphabet** (that is, the (finite) set of possible input symbols);

S is the **state set** or **state alphabet** (that is, the (finite) set of possible state symbols);

$\delta : S \times X \mapsto 2^S$ is the next state transition function[14];

$S_0 \subseteq S$ is a set of initial states.

If we compare this to to the definition of FSMs in Section 7.3, we can easily see that every FSM is built on embedded FST.

We shall refer to $\mathcal{T}$ as an "FST on X". The fact that X and S are *finite*, has profound ramifications. For example, it guarantees that the FST can be implemented as a physical sequential circuit. It is further guaranteed that two different FSTs, FSMs, or FAs can be proved equivalent or non-equivalent in time bounded from above by a function which is polynomial in the number of states and input symbols. If these were infinite sets, this important problem would be undecidable, and the FST would not be physically implementable.

We distinguish two types of FSTs: **Deterministic FSTs** and **Nondeterministic FSTs**. First consider the deterministic case. Given an input $x \in X$ and a

[14] As discussed in Section 3.1.1, each element of the set 2^S is one of the $2^{|S|}$ subsets of the finite set S.

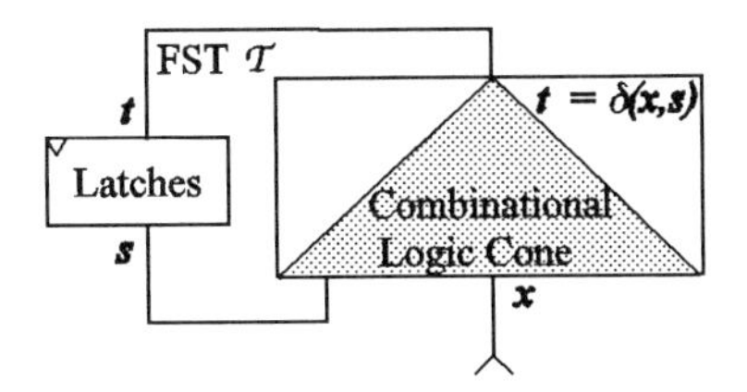

Figure 7.33: A Finite State Transition Structure.

present state $s \in S$, the next state transition function $\delta(s,x)$ maps s into a unique next state t of the FST $\mathcal{T}$, so that in this case

$$\begin{aligned} \delta(s,x) &: X \times S \mapsto S, \\ t &= \delta(s,x) \in S. \end{aligned}$$

However, in the general case,

$$\begin{aligned} \delta(s,x) &: X \times S \mapsto 2^S, \\ T &= \delta(s,x) \in 2^S, \\ T &\subseteq S. \end{aligned}$$

Definition 7.7.1 *FST $\mathcal{T} = \langle X, S, \delta, S_0 \rangle$, is* **deterministic** *if the image of all pairs (x,s) is a singleton next state. In this case, the specified δ does not map any given state into a set of two or more states. If in addition the initial state set S_0 of $\mathcal{T}$ is a singleton set, $\mathcal{T}$ is said to be* **strongly deterministic**. *If the mapping δ is specified for all pairs of states and input symbols, $\mathcal{T}$ is said to be* **complete**.

By this definition, we see that the FST of Figure 7.32a is strongly deterministic and complete. The FST of the NFA of Figure 7.32d is nondeterministic and complete. Observe that although we sometimes refer to incompleteness as a form of nondeterminism, by this definition an deterministic machine may yet be incomplete.

We see that the next state function $\delta(s,x)$ maps a state-input pair (s,x) into a subset $T_{s,x}$ of S. We say that $T_{s,x}$ is the image of the pair (s,x) (See Page 84). $\delta(s,x)$ is thus deterministic if and only if every pair in the Cartesian product $S \times X$ has a singleton image set, which would properly be written $T_{s,x} = \{t\}$, but usually we shorten this to just t.

Note that we have referred to nondeterminism as corresponding to non-physical behavior. Nevertheless, as discussed below, physical circuits can, through a process of encoding, represent both deterministic and nondeterministic FSTs [15]. After this special encoding process, the non-physical nature of the nondeterministic FST is manifested in a physical circuit with extra unspecified inputs. Thus, an FST $\mathcal{T}$ is an abstraction of a physical system comprised of a passive[16] sequential circuit with memory. The memory is modeled by the latch register shown in Figure 7.33.

Thus $\mathcal{T}$ stays in one of its initial states $s_0 \in S_0$ until it receives an input letter (or symbol) $x \in X$ from the external environment. The next state t of the FSM is jointly

[15] This was done in the development of the VIS verification tool suite [225]

[16] The circuit is passive in the sense that it has no outputs — hence it receives, but does not send, information to the external world.

δ	$\delta(s,x) : S \times X \mapsto 2^S$											
s, x	$A, 0$	$B, 0$	$C, 0$	$D, 0$	$E, 0$	$F, 0$	$A, 1$	$B, 1$	$C, 1$	$D, 1$	$E, 1$	$F, 1$
t	E	F	E	F	C	B	D	D	B	B	F	C

Table 7.2: The δ mapping for a deterministic FST with initial state A.

δ	$\delta(s,x) : S \times X \mapsto 2^S$											
s, x	$A, 0$	$B, 0$	$C, 0$	$D, 0$	$E, 0$	$F, 0$	$A, 1$	$B, 1$	$C, 1$	$D, 1$	$E, 1$	$F, 1$
$T_{s,x}$	$\{D, E\}$	F	E		C	B	D	D	B	$\{A, B, F\}$	F	C

Table 7.3: The δ mapping for a nondeterministic FST with initial state A.

determined by x and **present state** $s \in S$. On each cycle of the implicit clock, the next states t are stored in memory, which in a physical system is implemented by **latches**, which we shall also refer to as **flip-flops** or **registers**. A register is just a set of latches.

It is seen that once the input alphabet X and and state set S have been specified, the functionality of the FST is determined by the transition function $\delta(s, x)$. As discussed in Section 7.3 such a function can be variously specified. One method is to give a table specifying the set $T_{s,x}$ for each pair (s, x). A deterministic and complete FST, with $X = \{0, 1\}$, $S = \{A, B, C, D, E, F\}$, and $S_0 = \{A\}$, is given in Table 7.2. Note that each present state and input combination (s, x) is mapped into a unique next state t. Note also the specification is complete in the sense that all 12 of the elements in the Cartesian product $S \times X$ appear in the domain specification of δ.

A nondeterministic FST, defined on the same X, S, and S_0 is given in Table 7.3. Note that there are two present state and input combinations (s, x), namely $(A, 0)$ and $(D, 1)$, which are mapped into non-singleton sets of next states t.

The STGs of the two tables just given are shown in Figure 7.34. Note the second differs by the deletion of the $(D, 0)$ edge, and by the addition of extra edges on A and D.

The FST of Table 7.2 is deterministic and complete and the FST of Table 7.3 is

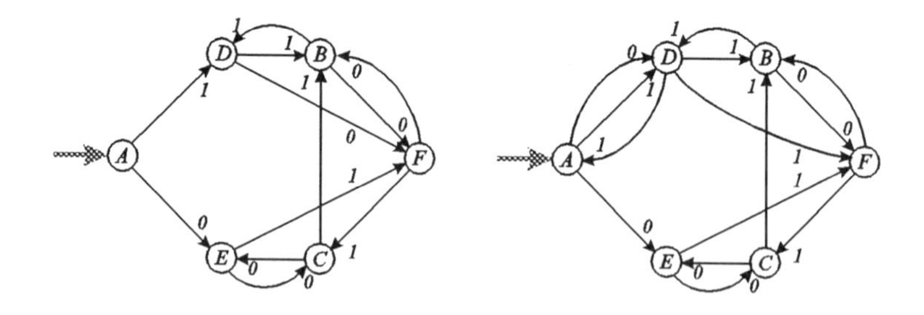

Figure 7.34: The STGs of Tables 7.2 and 7.3.

nondeterministic and incomplete.

Recall that such an incomplete STG can be converted into an equivalent complete STG by sending unspecified transition to a **trap states** $s_?$ (Cf., Section 7.3, Page 261).

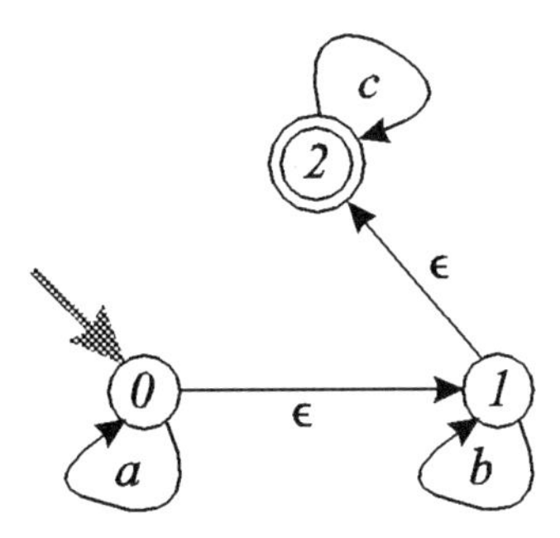

Figure 7.35: NFA example with ϵ-moves.

7.7.2 NFAs and ϵ-moves

Here we introduce the notion of **ϵ-moves** , which represent the further conceptual ability of an NFA to be in more than one state at a time. The concept of ϵ-moves is important in the procedure for deciding whether a specified string is accepted by a given FA (Section 9.1.1), and for constructing a DFA which accepts the same language as a given NFA (Section 9.2.2). An FST with ϵ-moves is illustrated in Figure 7.35, In this NFA, $S_0 = \{0\}$, $\delta(0, a) = 0$, $\delta(1, b) = \delta(0, \epsilon) = 1$, $\delta(2, c) = \delta(1, \epsilon) = 2$, and all the undefined transitions go to the unaccepting trap state (not shown). We interpret the ϵ-moves as follows. The initial state 0 can reach States 1 and 2 without any input, or, equivalently, with the input of the empty string ϵ. Thus if the NFA is in State 0, it can be regarded as effectively in States 1 and 2 as well. Similarly, if the NFA is in State 1, it can be regarded as effectively in State 2 as well. There are no ϵ-transitions from State 2.

7.7.3 FSTs as Labeled Digraphs

Sometimes an FST $\mathcal{T} = \langle X, S, \delta, S_0 \rangle$ is specified by giving its STG: $G = (S, E, L)$. Here

- The vertices $s \in S$ correspond to the states of $\mathcal{T}$.
- The edges $(s, t) \in E$ are directed from s to t and signify a possible **state transition:**
$$(s, t) \in E \Rightarrow t = \delta(s, x).$$
- For each edge $s, t \in E$ of G,
$$L_{(s,t)} = \{x \in X \mid t \in \delta(s, x)\} = \{x \in X \mid t = \delta(s, x)\}.$$

A well known example of an FST, which represents the Mead-Conway traffic controller [196], is given in Figure 7.36. This FST controls access to a busy highway from an infrequently used farm road. The capital letters refer to the busy highway and the lower case letters to the farm road. The input alphabet is an example of a compound alphabet[17], that is, the Cartesian product of two sub-alphabets $X^c =$

[17] Actually this input alphabet can be expressed as a product of Boolean algebras, as discussed in Chapter 3.

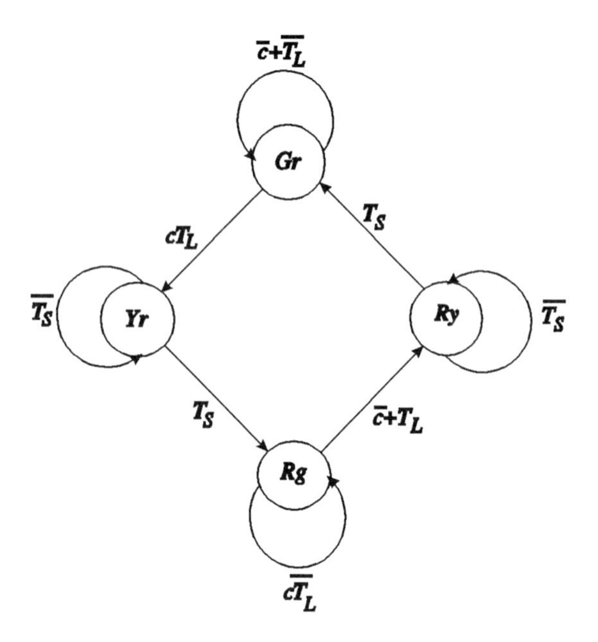

Figure 7.36: The FST of the Mead-Conway Traffic Controller.

$\{c, \overline{c}\}$, where the symbol c denotes a sensor output indicating the presence of cars on the farm road, and $X^T = \{T_L, \overline{T}_L, T_S, \overline{T}_S\}$ denotes the time-outs for the green and yellow lights. Thus there are a total of 8 possible input symbols in the alphabet X.

$$X = X^c \times X^T = \begin{array}{llll} \{cT_LT_S, & cT_L\overline{T}_S, & c\overline{T}_LT_S, & c\overline{T}_L\overline{T}_S, \\ \overline{c}T_LT_S, & \overline{c}T_L\overline{T}_S, & \overline{c}\overline{T}_LT_S, & \overline{c}\overline{T}_L\overline{T}_S\}. \end{array}$$

Thus the figure represents an FST $\mathcal{T} = \langle X, S, \delta, S_0 \rangle$, where $S = \{Gr, Yr, Rg, Ry\}$, $S_0 = \{Gr\}$, and δ is as indicated by the STG of the figure.

States (vertices) are labeled with a 2-letter code indicating the color of the traffic lights controlling the two roads. Long and short timers (not shown) control the waiting times (self-loops) in each state. While the short (long) timer is counting, input $\overline{T}_S$ ($\overline{T}_L$) is received, otherwise input T_S (T_L) is received.

These timers (not shown, so this is an example of abstraction at work) start counting when a car appears on the farm road. The input conjunction cT_L indicates that there is a car that has been waiting on the farm road for the specific length of time associated with the long counter, and the controller transitions then to the Yr state (highway Yellow, farm road red). This transition causes the short timer to start counting. While it is counting input $\overline{T}_S$ is received, and when it reaches the specified time, input T_S is received, and the controller transitions to the Rg state (highway Red, farm road green).

Other transitions are similar to those just described. In normal operation, the FST cycles counter-clockwise around the simple cycle connecting the four states. While there are no cars (state Gr) or while the timers are counting (all four states), the FST continually transitions back to its current state.

The input notation $(\bar{c} + T_L)$ is a shorthand for the following set of input symbols

$$\{\bar{c}T_LT_S,\ \ \bar{c}T_L\overline{T}_S,\ \ \bar{c}\overline{T}_LT_S,\ \ \bar{c}\overline{T}_L\overline{T}_S\}\cup$$
$$\{cT_LT_S,\ \ cT_L\overline{T}_S,\ \ \bar{c}T_LT_S,\ \ \bar{c}T_L\overline{T}_S\},$$

which can be seen as the union of the set of all inputs with sub-symbol $\bar{c}$ (meaning that no cars are present), and the set of all inputs with sub-symbol T_L (meaning that the long timer has reached its full count). Note the symbols $\bar{c}T_LT_S$, $\bar{c}T_L\overline{T}_S$ occur in both sets in the above union, so there are exactly six symbols represented by $(\bar{c}+T_L)$.

Note that this FST is **safe** in the sense that the each of the four states have the letter R or the letter r as part of their label, signifying, traffic is always stopped in one of the two directions. Further, notice that if the timer intervals controlling the X^T inputs are finite, any (non-simple) path of infinite length in the STG will have an *infinite number of states* with a G as part of their label, as well as an *infinite number of states* with a g as part of their label. That is, traffic in both directions will eventually be allowed to cross. In this sense, the FST is said to be **fair**.

We shall note in our discussion of Finite Automata (Chapter 9) that in practical formal verification packages like SMV [194] and HSIS/VIS [10], such temporal properties are often expressed in terms of Finite Automaton with design specific acceptance conditions. Alternatively, such properties are sometimes verified with temporal logic, for which the truth of a formula is decided by techniques similar to the reachability analysis of Section 7.9.

7.7.4 Strings, Tapes and Runs of FSTs

A **run** of an FST is a sequence of states which starts in an initial state $s_0 \in S$, occurs in response to some possible input sequence. A sequence of inputs is called a **string** if it is finite and a **tape** if it is infinite. The combination of a string/tape and a run is sometimes called a **chain**. In the example of Figure 7.35, *aabbbcccc* is a possible input string and 00111222 is the corresponding run. Similarly, $aabbbcc\cdots$ is a possible tape and $0011122\cdots$ is the corresponding run. Note that a tape is not a string, since strings are defined to be finite. Sequences, however, can be either finite or infinite.

Let X be the **alphabet** of the FST. A **string** $\mathbf{x}$ of finite length n on X is a sequence $\mathbf{x} = (x_0, \ldots, x_{n-1})$ of length n on X. A string of length $n = 0$ is called the **empty string**, and is denoted ϵ. We shall see in the discussion of NFAs that the response of an FST to an empty string should be considered carefully.

We shall initially focus on finite length sequences, that is, strings. Sequences of infinite length are discussed in detail in Section 9.4.1.

Let $\mathcal{T}$ be an FST $= \langle X, S, \delta, S_0\rangle$. Given a string or tape $\mathbf{x} = (x_0, x_1, \ldots,)$, a corresponding **run** is defined to be a **sequence** $\mathbf{s}$ on S. Thus $\mathbf{s}$ is an ordered set $(s_0, s_1, \ldots)$, where $s_0 \in S_0$, and $s_i \in \delta(s_{i-1}, x_{i-1})$, $1 \leq i$. Note that if the input sequence is a string of length n, the corresponding run is of length $n+1$.

A string (or a tape) represents a sequence of inputs to an FST. Each member x of the sequence causes a transition from the current state s to a next state $t \in \delta(s, x)$ (or $t = \delta(s, x)$ in the deterministic case) determined by the transition function $\delta(s, x)$.

In the nondeterministic case there is a set $R(\mathcal{T}, \mathbf{x})$ of runs, defined by

$$\begin{aligned} \mathbf{x} &= (x_0, x_1, \ldots,) \\ R(\mathcal{T}, \mathbf{x}) &= \{\mathbf{s} = (s_0, s_1, \ldots,) \mid s_0 \in S_0 s_i \in \delta(s_{i-1}, x_i), \ \ 1 \leq i\} \end{aligned} \tag{7.1}$$

Note that if the FST is strongly deterministic, $|R(\mathcal{T}, \mathbf{x})| = 1$, that is, there will be exactly one run for each input string or tape.

For example, in the deterministic FST of Table 7.2, the string $\mathbf{x} = 1101$ produces the run $ADBFC$, whereas in the nondeterministic FST of Table 7.3, the same string produces the run set

$$R(\mathcal{T}, \mathbf{x}) = \{ADBFC, ADFBD, ADAEF, ADADA, ADADB, ADADF\}.$$

Similarly, in the deterministic traffic controller example of Figure 7.36, the string $\mathbf{x} = (c\overline{T_L}\overline{T_S}, c\overline{T_L}\overline{T_S}, c\overline{T_L}\overline{T_S}, c\overline{T_L}T_S)$ produces the run set

$$R(\mathcal{T}, \mathbf{x}) = \{(Gr, Yr, Yr, Yr, Rg)\}.$$

7.7.5 Product of FSTs

In the sequel, it will be convenient and sometimes necessary to speak of the **product of FSTs**. This is necessitated by the fact that the FSTs that arise in practical, industrial strength applications simply have too many states to be conceived, designed, or specified as a single machine. One approach, sometimes referred to as *distributed control*, is a design style in which the component FSTs have manageable size, on the order of, say, just 10 states. However, if there are a hundred sub-FSTs, the overall FST will have on the order of 10^{100} states! Further, even if one is dealing with a very large *ab initio* specification, the first step in verification, synthesis, or test might be to try to decompose the FST into the product of component FSTs, again with the purpose of reducing the complexity of representation for synthesis or verification.

For simplicity, we use transition functions in our definition of product. Extension to the other representations (transition relations, STGs) is straightforward.

Definition 7.7.2 *Assume that the transitions in any sub-FST $\mathcal{T}_j$ depend on a global input alphabet X, but depend only on the local state set S_j. Then we may define the* **product of FSTs** *as follows.*

Given n [deterministic] FSTs $\mathcal{T}_j = \langle X, S_j, \delta_j, S_j^0 \rangle$, $1 \leq j \leq n$ on X, we define their product $\prod_j M_j$ as the [deterministic] FST

$$\begin{aligned} M &= \langle X, S, \delta(x, s), S^0 \rangle, \ \ \textit{where} \\ S &= \textstyle\prod_{j=1}^{j=n} S_j, \\ S^0 &= \textstyle\prod_{j=1}^{j=n} S_j^0, \\ \delta((s_1, \ldots, s_n), x) &= \textstyle\prod_{j=1}^{j=k} \delta_j(s_j, x) : X \times S \mapsto 2^S. \end{aligned}$$

A sub-FST $\mathcal{T}_j$ is deterministic if its transition function $\delta_j(s_j, x)$ is. If each sub-FST is deterministic, then the product is too.

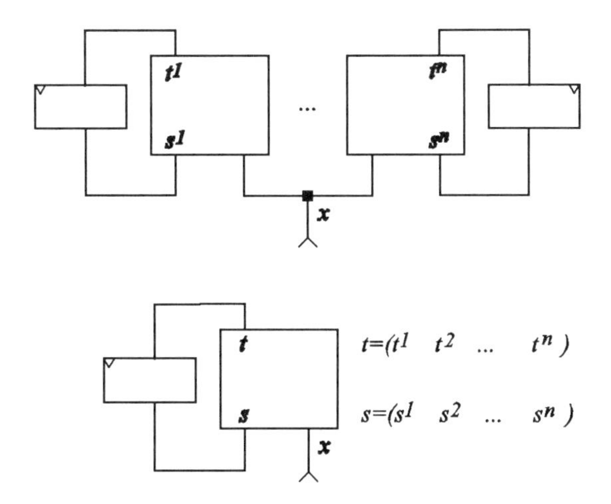

Figure 7.37: Product of FSTs.

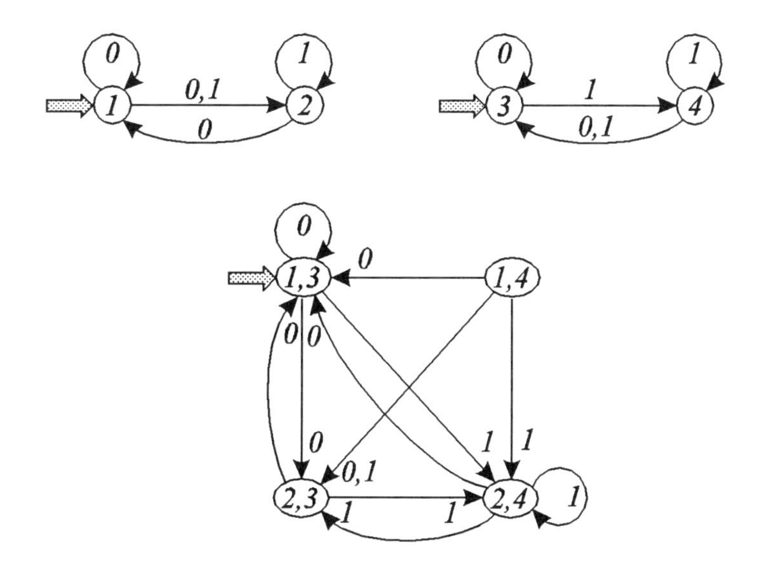

Figure 7.38: Product of Nondeterministic FSTs.

Note that the global state (state of the product) is a member $(s_1, \ldots, s_n) \in S$, where $s_j \in S_j$, $j = 1, \ldots, n$. Thus a single present state $s = s_1, s_2, \ldots, s_n$ is represented as a *catenation* (ordered set) of sub-states. Similarly the x-successor of s is determined by catenating the x-successors of each sub-FST.

The product of FSTs is illustrated for the deterministic case in Figure 7.37. Informally speaking, when we form the product of a given set of FSTs, we are in fact connecting them in parallel, letting the external input x enter them simultaneously, and allowing each FST to act on this input, independently of the others.

The product of nondeterministic STGs is illustrated in Figure 7.38. The upper left FST has nondeterministic transitions from state 1 with input 0, and the upper right FST has nondeterministic transitions from state 4, with input 1. The product of the two machines is shown at the bottom. Note

$$\begin{aligned}\delta((1,3),0) &= \delta_1(1,0)\times\delta_2(3,0) &= \{1,2\}\times\{3\} &= \{(1,3),(2,3)\},\\ \delta((2,4),1) &= \delta_1(2,1)\times\delta_2(4,1) &= \{2\}\times\{4,3\} &= \{(2,4),(2,3)\}.\end{aligned}$$

7.8 FSM Equivalence Checking

Here we address the question of how to determine if two FSMs,

$$\mathcal{M}_1 = \langle X, S^1, \delta^1, S_0^1, O, \lambda^1\rangle, \quad \mathcal{M}_2 = \langle X, S^2, \delta^2, S_0^2, O, \lambda^1\rangle$$

are equivalent. Note that for the equivalence question to make sense, the FSMs must share the same input and output alphabet. In practice, the two given machines $\mathcal{M}_1$ and $\mathcal{M}_2$ are typically comprised of one derived from a high level HDL specification and one derived from a lower level circuit implementation obtained by manual design or synthesis.

We shall consider only the special case in which both FSMs are strongly deterministic, meaning that each state transitions to a unique next state and there is a unique initial state. Note that an FSM is (strongly) deterministic if and only if its underlying FST is (Cf., the definition of Page 293). The more general case of multiple initial states can be similarly handled[18]. In this case, the equivalence problem is: "Find a string $\mathbf{x}$ which distinguishes the initial states s_0^1 of $\mathcal{M}^1$ and s_0^2 of $\mathcal{M}^2$". Methods for solving this problem are part of the foundation of the field of formal verification. Since the FSMs have finite state, it suffices to traverse the STG of their product, and show that no product state is reached which gives different outputs for the two submachines. This can be done in time polynomial in the size of the STG.

Note that whereas in Section 7.4.1 we considered distinguishing input sequences for pairs of states in the same machine, we are interested here in testing for the existence of a sequence which distinguishes between initial states of two different machines. We begin our discussion by elaborating on the idea of distinguishing sequences, and then treat the construction and traversal of the product machine.

7.8.1 Strings which Distinguish Two Machines

For the purposes of synthesis and verification, for example, for the FSM equivalence problem discussed below, it will be necessary to consider (implicitly, for the sake of efficiency) all possible input strings and their corresponding runs. Recall that strings and runs were defined in Definition 7.4.1 on Page 266. Given a starting state $s_0 \in S_0$,the k-string $\mathbf{x} = x_0, \ldots, x_{k-1}$ produces a run $\mathbf{s} = s_0, \ldots, s_k$, where $s_i = \delta(s_{i-1}, x_{i-1})$, and an **output string** $\mathbf{z} = (z_1, \ldots, z_k)$, where $z_i = \lambda(s_{i-1}, x_{i-1})$. We show below that two FSMs $\mathcal{M}_1$ and $\mathcal{M}_2$ are proved equivalent by showing that no (common) input string $\mathbf{x}$ produces different output strings in the two machines. A key concept in equivalence testing is reachability, defined as follows.

Definition 7.8.1 *A state t is* **reachable from state** *s if there exists a string $\mathbf{x}$ which produces a run $\mathbf{s}$, ending in state t. A state t is simply* **reachable** *if it is reachable from any state in S^0.*

[18]Pixley has shown ([215], *et. seq.*) how to define and test for equivalence of two machines that do not have initial state specifications.

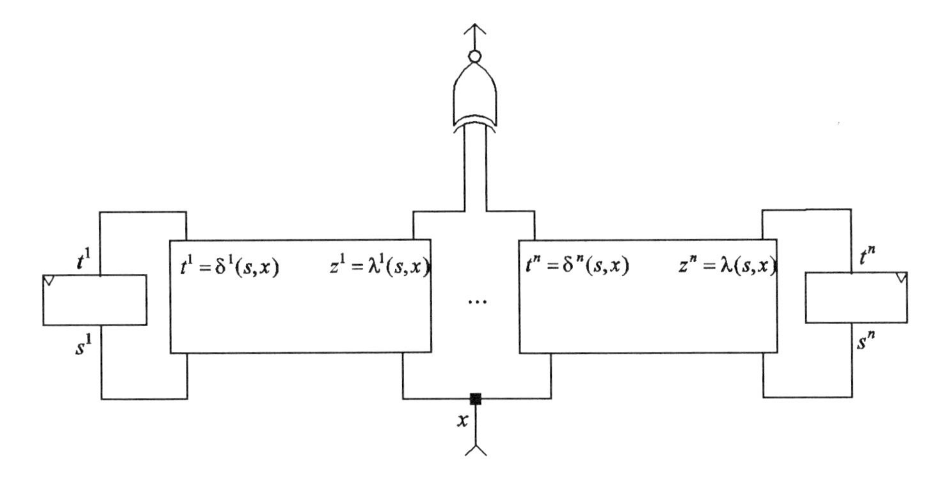

Figure 7.39: Product Machine for Equivalence Checking.

The problem of traversing a product FSM (or an FA, as discussed below) is to enumerate all reachable states in the underlying product FST. Note that in reachability analysis we are only concerned with the next-state function $\delta(s,x)$. Hence, when discussing reachability, we focus on the FST, thus ignoring the output function $\lambda(s,x)$. Once the reachable product states are enumerated, we then check whether the outputs of the submachines are equal in each such state.

7.8.2 Building the Product Machine

Typically one solves the equivalence problem by building a **product machine** $\mathcal{M}^{12} = \mathcal{M}_1 \times \mathcal{M}_2$, as illustrated in Figure 7.39. The two components $\mathcal{M}^1$ and $\mathcal{M}^2$ of the product machine $\mathcal{M}^{12}$ share the same primary input vector x, and have their corresponding outputs fed through a comparator, so that $\mathcal{M}^{12}$ has a single output The product machine $\mathcal{M}_{12} = \langle X, S^{12}, \delta^{12}, S_0^{12}, O^{12}, \lambda^{12} \rangle$ is defined as follows.

$$\begin{aligned}
s^{12} &= (s^1, s^2), \\
t^{12} &= (t^1, t^2), \\
&= \delta^{12}(s^{12}, x) : (S^1 \times S^2) \times X \mapsto (S^1 \times S^2), \\
&= (\delta^1(s^1, x), \delta^2(s^2, x)), \\
z^{12} &= \lambda^{12}(s^{12}, x) : (S^1 \times S^2) \times X \mapsto \{0, 1\},
\end{aligned}$$

where

$$\lambda^{12}(s^{12}, x) = \begin{cases} 1 & \textbf{if } \lambda^1(s^1, x) = \lambda^2(s^2, x) \\ 0 & \text{otherwise} \end{cases}$$

In words, this means that states of the product are formed by simple concatenation of the states, as discussed above for the product of FSTs. That is, the next states of the two component machines are computed independently, and the results catenated to form the next state of the product. The outputs specified for these two transition are then compared, and if they are equal, the product machine outputs a 1, else, it outputs a 0.

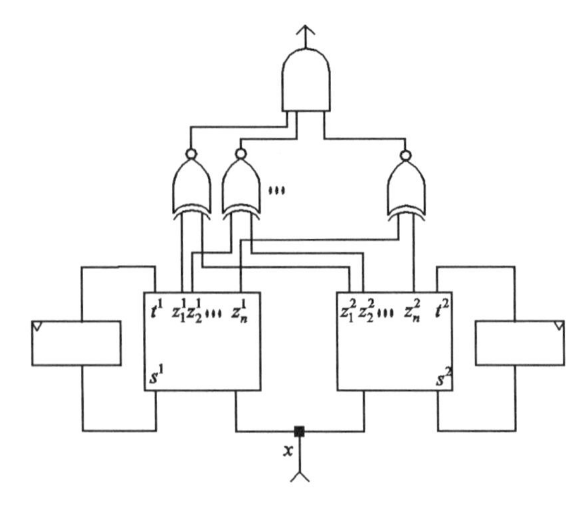

Figure 7.40: Encoded Product Machine for Equivalence Checking.

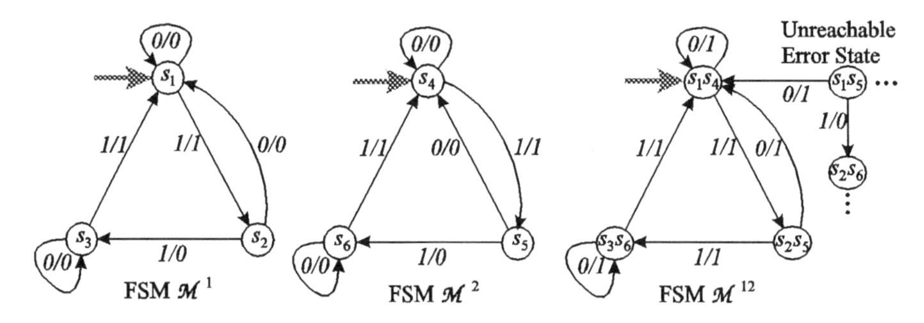

Figure 7.41: Product of two equivalent FSMs.

If the two FSMs are already encoded, and the output alphabet is given an n-bit code, or are sequential circuits with latches, the comparator operation is the AND of XNOR gates, one for each output code bit z_i, $i = 1, \ldots, n$. Figure 7.40.

To solve the equivalence problem, one traverses the STG of $\mathcal{M}_{12}$ starting from the initial state $s_0^{12} = (s_0^1, s_0^2)$ and checks for the output being identically 1 in all states that can be reached.

Example:

An example of equivalence checking based on the product operation is given in Figure 7.41. Here $s_0^1 = s_1$, $s_0^2 = s_4$, $X = O^1 = O^2 = O^{12} = \{0, 1\}$, $S^1 = \{s_1, s_2, s_3\}$, and $S^2 = \{s_4, s_5, s6\}$. In this example, the product machine has $|S^1 \times S^2| = 9$ states. However, only 3 of these states, s_1s_4, s_2s_5, and s_3s_6, are reachable from the initial state. For brevity, only two ((s_1, s_5) and (s_2, s_6)) of the six unreachable states of the product are shown. By comparing the two sub-outputs on all the reachable transitions, we see that they are always equal, so the product machine always outputs a 1. However, it is interesting to observe that transitions from the unreachable states of the product machine may produce error outputs ($z^{12} = 0$). For example, the transition (s_1, s_5) to (s_2, s_6) in response to input $x = 1$ produces an error output.

```
Line Procedure BFS_FSM(S, X, δ, λ, S_0) {
0        Reached = From = New^0 = S_0; k = 0
         do {
            k = k + 1
1           To = Img(δ, From)
2           New^k = To − Reached
            for(s ∈ New^k) {
               for(x ∈ X) {
3                 if(λ(s, x) = 0) {
                     (s, x) = ERROR_TRACE(s, x, k, δ, X, New^j, j = 0, ..., k − 1)
4                    return(FALSE, s, x)
                  }
               }
            }
            From = New^k
            Reached = Reached ∪ New^k     Reached is augmented
5        } while (New^k ≠ ∅)               with vertices newly reached.
6        return(TRUE, ε, ε)
     }
```

Figure 7.42: Procedure for equivalence checking a product machine.

■

In this simple example, the reachable states are visually identifiable, However, in practice this is not usually the case, and automatic procedures must be used, based on the graph traversal procedures BFS (Section 7.5.2) and DFS (Section 7.5.3). Such a modification is presented here as Procedure BFS_FSM of Figure 7.42, which traverses the STG of a given FSM, and on finding an error state, performs a backtrace operation showing how the error state may be reached from the initial state. The backtrace operation is performed by Procedure ERROR_TRACE, given in Figure 7.43. Procedure ERROR_TRACE uses the supplied New^k sets to produce a string **x** and corresponding run **s** which takes the initial state to the error transition.

Example:

Another example of equivalence checking based on the product operation uses a slight alteration of Figure 7.41. The alteration consists changing the output label of the transition from s_6 to s_6 (under input 0) from 0 to 1. This insures that the two machines are no longer equivalent, but retains the same set of reachable states in the product machine. The alteration also changes the output label of the product machine transition from (s_3, s_6) to (s_3, s_6) (under input 0) from 1 to 0. The following is a

```
    Procedure ERROR_TRACE(s_E, x_E, k_e, δ, X, New^j, j = 0,...,k_E − 1){
       /* Error state: λ(s_E, x_E) = 0
0      s = (s_E); x = (x_E)
1      for(k = k_E − 1; k ≥ 1; k++) {
2         for(s ∈ New^k) {
3            for(x ∈ X) {
                if(s_E = δ(s, x)) {
4                  s_E = s; x_E = x
5                  s = (s_E, s); x = (x_E, x)
                }
             }
          }
       }
6      return(s, x)
    }
```

Figure 7.43: Procedure for finding a shortest error trace.

data trace of Procedure BFS_FSM applied to this altered product machine.

$$\begin{array}{llcl} k = 0 & New^0 & = & \{(s_1, s_4)\}, \\ k = 1 & New^1 & = & \{(s_2, s_5)\}, \\ k = 2 = K_E & New^2 & = & \{(s_3, s_6)\}. \end{array}$$

On the kth pass through **do-while** loop, we look at all transitions out of states in the set New^k. This is done by the **if** body of Line 3. On the second pass through the **do-while** loop, we encounter the case $\lambda(s, x) = 0$, signifying that an error state $s = (s_3, s_6)$ has been reached, from which their is a transition for input $x = 0$ with product machine output label 0.

Then Subprocedure ERROR_TRACE is called, which works backward through the provided sets New^k to find string $\mathbf{x}$, which is a distinguishing sequence for the initial pair of states s_1 of machine $\mathcal{M}_1$ and s_4 of machine $\mathcal{M}_2$. This is done by the nested **for** loops of Lines 1, 2, and 3. Note in Line 4 how the new state s_E and input x_E are pushed onto the from of the accumulated run $\mathbf{s}$ and string $\mathbf{x}$, which are thus handled like stacks.

The corresponding data trace of Subprocedure ERROR_TRACE, for $k = 2, 1, 0$ ($k_E = 2$) is

$$\begin{array}{llll} s_E = (s_3, s_6) & x_E = 0 & \mathbf{s} = ((s_3, s_6)) & \mathbf{x} = (0), \\ s_E = (s_2, s_5) & x_E = 1 & \mathbf{s} = ((s_2, s_5), (s_3, s_6)) & \mathbf{x} = (1, 0), \\ s_E = (s_1, s_4) & x_E = 0 & \mathbf{s} = ((s_1, s_4), (s_2, s_5), (s_3, s_6)) & \mathbf{x} = (0, 1, 0). \end{array}$$

■

7.8.3 Equivalence Identification by Isomorphism

In the simple case of the unmodified Figurefi:fsmmin3, the STGs of the two machines are identical except for labeling, that is, the two STGs are *isomorphic*[19], defined as follows. In such a case the reachable states part of the STG of the product machine will also be isomorphic.

Definition 7.8.2 *Two graphs* $G^a = (V^a, E^a)$ *and* $G^b = (V^b, E^b)$ *are said to be* **isomorphic** *if* $|V^a| = |V^b|$ *if the nodes of* V^a *can be relabeled so that the two graphs are identical. A "relabeling function" is a function* $\rho : V^a \longmapsto V^b$ *such that for each node* $v^b \in V^b$ *there is exactly one node* $v^a \in V^a$ *such that* $v^b = \rho(v^a)$ *(Such a function is said to be "1 to 1 and onto" and is called an "isomorphism").*

The relabeling function $\mu : S^1 \longmapsto S^2$ for the FSMs $\mathcal{M}^1$ and $\mathcal{M}^2$ of Figure 7.41 is given by

$$\begin{array}{ccc} s & \leftrightarrow & \mu(s) \\ \\ s_1 & \leftrightarrow & s_4 \\ s_2 & \leftrightarrow & s_5 \\ s_3 & \leftrightarrow & s_6 \end{array}$$

Note that the STGs of two FSMs do not have to be isomorphic for the two to be equivalent. For example, consider the the FSMs of Figure 7.11 and Figure 7.12, which are equivalent, even though their FSMs are not isomorphic.

7.9 Reachability Analysis

The essence of the above method for equivalence checking is to reduce the verification of a global property of one or more machines (in our example, equivalence) to the verification of a local property (identical outputs) that must hold for all transitions from states which are reachable from the initial states. Therefore, **reachability analysis** plays a central role in the verification of finite-state systems and we focus now on this problem. In particular, we shall concentrate on finite state machines, for which the reachability analysis is often called **FSM traversal.**

We set the stage for the discussion of reachability analysis for large state machines by first presenting the notion of using BDDs to represent encoded transition functions, transition relations and the characteristic functions of state sets (BDDs were discussed in Chapter 6).

7.9.1 FSM Traversal Using Binary Decision Diagrams

It is clear from examination of Procedure BFS_FSM that the tractability of FSM traversal rests on two requirements: first, the image of each *From* set must be computable (Line 1) in affordable time; second, the sets *From*, *To*, New^k, and *Reached* must be storable in reasonable disk space. These time and space requirements may be severe — often BFS_FSM must deal with sets with more (or even vastly more) than the sets of 10^7 to 10^8 states that might conceivably be storable on today's largest computers (in

[19] If two FSMs have isomorphic STGs, they are guaranteed to be equivalent [162, Chapter 10]

which 1-2GB of RAM is sometimes available). In many cases, FSM traversal techniques [193, 71, 24] have been able to go far beyond these limitations of explicit set representation, compactly representing sets of vast proportion using BDDs [60, 262].

Once we have formed the product FSM, we know its transition function $t = \delta(s, x) : S \times X \longmapsto S$, and the corresponding **transition relation** $T(s, x, t) = (t \equiv \delta(s, x))$. For efficient traversal of the large product machines which arise in practice, it is necessary to encode the states and inputs of the FSM, as well as the transition function δ itself. Examples of such encodings were given in Section 7.2. We shall treat methods for effectively encoding the inputs and states later, in Section 8.3. For now we focus attention on efficient traversal of encoded FSMs obtained by encoding the states with an n-bit code $s = (s_1, \ldots, s_n)$, and the inputs with a p-bit code $x = (x_1, \ldots, x_n)$.

Suppose the product machine had $|S| = m$ states. S may be given an n-bit binary encoding $s = (s_1, \ldots, s_n)$, where $n = \lceil \log_2(m) \rceil$[20]. The characteristic function of any such encoded set S, is denoted $f_S(s) : \{0,1\}^n \mapsto \{0,1\}$ is a Boolean function (Cf., Section 3.3, Page 95).

Now suppose the states of a set S were given binary integer labels. That is, the first element may be given the code $00 \cdots 0$, the second $00 \cdots 1$, and so on. For example, let

$$S = \{[0,3], [8,15], [16,23]\},$$

where the closed interval notation $[0,3]$ stands for the set of integers between 0 and 3, inclusive. Thus $|S| = 20$, and $n = \lceil \log_2(20) \rceil = 5$. The natural encoding of this set would assign the code 00000 to element 0, the code 00001 to 1 and the code 10111 to

$$23 = 1 \times 2^4 + 0 \times 2^3 + 1 \times 2^2 + 1 \times 2^1 + 1 \times 2^0.$$

The cited literature typically employs functional notation such as $E(u, v)$ for an edge relation E encoded with binary variable vectors u and v, or $S(v)$ for a similarly encoded the characteristic functions of a state set S. This emphasizes the dramatic algorithmic efficiencies that are often available when a combinatorially large set S, $|S| > 10^{10}$ is represented by BDDs [47].

A BDD is a DAG (Directed Acyclic Graph) in which every vertex has exactly two successors, and there is exactly one source vertex and exactly two sink vertices, labeled 0 and 1. As we shall see in Section 7.9 ahead, BDDs are of great importance in the verification and synthesis of sequential circuits. Their utility derives from two sources: (i) their remarkable power in compactly representing very large sets, and (ii) under certain conditions, BDDs are a canonical representation of logic functions (Cf., Section 6.1 on Page 220).

When the members of a set S are given binary encodings, an encoding of length only $\log |S|$ suffices. We shall see in Section 3.3 below, how a BDD is a canonical directed graph representation of the characteristic function of S and in many cases has only on the order of $\log |S|$ vertices. Examples of BDDs are given in Section 6.1 ahead.

We can easily find a BDD which represents the characteristic function $f_S(s)$ of the 20 state S specified above. Such a BDD is shown in Figure 7.44. It is of interest

[20]The notation $\lceil \log_2(k) \rceil$ means "the smallest integer greater than the real number $\log_2(k)$.

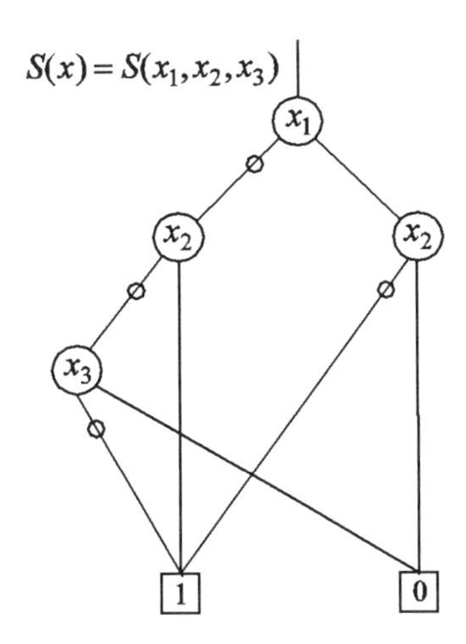

Figure 7.44: A simple BDD representing the characteristic function of the set S.

to observe that a characteristic function of a 20-element set can be represented by a BDD with just 3 nodes. This is not an isolated case but is in fact a powerful general rule — sets of huge cardinality can often be represented by BDDs which are quite small — frequently the node count is on the order of $\log(|S|)$.

Now let us explore in some detail just how the BDD of Figure 7.44 represents the characteristic function f_S of S. With $n = 5$ there are $2^5 = 32$ unique codes. Note that the BDD does not depend on variables s_4 or s_5. The truth table of $f_S(s)$ will have 12 of the 32 possible minterms evaluating to 0. We can enumerate the minterms evaluating to 1 as follows.

x_1	x_2	x_3	x_4	x_5	f	Elements	Count
0	0	0	–	–	1	[0 – 3]	4
0	1	–	–	–	1	[8 – 15]	8
1	0	–	–	–	1	[16 – 23]	8

Thus there are $4 + 4 + 5 = 20$ elements total, each recognized by an path (for the indicated input value assignments) in the BDD to from the root to the 1-node.

If we increase n to 6 but keep, essentially, the same BDD (unchanged except for the inclusion of one additional (unreferenced) input s_6), the same BDD now represents the characteristic function of a larger set

$$S_6 = \{[0,7],[16,31],[32,47]\}.$$

Although the BDD representing the characteristic function stays the same, the set elements represented become

x_1	x_2	x_3	x_4	x_5	x_6	f	Elements	Count
0	0	0	–	–	–	1	[0 – 7]	8
0	1	–	–	–	–	1	[16 – 31]	16
1	0	–	–	–	–	1	[32 – 47]	16

Thus there are $8 + 16 + 16 = 40$ elements total, each recognized by an path (for the indicated input value assignments) in the BDD to from the root to the 1-node. Since there are more (unreferenced) variables, there are more paths to the 1-node.

In fact, depending on the context, the same BDD can represent sets with $5 \times 2^{n-3}$ elements. For example, with $n = 32$, the same 3-node BDD also represents the characteristic function of a corresponding set of about 2.5 billion elements.

It is now becoming common and important practice in the field CAD software development to represent such characteristic functions with called **Binary Decision Diagrams** or just **BDDs**.

7.10 Symbolic FSM State Traversal

Many problems in the formal verification of hardware are based on such reachable states computations for FSMs (Finite State Machines). A reachable state is just one that is reachable for some tape (input sequence) from a given set of initial states. This type of computation uses a "symbolic" breadth first search approach to reach all reachable states by *shortest paths.*

In the context of FSMs, reachable states computations are based on implicit traversal of the STG (State Transition Graph) of the underlying FST. The key step in reachable states computations (and a host of other related computations of formal verification) is the computation of the **image**, $\text{Img}(\delta(s,x), C(s))$, of a given set of points C in the domain of a specified transition function $\delta(s,x)$. Image computation plays a key role in formal verification, especially in FSM verification based on *symbolic* traversal of the STG. Preimage computation plays a similar, perhaps even more important role.

In symbolic traversal one computes *sets of states*, which are reachable in one FSM transition from a specified *set* of states [71, 73, 60, 262, 74, 79]. The breadth first search method of Procedure BFS_FSM allows one to deal naturally with multiple states simultaneously and has thus become the method of choice for the traversal of large machines[21]. The full power of this approach is realized when BDDs are used to represent the characteristic functions of these sets (See Section 6.1). This process is sometimes called **symbolic simulation** of λ.

7.10.1 Transition Relations and Symbolic Image Computation

Let $t_i = \delta_i(s,x), i = 1, \ldots, n$ be the ith encoded next state transition function of a given encoded FSM, and let s and x the coding vectors for states and inputs. A given symbolic state set $C(s)$ (characteristic function) is mapped by $\delta(s,x)$ into a state set $I \subseteq \{0,1\}^n$ in the range, (or co-domain) of the functional vector δ. The set of such co-domain points is called the image of δ under C (Cf., the definitions following Page 84). In the symbolic approach, the image is typically computed using **transition relations**.

[21] It is also possible to extend the depth-first method to deal with groups of states simultaneously [7], yet in a less general and satisfactory way.

Definition 7.10.1 *Given a deterministic transition function $\delta(s,x)$, the corresponding transition relation $T(s,x,t)$ is defined*[22] *by:*

$$T(s,x,t) = \prod_{i=1}^{i=n} (t_i \equiv \delta_i(s,x)).$$

The equation $T(s,x,t) = 1$, denotes a set of encoded triples s, x, t of state s, input x, and x-successor t of s, each representing a transition in the FST of the given FSM.

Given the transition relation it is straightforward to compute the image by Boolean manipulations, but for this we need to define a new Boolean operation called **existential abstraction**.

Definition 7.10.2 *Given an m-variable Boolean function $f(x_1, \ldots, x_m)$, the* **existential abstraction**[23] *of f with respect to x_i is:*

$$\exists_{x_i} f = f_{x_i} + f_{\overline{x}_i}.$$

Here $f_{x_i} = f(x_1, \ldots, x_{i-1}, 1, x_{i+1}, \ldots, x_m)$ stands for the positive cofactor of f with respect to x_i and $f_{\overline{x}_i} = f(x_1, \ldots, x_{i-1}, 0, x_{i+1}, \ldots, x_m)$ stands for the corresponding negative cofactor.

The name derives from the following property, which can be easily verified.

$$(\exists_{x_i} f)(x_1, \ldots, x_{i-1}, x_{i+1}, \ldots, x_n) = 1 \Leftrightarrow \exists x_i \ni f(x_1, \ldots, x_n) = 1.$$

It can be shown that f_{x_i} is the smallest (fewest minterms) function that contains all minterms of f, and is independent of x_i.

We illustrate this definition by computing $\exists_x f$ for the following function.

$$f = x'y'z + xz' + xy.$$

The two cofactors are:

$$f_z = x'y' + xy \qquad \text{and} \qquad f_{z'} = x.$$

Hence, $\exists_z f = x + y'$.

Given $f(s,x) = f(s_1, \ldots, s_n, x_1, \ldots, x_m)$, the existential and universal abstractions with respect to a set of variables are easily defined.

$$\exists_x f(s,x) = \exists_{x_1}(\cdots(\exists_{x_{m-1}}(\exists_{x_m} f(s,x)))).$$

That is, we first abstract x_m from the original function f. Then we abstract x_{m-1} from the result of this abstraction. From this second result we abstract x_{m-2}, and so on. The order in which we do these abstractions is immaterial. The final result $g(s) = \exists_x f(s,x)$ is the smallest (fewest minterms) function that contains all minterms of $f(s,x)$, and is independent of x.

[22] Recall that $a \equiv b = ab + \overline{a}\overline{b}$.

[23] There is a corresponding dual operation, called **universal abstraction** which takes the conjunction rather than the disjunction of the cofactors.

Given the above definition we can easily compute the image of the set $C(s)$ as

$$I(t) = \text{Img}(T, C) = \exists_x \exists_s C(s) \cdot T(s, x, t). \tag{7.2}$$

In words, the image computation proceeds as follows. First compute the transition relation $T(s, x, t)$, and then compute the conjunction of this function and the function $C(s)$ and call the result $f(s, x, t)$. Then existentially abstract all the s-variables and all the x-variables to obtain the result $I(t)$. The result is the smallest function independent of s and x which contains all the triples in $f(s, x, t)$.

Like this method of image computation, all the other steps of Procedure BFS_FSM can be converted into symbolic, BDD-based procedures.

We conclude our treatment of the symbolic approach with a simple example. Consider the two FSMs of Figure 7.45.

Example:

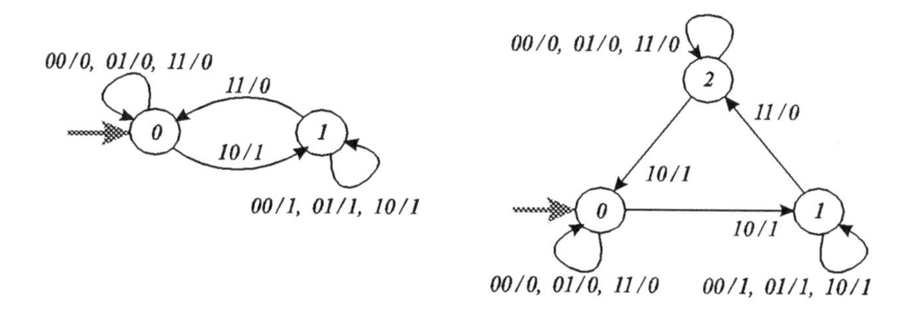

Figure 7.45: Two non-equivalent FSMs

These two machines are not equivalent. Their product is illustrated in Figure 7.46. Note in Figure 7.46 and other computer generated graphics,

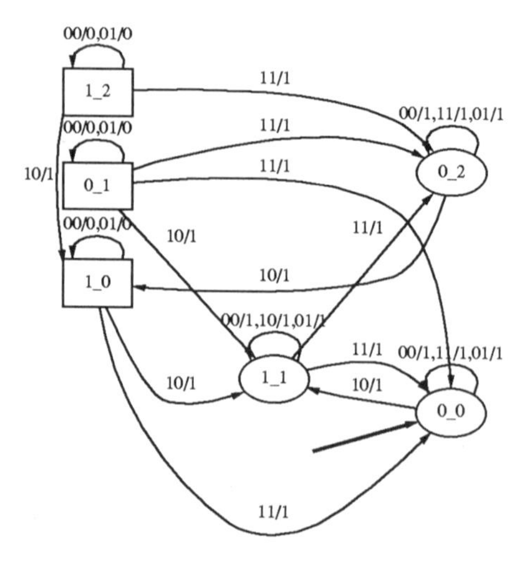

Figure 7.46: Product of the Two FSMs of Figure 7.45.

in the sequel, the octagonal state indicates the initial state of the FST.

Here the product states are given by the left-machine state (0 or 1) followed by an underscore and the right-machine state (0, 1, or 2). The error states of the product are $\{1\_0, 0\_1, 1\_2\}$, are distinguished by box shapes rather than ellipses. Note the strings $\mathbf{x^1} = (10, 11, 10, 00)$ and $\mathbf{x^2} = (10, 11, 10, 01)$ produce the corresponding runs $\mathbf{r^1} = \mathbf{r^2} = (0\_0, 1\_1, 0\_2, 1\_0, 1\_0)$, which have the output strings $\mathbf{o^1} = \mathbf{o^2} = (1, 1, 1, 0)$. Thus these two strings are distinguishing sequences for the pair of initial states of the two machines. Examination of the product FSM shows that these are the shortest error traces possible.

We now show how the above error traces can be obtained with symbolic, BDD-based computation. We illustrate only the first image computation in the application of Procedure BFS_FSM of Page 303. We first treat the 2-state FSM on the left of the figure, using the natural encoding with code bit s_1 to encode the two states, and the natural encoding for the four input symbols, which leads, as discussed in Section 7.1 of Page 255, to

$$\begin{aligned} t_1 = \delta_1^1 &= \overline{s}_1 x_1 \overline{x}_2 + s_1(\overline{x}_1 + \overline{x}_2) \\ \lambda^1 &= \delta_1^1. \end{aligned}$$

We then treat the 3-state FSM on the right. We begin by encoding the states — we use the natural binary encoding of the digits, so that states $0, 1, 2$ of the left FSM are encoded as $(s_1, s_2) = 00, 01, 10$. Since two code bits are required we must have two latches, that is, one next state transition function for each code bit. We then construct a truth table (not shown) to realize the three functions $\delta_3^2(s_2, s_3, x_1, x_2)$, $\delta_2^2(s_2, s_3, x_1, x_2)$, and $\lambda(s_2, s_3, x_1, x_2)$.

By collecting the terms of the truth table where functions evaluate to 1, we get, after some simplification, the following expressions for $\delta_1(s, x)$, $\delta_2(s, x)$ and $\lambda(s, x)$.

$$\begin{aligned} t_2 = \delta_2^2 &= s_3 x_1 x_2 + s_2(\overline{x}_1 + x_2) \\ t_3 = \delta_3^2 &= \overline{s}_2 x_1 \overline{x}_2 + s_3(\overline{x}_1 + x_2) \\ \lambda^1 &= s_3 \overline{x}_1 + x_1 \overline{x}_2. \end{aligned}$$

We then form the product machine transition relation as the conjunction of the transition relations for the two submachines, leading to

$$T(s, x, t) = (t_1 \equiv \delta_1^1) \cdot (t_2 \equiv \delta_2^2) \cdot (t_3 \equiv \delta_3^2).$$

The first *From* set is the $C(s) = \overline{s}_1 \overline{s}_2 \overline{s}_3$, which is the characteristic function of the set consisting of only the initial state of the product machine. We now wish to compute the image of $C(s)$, using Equation 7.2. First we observe that

$$\begin{aligned} &T(s, x, t) \cdot C(s) = T(s, x, t) \cdot \overline{s}_1 \overline{s}_2 \overline{s}_3 = \\ &\qquad (t_1 \equiv \overline{s}_1 x_1 \overline{x}_2) \cdot (t_2 \equiv 0) \cdot (t_3 \equiv \overline{s}_2 x_1 \overline{x}_2) \cdot \overline{s}_1 \overline{s}_2 \overline{s}_3. \end{aligned}$$

Since this conjunction evaluates to 1 for just one s-minterm $(\overline{s}_1 \overline{s}_2 \overline{s}_3)$, it should be clear that

$$g(x, t) = \exists_s (T(s, x, t) \cdot C(s)) = (t_1 \equiv x_1 \overline{x}_2) \cdot (t_2 \equiv 0) \cdot (t_3 \equiv x_1 \overline{x}_2)$$

If we observe that $g_{x_1\overline{x}_2} = (t_1 \equiv 1) \cdot (t_2 \equiv 0) \cdot (t_3 \equiv 1)$, and $g_{x_1x_2} = g_{\overline{x}_1x_2} = g_{\overline{x}_1\overline{x}_2} = (t_1 \equiv 0) \cdot (t_2 \equiv 0) \cdot (t_3 \equiv 0)$, we can easily show that

$$I(t) = \text{IMG}(T, C) = \exists_x g(x,t) = \overline{t}_1\overline{t}_2\overline{t}_3 + t_1\overline{t}_2t_3.$$

Observe that this is consistent with the fact that in the STG of Figure 7.45, the image of the initial state of the product machine is the set $\{0\_0, 0\_1\}$, which is encoded as the set with the characteristic function $I(t) = \overline{t}_1\overline{t}_2\overline{t}_3 + t_1\overline{t}_2t_3$.

This example demonstrates how the symbolic method correctly computes the image of the product machine by Boolean function manipulation.
■

The great practical importance of this symbolic approach derives from the fact that FSMs of interest may have 10^{10} states (about 34 latches) to 10^{100} states (about 333 latches). or an even greater number of states. This makes it impractical to represent states as individual entities in a computer program. Instead, the state sets of algorithm BFS_FSM are represented symbolically as BDDs. Often billions of states may be stored by a BDD with just thousands of nodes — leading to a cpu storage cost of less than one megabyte. Coudert and Madre [70] are generally credited with this extremely important discovery, although McMillan [194] did earlier, but unpublished, similar work.

Another major advantage of BDDs over other representations of functions such as logic functions and the characteristic functions of sets is that BDDs are canonical. That is, given the ordering restriction discussed in Section 6.1, there is just one unique BDD graph structure for each such function. As a result to FSMs with the same transition relation BDD are equivalent. This aspect is crucial in many synthesis and verification applications.

Unfortunately, however, BDDs are not the panacea. Multipliers, for example, have no efficient BDD representation, and some smallish sequential circuits with less than 100 latches require more than 1 gigabyte of BDD memory to store their reachable state sets. After a proper introduction of Boolean algebras, we shall present, in Section 6.1, the details of how BDDs are used in the representation of large sets and of logic functions.

7.11 Notes

State minimization of completely specified machines such as large sequential circuits has been studied in [177]. The complexity of state minimization in this context has been studied in [146].

Coudert *et al.* [71] have shown that breadth first traversal is more amenable to symbolic treatment than depth first traversal, and hence can deal with sequential machines with many more states. Although quite successful, the symbolic methods developed so far (see, for instance, [50, 60, 153]) cannot complete the reachability analysis for many large finite state machines, because they require too much memory, or are computationally intensive. Depth first traversal was combined with BDD-based symbolic methods in [226].

7.12 Summary

In this chapter we have briefly reviewed some important models that have been used to characterize sequential digital systems. These models were illustrated in Figure 7.32:

- **Finite State Transition Structures**, which we have called FSTs — In a given state, FSTs receive an input symbol, and make a transition to a new state;
- **Finite Automata**, which we shall call FAs — FAs are FSTs which also take notice when a favorable state (called *accepting*) is entered;
- **Finite State Machines**, which we shall call FSMs — FSMs are FSTs which emit a specified output symbol when they make a transition;

The treatment has emphasized the role of the underlying FSTs upon which automata and state machines are based.

We have presented in some detail the subject of state equivalence, beginning by defining the concept of "k-equivalence". Two states s and t of an FSM are k-equivalent if there exists no distinguishing input sequence of length k. If s and t are not k-equivalence, they are k-distinguishable. This means that when the length-k distinguishing sequence is applied, the kth output obtained when the machine is started in state s, is different than it would be if started in state t.

If there are n states in an FSM, two states were shown to be equivalent if and only if they are $n-1$ equivalent. This has led to the partition/refinement algorithm for state minimization of an FSM, which first identifies all pairs of equivalent states, and then collapses sets of pairwise equivalent states into single states which are "representatives" of their equivalence classes.

By the definition of product machine, the concept of state equivalence was then used to define the behavioral equivalence of two FSMs. The bottom line is (roughly speaking) that two FSMs are equivalent if their initial states are equivalent.

We concluded by describing the revolutionary effect that BDDs have had on the nascent field of FSM equivalence checking in particular, and on Formal Verification in general.

7.13 Problems

1. Give the flow table which corresponds to the STG of Figure 7.11.

   Solution.

Present State	Next State/Output	
	$x = 0$	$x = 1$
A	E/0	D/1
B	D/0	F/0
C	E/0	B/1
D	B/0	F/0
E	C/0	F/1
F	B/0	C/0

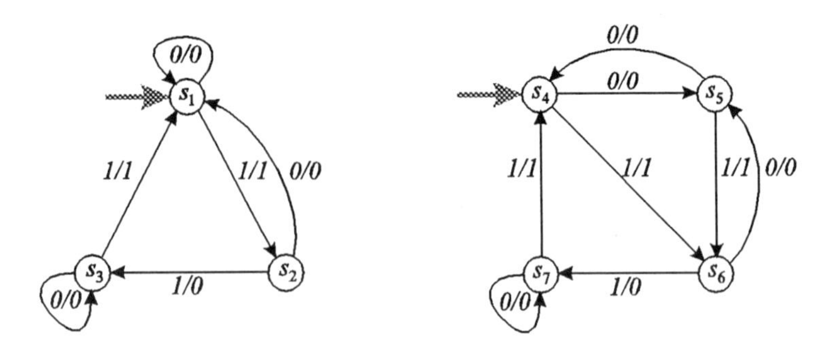

Figure 7.47: The STGs of two equivalent FSMs.

□

2. (a) Find the STG of the flow table of Figure 7.15.

   (b) Give two shortest possible distinguishing sequences for the pairs (A, D) and (D, G). For each sequence, give the corresponding run and output sequence.

3. Consider the STGs of two FSTs of Table 7.2 and Table 7.3, given in Figure 7.34 of Page 294. Note the second table differs by the deletion of the $(D, 0)$ edge, and by the addition of extra edges on A and D .
   (a) State why these FSTs are or are not deterministic and complete.
   (b) Give the run sets $R(\mathcal{T}, \mathbf{x})$ for the FA's defined on the above two FSTs, for the string $\mathbf{x} = (0, 0, 0, 1, 0, 1, 0)$, and for the initial state A.
   Solution. (a) The FST of Table 7.2 is deterministic and complete because the image of every state-input pair is specified as a singleton set. The FST of Table 7.3 is nondeterministic and incomplete, because the images of $(A, 0)$ and $(D, 1)$ are not singleton sets, and the image of $(D, 0)$ is not specified.

   (b) For the string $\mathbf{x} = (0, 0, 0, 1, 0, 1, 0)$, the run set of The FST of Table 7.2 is $\{A, E, C, E, F, B, D, F\}$. For the string $\mathbf{x} = (0, 0, 0, 1, 0, 1, 0)$, the run set of The FST of Table 7.3 is

   $$\{A, E, C, E, F, B, D, s_?\}$$
   $$\{A, D, s_?\}$$

   Note the first run starts out the same as for the first table, but stops at D, since D has no specified 0-successor. The second run exists because D is an alternate nondeterministic 0-successor of A. However, the run stops there since D has no 0-successor. Thus the extra 1-successors of D have no effect for this string. □

4. Give the STG of the product machine corresponding to the two FSMs of Figure 7.47. Show all transitions from all states, including unreachable states.

5. Show in what sense the STG of Figure 7.48 represents a modulo 3 counter,

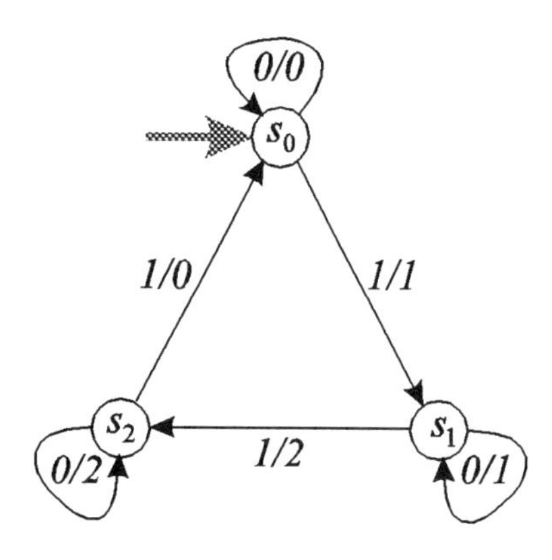

Figure 7.48: The STG of a modulo 3 counter.

Solution. First, it returns to the initial state after receiving a total of 3 1-inputs (a counter counts up one for each 1-input). Second, it outputs the residue modulo 3 of the number of 1's in the input string. □

6. Apply the minimization procedures of Section 7.4 to the FSM of Figure 7.49.

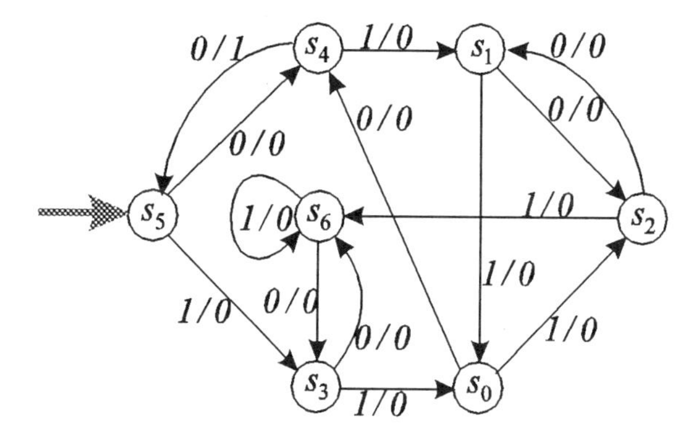

Figure 7.49: The STG of an FSM to be minimized.

List the set of k-equivalent state pairs, for $k = 0, 1, 2, 3, 4, 5$ (to save space, you may just list the sets of states that are pair-wise k-equivalent). Give the undirected graph representing the equivalence classes.

7. In this problem we present an alternative algorithm for state equivalence checking, consisting of formulate two procedures, EQUIV_1 and EQUIV_N. Working in concert, these procedures determine the set $\xi^n(s)$ of all equivalent state pairs of a given n-state FSM.

   After reading the discussion given below, apply Procedure EQUIV_N to the STG of Figure 7.11.

```
EQUIV_1 (X, S, δ(s, x), λ(s, x)) {
   ξ^0 = S × S;  ξ^= = {(s, t) | s = t ∈ S}
   ξ^1 = ξ^=
   for((s, t) ∈ ξ^0 − ξ^=) {
      test = 1
      for(x ∈ X) {
         if(λ(s, x) ≠ λ(t, x)) {
            test = 0; break
         }
      }
      if(test = 1)  ξ^1 = ξ^1 ∪ {(s, t), (t, s)}
   }
   if(ξ^1 = ξ^0) return(−1, ξ^0)}          Combinational case.
   else if  (ξ^1 = ξ^=) return(0, ξ^1)}  No 1-equivalent pairs.
   else return  (1, ξ^1)}                 Normal termination.
}
```

Figure 7.50: Procedure for finding 1-equivalent states of an FSM.

First, consider Procedure EQUIV_1 of Figure 7.50, which determines the set $\xi^1(s)$ of all 1-equivalent state pairs. The procedure begins by computing ξ^0 as the set of all pairs and the set $\xi^=$ as the set of identity pairs s, s (by the above definition, every pair is 1-equivalent (and equivalent) to itself). The desired set ξ^1 is initialized to $\xi^=$. Then, for each pair $(s,t) \in \xi^0 - \xi^=$, we test all inputs $x \in X$ to see if the outputs on the corresponding transition are always identical. If so, then the pair (s,t) (and (t,s) as well, since the equivalence relation is symmetric) is added to ξ^1.

If Procedure EQUIV_1 returns $(1, \xi^1)$ the first return argument indicates normal termination. If Procedure EQUIV_1 returns $(0, \xi^=)$ the first return argument indicates that no state is 1-equivalent to any state besides itself. If Procedure EQUIV_1 returns $(0, \xi^0)$ the first return argument indicates that all the states of the specified machine are pairwise equivalent

Once we know the set of 1-equivalent state pairs, we can find the equivalent pairs without further reference to the output function $\lambda(s,x)$, by calling Procedure EQUIV_N of Figure 7.51. The arguments of EQUIV_N include ξ^1, as computed by EQUIV_1. The procedure consists of **for**-loop for k from 2 through n. In each loop the set ξ^k of k-equivalent pairs is computed according to Definition 7.4.2 above. Loop processing begins by initializing ξ^k to $\xi^=$, since we know that every state is k-equivalent to itself.

Then, each state pair $(s,t) \in \xi^{k-1} - \xi^=$ is tested to see if, for all $x \in X$, the x-successors s_x, t_x, where $s_x = \delta(s,x)$ and $t_x = \delta(t,x)$,

```
EQUIV_N (X, S, ξ¹) {
/*  ξ¹ is the set of 1-equivalent state pairs */
        ξ⁼ = {(s,t) | s = t ∈ S}
        for(k = 2, ···, |S|) {
                ξᵏ = ξ⁼
                for((s,t) ∈ ξᵏ⁻¹ − ξ⁼) {
                        if ((s,t) ∈ ξᵏ) continue
                        test = 1
                        for(x ∈ X) {
                                s_x = δ(s,x);  t_x = δ(t,x)
                                if ((s_x,t_x) ∉ ξᵏ⁻¹) {
                                        test = 0; continue
                                }
                        }
                        if (test = 1)ξᵏ = ξᵏ ∪ {(s,t),(t,s)}
                }
                if (ξᵏ = ξᵏ⁻¹) break
        }
        return (k, ξᵏ)
}
```

Figure 7.51: Procedure for finding equivalent states of an FSM.

are both members of the previously computed ξ^{k-1}. If so, both (s,t) and (t,s) can be added to ξ^k as shown by Theorem 7.4.1.

As per Theorem 7.4.2, and its preceding proposition, the procedure terminates when it is observed that $(\xi^k = \xi^{k-1})$, which is bound to occur for some $k \leq n = |S|$.

Solution. In the example of Figure 7.11 we summarize the action of EQUIV_1 and EQUIV_N as follows. For EQUIV_1 we have

$$\begin{array}{lcccccc} \text{states:} & A & B & C & D & E & F, \\ \text{0-outputs:} & 0 & 0 & 0 & 0 & 0 & 0, \\ \text{1-outputs:} & 1 & 0 & 1 & 0 & 1 & 0. \end{array}$$

Thus the 1-equivalence classes are $B_1^1 = \{A, C, E\}$, $B_2^1 = \{B, D, F\}$. Similarly, for For EQUIV_N

$$\begin{array}{lccccccc} \text{states:} & A & C & E & & B & D & F \\ \text{0-successors:} & E & E & C & & D & B & B \\ \text{1-successors:} & D & B & F & & F & F & C \end{array}$$

Note the 0-successors of A, C, and E are all 1-equivalent, and similarly for the 1-successors. However, The 1-successor of F is $C \not\equiv_1 F$, so $F \not\equiv_1 B$. Thus the 2-equivalence classes are $B_1^2 = \{A, C, E\}$, $B_2^2 = \{B, D\}$. and $B_3^2 = \{F\}$. Continuing, we have

$$\begin{array}{lcccccccc} \text{states:} & A & C & E & & B & D & & F \\ \text{0-successors:} & E & E & C & & D & B & & B \\ \text{1-successors:} & D & B & F & & F & F & & C \end{array}$$

This time, the 0-successors of A, C, and E are all 2-equivalent, but the 1-successors are not. The 1-successor of E is $F \not\equiv_2 D$, so $E \not\equiv_3 A$. Thus the 3-equivalence classes are $B_1^3 = \{A, C\}$, $B_2^3 = \{E\}$, $B_3^3 = \{B, D\}$. and $B_4^3 = \{F\}$. Continuing, we have $\xi^4 = \xi^3$, so by the above theorem, these are also the n-equivalence classes of the given FSM. □

8. Analyze the upper bound asymptotic complexity of Phase 1 (Lines 1-4) of STATE_EQUIVALENCE, according to the definitions given in Section 1.6.1 of Page 34. Then compare this to the complexity of Procedure EQUIV_1 of Problem 7

9. For the flow table of Figure 7.52, draw the state transition graph.
   Solution. The STG is given in Figure 7.53. □

10. This problem and the following one explore the limits of finite state machines. Consider a completely specified, Moore-type FSM $\mathcal{M} = \langle I, S, O, \delta, \lambda, \{s_1\}\rangle$ and suppose $I = \{0\}$, $O = \{0, 1\}$, $S = \{s_1, \ldots, s_n\}$, for n finite, but otherwise arbitrary. No assumptions are made for δ and λ.

	0	1	
1	2	4	0
2	1	3	0
3	6	7	1
4	5	8	1
5	1	8	0
6	2	7	0
7	3	5	1
8	4	6	1

Figure 7.52: A Completely Specified Flow Table.

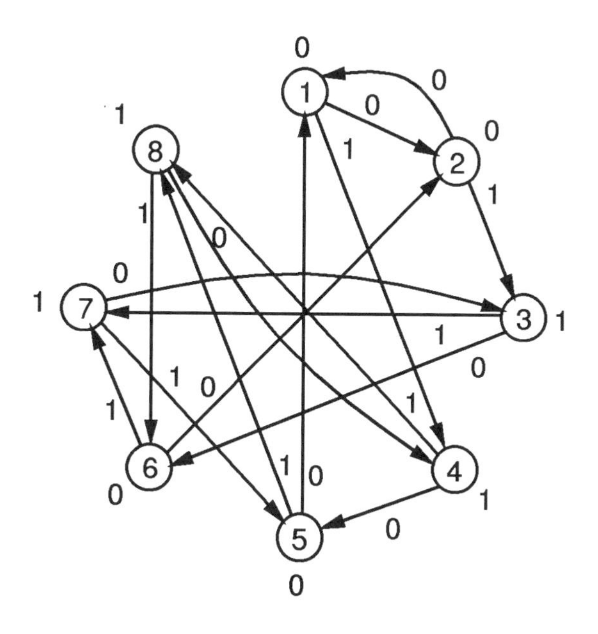

Figure 7.53: STG for the Flow Table of Figure 7.52.

The only possible input sequence is the one composed of all zeroes. Suppose this sequence is applied to $\mathcal{M}$. Show that the resulting output sequence is eventually periodic and that the period is at most n.

11. Based on the result of the previous problem, say whether it is possible to design a finite state machine that, upon receipt of a string of all zeroes, produces an output strings of zeroes and ones, such that the i-th element of the string is 1 if and only if i is a prime number.

Solution. In view of the result of Problem 26, it is sufficient to show that the prime numbers do not eventually form a periodic sequence. Suppose they did. Then it would be possible to write periodic numbers as:

$$p_k = m + kn,$$

for $k > k_0$. (We make this assumption, because we can accept a finite prefix of the sequence of primes that is not periodic.) However, if we now take l such that $ml > k_0$, we have:

$$p_{ml} = m + mln = m(1 + ln).$$

Hence, p_{ml} is not a prime, if $m \neq 1$. If $m = 1$, then:

$$p_k = 1 + kn.$$

Take l such that $l(n+1) + 1 > k_0$. Then,

$$p_{l(n+1)+1} = 1 + ln^2 + ln + n = (n+1)(ln+1).$$

Hence, $p_{l(n+1)+1}$ is not prime. This concludes our proof. □

12. For the flow table of Figure 7.52, find the minimal equivalent flow table, by applying the partition refinement algorithm.

13. For the flow table of Figure 7.54, find the minimal equivalent flow table, by applying the partition refinement algorithm, Procedure STATE_EQUIVALENCE, to identify the equivalent states.

Solution. The initial partition, based on the output values, is:

$$\{\{1,2,4,6\},\{3,5,7,8\}\}.$$

Refining the first block, we get:

$$\{\{1,2\},\{4,6\},\{3,5,7,8\}\}.$$

Finally, refining the last block, we get:

$$\{\{1,2\},\{4,6\},\{3,7\},\{5,8\}\}.$$

The reduced flow table is shown in Figure 7.55. □

	0	1	
1	5	4	1
2	5	6	1
3	5	7	0
4	1	7	1
5	2	1	0
6	2	7	1
7	8	3	0
8	1	2	0

Figure 7.54: A Completely Specified Flow Table.

		0	1	
{1, 2}	a	d	b	1
{4, 6}	b	a	c	1
{3, 7}	c	d	c	0
{5, 8}	d	a	a	0

Figure 7.55: Minimized Flow Table for Figure 7.54.

	00	01	11	10	
1	1	2	3	5	0
2	4	6	2	1	0
3	3	2	1	5	0
4	2	2	1	6	1
5	6	3	3	1	1
6	4	2	6	3	0

Figure 7.56: Flow Table for Problem 14.

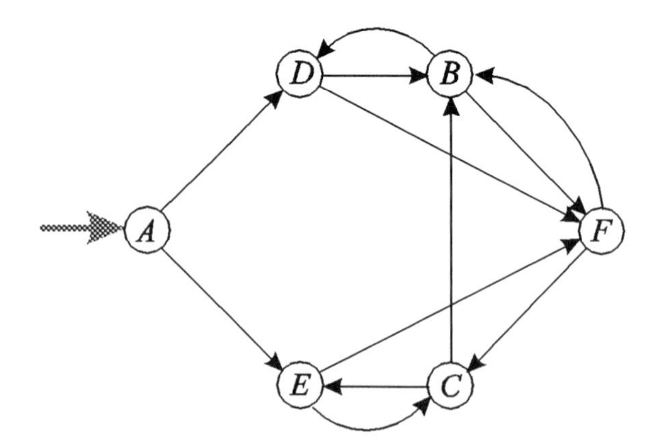

Figure 7.57: A simple directed graph.

14. Find the state-minimal FSM equivalent to the one of Figure 7.56, using Procedure STATE_EQUIVALENCE to identify the equivalent states.

15. Enumerate the simple cycles in the graph of Figure 1.11 on Page 23.
Solution.
Simple cycles are,
$((D,B)(B,D)), ((B,F)(F,B)), ((C,E)(E,C)), ((B,F)(F,C)(C,B)),$
$((D,F)(F,B)(B,D)), ((C,E)(E,F)(F,C)), ((C,B)(B,D)(D,F)(F,C))$
□

16. Find the CCCs of the graph of Figure 7.58.

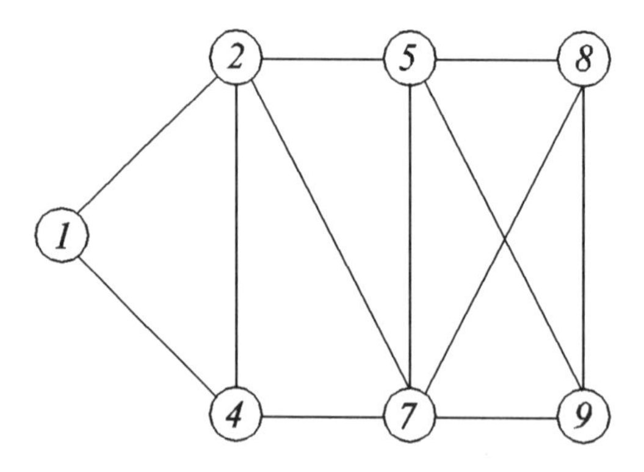

Figure 7.58: Another simple undirected graph.

17. Find the SCCs of the graph if Figure 1.11.
Solution. The SCCs of the graph of Figure 1.11 are, $\{A\}, \{B, C, D, E, F\}$ □

18. Apply algorithm DFS to the graph of Figure 7.31, starting at vertex 0. Give the pre-order and post-order search tags as in Figure 7.24.

19. Apply algorithm BFS to the graph of Figure 7.31, with starting vertex set $V_0 = \{1, 10\}$.

Solution.

(a) $Reached = From^0 = \{1, 10\}$

(b) $To^1 = New^1 = From^1 = \{6, 9\}$

(c) $To^2 = New^2 = From^2 = \{7, 8\}$

(d) $To^3 = \{8\}$ $New^3 = \{\emptyset\}$

$Reached = \{1, 6, 7, 8, 9, 10\}$ □

20. Apply algorithm DFS to the graph of Figure 7.59, starting at vertex 0, and always searching first the fanout vertex of maximal index. For example, when

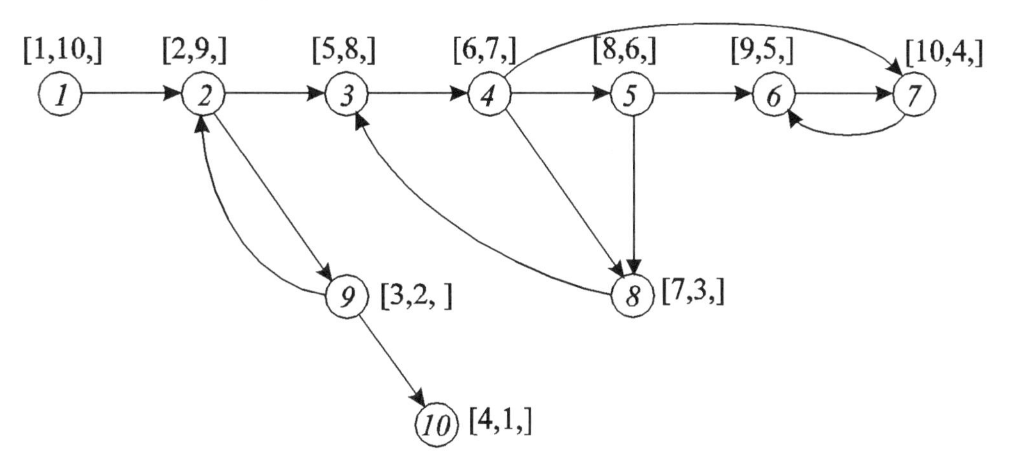

Figure 7.59: Partial labeling of directed acyclic graph 7.59.

searching vertex 2, recursively search fanout vertex 9 before vertex 3. Label each vertex with 3 integers: first, the pre-order and post-order search tags as in Figure 7.24, and third, the **lowlink**, defined as follows. When a search of vertex u completes, the lowlink value of the node is the minimum of the lowlink values of its fanout edges. Edge lowlink values are defined only for two cases.

First, when traversing any edge from the active vertex u to any previously visited vertex v whose DFS is currently incomplete, the edge lowlink value is defined as the pre-order search index of v.

Second, when a search of vertex v is completed, the lowlink value of the edge (u, v) traversed to initiate that search is defined as the lowlink value of v.

As you traverse the graph of Figure 7.59, complete the triple vertex label with the lowlink value of the vertex.

Note that the SCCs of the DAG are identified by the condition pre-order[v]=lowlink[v].

21. Complete the data table of Table 7.1. That is, continue adding lines to the table (above the "final" row), until all reachable vertices have become active.

Solution.

$v/\lambda^*/$ adj(v) Line 6	$(\lambda_v/(v \in \mathit{Reached}))$ Line 10										S Line 10
	1	2	3	4	5	6	7	8	9	10	
0/0/ {1,2,3,4}	1/0	2/0	3/0	4/0	∞/0	∞/0	∞/0	∞/0	∞/0	∞/0	{1,2,3,4}
1/1/{6}	1/1	2/0	3/0	4/0	∞/0	2/0	∞/0	∞/0	∞/0	∞/0	{2,6,3,4}
2/2/{6}	1/1	2/1	3/0	4/0	∞/0	2/0	∞/0	∞/0	∞/0	∞/0	{6,3,4}
6/2/{7,8}	1/1	2/1	3/0	4/0	∞/0	2/1	6/0	3/0	∞/0	∞/0	{3,8,4,7}
3/3/{5,6}	1/1	2/1	3/1	4/0	8/0	2/1	6/0	3/0	∞/0	∞/0	{4,5,7,8}
8/3{$\emptyset$}	1/1	2/1	3/1	4/0	8/0	2/1	6/0	3/1	∞/0	∞/0	{4,5,7}
4/4/{5,7,9}	1/1	2/1	3/1	4/1	8/0	2/1	6/0	3/1	7/0	∞/0	{5,7,9}
7/6/{8,9}	1/1	2/1	3/1	4/1	8/0	2/1	6/1	3/1	7/1	∞/0	{5,9}
5/8/{6}	1/1	2/1	3/1	4/1	8/1	2/1	6/1	3/1	7/1	∞/0	{9}
9/7/{$\emptyset$}	1/1	2/1	3/1	4/1	8/1	2/1	6/1	3/1	7/1	∞/0	{$\emptyset$}
final λ:	1	2	3	4	8	2	6	3	7	∞	

Table 7.4: Data trace for procedure shortest_path

□

22. (A) Consider the logic graph of Figure 1.12 to be a graph $G = (V, E)$, where the 11 vertices in the set V are in 1-1 correspondence with the logic gates number 1 to 11 (including the input and output buffers). Vertex pair $(u, v) \in E$ if and only if the output of gate u is an input of gate v

    (a) Is the edge relation E an equivalence relation?

    (b) Is it a partial order?

    Explain.

    (B) Now modify the circuit with a wire connecting output buffer 10 with input buffer 1, so that a new edge (10,1) is added to the edge relation of part (a).

    (a) Is the modified edge relation E^+ an equivalence relation?

    (b) Is it a partial order?

    (C)
    Identify the strongly connected components of the graph of part (b). Assume that any graph with just 1 vertex and no edges (or one self-loop edge) is strongly connected.

Chapter 8

Synthesis and Verification of Finite State Machines

In Chapter 7 we established a foundation for understanding the operation of FSTs, FAs, and FSMs as abstract transition systems, and also treated the graph models which are used in their characterization. In Section 7.2 we outlined FSM state minimization and equivalence checking. These are two of the most crucial steps crucial steps in automated logic design, since they establish the basis for equivalence preserving optimization steps. However, the treatment was limited to completely specified FSMs.

In the present chapter, we extend the treatment of state minimization to the much more complicated case of incompletely specified machines. Then, in Section 7.4 we discuss the details of state minimization of FSMs. We also include a discussion of the practical algorithms for state encoding, which is required to synthesize a physical sequential circuit from an abstract FSM. We conclude with a brief treatment of the "partition with substitution property" theory of Hartmanis and Stearns [138].

Although the entire treatment is in terms of FSMs, we emphasize that all the presented techniques are equally applicable to DFAs and NFAs (recall that in Definition 7.4.5 of Page 272, we described how an equivalent FSM can be derived for any given FA). This broad applicability is due to the fact that FSMs and FAs are both built on top of FSTs.

8.1 Minimization of Incompletely Specified Machines

The state minimization problem becomes more difficult when we move from completely specified to incompletely specified machines. For starters, the simplistic approach of trying all possible assignments of states and output values to the *don't cares* and choosing the best result does not guarantee the minimum cost solution. (We shall see a counterexample.) This situation should be contrasted with the one encountered in the minimization of incompletely specified switching functions. In the case of a switching function with don't cares, it is possible—though inefficient—to obtain the minimum cost solution by trying all possible assignments of zeroes and ones to the don't cares. As a result, the modification to the minimization algorithm to accommodate don't cares are relatively minor. This is not the case for FSMs,

	0	1	
1	3	2	0
2	-	1	-
3	1	-	1

Figure 8.1: An incompletely specified Moore machine.

	0	1	
1	2	3	-
2	-	1	0
3	1	-	1

Figure 8.2: Another incompletely specified Moore machine.

where the minimization algorithm for incompletely specified machines is quite different from and noticeably more complex than the partition refinement method that we have examined in the previous section.

We still want to base state minimization on the notion of indistinguishability, but we need to define it carefully, in the presence of don't cares. Consider the flow table of Figure 8.1. Obviously, States 1 and 3 are distinguishable because they produce different outputs. However, State 2 could be made indistinguishable from, say, 1, by judiciously filling the don't care entries. This possibility leads us to declare States 1 and 2 *compatible*. Two states are compatible, in general, if they have the same output values, wherever they are both specified, and their pairs of successors are also compatible, when they are both specified.

From Figure 8.1 we see that 2 and 3 are also compatible. This illustrates the fundamental difference with respect to the case of completely specified machines: Compatibility is not transitive. Two equivalent states can be merged into one. Similarly, two compatible states can be merged into one. However, in the case of equivalent states, each state belongs exactly to one class and no decisions are required as to what states should be merged. In the presence of don't care entries, one has to make choices. In our simple example, one could merge 2 with 1, with 3, or with both. In all cases a two-state solution would be obtained, but in general the choices will affect the number of states and will not be independent. Furthermore, the simple partition refinement algorithm cannot be used to derive the sets of compatible states, since they do not form a partition.

Let us consider now the example of Figure 8.2. There are three don't care entries in the flow table: Two next state don't cares and one output don't care. Let us examine what happens when the output don't care is replaced by a 0. We see that 1 and 3 are not compatible and so are 2 and 3, because of the different outputs they produce. Compatibility between 1 and 2 would require that between 1 and 3, and

	0	1	
a	a	b	0
b	a	b	1

Figure 8.3: Reduced machine obtained from the one of Figure 8.2.

	0	1	
1'	2	3	-
1"	2	3	-
2	-	1"	0
3	1'	-	1

Figure 8.4: Machine obtained from the one of Figure 8.2 by state splitting.

therefore no pair of compatible states exists. We can repeat the experiment, this time substituting a 1 for the output don't care: We come again to the conclusion that there are no pairs of compatible states.

We may be tempted to conclude that the machine of Figure 8.2 is already minimal, but that would be wrong. According to our definition of compatibility, 1 and 2 are compatible if 1 and 3 are compatible and vice versa. We can then merge 1 with both 2 and 3, to obtain the two-state machine of Figure 8.3.

One may argue that in merging 1 with 2 we have assumed that the output for 1 was 0, and in merging 1 and 3, we have made the opposite assumption. Therefore, we appear to have violated the law of excluded middle. The violation is, however, only apparent. We can convince ourselves of the correctness of our procedure in at least two ways.

On the one hand, we can try to find a sequence of inputs that will not cause the original machine to go through an incompletely specified next state entry, while at the same time causing a mismatch on the outputs of the two machines. Our attempts will fail, suggesting that the two machines are actually compatible. This procedure, however, only offers circumstantial evidence.

The other way of proving that what we are doing is correct is to realize that we can split State 1 in two identical states, say $1'$ and $1''$. We can then change the next state entries referring to 1 so as to refer to either $1'$ or $1''$. The result of this is shown in Figure 8.4. A moment's thought shows that the new machine expresses the same behavior as the original one[1]. It is also clear that now we can assign 0 to the output of $1'$ and 1 to the output of $1''$ and obtain the solution of Figure 8.3. In summary, when we merge a state with two other states having conflicting requirements, we can always think of splitting this state first, and then merging the newly generated states.

Having seen how the new problem differs from the previous one, we now concen-

[1]These considerations can be made more precise.

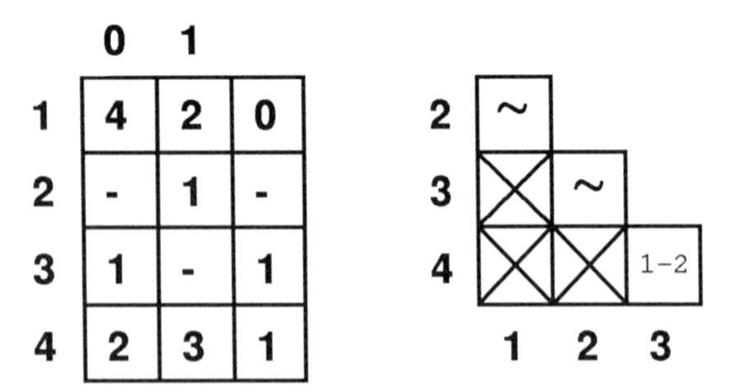

Figure 8.5: A flow table and its compatibility table.

trate on solving it. The outline of the algorithm is as follows. We first find the pairs of compatible states. We then find the maximal compatibles, i.e., sets of compatible states that are not strictly contained in any set of compatible states. It would be nice if we could restrict our attention to maximal compatibles, but unfortunately that would preclude sometimes the achievement of the minimum number of states. Therefore we need to find a larger set of compatibles that we shall call *prime compatibles*, since they play a role similar to that of prime implicants in the minimization of two-level circuits. Finally, we shall select a subset of the prime compatibles that covers all the states of the original machine. In doing so, we shall obey constraints that guarantee the consistency of the choices made. Hence, we shall solve a more complex covering problem than in the minimization of two-level circuits, namely, a *binate* covering problem. We now describe each phase of the algorithm in detail.

8.1.1 Finding the Compatible Pairs.

Compatible pairs can be found with the help of a *compatibility table* or *compatibility graph*. (They represent the same information in slightly different ways.) A simple flow table and the corresponding compatibility table are shown in Figure 8.5. In the compatibility table, the $\sim$ symbol in the $(1,2)$ entry specifies that states 1 and 2 are unconditionally compatible. The $(3,4)$ entry, on the other hand, says that states 3 and 4 are compatible only if states 1 and 2 are merged. Finally, the crosses indicate incompatibility. For the $(1,3)$ and $(1,4)$ entries the incompatibility is due to the different output values. For the $(2,4)$ entry, the outputs are compatible, but states 1 and 3 should be compatible, which is not the case.

In general, we start by examining the compatibility of the first state with all the remaining states; then we compare the second state to all the states but the first, and so on. For each pair of states, the following cases may occur.

1. There are conflicts in the outputs: The two states are incompatible.

2. Some next states for the same input are known to be incompatible: The two states are incompatible.

3. Some next states for the same input are different, but are not known to be incompatible: The compatibility of the two states is conditional on the compatibility of their next states; the pair of next states is entered in the compatibility table.

4. No conflicts occur and no different states occur. The two states are unconditionally compatible.

Every time a pair of states is found to be incompatible, we check the table entries containing this pair. The corresponding pairs of states are declared incompatible. Each such pair is checked in turn.

8.1.2 Finding the Maximal Compatibles

Among all the methods for the generation of the maximal compatibles, we shall present the one due to Marcus [184]. This method works with the incompatible pairs, that are easily found from the compatibility table. If we know that s_i and s_j are two incompatible states, then we know that no (maximal) compatible will contain both s_i and s_j. More precisely, every set of states that does not contain any incompatible pair is a compatible set. We can thus write a Boolean formula that expresses the conditions for a set of states to be compatible. We associate a variable x_i to state s_i and stipulate that $x_i = 1$ means that s_i is an element of the set. We can then write a conjunction of two-literal clauses. Each clause corresponds to an incompatible pair and expresses the condition that the two states cannot both be present. For the example of Figure 8.5, we have:

$$(x_1' + x_3')(x_1' + x_4')(x_2' + x_4').$$

Note that $(x_1' + x_3') \Longleftrightarrow x_1 \Rightarrow x_3' \Longleftrightarrow x_3 \Rightarrow x_1'$. Thus each clause in the product of sums expresses the exclusivity of incompatible states s_1 and s_3.

We can convert the above expression into sum-of-product form by multiplying it out and eliminating the absorbed terms. From what we know of prime generation (see Section 4.6 of Page 138), this procedure produces a complete sum:

$$(x_1' + x_3'x_4')(x_2' + x_4') = \\ x_1'x_2' + x_1'x_4' + x_3'x_4'.$$

Every term of this sum of products identifies a maximal compatible set. For instance, $x_1'x_4'$ identifies the maximal compatible $\{s_2, s_3\}$. In general, one takes the states corresponding to the variables that do not appear in the prime implicant. Since the complete sum is the set of all prime implicants, we know that there is no smaller product expressing the exclusivity of the incompatible pairs. Hence the compatibles formed with this procedure are maximal compatibles.

This example may be generalized as follows. First, form the product of sums as above, with one sum $(x_i' + x_j') \Longleftrightarrow x_i \Rightarrow x_j' \Longleftrightarrow x_j \Rightarrow x_i'$ for each incompatible pair denoted by a cross in the compatibility table (for example that of Figure 8.5) derived as in Section 8.1.1.

8.1.3 Finding the Prime Compatibles.

Suppose a set of compatibles is given and we start selecting some of these compatibles to cover all the states. We have seen that some pairs are compatible only if other pairs are merged into a single state. This means that the selection of one compatible may imply the selection of other compatibles. The set of compatibles implied by

	x_1	x_2	x_3	x_4	x_5	x_6	x_7
a	a,0	–	d,0	e,1	b,0	a,–	–
b	b,0	d,1	a,–	–	a,–	a,1	–
c	b,0	d,1	a,1	–	–	–	g,0
d	–	e,–	–	b,–	b,0	–	a,–
e	b,–	e,–	a,–	–	b,–	e,–	a,1
f	b,0	c,–	–,1	h,1	f,1	g,0	–
g	–	c,1	–	e,1	–	g,0	f,0
h	a,1	e,0	d,1	b,0	b,–	e,–	a,1

Figure 8.6: A flow table to illustrate the computation of prime classes.

b	a,d						
c	×	∼					
d	b,e	a,b d,e	d,e a,g				
e	a,b a,d	d,e a,b a,e	×	∼			
f	×	×	c,d	×	×		
g	∼	×	c,d f,g	×	×	e,h	
h	×	×	×	∼	a,b a,d	×	×
	a	b	c	d	e	f	g

Figure 8.7: Compatibility table for the flow table of Figure 8.6.

one compatible is easily found from the flow table and is called the *class set* of the compatible. The selection of the implied compatibles guarantees the *closure* of the solution. Observing Figure 8.5, we see that 3 and 4 are compatible only if 1 and 2 are compatible. This means that we can only use the compatible $\{3, 4\}$ in the solution, if we also include $\{1, 2\}$. The class set of $\{3, 4\}$ is therefore $\{\{1, 2\}\}$.

Let C_1 and C_2 be two compatibles and let Γ_1 and Γ_2 be their respective class sets. Suppose $C_1 \subset C_2$. We may be tempted to choose C_2 over C_1, since it covers more states. However, it may be that $\Gamma_1 \not\supseteq \Gamma_2$. In other words, C_2 may imply the choice of some other compatibles that would not be required, were C_1 selected instead. For this reason, we cannot restrict ourselves to maximal compatibles.[2] One possible solution to the problem is to consider all possible sets of compatible states. However, this could be very inefficient. The previous discussion indicates how to avoid that inefficiency. It tells us that if $C_1 \subset C_2$ and $\Gamma_1 \supseteq \Gamma_2$, then C_1 is inferior and should be discarded. More specifically we say that C_1 is not prime and that a solution with the minimum number of states can be built also without C_1.

Consider the flow table of Figure 8.6, taken from [118]. The corresponding compatibility table is shown in Figure 8.7. From the compatibility table, we can derive the set of maximal compatibles as described above in Section 8.1.2. To do this, we

[2]There are however machines for which that is possible [95].

form the following expression:

$$(a' + c')(a' + f')(a' + h')(b' + f')(b' + g')(b' + h')(c' + e')$$
$$(c' + h')(d' + f')(d' + g')(e' + f')(e' + g')(f' + h')(g' + h').$$

Here we formed a product of sums, with one sum $(x'_i + x'_j) \Longleftrightarrow x_i \Rightarrow x'_j \Longleftrightarrow x_j \Rightarrow x'_i$ for each incompatible pair denoted by a $\times$ in the compatibility table of Figure 8.7. The prime implicants of this negative unate function are the smallest products satisfying the incompatibility function.

This can be translated into a sum of products efficiently by applying the distributivity property to all the terms containing f', then all the terms containing h' and so on—at every step choosing one of variables appearing in the largest number of terms. We get:

$$(f' + a'b'd'e'h')(h' + a'b'c'g')(g' + b'd'e')(c' + a'e').$$

Now, multiplying out and eliminating absorbed terms, we get:

$$(f'h' + a'b'c'f'g' + a'b'd'e'h')(c'g' + a'e'g' + b'c'd'e' + a'b'd'e'),$$

and finally:

$$c'f'g'h' + a'e'f'g'h' + b'c'd'e'f'h' + a'b'c'f'g' + a'b'd'e'h'.$$

From this complete sum, we extract the following maximal compatibles:

$$abde, bcd, ag, deh, cfg.$$

(For instance, a, b, d, and e are exactly the variables not appearing in $c'f'g'h'$.) The maximal compatibles and the remaining prime compatibles are shown in Figure 8.8.

The maximal compatibles *MAXCOMPS* are prime by definition. The algorithm for computing the entire set of prime compatibles is given as Procedure 8.9. The inputs are *MAXCOMPS*, the set of maximal compatibles, and *CM*, the compatibility matrix (see Section 8.1.3). Initially, the largest maximal compatible, of size k_{max}, is identified. Then, on each pass through the outer **for** loop (Line 1) prime compatibles of size k are identified and added to the list of primes, which are stored in a queue denoted by P (A maximal compatible is prime by definition). This ordering by size is meant to minimize unnecessary computation. One can verify that in this way no compatible added to the list has to be removed later.

Then in the for loop of Line 2, all prime compatibles p, of size k, are treated by generating and processing all sub-compatibles (subsets of size $k-1$). However, if a compatible has an empty class set, its subsets are not generated, because they are known not to be prime. In this case the **foreach** loop continues at Line 3 by skipping to the next p of size k.

Otherwise, all subsets $s \in S_p$, each of size $k-1$, are subjected to to a primality test. In case a previously treated subset s is encountered, the **foreach** loop continues at Line 4.

The primality test is executed in the **foreach** loop of Line 5. It consists of looking at all larger primes q on the list P. If $s \subset q$, the class set CS_s of s is tested to see

	maximal compatibles	class set
1	{a,b,d,e}	∅
2	{b,c,d}	{{a,b}, {a,g}, {d,e}}
3	{c,f,g}	{{c,d}, {e,h}}
4	{d,e,h}	{{a,b}, {a,d}}
11	{a,g}	∅

	remaining prime compatibles	class set
5	{b,c}	∅
6	{c,d}	{{a,g}, {d,e}}
7	{c,f}	{{c,d}}
8	{c,g}	{{c,d}, {f,g}}
9	{f,g}	{{e,h}}
10	{d,h}	∅
12	{f}	∅

Figure 8.8: Prime compatibles for the flow table of Figure 8.6.

if it contains the class set CS_q of q. If this test passes (Line 6), the inner **for** loop is exited at Line 6, after setting *prime* to 0. If no such prime q is found in the inner **for** loop, s is added to the list of primes at Line 7.

With reference to Figure 8.8, the numbers on the left give the order in which primes are entered in the list. The first element of the list is $\{a, b, d, e\}$, the only maximal compatible of size 4. Since its class set is empty, no subsets are generated and p is decreased from 4 to 3. All maximal compatibles of size 3 are then appended to the list. The subsets of $\{b, c, d\}$ of size 2 are then considered. There are three such subsets. Only two are prime, since $\{b, d\} \subset \{a, b, d, e\}$ and $\{a, b, d, e\}$ is already on the list with an empty class set. The process continues, until the last prime, $\{f\}$ is added to the list.

8.1.4 Setting up the Covering Problem.

Once the prime compatibles and their class sets are available, we can formulate the conditions under which a collection of primes forms a valid solution. Such a collection must be a *closed cover*, i.e., all states must be contained is some compatibles (covering) and all compatibles implied by any compatible of the solution must be contained in some other compatible of the solution (closure).

We associate a variable c_i to the i-th prime compatible and stipulate that $c_i = 1$ implies that the i-th compatible is part of the solution. We then write a Boolean formula that expresses the conditions for a set of compatibles to represent a closed cover. This formula is in conjunctive form (product of sums). The terms will be divided in two groups, depending on whether they express covering or closure constraints.

Let us consider the covering constraints first. We need one clause for each state,

```
  Procedure PRIME_COMPATIBLES(MAXCOMPS, CM){
     p =LARGEST(MAXCOMPS); k_max = |p|
1    for(k = k_max; k >= 1; k - -) {                 Candidate primes of size k.
        Q = SELECT_BY_SIZE(MAXCOMPS, k)
        for(q ∈ Q) ENQUEUE(P, p)                    Maximal compatibles of of size k.
2       foreach(p ∈ P; |p| = k) {                   Process subsets of size k - 1.
           CS_p = CLASS_SET(CM, q)
3          if(CS_p = ∅) continue
           S_p = MAX_SUBSETS(p)
           for(s ∈ S_p) {
4             if(DONE(s)) continue
              CS_s = CLASS_SET(CM, s)
              prime = 1
5             foreach(q ∈ P; |q| >= k) {
                 if (s ⊂ q) {
                      CS_q = CLASS_SET(CM, q)
                      if (CS_s ⊇ CS_q) {
6                          prime = 0; break
                      }
                 }
              }
7             if(prime = 1) ENQUEUE(P, s)
              DONE(s) = 1
           }
        }
     }
  }
```

Figure 8.9: Algorithm for computing prime compatibles.

expressing the condition that the state is part of the solution. This condition is satisfied if at least one compatible including the state is selected. For state a of Figure 8.6, for instance, the only two compatibles that cover it are 1 and 11; hence:

$$(c_1 + c_{11})$$

will be one term of the formula. The terms for the other states are similar.

We now discuss the closure constraints with the help of an example. Consider Figures 8.6 and 8.8 again. Prime 2 ({b,c,d}) requires {(a,b),(a,g),(d,e)}. If we set $c_2 = 1$, then we need also to select compatibles that contain these implied compatibles. By inspection of Figure 8.8, we see that

(a,b)	is found in	{a,b,d,e} (c_1)
(a,g)	is found in	{a,g} (c_{11})
(d,e)	is found in	{a,b,d,e} (c_1) and {d,e,h} (c_4).

Hence, we have:

$$\begin{aligned} c_2 &\Rightarrow c_1 \\ c_2 &\Rightarrow c_{11} \\ c_2 &\Rightarrow (c_1 + c_4) \end{aligned}$$

that can be written as

$$(c_2' + c_1)\ (c_2' + c_{11})\ (c_2' + c_1 + c_4)$$

Similar expressions can be written for all the prime classes.

The conjunction of all the constraints forms an expression that must be satisfied for the selection to be valid. We are interested in a selection of minimum cost, that can be found by solving a binate covering problem. In our case we need to find a solution to:

$$\begin{aligned} &(c_1 + c_{11})(c_1 + c_2 + c_5)(c_2 + c_3 + c_5 + c_6 + c_7 + c_8)(c_1 + c_2 + c_4 + c_6 + c_{10}) \\ &(c_1 + c_4)(c_3 + c_7 + c_9 + c_{12})(c_3 + c_8 + c_9 + c_{11})(c_4 + c_{10}) \\ &(c_2' + c_1)(c_2' + c_{11})(c_2' + c_1 + c_4)(c_3' + c_2 + c_6) \\ &(c_3' + c_4)(c_4' + c_1)(c_4' + c_1)(c_6' + c_{11})(c_6' + c_1 + c_4) \\ &(c_7' + c_2 + c_6)(c_8' + c_2 + c_6)(c_8' + c_3 + c_9)(c_9' + c_4) = 1. \end{aligned}$$

One can verify that $c_1 = c_4 = c_5 = c_9 = 1$ and all other variables equal to 0 is a solution. It is actually a minimum cost solution as we shall see shortly.

8.1.5 Forming the Reduced Table

Given a solution, we can build the minimized cover for the FSM as follows. For each compatible in the solution there will be a state in the reduced machine. The next state entry for given present state of the reduced machine and input is determined by finding a compatible in the solution that contains all the states that are next states for the given present state compatible and input.

In our example, for instance, let us consider the next state of compatible 1 under input x_1. Since compatible 1 is $\{a, b, d, e\}$, we consider the entries in Column x_1 of

	x_1	x_2	x_3	x_4	x_5	x_6	x_7
1	1,0	$\{1,4\}$,1	1,0	1,1	1,0	1,1	1,1
4	1,1	$\{1,4\}$,0	1,1	$\{1,5\}$,0	$\{1,5\}$,0	$\{1,4\}$,–	1,1
5	$\{1,5\}$,0	$\{1,4\}$,1	1,1	—	1,–	1,1	9,0
9	$\{1,5\}$,0	5,1	–,1	4,1	9,1	9,0	9,0

Figure 8.10: Reduced flow table obtained from the one of Figure 8.6.

	x_1	x_2	x_3	x_4	x_5	x_6	x_7
1	1,0	1,1	1,0	1,1	1,0	1,1	1,1
4	1,1	1,0	1,1	1,0	1,0	1,–	1,1
5	1,0	1,1	1,1	—	1,–	1,1	9,0
9	1,0	5,1	–,1	4,1	9,1	9,0	9,0

Figure 8.11: Reduced flow table obtained from the one of Figure 8.10 by heuristic choices of the next state entries.

the table of Figure 8.6 corresponding to states a, b, d, and e. We find that a and b are the only states appearing in those entries. Hence we look for a compatible containing $\{a, b\}$. The only such compatible is compatible 1 and therefore we choose 1 as next state of 1 for input x_1. The output in this case must be 0, because that is the value for the entries where it is not unspecified.

We consider next the entry for present state 1 and input x_2. We find that states d and e appear in the appropriate entries in Figure 8.6. These two states are contained in two of the compatibles forming the solution, namely 1 and 4. Hence we have a choice and we enter $\{1,4\}$ in the reduced table to indicate that. The complete reduced table is given in Figure 8.10.

The choices for the next state entries are arbitrary as far as the number of states are concerned. However, they may impact the number of literals after state encoding and logic optimization. Finding the exact solution to this problem is beyond our present scope and is computationally very expensive [178]; we just mention that one can make heuristically good choices by creating uniformity, rather than disuniformity. For instance, in our problem, a good heuristic solution would consist of selecting 1 wherever there is a choice. The result is shown in Figure 8.11.

8.2 The Binate Covering Problem

We recall that the Binate Covering Problem (BCP) is the problem of finding the minimum cost assignment that satisfies a Boolean formula in conjunctive form. The difference with respect to ordinary (unate) covering is that the formula is not supposed to be unate. As a consequence, some simplifying assumptions that we implicitly made when dealing with unate covering are not valid in the more general case. However, we shall see that the similarities between the two problems are much stronger than

```
BCP(F, U, currentSol) {
1     (F, currentSol) = REDUCE(F, currentSol)
      if (terminalCase(F)) {
          if (F ≠ 0 and COST(currentSol) < U) {
              U = COST(currentSol)
2             return (currentSol)
          }
3         else return ("no solution")
      }
4     L = LOWER_BOUND(F, currentSol)
      if (L ≥ U) return ("no solution")
5     x_i = CHOOSE_VAR(F)
6     S^1 = BCP(F_{x_i}, U, currentSol ∪ {s_i})
7     if (COST(S^1) = L) return (S^1)
      S^2 = BCP(F_{x_i'}, U, currentSol)
8     return BEST_SOLUTION(S^1, S^2)
}
```

Figure 8.12: Branch and bound algorithm for binate covering.

the differences, so that it will be possible to adapt the branch-and-bound algorithm for unate covering to BCP without drastic changes.

In particular, our general scheme will be the same. We shall put the formula in matrix form and try to identify singleton rows. We shall then try to apply row and column dominance to simplify the problem. As in the unate case, we shall sometimes find the solution by exclusive application of these reduction techniques. In the general case, though, we shall select a branching variable and decompose the problem in two sub-problems. As in the unate case, we shall keep an upper and a lower bound on the cost of the solutions, so as to prune as much of the search tree as possible. (See Figure 8.12.)

Unlike the unate case, BCP may not have a solution. This occurs when the given formula represents the function that is identically 0. One such formula is:

$$(x_1 + x_2)(x_1 + x_2')(x_1' + x_2)(x_1' + x_2').$$

Likewise, some of the subproblems generated by branching may not have a solution. In the case of FSM minimization, the initial problem is guaranteed to have a solution, but some of the subproblems may not have one. Note that the algorithm of Figure 8.12 tests for unsatisfiable subproblems when a terminal case is reached. We are now ready to present the problem more formally.

8.2.1 Formulation of BCP

We are given a Boolean formula F in conjunctive form and an objective function:

$$\min \sum_{j=1}^{t} w_j x_j \tag{8.1}$$

It is convenient to represent F in matrix form, by associating a column to every variable and a row to every clause. The generic element f_{ij} is $-$ if variable x_j does not appear in clause i, is 0 if variable x_j appears complemented in clause i, and is 1 otherwise. We denote the i-th row of the matrix F by f_i and the j-th column by F_j.
Example:

For this formula

$$F = (x_1 + x_3)(x_2 + x_4 + x_6)(x_3' + x_4 + x_5)(x_6')(x_1' + x_6')(x_3 + x_4' + x_5)$$

we have in matrix form

$$F = \begin{array}{c} \begin{matrix} x_1 & x_2 & x_3 & x_4 & x_5 & x_6 \end{matrix} \\ \begin{bmatrix} 1 & - & 1 & - & - & - \\ - & 1 & - & 1 & - & 1 \\ - & - & 0 & 1 & 1 & - \\ - & - & - & - & - & 0 \\ 0 & - & - & - & - & 0 \\ - & - & 1 & 0 & 1 & - \end{bmatrix} \end{array}$$

■

With F in matrix form, the binate covering problem can be formulated as follows:

Find a subset S of columns of minimum cost according to (8.1), such that for every row f_i, either

1. $\exists j : (f_{ij} = 1) \wedge (F_j \in S)$; or,
2. $\exists j : (f_{ij} = 0) \wedge (F_j \notin S)$.

By contrast, in UCP the coefficients of F are only ones and dashes and only condition 1 may hold. **Example:**

In Example 8.2.1, if the cost of all columns is the same, a solution is given by Columns 1 and 2 or, in other words, by setting $x_1 = x_2 = 1$ and $x_3 = x_4 = x_5 = x_6 = 0$. This solution is not unique: Columns 3 and 4 form another solution. ■

8.2.2 Reduction Techniques

As mentioned in the introductory remarks, the outline of the branch-and-bound algorithm is the same as in the ordinary unate covering problem. (See Figure 8.12.) In the following, we extend the reduction techniques from the unate to the binate case, starting from the detection of essential variables.

Essential and Unacceptable Variables

Definition 8.2.1 *An* essential *row of F is a row f_i where only one coefficient is different from $-$. Alternatively, it is a clause of F consisting of a single literal.*

Essential rows correspond to clauses of the form (x_j) or (x_j'). They cause the value of x_j to be 1 or 0, respectively, in any assignment satisfying F. Therefore, in the solution of an instance of BCP, we proceed to the identification of the literals corresponding to the essential rows. These are called essential literals. The variables corresponding to the positive essential literals are *essential variables*. The variables corresponding to negative essential literals are *unacceptable variables*. When an essential row is found, F is cofactored with respect the the essential literal. **Example:**

In Example 8.2.1, x_6' is essential (x_6 is unacceptable). The cofactored formula is:

$$F = \begin{array}{c} \begin{array}{ccccc} x_1 & x_2 & x_3 & x_4 & x_5 \end{array} \\ \left[\begin{array}{ccccc} 1 & - & 1 & - & - \\ - & 1 & - & 1 & - \\ - & - & 0 & 1 & 1 \\ - & - & 1 & 0 & 1 \end{array}\right] \end{array}$$

■

As a concluding remark, we notice that, if x_i is essential, then $F \leq x_i$. Similarly, if x_i is unacceptable, $F \leq x_i'$. These two inequalities actually give the most general definition of essentiality.

Row Dominance

Definition 8.2.2 *A row f_i* dominates *another row f_l if f_i is satisfied whenever f_l is satisfied, i.e., if $f_l \leq f_i$.*

Dominating rows can be eliminated without affecting the set of solutions, because $f_i f_l = f_l$. In practice, the dominating row has all the ones and zeroes of the dominated rows, plus possibly some.

Example:

Consider the following matrix:

$$F = \begin{array}{c} \begin{array}{cccc} x_1 & x_2 & x_3 & x_4 \end{array} \\ \left[\begin{array}{cccc} 0 & 1 & 0 & - \\ - & 1 & 0 & - \\ 1 & - & - & 1 \\ 1 & 0 & 1 & 0 \end{array}\right] \end{array}$$

The first row, f_1, dominates f_2. The ensuing simplification yields:

$$F = \begin{array}{c} \begin{array}{cccc} x_1 & x_2 & x_3 & x_4 \end{array} \\ \left[\begin{array}{cccc} - & 1 & 0 & - \\ 1 & - & - & 1 \\ 1 & 0 & 1 & 0 \end{array}\right] \end{array}$$

This simplification is based on the identity $(x_1' + x_2 + x_3')(x_2 + x_3') = (x_2 + x_3')$.

■

Column (Variable) Dominance

Definition 8.2.3 *Let F_j, F_k be two columns of the matrix F. F_j* dominates *$F_k (F_j \geq F_k)$ if, for each clause f_i of F, one of the following occurs:*

- $f_{ij} = 1$;
- $f_{ij} = -$ *and* $f_{ik} \neq 1$;
- $f_{ij} = 0$ *and* $f_{ik} = 0$.

Theorem 8.2.1 *Let F be satisfiable. If $F_j \geq F_k$ and $w_j \leq w_k$, then there is at least one minimum solution with $x_k = 0$.*

Proof. Note that every clause in F that is set to 1 by $x_k = 1$ is also set to 1 by $x_j = 1$, and every clause set to 0 by $x_j = 1$ is also set to 0 by $x_k = 1$. Suppose there is a solution with $x_k = 1$. The assignment to the x's obtained from that solution by setting $x_j = 1$ and $x_k = 0$ is still a solution, not higher in cost than the original one. □

Example:

Continuing Example 8.2.2, let us assume that all variables have unit costs. We see that after simplification by row dominance, F_4 is dominated by F_1. Hence, we set $x_4 = 0$ and we get the following matrix:

$$F = \begin{array}{c} \begin{matrix} x_1 & x_2 & x_3 \end{matrix} \\ \begin{bmatrix} - & 1 & 0 \\ 1 & - & - \end{bmatrix} \end{array}$$

Variable x_1 has now become essential. Setting $x_1 = 1$, we get:

$$F = \begin{array}{c} \begin{matrix} x_2 & x_3 \end{matrix} \\ \begin{bmatrix} 1 & 0 \end{bmatrix} \end{array}$$

We have another case of column dominance. Specifically, F_2 dominates F_3. Setting $x_3 = 0$ covers the only remaining row and hence solves the problem. All unassigned variables (in this case, only x_2) are set to 0. The final solution is $x_1 = 1$, $x_2 = x_3 = x_4 = 0$. Notice that there is another solution of the same cost, namely, $x_4 = 1$, $x_1 = x_2 = x_3 = 0$. This solution cannot be found if column dominance is applied. As in the unate case, column dominance only guarantees that at least one minimum cost solution will be found. ■

As a special case of column dominance, we may consider those columns that do not contain any ones (i.e., they only contain zeroes and dashes). The variables corresponding to these columns can be set to 0 safely.

The definition of column dominance that we have given is tailored to our formulation of the problem, where F is given in POS form. A more general definition can be found in [155].

Row dominance has to do with the identification of a subset of clauses of F that is sufficient to express all constraints. Column dominance, on the other hand, is used to extract a subset of the variables that contains a minimum cost solution, if one exists.

It should be noted that the set of reduction rules presented here is not exhaustive and the reduced form of a given problem may or may not be cyclic depending on the set of reduction rules and their order of application. Some additional reduction techniques are extensions of techniques available for unate covering [253]. Others are more specific to BCP [119, 147]. Their usefulness varies depending on the type of problems that need to be solved. Finally, we notice that a binate covering problem can be partitioned—exactly as a unate covering problem—whenever the matrix has a block-diagonal structure.

8.2.3 Choice of the Splitting Variable and Bounding

The application of the reduction techniques sometimes reduces the problem to a trivial one. In general, however, the reduced problem is *cyclic* and it is necessary to "branch" by temporarily assigning a chosen variable to 0 (or 1) and finding the minimum solution, and then repeating this when the variable is assigned to 1 (or 0). The choice of the branching variable is usually a heuristic, but a very important one.

Also, we keep a lower and an upper bound. Whenever the lower bound exceeds the upper bound, the current node is pruned, since there is provably no optimal solution to be found by expanding this node. We now discuss these two important aspects of the algorithm, by discussing the few changes required to adapt the algorithm for unate covering to the binate case. We also discuss how infeasible (or unsatisfiable) subproblems are identified. The strategy we follow is the one of [41].

8.2.4 Maximal independent set.

A maximal independent set of clauses provides a lower bound to the cost of the solution and also helps in selecting the branching variable. Two clauses are independent if it is not possible to satisfy both by setting at most one of the variables to 1. For the binate case, we note that a clause including a complemented variable is dependent on any other clause, according to this definition. Indeed, it can be satisfied by setting to 0 one of the complemented variables. Therefore, in the binate case we only consider the clauses that do not contain any complemented literal and find an independent set for those. Consider, as an example, the following constraint matrix:

$$F = \begin{array}{c} \begin{array}{cccc} x_1 & x_2 & x_3 & x_4 \end{array} \\ \left[\begin{array}{cccc} 1 & 1 & - & - \\ - & - & 0 & 1 \end{array} \right] \end{array}$$

The two rows have no columns in common. However, f_2 can be satisfied by $x_3 = 0$. Hence, the size of the independent set is 1, and the lower bound is the minimum of the costs of x_1 and x_2. One can easily verify that in this case this lower bound is exact, i.e., the lower bound is precisely the cost of the cheapest solution.

Clearly the lower bound we get by ignoring the clauses with complemented literals is less tight than in the unate case. Consider the following matrix as an example:

$$F = \begin{array}{c} \begin{array}{cccc} x_1 & x_2 & x_3 & x_4 \end{array} \\ \left[\begin{array}{cccc} 1 & 1 & - & - \\ - & 1 & 1 & - \\ - & 0 & - & 1 \end{array} \right] \end{array}$$

The size of the independent set is one, but a minimum cost solution requires at least two variables set to 1. On the other hand, for the unate covering problem obtained by considering only f_1 and f_2, the bound is exact.

Unlike the unate case, it is possible for the largest independent set to be empty. This occurs when all clauses contain at least one complemented literal. In such a case, the problem admits the solution where all variables are set to zero, as in the following example:

$$F = \begin{array}{c} \begin{array}{cccc} x_1 & x_2 & x_3 & x_4 \end{array} \\ \left[\begin{array}{cccc} 1 & 0 & - & - \\ 0 & 1 & - & - \\ - & 0 & 1 & - \\ - & - & 0 & 1 \end{array} \right] \end{array}$$

Notice that this example is cyclic. Hence it also shows that in binate covering a cyclic constraint matrix may be covered by fewer than two columns. As we saw, that is not the case for unate covering.

8.2.5 Choice of the branching column.

As in the case of unate covering, columns that intersect many short rows are favored. However, a given row, f_i, is covered by a selected column, F_j, only if $f_{ij} = 1$. Therefore, treating all intersected rows the same is not necessarily accurate. One looks for a good trade-off between the complexity and the quality of the choice criterion. Ignoring the rows f_i such that $f_{ij} = 0$, while computing the relative merit of F_j leads to a simple, yet effective selection method.

8.2.6 Infeasible problems.

Whenever we find two essentials x_j and x_j', we have an infeasible problem and we prune the current branch of the search tree. Consider the following cyclic matrix:

$$F = \begin{array}{c} \begin{array}{cc} x_1 & x_2 \end{array} \\ \left[\begin{array}{cc} 1 & 1 \\ 0 & 1 \\ 1 & 0 \\ 0 & 0 \end{array} \right] \end{array}$$

If we split on x_1, we obtain for both subproblems that x_2 is both essential and unacceptable. Hence F is not satisfiable.

Example:

In Example 8.2.2, after the identification of the unacceptable variable, we are left with a cyclic problem and we need to split. We start with an upper bound of 7 and a lower bound of 2 (computed after elimination of the essentials). We choose $x_3 = 1$ and we obtain

$$F = \begin{array}{c} \begin{matrix} x_1 & x_2 & x_4 & x_5 \end{matrix} \\ \begin{bmatrix} - & 1 & 1 & - \\ - & - & 1 & 1 \end{bmatrix} \end{array}$$

From which we easily see that the optimal solution is obtained by selecting x_4. Since the cost is 2, i.e., it equals the lower bound at the root of the tree, this solution is globally optimal and we don't need to consider the branch corresponding to $x_3 = 0$. ■

8.2.7 An Example of Reductions

We conclude with an example of application of reduction techniques. Let us consider the following problem (we assume unit costs).

$$F = \begin{array}{c} \begin{matrix} x_1 & x_2 & x_3 & x_4 & x_5 & x_6 \end{matrix} \\ \begin{bmatrix} - & - & 0 & 0 & - & - \\ - & - & - & 0 & 0 & - \\ - & - & 1 & 1 & 1 & - \\ - & 1 & - & 1 & - & 1 \\ - & 1 & - & - & 1 & - \\ - & - & - & - & - & 1 \end{bmatrix} \end{array}$$

We find $x_1 = 0, x_6 = 1$. Cofactoring, we get:

$$F^1 = \begin{array}{c} \begin{matrix} x_2 & x_3 & x_4 & x_5 \end{matrix} \\ \begin{bmatrix} - & 0 & 0 & - \\ - & - & 0 & 0 \\ - & 1 & 1 & 1 \\ 1 & - & - & 1 \end{bmatrix} \end{array}$$

From which, by column dominance, $x_4 = 0$. Hence,

$$F^2 = \begin{array}{c} \begin{matrix} x_2 & x_3 & x_5 \end{matrix} \\ \begin{bmatrix} - & 1 & 1 \\ 1 & - & 1 \end{bmatrix} \end{array}$$

This gives $x_2 = x_3 = 0$. (Again, by column dominance.) Finally,

$$F^3 = \begin{array}{c} x_5 \\ \begin{bmatrix} 1 \\ 1 \end{bmatrix} \end{array}$$

that yields $x_5 = 1$. The complete solution is therefore $x_1 = x_2 = x_3 = x_4 = 0$ and $x_5 = x_6 = 1$.

8.3 State Encoding

We shall focus on the problem of state assignment for synchronous machines, where the primary objective is the reduction of the cost of the implementation. Other objectives may be speed and testability.[3]

In this section we consider the number of literals as the cost figure. As pointed out earlier, this is an approximation, since the effects of technology mapping and the cost of interconnections are not modeled, but it is usually a good one. We also mentioned that the number of possible assignments is very high. If one uses k bits to encode p states, there are $(2^k)!/(2^k - p)!$ possible assignments. If one considers two assignments obtained by permutation or complementation of some of the bits as essentially the same assignment, then there are

$$\frac{(2^k - 1)!}{(2^k - p)!k!}$$

distinct assignments [189]. Therefore, practical encoding methods are heuristic. We shall present one such algorithm, called MUSTANG [87], WHOSE PURPOSE IS TO ANTICIPATE THE EFFECTS OF LOGIC OPTIMIZATION, AS PERFORMED BY SIS. Other more sophisticated evolutions of the ideas of MUSTANG have been incorporated in later programs [176, 92, 94], and there is a rich earlier literature on the subject (see, for instance [6, 89, 84]), but for our illustrative purposes MUSTANG will suffice. In the next classes we shall consider another approach to state assignment, that originates from the work in [137].

8.3.1 Practical Encoding Algorithms

As many other approaches to state encoding, the one of MUSTANG tries to identify pairs of states that should receive adjacent codes. Two codes are adjacent if they only differ in one bit, like 001 and 101. Throughout our discussion of MUSTANG, we shall refer to the example of Figure 8.13. Our first objective is to build a graph representing the *attraction* between each pair of states. Such a graph is presented in Figure 8.14. There is a node in the graph for each state of the FSM. There is an edge in the graph for each pair of nodes, labeled with the attraction between the corresponding states. Two states that have a strong attraction should be given adjacent codes. Notice that it will not always be possible to take into account all the suggestions that come from the attraction graph, so that we shall have to heuristically select what adjacencies to actually implement.

In MUSTANG, the attractive force between two states is related to the ability of extracting common cubes from the next state and output functions, if the two states are given adjacent codes. Consider the following fragment of an encoded cube table.

x_1x_2	y_1y_2	Y_1Y_2	O_1
01	01	11	1
01	00	11	1

[3] The task of assigning binary codes to states has different objectives for synchronous and asynchronous circuits. For the latter, the main goal is to achieve correct behavior by avoiding or controlling the effects of races.

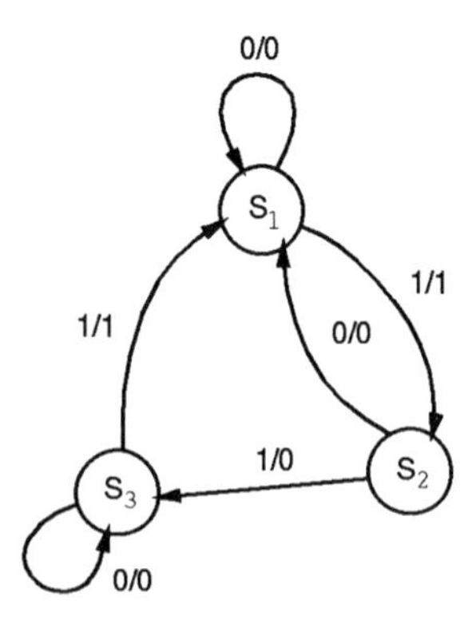

Figure 8.13: Example FSM for the discussion of state encoding.

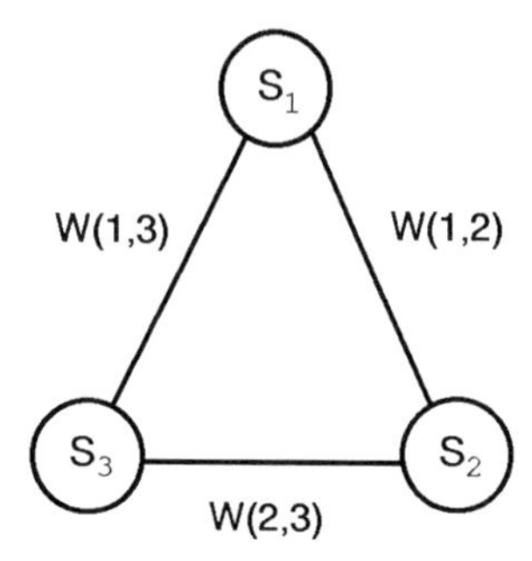

Figure 8.14: Attraction graph for the FSM of Figure 8.13.

We have two states (01 and 00) that are mapped into the same state (11) when the input is 01. These two states are said to have a common fanout state. In the equations for the next state variables and the output, before minimization, we shall find terms corresponding to these lines of the cube tables:

$$\begin{aligned} Y_1 &= \cdots + y_1' y_2 x_1' x_2 + y_1' y_2' x_1' x_2 + \cdots \\ Y_2 &= \cdots + y_1' y_2 x_1' x_2 + y_1' y_2' x_1' x_2 + \cdots \\ O_1 &= \cdots + y_1' y_2 x_1' x_2 + y_1' y_2' x_1' x_2 + \cdots \end{aligned}$$

but, since 01 and 00 are adjacent, we can simplify those expressions as follows:

$$y_1' y_2 x_1' x_2 + y_1' y_2' x_1' x_2 = y_1' x_1' x_2$$

and therefore the common cube $y_1' x_1' x_2$ can be extracted from the three equations. In a similar way we can analyze the following fragment of cube table. Here we see that states 10 and 11 have one fanin state in common.

x_1x_2	y_1y_2	Y_1Y_2	O_1
10	01	10	1
11	01	11	0

We focus in this case on the equation for Y_1, that will contain, before minimization, terms corresponding to both lines of the fragment. (The equations for Y_2 and O_1 will contain only a term corresponding to one of these lines.)

$$Y_1 = \cdots + y_1' y_2 x_1 x_2' + y_1' y_2 x_1 x_2 + \cdots$$

In this case we can also simplify the expression for Y_1 to:

$$Y_1 = \cdots + y_1' y_2 x_1 + \cdots$$

No common cube can be extracted in this particular example, but in general it may be possible.

Each of these two examples illustrates one of the two ways in which MUSTANG builds the attraction graph. In the *fanout-oriented* algorithm, whenever two states, s_i and s_j, have a common fanout state, the weight of the edge (s_i, s_j) of the attraction graph is increased. Similarly, for the *fanin-oriented* algorithm, if s_i and s_j have a common fanin state, the weight of the edge (s_i, s_j) of the attraction graph is increased. Once the graph of the attractions is found, we try to assign adjacent codes to pairs of states that have strong attractions.

Fanout-Oriented Algorithm

In the fanout-oriented method, we build two matrices: The first with one row for each present state and one column for each next state and the second with one row for each present state and one column for each output. For the example of Figure 8.13, we obtain the following matrices, where the superscripts p and n stand for present state and next state.

$$S = \begin{array}{c|ccc} & s_1^n & s_2^n & s_3^n \\ s_1^p & 1 & 1 & 0 \\ s_2^p & 1 & 0 & 1 \\ s_3^p & 1 & 0 & 1 \end{array} \qquad Z = \begin{array}{c|c} & z \\ s_1^p & 1 \\ s_2^p & 0 \\ s_3^p & 1 \end{array}$$

A 1 in row s_1^p and column s_2^n indicates that there is one arc of the state transition graph going from s_1 to s_2. A 1 in row s_1^p and column z indicates that there is an arc going out of s_1 that asserts that output z should be 1. In general, the entries of the matrix are non-negative integers that give the number of arcs connecting two states or the number of arcs going out of a state and asserting a given output.

Let S_i be the i-th row of S and Z_i the i-th row of Z. Let also N_b be number of encoding bits. Then the attraction between states s_i and s_j is given by:

$$N_b \cdot S_i \cdot S_j^T + Z_i \cdot Z_j^T,$$

where the operations are on integers and T means transpose. In our example, assuming $N_b = 2$, the attraction of states 2 and 3 is computed as:

$$W(2,3) = 2 \cdot \begin{bmatrix} 1 & 0 & 1 \end{bmatrix} \cdot \begin{bmatrix} 1 \\ 0 \\ 1 \end{bmatrix} + \begin{bmatrix} 0 \end{bmatrix} \cdot \begin{bmatrix} 1 \end{bmatrix} = 4$$

Proceeding similarly for the other pairs of states, we get the attraction graph of Figure 8.15. The reason for the N_b factor is simple: We want to take into account that the cube that will be generated will be shared by multiple next state equations, possibly all of them. A more conservative estimate would use $N_b/2$.

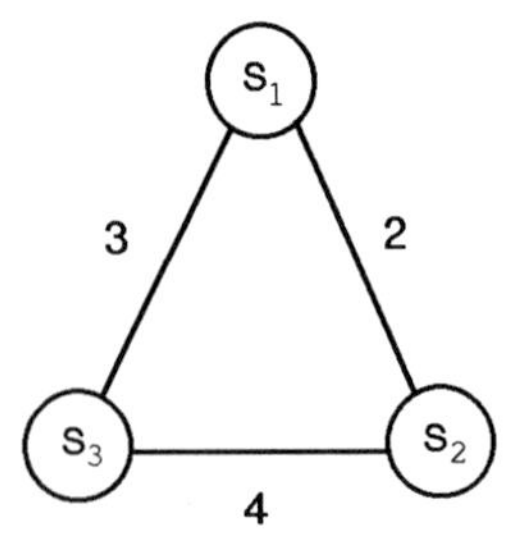

Figure 8.15: Attraction graph produced by the fanout-oriented algorithm of MUSTANG.

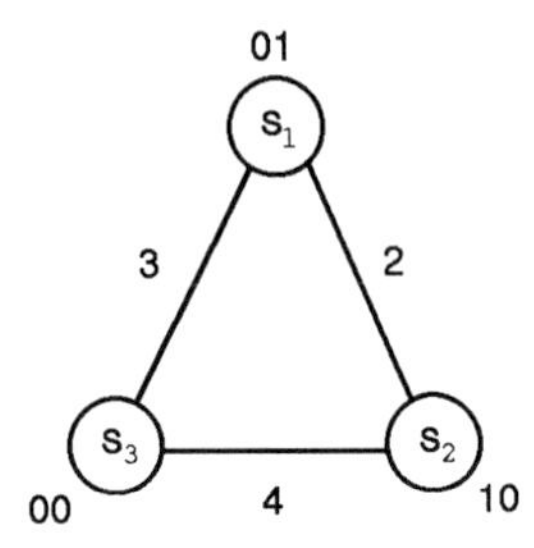

Figure 8.16: An assignment derived by the fanout-oriented algorithm.

The Embedding Algorithm

We are now ready to assign codes to states. We select first the node for which the sum of the weights of the N_b heaviest incident edges is maximum. (In our case s_3, with a sum of 7.) We only consider the N_b heaviest edges, because there are only N_b adjacent codes to any given code of N_b bits. We arbitrarily assign a code to it and *adjacent* codes to the N_b adjacent states connected by the heaviest edges. In our example this completes the process. The result is shown in Figure 8.16. In general, the states that have been assigned codes are removed and the procedure is repeated. For the assignment of Figure 8.16, the resulting logic is:

$$\begin{aligned} Y_1 &= y_1' y_2 x \\ Y_2 &= y_1' y_2 x' + y_1 y_2' x' + y_1' y_2' x \\ z &= y_1' y_2 x + y_1' y_2' x = y_1' x \end{aligned}$$

In the arbitrary choice of the initial code, it is not uncommon to choose the all-zero code. This has the effect of heuristically minimizing the number of ones in the output parts of the cubes, which normally leads to simplification of the equations.

Fanin-Oriented Algorithm

In the fanin-oriented method, we again build two matrices: The first with one row for each next state and one column for each present state and the second with one row for each next state and two columns for each input; one column is for the true input and the other is for the complement. For our example, we get the following matrices.

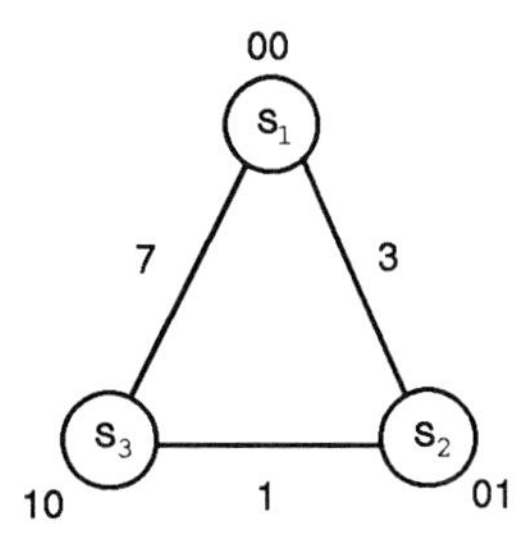

Figure 8.17: An assignment derived by the fanin-oriented algorithm.

$$\hat{S} = \begin{array}{c|ccc} & s_1^p & s_2^p & s_3^p \\ s_1^n & 1 & 1 & 1 \\ s_2^n & 1 & 0 & 0 \\ s_3^n & 0 & 1 & 1 \end{array} \qquad X = \begin{array}{c|cc} & x & x' \\ s_1^n & 1 & 2 \\ s_2^n & 1 & 0 \\ s_3^n & 1 & 1 \end{array}$$

A 1 in row s_1^n and column s_2^p of $\hat{S}$ indicates that there is one arc of the state transition graph going from s_2 to s_1. A 2 in row s_1^n and column x' of X indicates that there are two arcs entering s_1 that are labeled x'. Notice that $\hat{S} = S^T$. The attraction between states s_i and s_j is now given by:

$$N_b \cdot \hat{S}_i \cdot \hat{S}_j^T + X_i \cdot X_j^T .$$

The resulting graph and a possible encoding are shown in Figure 8.17. The embedding procedure is the same as before. For our example, the resulting logic is:

$$\begin{aligned} Y_1 &= y_1 y_2' x' + y_1' y_2 x \\ Y_2 &= y_1' y_2' x \\ z &= y_1' y_2' x + y_1 y_2' x = y_2' x . \end{aligned}$$

8.4 Decomposition and Encoding

In Section 7.1 we mentioned that one of the approaches to state encoding aimed at reducing the mutual dependence of the state variables. If we have n state variables, the next-state functions take the form

$$\begin{aligned} Y_1 &= f_1(y_1, \ldots, y_n, x_1, \ldots, x_m) \\ &\vdots \\ Y_n &= f_n(y_1, \ldots, y_n, x_1, \ldots, x_m), \end{aligned}$$

where x_i is the i-th input variable. Rather than aiming directly at minimizing the number of literals in the next-state functions, one may actually try to minimize the *support* of the functions. In so doing, he or she could achieve two objectives: Reduction of the number of literals and simplification of the interconnections.[4]

[4]The testability of the circuit may also be improved.

Consider the case of a machine with four state variables and suppose that, thanks to a careful encoding, we have:

$$\begin{aligned} Y_1 &= f_1(y_1, y_2, x_1, \ldots, x_m) \\ Y_2 &= f_2(y_1, y_2, x_1, \ldots, x_m) \\ Y_3 &= f_3(y_3, y_4, x_1, \ldots, x_m) \\ Y_4 &= f_4(y_3, y_4, x_1, \ldots, x_m). \end{aligned}$$

We see that, for instance, Y_1 does not depend on y_3 and y_4. If we ignore the output function for the moment, we have two circuits that share only the primary inputs and thus actually work in parallel. In general, the reduction of the dependences among the state variables may lead to the decomposition of the FSM into two or more components.

We shall start our investigation of reduced dependence from the study of the decomposition of a FSM and we shall then see the reduction of the dependence as a generalization of decomposition. We shall come to the conclusion that decomposition and state encoding are essentially the same problem, though from a practical standpoint it may be convenient to separate them. Our treatment of the subject will be largely based on [138] and will be restricted to completely specified machines.

8.4.1 Partitions

A **partition** π on a set S is a collection of disjoint subsets of S whose set union is S, i.e.,

$$\pi = \{B_\alpha\}$$

such that

$$B_\alpha \cap B_\beta = \emptyset \quad \text{for} \quad \alpha \neq \beta$$

and

$$\bigcup \{B_\alpha\} = S$$

Each subset is called a *block* of the partition. We use a bar over the elements of a block in our notation, as shown in the following example.

$$S = \{1, 2, 3, 4, 5, 6, 7, 8\}$$

$$\pi = \{B_1; B_2; B_3\} = \{\overline{1, 2, 4, 5}; \overline{3, 6}; \overline{7, 8}\}$$

We write $s \equiv t(\pi)$ to indicate that s and t are in the same block of π.

Partitions play a central role in the decomposition theory: Hence we introduce some definitions. We indicate with $\mathbf{0}$ and $\mathbf{1}$ the trivial partitions. For instance, if

$$S = \{1, 2, 3, 4\}$$

then

$$\mathbf{0} = \{\bar{1}; \bar{2}; \bar{3}; \bar{4}\} \quad \text{and} \quad 1 = \{\overline{1, 2, 3, 4}\}.$$

For π_1 and π_2 on S, we say that π_2 is *greater than or equal* to π_1, and write

$$\pi_1 \leq \pi_2$$

if and only if every block of π_1 is contained in a block of π_2. We immediately recognize that $\leq$ is a partial order. As an example, we have:

$$\{\overline{1}; \overline{2,3}; \overline{4}\} \leq \{\overline{1,4}; \overline{2,3}\}.$$

If π_1 and π_2 are partitions on S, then

$$\pi_1 \cdot \pi_2$$

is the partition on S such that

$$s \equiv t(\pi_1 \cdot \pi_2) \quad \text{if and only if} \quad s \equiv t(\pi_1) \quad \text{and} \quad s \equiv t(\pi_2),$$

whereas,

$$\pi_1 + \pi_2$$

is the partition on S such that

$$s \equiv t(\pi_1 + \pi_2) \quad \text{if and only if there exists a sequence in } S$$

$$s = s_0, s_1, s_2, \ldots, s_n = t$$

for which either

$$s_i \equiv s_{i+1}(\pi_1) \quad \text{or} \quad s_i \equiv s_{i+1}(\pi_2),$$

$0 \leq i \leq n-1$. For instance, if

$$S = \{1,2,3,4,5,6,7,8,9\}$$

$$\pi_1 = \{\overline{1,2}; \overline{3,4}; \overline{5,6}; \overline{7,8,9}\}$$

$$\pi_2 = \{\overline{1,6}; \overline{2,3}; \overline{4,5}; \overline{7,8}; \overline{9}\}$$

then

$$\pi_1 \cdot \pi_2 = \{\overline{1}; \overline{2}; \overline{3}; \overline{4}; \overline{5}; \overline{6}; \overline{7,8}; \overline{9}\}$$

and

$$\pi_1 + \pi_2 = \{\overline{1,2,3,4,5,6}; \overline{7,8,9}\}.$$

It can be seen that these two operations are *meet* and *join* for partitions and that the partitions on a set form a *lattice.*

The importance of partitions is best understood by reminding that an equivalence relation induces a partition on a set and, vice versa, a partition on a set identifies an equivalence relation. We have used one equivalence relation, namely indistinguishability, in minimizing the number of states of a completely specified machine. It is clear that indistinguishability is an equivalence relation, but in general it is sufficient to have a relation that is reflexive, symmetric, and transitive.

Two elements in the same block have some property in common, such that for some purpose, we may not need to distinguish them. Partitions may then be seen as a mechanism for summarizing information or abstracting detail. How exactly we do that is the subject of the next section.

	0	1	
0	3	2	0
1	5	2	0
2	4	1	0
3	1	4	1
4	0	3	0
5	2	3	0

Figure 8.18: Example of FSM with parallel decomposition.

8.4.2 Partitions with Substitution Property

So far, we have considered partitions on generic sets. We now focus on partitions of the set of states of a machine. Let us consider a machine M, composed by the interconnection of two components, M_1 and M_2. A state of M is given by a pair of states, one from M_1 and the other from M_2. If we consider all the states of M with the same first component as equivalent, we have a partition. So, we see that a decomposition identifies a partition. Suppose we are given a machine and we are asked to decompose it: We are then interested in seeing what partitions correspond to nice decompositions. To this end we introduce the notion of partition with substitution property. A partition π on the set of states of the machine

$$M = (S, I, O, \delta, \lambda)$$

is said to have the *substitution property* if an only if

$$s \equiv t(\pi)$$

implies that

$$\delta(s, a) \equiv \delta(t, a)\ (\pi)$$

$\forall a \in I$. Intuitively, a partition with substitution property (an S.P. partition) represents an amount of uncertainty on the current state of a machine that 'does not spread.' If π has S.P. and we know the block in which to which the current state of the machine belongs, then we can compute what block the next state will be in, without knowing exactly the state. Thus, an S.P. partition represents a useful way of summarizing the behavior of the machine. As a first example of the importance of S.P. partitions we offer the following result.

Theorem 8.4.1 *A sequential machine M has a non-trivial parallel decomposition of its state behavior if and only if there exist two nontrivial S.P. partitions π_1 and π_2 on M such that*

$$\pi_1 \cdot \pi_2 = \mathbf{0}$$

A similar theorem can be stated for serial decomposition: If a machine has one nontrivial S.P. partition, then it has a nontrivial serial decomposition.

Let us see how to perform the decomposition on the FSM of Figure 8.18. One

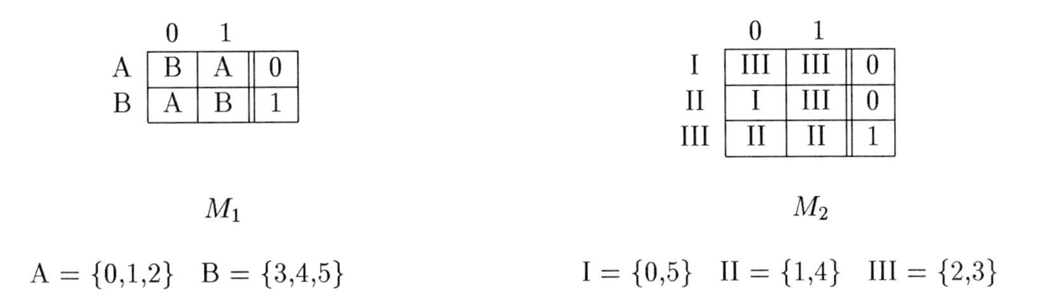

	0	1	
A	B	A	0
B	A	B	1

M_1

	0	1	
I	III	III	0
II	I	III	0
III	II	II	1

M_2

A = {0,1,2} B = {3,4,5}   I = {0,5} II = {1,4} III = {2,3}

Figure 8.19: Components of the FSM of Figure 8.18.

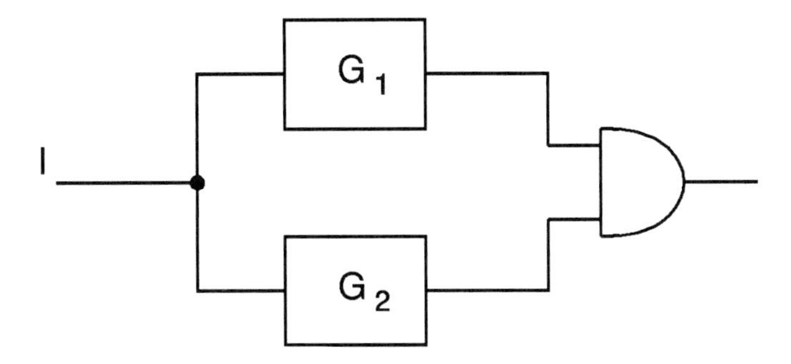

Figure 8.20: Structure of the parallel decomposition.

can easily verify that

$$\pi_1 = \{\overline{0,1,2}; \overline{3,4,5}\} \quad \text{and} \quad \pi_2 = \{\overline{0,5}; \overline{1,4}; \overline{2,3}\}$$

are 2 S.P. partitions and that $\pi_1 \cdot \pi_2 = \mathbf{0}$. The basic idea is that each of the two parallel components corresponds to one of the two S.P. partitions. The states correspond to the blocks of the partitions. Each component machine has approximate information on the state of the overall machine; however, its state can be reconstructed because the product of the two partitions is **0**. The result of the decomposition for our example is illustrated in Figure 8.19. Notice that in this case we can assign the outputs of the two components so as to have

$$\lambda(s) = \lambda_1(s_1) \cdot \lambda_2(s_2),$$

and hence the overall structure of the result is the one depicted in Figure 8.20. In general, the outputs of the parallel composition will be a function of the states of the two machines and of the primary inputs.

We now consider the serial decomposition of a machine. If we only have one S.P. partition, we can find only one component that can independently compute what block the next state of the machine will be in. We call this sub-machine the *independent component*. It is possible to build a second component that computes the missing information, based on the state of the independent component. This second machine is called the *dependent component*. The dependent component corresponds

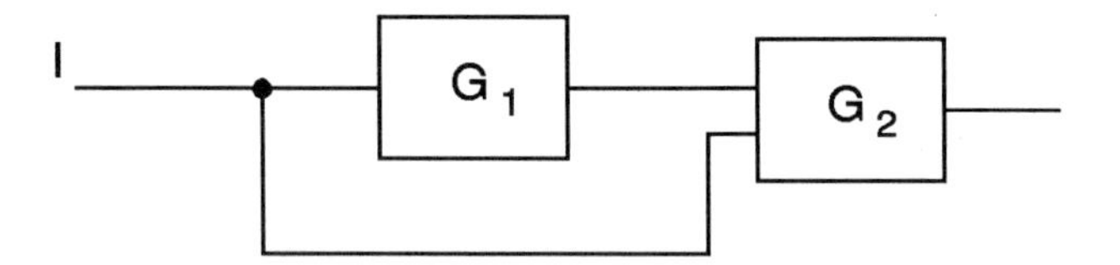

Figure 8.21: Structure of the serial decomposition.

	0	1	0	1
1	1	4	1	0
2	3	5	1	0
3	2	4	0	0
4	3	1	1	1
5	2	2	0	1

Figure 8.22: Example of FSM with serial decomposition.

to a partition of the states such that its product with the S.P. partition is **0**. This guarantees that the state of the original machine is computed exactly. The general structure of the serially decomposed machine is illustrated in Figure 8.21. We examine the details of the decomposition for the FSM of Figure 8.22. It can be verified that $\pi = \{\overline{1,2,3};\overline{4,5}\}$ is the only S.P. partition. We choose $\tau = \{\overline{1,4};\overline{2,5};\overline{3}\}$ as partition for the dependent component. Clearly, $\pi \cdot \tau = \mathbf{0}$.

The flow table of the independent component is computed as in the case of parallel decomposition. The result is shown in Figure 8.23, where state A corresponds to states 1, 2, and 3 of the original machine, and state B corresponds to states 4 and 5. Notice that there are no outputs indicated, since the outputs are the state variables. For the dependent component, we first rewrite the flow table of the original machine, substituting the partition block names for the original states, as shown in Figure 8.24. We then rearrange the table so as to show the inputs to the dependent component across the top of the table, as usual. This is illustrated in Figure 8.25.

8.4.3 Computation of the S.P. Partitions

The computation of the S.P. partitions is based on the partition algebra that we have introduced. The S.P. partitions form a sub-lattice of the lattice of all partitions. This suggests the following generation strategy. We first generate the *minimal* S.P. partitions and then we sum them until we have considered all possible sums.

	0	1
A	A	B
B	A	A

Figure 8.23: Independent component for the FSM of Figure 8.22.

		0	1	0	1
	I	I	I	1	0
A	II	III	II	1	0
	III	II	I	0	0
B	I	III	I	1	1
	II	II	II	0	1

Figure 8.24: First step in the construction of the dependent component.

	A,0	A,1	B,0	B,1	A,0	A,1	B,0	B,1
I	I	I	III	I	1	0	1	1
II	III	II	II	II	1	0	0	1
III	II	I	—	—	0	0	—	—

Figure 8.25: Second step in the construction of the dependent component.

The minimal partitions are those obtained by requiring that two states only are included in a block. Consider the machine of Figure 8.26. (The outputs are not shown, because the computation of the S.P. partitions does not depend on them.) In order to compute the minimum partition such that states 1 and 2 belong to the same block, we consider the successors of 1 and 2. They are states 1 and 4 under input 0 and states 2 and 3 under input 1. Hence, if 1 and 2 are to be in the same block, so must be states 1 and 4 and states 2 and 3. However, this implies, due to transitivity, that states 1, 2, 3, and 4 must all belong to the same block if states 1 and 2 are to belong to the same block. Considering the successors of these four states, we rapidly come to the conclusion that there is no S.P. partition other than the trivial partition **1** that contains 1 and 2 in the same block.

As a second example, we consider states 1 and 4. These two states have the same successors for both input values. Hence there are no applications of transitivity and $\{\overline{1,4};\overline{2};\overline{3};\overline{5}\}$ is the minimal S.P. partition such that states 1 and 4 belong to the same block.

Finally, we consider states 4 and 5. Looking at their successors, we see that states

	0	1
1	1	2
2	4	3
3	5	2
4	1	2
5	1	3

Figure 8.26: Example FSM for the computation of the S.P. partitions.

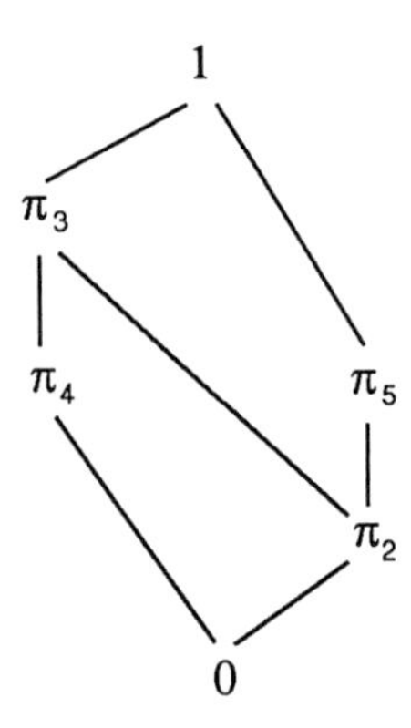

Figure 8.27: S.P. partition lattice for the example of Figure 8.26.

2 and 3 must be in the same block. No further implications occur and we get the S.P. partition $\{\overline{1}; \overline{2,3}; \overline{4,5}\}$.

If we carry out the computation for all pairs of states, we end up with the following list of S.P. partitions.

$$\begin{aligned}
\pi_1 &= \mathbf{1} \\
\pi_2 &= \{\overline{1,4}; \overline{2}; \overline{3}; \overline{5}\} \\
\pi_3 &= \{\overline{1,4,5}; \overline{2,3}\} \\
\pi_4 &= \{\overline{1}; \overline{2,3}; \overline{4,5}\} \\
\pi_5 &= \{\overline{1,4}; \overline{2,5}; \overline{3}\}
\end{aligned}$$

One can easily verify that no new S.P. partition can be generated by summing the partitions of this list. The resulting lattice is shown in Figure 8.27.

8.4.4 General Decomposition and State Encoding

S.P. partitions allow one to find serial/parallel decompositions. If we can repeatedly apply serial and parallel decompositions until each component machine has two states, then we have reduced the encoding problem to the one of selecting which state of each sub-machine will get the code 0. If we assume that the cost of the final implementation will not depend much on this choice, we then conclude that we have practically solved the state assignment problem.

However, relatively few circuits can be decomposed that way. We need to resort to something more general than S.P. partitions, namely, *partition pairs.* We shall only define partition pairs and see an application to state encoding. The reader interested in more detail should refer to [138].

A partition pair (π, π') on the machine

$$M = (S, I, O, \delta, \lambda)$$

is an ordered pair of partitions on S such that

$$s \equiv t(\pi) \text{ implies that } \delta(s,a) \equiv \delta(t,a)\ (\pi') \qquad \forall a \in I.$$

	00	01	10	11	
1	1	2	3	4	1
2	3	4	1	2	1
3	2	1	4	3	0
4	4	3	2	1	0

Figure 8.28: Example FSM for encoding based on partition pairs.

According to this definition, the knowledge of the block of π containing the present state and of the current input allows one to compute the block of π' that will contain the next state. It is evident that if (π, π) is a partition pair, then π has substitution property. Hence, partition pairs generalize S.P. partitions.

We are now interested in using partition pairs to encode the machine of Figure 8.28. One can easily verify that

$$\begin{aligned}(\pi_1, \pi_2) &= (\{\overline{1,2}; \overline{3,4}\}, \{\overline{1,3}; \overline{2,4}\}) \\ (\pi_2, \pi_1) &= (\{\overline{1,3}; \overline{2,4}\}, \{\overline{1,2}; \overline{3,4}\})\end{aligned}$$

are two partition pairs and that

$$\pi_1 \cdot \pi_2 = \mathbf{0}.$$

Suppose we use a minimum-length encoding of two bits, y_1 and y_2. We encode y_1 according to π_1 and y_2 according to π_2, that is, we use the encoding bits to designate the blocks of the partitions. One possible way to do that is:

$$\begin{array}{ccc} & & y_1 y_2 \\ 1 & \longrightarrow & 0\,0 \\ 2 & \longrightarrow & 0\,1 \\ 3 & \longrightarrow & 1\,0 \\ 4 & \longrightarrow & 1\,1. \end{array}$$

(Notice that each state has a unique code.) The key observation here is that computing the next value of y_1 only requires the knowledge of the current value of y_2, and vice versa. This follows from (π_1, π_2) and (π_2, π_1) being partition pairs. In particular, one obtains the following equations for the machine:

$$\begin{aligned}Y_1 &= x_1' y_2 + x_1 y_2' \\ Y_2 &= x_2' y_1 + x_2 y_1' \\ z &= y_1'.\end{aligned}$$

This example illustrates the general principle that partition pairs can be used to reduce the dependence among the state variables. It is also possible to reduce the dependence on the inputs, so that only a subset of the inputs is required to compute a given next-state function. A look at the block diagram of the encoded machine,

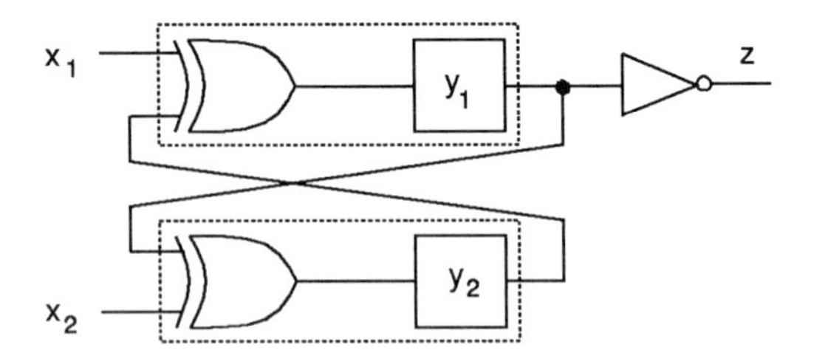

Figure 8.29: Schematic for the encoding of the machine of Figure 8.28.

depicted in Figure 8.29, shows how we have decomposed the machine in the process of encoding it. (We may also say: "How we encoded the machine in the process of decomposing it." It is largely a matter of interpretation.)

8.5 Notes

Books that cover finite state machines are [136, 28, 140, 162, 187]. The method outlined in Section 8.1 is that of Grasselli and Luccio [118]. This method is used in the program STAMINA [130, 228] that will be used in some of the homework problems. The first studies on the minimization of incompletely specified machines are due to Ginsburg [115, 113] and Paull and Unger [211].

State minimization is an important step in the design of FSM-based circuits. Though the problem has received considerable attention in the past [211, 114, 118, 220, 170, 9, 8] (see [227] for an extensive bibliography), it is the recent development of sequential synthesis systems that has created the need for efficient algorithms that can minimize large FSM's.

The current state of the art in FSM state minimization is that established in [228]. This paper showed that exact state minimization is feasible for a large class of practical examples, certainly including most hand-designed FSM's. However, FSM's generated by sequential synthesis systems may have many states and, in particular, many compatible states. Heuristic techniques are therefore of interest. The ones we present in this paper have been very successful in reducing time and memory requirements, without appreciably affecting the optimality of the solution.

Normally a reduction in the number of states is attempted in the hope of reducing the complexity of the resulting FSM, as measured, for instance, by the gate count of a multilevel implementation after technology mapping.[5] However, solutions with the same number of states may have different gate counts. Several steps in the algorithms influence the cost of the resulting implementation. These are analyze in detail in [228].

BDD-Based Symbolic Methods are beginning to appear [157]. BDDs are well suited for representing the characteristic functions of huge sets of prime compatibles that arise in practical applications. BDDs have also led to some breakthroughs in the unate and binate covering problems [178].

Significant problems in controlling power dissipation and/or traversal tractability has caused a recent renaissance of sorts in the field of encoding (or re-encoding) digital circuits. Early work on encoding for low power [234, 19, 124, 20] was limited to

[5] Another objective may be increased testability.

	00	01	11	10
1	3,0	—	2,–	—
2	—	4,0	6,–	—
3	5,1	—	—	–,0
4	—	4,1	—	–,0
5	1,–	—	3,1	6,–
6	4,–	5,–	6,1	6,–

Figure 8.30: An incompletely specified flow table.

relatively small state machines, where the states may be referred to *explicitly*. In this sense, these techniques follow the qualitative paradigm of traditional FSM encoding paradigm discussed in the present chapter. The algorithms discussed derive from the treatments in [87, 176, 93]. Other significant publications in this area include [89, 6, 84].

Recent literature on the encoding problem includes [267, 243, 181, 88, 216, 125, 124].

8.6 Notes

8.7 Summary

In this chapter we have discussed in detail the method due to Rho, et al, [228] for minimizing incompletely specified FSMs. This method is applicable in the most significant FSM synthesis scenario, in which the FSM is initially designed and simulated in some high level language, and then synthesized down to the RTL (Register Transfer Level) and logic levels of abstraction, which naturally leads to incomplete specification.

We further introduced an approach to FSM encoding that is used in the most widely available CAD tools. The basic notion is that of *anticipating*, in the encoding step, the the logic minimization operations that will likely be applied at later steps in the design cycle.

We concluded with a brief treatment of Hartmanis and Stearns' elegant work on encoding via partitions with the substitution property.

8.8 Problems

1. For the flow table of Figure 8.30, draw the compatibility table.
   Solution. The compatibility table is shown in Figure 8.31.

   □

2. For the flow table of Figure 8.30, compute the maximal compatibles, using the compatibility table derived in Problem 1.

2	2,6				
3	×	∼			
4	∼	×	∼		
5	×	3,6	×	∼	
6	3,4 2,6	4,5	4,5	4,5	1,4 3,6
	1	2	3	4	5

Figure 8.31: Compatibility table for the flow table of Figure 8.30.

3. For the flow table of Figure 8.30, compute a minimal closed cover *using maximal compatibles only*. Use the maximal compatibles derived in Problem 2. Set up and solve the binate covering problem. Draw your search tree. Finally, draw the reduced flow table. [Hint: In the covering problem, split on the maximal compatible composed of States 4, 5, and 6.]

Solution. We begin by listing the maximal compatibles with their class sets.

	maximal compatibles	class set
$m_1 =$	{4,5,6}	{{1,4}, {3,6}}
$m_2 =$	{3,4,6}	{{4,5}}
$m_3 =$	{2,5,6}	{{1,4}, {4,5}, {3,6}}
$m_4 =$	{2,3,6}	{{4,5}}
$m_5 =$	{1,4,6}	{{3,4}, {4,5}, {2,6}}
$m_6 =$	{1,2,6}	{{3,4}, {4,5}}

This list will guide us in writing the constraint matrix.

$$F = \begin{array}{c} \begin{matrix} x_1 & x_2 & x_3 & x_4 & x_5 & x_6 \end{matrix} \\ \left[\begin{matrix} - & - & - & - & 1 & 1 \\ - & - & 1 & 1 & - & 1 \\ - & 1 & - & 1 & - & - \\ 1 & 1 & - & - & 1 & - \\ 1 & - & 1 & - & - & - \\ 1 & 1 & 1 & 1 & 1 & 1 \\ 0 & - & - & - & 1 & - \\ 0 & 1 & - & 1 & - & - \\ 1 & 0 & - & - & - & - \\ - & - & 0 & - & 1 & - \\ 1 & - & 0 & - & - & - \\ - & 1 & 0 & 1 & - & - \\ 1 & - & - & 0 & - & - \\ - & 1 & - & - & 0 & - \\ 1 & - & - & - & 0 & - \\ - & - & 1 & 1 & 0 & 1 \\ - & 1 & - & - & - & 0 \\ 1 & - & - & - & - & 0 \end{matrix}\right] \end{array} \begin{matrix} \\ 1 \\ 2 \\ 3 \\ 4 \\ 5 \\ 6 \\ 7 \\ 8 \\ 9 \\ 10 \\ 11 \\ 12 \\ 13 \\ 14 \\ 15 \\ 16 \\ 17 \\ 18 \end{matrix}$$

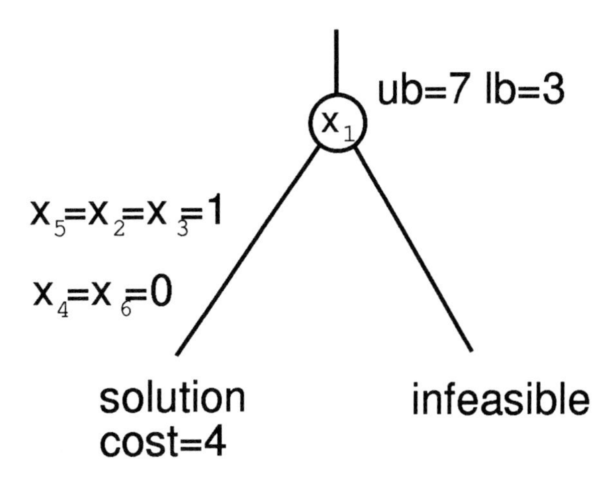

Figure 8.32: Search tree for Problem 3.

	00	01	11	10
1	5,–	1,1	2,1	1,–
2	1,1	1,1	1,1	1,0
3	5,–	1,0	2,1	1,–
5	2,0	1,1	3,1	1,0

Figure 8.33: Result of minimizing the flow table of Figure 8.30 using maximal compatibles only.

The first six rows of the matrix give the covering constraints and the remaining rows give the closure constraints. One can easily see that Rows 6, 8, and 16 dominate other rows and therefore can be dropped. The resulting matrix is cyclic and splitting is necessary. First of all, we compute the lower bound, which is 3. (Rows 1, 3, and 5 are independent.) The upper bound is of course 7. Then, following the hint, we split on x_1. Selecting x_1 causes x_5 to become essential and the choice of x_5 causes x_2 to become essential. After all covered rows are removed, only Row 2 is left. By dominance, Columns 4 and 6 are eliminated and we obtain the solution of cost 4:

$$x_1 = x_2 = x_3 = x_5 = 1 \quad x_4 = x_6 = 0.$$

Since this cost is higher than the lower bound, we need to consider also $x_1 = 0$. However, imposing $x_1 = 0$ causes x_5 and x_6 (among others) to become unacceptable. However, if $x_5 = x_6 = 0$, Row 1 is not covered. Hence, the subproblem is infeasible and the final solution is the one of cost 4 that we found. The search tree is shown in Figure 8.32.

Figure 8.33 gives a minimized FSM for the solution that we have found. In all cases where multiple choices existed for the next state, State 1 was selected. □

4. For the flow table of Figure 8.30, compute the prime compatibles. Use the results of Problem 2.

5. Use MINCOV to solve the covering problem based on the prime compatibles found in Problem 4. Include the output from the program, showing the echo of the input matrix.

   Solution. Here is the output of MINCOV, including the echo of the input matrix.

```
# mincov -v 10 -i pb7-12.t
# Mincov Version #1.0, Release date 10/01/92
     00000000011111
     12345678901234
     --------------
   1:....++....+++.
   2:..++.+..++..+.
   3:.+.+...+.+....
   4:++..+.++..+...
   5:+.+...+.+.....
   6:++++++.....+.+
   7:-...+.....+...
   8:-+.+..........
   9:+-....+.......
  10:..-.+.....+...
  11:+.-...+.......
  12:.+-+..........
  13:+..-..+.......
  14:.+..-..+......
  15:+...-.+.......
  16:..++-+........
  17:.+...-.+......
  18:+....-+.......
  19:.+.+....-.....
  20:.+.....+...-..
  21:..++.+.....-..
  22:..++.+......-.
Options:  Independent Set  Beta Dominance
ABSMIN[ 0]    22x 14 sel=  0 bnd= 15 lb=  3     0.01 sec pick=1
ABSMIN[ 1]     0x  0 sel=  3 bnd= 15 lb=  3     0.01 sec BEST
new 'best' solution 3 at level 1 (time is 0.01 sec)
matrix      = 22 by 14 with 84 elements (27.273\%)
cover size = 3 elements
cover cost = 3
time        = 0.00 sec
components = 0
gimpel      = 0
nodes       = 2
max_depth   = 1
```

	00	01	11	10
1	2,0	—	5,1	—
2	1,0	3,–	—	—
3	3,–	1,1	5,–	4,1
4	—	2,–	1,–	—
5	3,–	2,1	–,0	6,–
6	6,1	1,–	2,–	5,–

Figure 8.34: Flow table for Problem 7.

```
**** 0 rows deleted -- 0 were not covered
Solution is  1 4 11
```

The cost of the first solution found equals the lower bound this time and therefore the branch corresponding to $x_1 = 0$ need not be explored. Notice that this problem has many solutions of cost 3, besides the one found by the program. For instance, 6, 7, and 8 form a solution as well as 4, 7, and 11. □

6. Use SIS to verify your answers are correct. Write a `.kiss2` file describing the flow table of Figure 8.30. Use SIS to invoke STAMINA. (Check the command `state_minimize`.) Alternatively, you may use STAMINA directly. Programs and man pages are in the usual places.

7. Minimize the incompletely specified FSM of Figure 8.34. Specifically, do the following:

   (a) Draw the compatibility table;

   (b) compute the maximal compatibles;

   (c) compute the prime compatibles;

   (d) set up and solve the binate covering problem, showing the search tree; and

   (e) form the reduced table.

   In addition, also compute the cost of the cheapest solution composed of maximal compatibles only, by setting up and solving another binate covering problem.

   You may want to verify your results running STAMINA on this FSM. As an example, the description of the flow table of Figure 8.34 in `.kiss2` format is as follows.

```
.i 2
.o 1
.p 18
.s 6
00 1 2 0
11 1 5 1
```

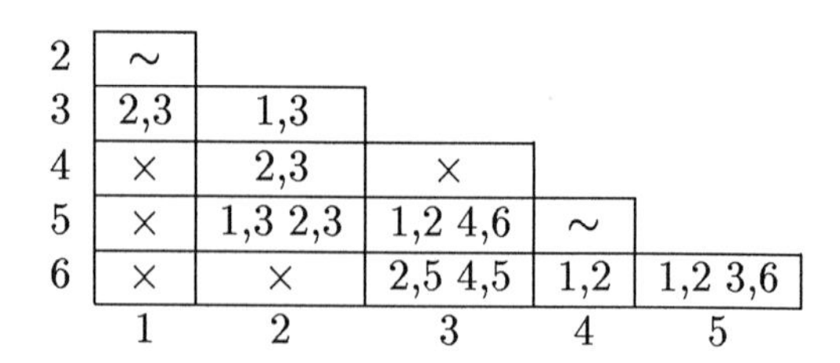

2	~				
3	2,3	1,3			
4	×	2,3	×		
5	×	1,3 2,3	1,2 4,6	~	
6	×	×	2,5 4,5	1,2	1,2 3,6
	1	2	3	4	5

Figure 8.35: Compatibility table for the flow table of Figure 8.34.

```
00 2 1 0
01 2 3 -
00 3 3 -
01 3 1 1
10 3 4 1
11 3 5 -
01 4 2 -
11 4 1 -
00 5 3 -
01 5 2 1
10 5 6 -
11 5 ANY 0
00 6 6 1
01 6 1 -
10 6 5 -
11 6 2 -
.e
```

Every line corresponds to a transition and is divided in four fields: primary input, present state, next state, and primary output. Notice the keyword ANY that allows you to specify the output, while leaving the next state unspecified. Also notice that no line corresponds to entries for which both next state and output are unspecified. The directive `.s 6` says that there are six states. Everything else is as in the input format to ESPRESSO.

Solution. The compatibility table for the flow table of Figure 8.34 is shown in Figure 8.35. The maximal compatibles can be derived from the compatibility table as follows:

$$(x_1' + x_4')(x_1' + x_5')(x_1' + x_6)(x_2' + x_6')(x_3' + x_4')$$

$$(x_1' + x_4'x_5'x_6')(x_2'x_3' + x_2'x_4' + x_3'x_6' + x_4'x_6')$$

$$x_1'x_2'x_3' + x_1'x_2'x_4' + x_1'x_3'x_6' + x_1'x_4'x_6' + x_4'x_5'x_6'$$

From which, the following maximal compatibles result:

$$\{4,5,6\}, \{3,5,6\}, \{2,4,5\}, \{2,3,5\}, \{1,2,3\}.$$

We can now derive the prime classes and their class sets as follows. We begin by computing the class sets of the maximal compatibles. Since all maximal compatibles are of the same size, they are all entered in the prime list.

	prime compatibles	class set
$p_1 =$	{4,5,6}	{{3,6}, {1,2}}
$p_2 =$	{3,5,6}	{{1,2},{2,5},{4,5,6}}
$p_3 =$	{2,4,5}	{{1,3}, {2,3}}
$p_4 =$	{2,3,5}	{{1,2,3},{4,6}}
$p_5 =$	{1,2,3}	$\emptyset$

We now add the primes of size two.

	prime compatibles	class set
$p_6 =$	{4,5}	$\emptyset$
$p_7 =$	{4,6}	{{1,2}}
$p_8 =$	{3,5}	{{1,2}, {4,6}}
$p_9 =$	{3,6}	{{2,5},{4,5}}
$p_{10} =$	{2,4}	{{2,3}}

Finally, we add one prime of size one.

	prime compatibles	class set
$p_{11} =$	{6}	$\emptyset$

There are a total of eleven prime compatibles. We can now proceed to build the covering table as shown in Figure 8.36. The search tree for this covering problem is in Figure 8.37. The best solution is given by prime compatibles 5, 6, and 7. Another solution of cost 3 is composed of 3, 5, and 7. The minimized machine derived from our solution is given in Figure 8.38. We now consider the restriction of the covering problem to maximal compatibles. This is the matrix:

$$F = \begin{array}{c} \begin{array}{ccccc} x_1 & x_2 & x_3 & x_4 & x_5 \end{array} \\ \left[\begin{array}{ccccc} - & - & - & - & 1 \\ - & - & 1 & 1 & 1 \\ - & 1 & - & 1 & 1 \\ 1 & - & 1 & - & - \\ 1 & 1 & 1 & 1 & - \\ 1 & 1 & - & - & - \\ 0 & 1 & - & - & - \\ 0 & - & - & - & 1 \\ - & 0 & - & - & 1 \\ - & 0 & 1 & 1 & - \\ 1 & 0 & - & - & - \\ - & - & 0 & - & 1 \\ - & - & 0 & 1 & 1 \\ - & - & - & 0 & 1 \\ 1 & - & - & 0 & - \end{array}\right] \end{array} \begin{array}{c} \\ 1 \\ 2 \\ 3 \\ 4 \\ 5 \\ 6 \\ 7 \\ 8 \\ 9 \\ 10 \\ 11 \\ 12 \\ 13 \\ 14 \\ 15 \end{array}$$

$F =$

x_1	x_2	x_3	x_4	x_5	x_6	x_7	x_8	x_9	x_{10}	x_{11}	
–	–	–	–	1	–	–	–	–	–	–	1
–	–	1	1	1	–	–	–	–	1	–	2
–	1	–	1	1	–	–	1	1	–	–	3
1	–	1	–	–	1	1	–	–	1	–	4
1	1	1	1	–	1	–	1	–	–	–	5
1	1	–	–	–	–	1	–	1	–	1	6
0	1	–	–	–	–	–	–	1	–	–	7
0	–	–	–	1	–	–	–	–	–	–	8
–	0	–	–	1	–	–	–	–	–	–	9
–	0	1	1	–	–	–	–	–	–	–	10
1	0	–	–	–	–	–	–	–	–	–	11
–	–	0	–	1	–	–	–	–	–	–	12
–	–	0	1	1	–	–	–	–	–	–	13
–	–	–	0	1	–	–	–	–	–	–	14
1	–	–	0	–	–	1	–	–	–	–	15
–	–	–	–	1	–	0	–	–	–	–	16
–	–	–	–	1	–	–	0	–	–	–	17
1	–	–	–	–	–	1	0	–	–	–	18
–	–	1	1	–	–	–	–	0	–	–	19
1	–	1	–	–	1	–	–	0	–	–	20
–	–	–	1	1	–	–	–	–	0	–	21

Figure 8.36: Covering table for Problem 7.

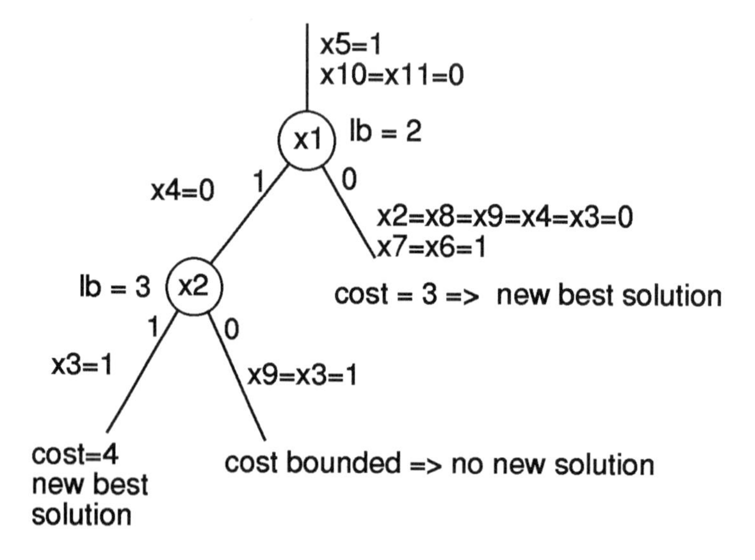

Figure 8.37: Search tree for the covering problem of Figure 8.36.

	00	01	11	10
5	5,0	5,1	6,1	6,1
6	5,–	5,1	5,0	7,–
7	7,1	5,–	5,–	6,–

Figure 8.38: Result of minimizing the flow table of Figure 8.34.

	00	01	11	10
1	1,1	1,1	5,0	2,1
2	1,1	4,0	3,0	3,0
3	3,1	4,0	4,0	4,0
4	1,0	5,0	5,0	1,0
5	3,1	5,0	5,0	1,0

Figure 8.39: Flow table for Problem 9.

We notice that x_5 is essential. Once that is taken into account, we obtain a cyclic reduced matrix. Splitting on $x_1 = 1$, we find a solution of cost 4 (1,2,3,5). For $x_1 = 0$ the subproblem is infeasible. Hence, in this case, the solution obtained by considering the maximal compatibles only is inferior to the one obtained using all prime compatibles. □

8. Solve the following binate covering problem, assuming unit cost for all variables. Show the search tree.

$$F = \begin{array}{c} \begin{array}{ccccc} x_1 & x_2 & x_3 & x_4 & x_5 \end{array} \\ \left[\begin{array}{ccccc} 1 & 1 & - & - & - \\ 1 & 0 & 0 & 1 & - \\ - & 1 & 1 & - & 1 \\ - & - & - & 0 & 1 \\ - & - & 1 & 1 & - \end{array} \right] \end{array} \begin{array}{c} \\ 1 \\ 2 \\ 3 \\ 4 \\ 5 \end{array}$$

9. Apply the fan-out oriented algorithm and the fan-in oriented algorithm for state encoding to the FSM of Figure 8.39. Use minimum-length codes. Show the matrices (S, Z, $\hat{S}$, and X), the attraction graphs, the ranking of the states for the embedding algorithm, and the codes derived by the embedding algorithm in both cases.

Solution. We start with the fanout-oriented algorithm. The matrices S and Z are shown in Figure 8.40.

The corresponding attraction graph is shown in Figure 8.41. States are ranked as follows:

$$S = \begin{array}{c|ccccc} & s_1^n & s_2^n & s_3^n & s_4^n & s_5^n \\ s_1^p & 2 & 1 & 0 & 0 & 1 \\ s_2^p & 1 & 0 & 2 & 1 & 0 \\ s_3^p & 0 & 0 & 1 & 3 & 0 \\ s_4^p & 2 & 0 & 0 & 0 & 2 \\ s_5^p & 1 & 0 & 1 & 0 & 2 \end{array} \qquad Z = \begin{array}{c|c} & z \\ s_1^p & 3 \\ s_2^p & 1 \\ s_3^p & 1 \\ s_4^p & 0 \\ s_5^p & 1 \end{array}$$

Figure 8.40: Matrices S and Z for Problem 9.

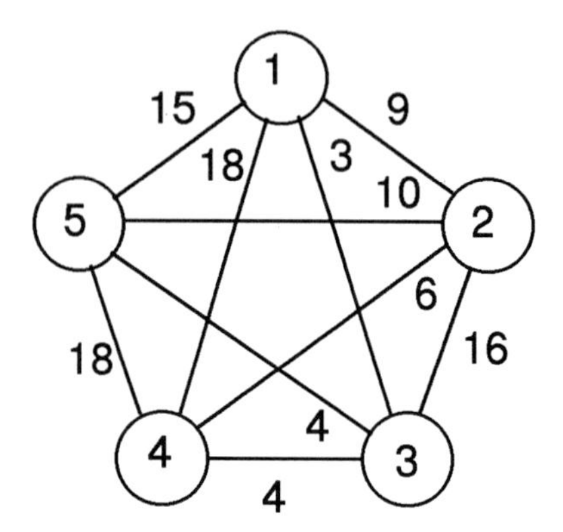

Figure 8.41: Attraction graph for the fanout-oriented algorithm.

1: 42
2: 35
3: 24
4: 42
5: 43

Hence, we initially place on the cube State 5. The three strongest attractions are for States 4, 1, and 2. The strongest attractions of the remaining state (3) are towards States 2 and 4. This gives the encoding shown in Figure 8.42.

We now turn our attention to the fanin-oriented algorithm. The matrices $\hat{S}$ and X are shown in Figure 8.43. The attraction graph is in Figure 8.44 and the encoding is in Figure 8.45. State 1 has the largest weight (48).

□

10. Use SIS to design and optimize the FSM of Figure 7.2. In particular, describe the state transition graph in `blif` format, using the `.start_kiss` and `.end_kiss` directives. Use the `state_assign` command to encode the states; optimize the combinational logic; and map using `lib2.genlib` and `lib2_latch.genlib`. Choose the options and the optimization commands that yield the circuit of minimum area.

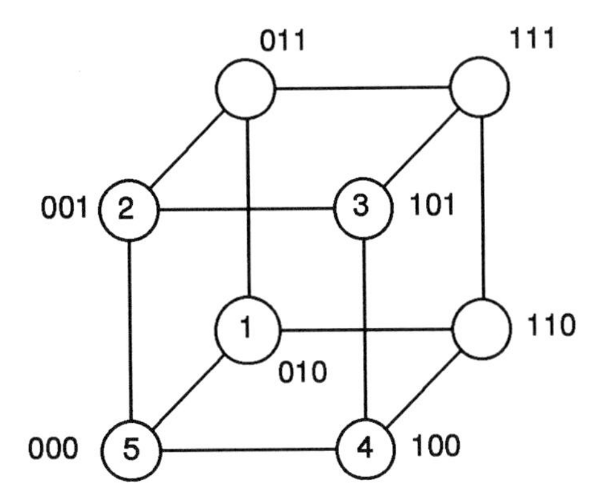

Figure 8.42: Encoding for the fanout-oriented algorithm.

$\hat{S} =$

	s_1^p	s_2^p	s_3^p	s_4^p	s_5^p
s_1^n	2	1	0	2	1
s_2^n	1	0	0	0	0
s_3^n	0	2	1	0	1
s_4^n	0	1	3	0	0
s_5^n	1	0	0	2	2

$X =$

	$x_1'x_2'$	$x_1'x_2$	x_1x_2	x_1x_2'
s_1^n	3	1	0	2
s_2^n	0	0	0	1
s_3^n	2	0	1	1
s_4^n	0	2	1	1
s_5^n	0	2	3	0

Figure 8.43: Matrices $\hat{S}$ and X for Problem 9.

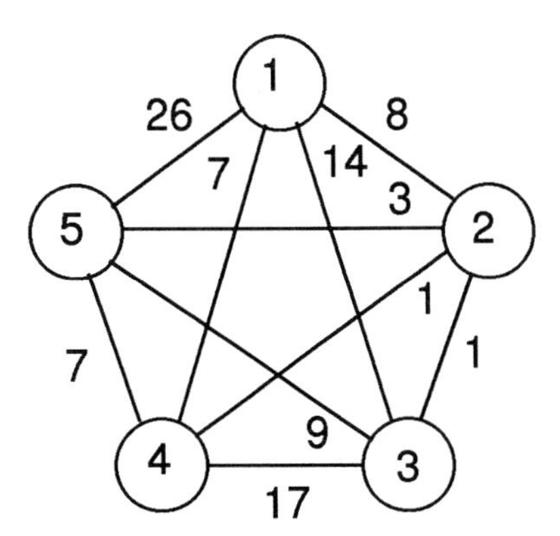

Figure 8.44: Attraction graph for the fanin-oriented algorithm.

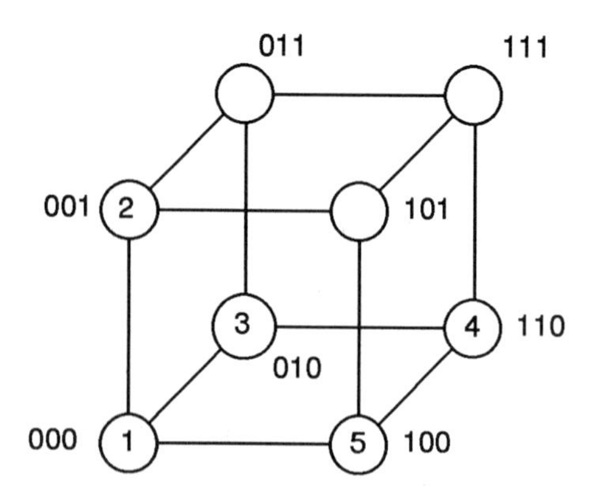

Figure 8.45: Encoding for the fanin-oriented algorithm.

Report the commands used to obtain your result, including the output of `ps`, `print_map_stats`, `pg`, and `p` for your mapped circuit.

[The area of the circuit of Figure 7.1 is 24592.]

Chapter 9

Finite Automata

In this chapter we will review some of the properties of finite automata that are pertinent to VSLI CAD, and are assumed as basic knowledge on the part of users of synthesis and verification programs such as HSIS/VIS [10, 225] and SMV [194]. Our goal is provide a set of basic notions which will enable readers to effectively employ the aforementioned CAD programs, and enable him/her to comfortably access the material in the reference texts cited above. Thus we will describe

- Deterministic and nondeterministic finite automata;
- Define regular expressions and languages (sets of finite strings) and their one to one correspondence to deterministic finite automata and nondeterministic finite automata ;
- Define regular and ω-regular languages (sets of tapes) and the corresponding finite automata;
- Define product, union and $*$ operations on sets of strings;
- Define the complementation of languages and automata.
- Discuss the following method for checking if two regular expressions $\mathcal{R}_1$ and $\mathcal{R}_2$ represent the same language:

  1. Directly build the NFAs corresponding to $\mathcal{R}_1$ and $\mathcal{R}_2$;
  2. Build equivalent DFAs to these DFAs using the subset construction of Section 9.2.2;
  3. Minimize these two DFAs as described in Section 7.4;
  4. Check that the resulting minimized STGs, whose form is canonical, are either equivalent (Cf., Section 7.8) or that they are effectively identical[1].

  All of these steps can be done efficiently and in time polynomial in the size of the various STGs.

[1]Here identical means identical up to a possible isomorphism, defined in Definition 7.8.2, given on Page 305. This is a polynomial time problem in general, and especially simple when the initial states of the two STGs are known in advance, which is the case here.

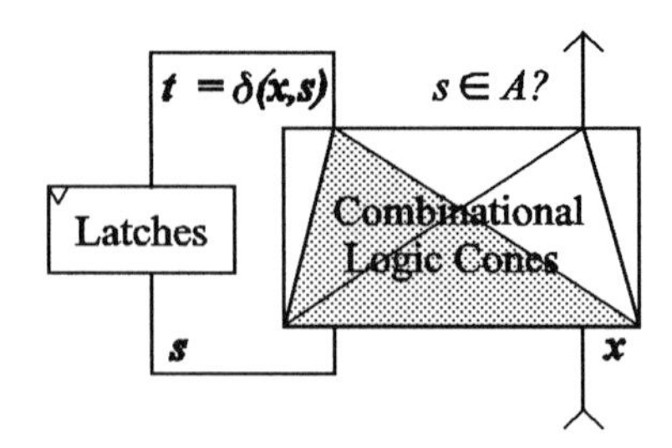

Figure 9.1: Physical implementation of a Finite Automaton.

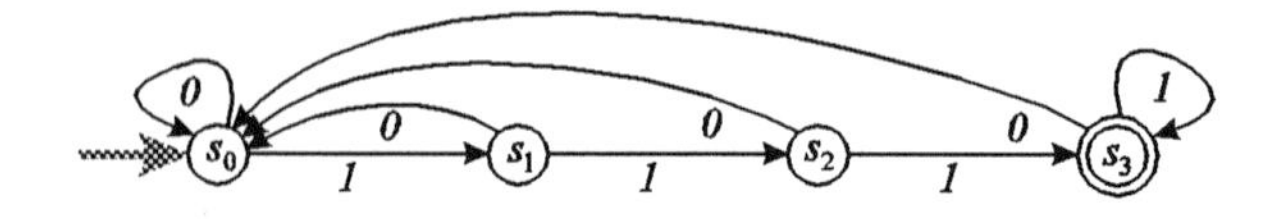

Figure 9.2: A DFA accepting all strings ending in 111.

9.1 Finite Automata and Regular Languages

Given a finite state transition structure $\mathcal{T}$ (see Section 7.7.1 on Page 292), we define a finite automaton as follows.

Definition 9.1.1 *A* **Finite Automaton** $\mathcal{A}$ *is an FST with an acceptance set A appended to the specification:*

$$\mathcal{A} = \langle \mathcal{T}, A \rangle = \langle X, S, \delta, S_0, A \rangle,$$

where the set A defines a set of accepting or final states, determined as follows. Consider an input string $\mathbf{x}$ *and corresponding run* $\mathbf{s}$*, where* $\mathbf{x} = (x_1, \cdots, x_n)$*, and* $x_i \in X,\ \ \forall 0 \leq i \leq n$*. Given a starting state* $s_0 \in S_0$*, the string* $\mathbf{x}$ *produces a run* $\mathbf{s}^{\mathbf{x}} = s_0^x, \cdots, s_n^x$*, where* $s_i \in \delta(s_{i-1}, x) \subseteq S$*. The string* $\mathbf{x}$ *is* accepted *if and only if* $s_n^x \in A$*.*

The automaton $\mathcal{A}$ *is deterministic if the transition function* δ *of the underlying FST is, and complete if* δ *is.*

We shall initially focus on deterministic FAs. A DFA, that is, a **Deterministic Finite Automaton** $\mathcal{A}$ is just a deterministic FST with a string acceptance set A appended to the specification.

A physical implementation of a DFA is depicted in Figure 9.1. The combinational logic cones of their implementations are similar to those of FSMs, except in the FA case, the output logic cone will simply implement the characteristic function $A(s)$ of the acceptance set A. That is, an implementation would output a 1 whenever the FA was in an accepting state.

A specific example of a DFA is $\mathcal{A} = \langle \mathcal{T}, A \rangle = \langle X, S, \delta, S_0, A \rangle$, the so-called 111-recognizer, of Figure 9.2. We define a $\mathbf{x}$-recognizer to be an automaton which accepts strings ending in $\mathbf{x}$, and only such strings. Here we have $X = \{0, 1\}$, $S = \{s_1, s_2, s_3\}$, and $A = \{s_3\}$. An input string $\mathbf{x} = (x_1, \cdots, x_n)$ with corresponding run $\mathbf{s} = s_0^x, \cdots, s_n^x$ is accepted if and only if $s_n^x = s_3$. Note that for any $\mathbf{x}$, the DFA $\mathcal{A}$ stays in its initial state until the first 1 in $\mathbf{x}$ occurs. If the string *ends* with 3 consecutive

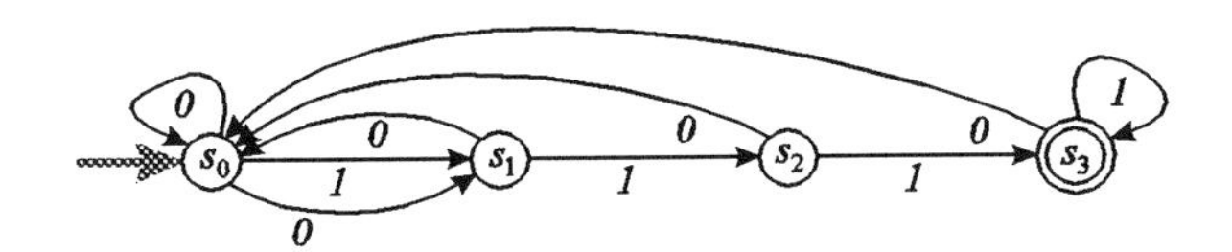

Figure 9.3: An NFA (Nondeterministic Finite Automaton).

(1)s, the string is accepted. This is why accepting states are sometimes called final states. Note that a string can have finitely many sub-strings of 3 or more consecutive (1)s, but is accepted only if it ends with 3 consecutive (1)s.

If the STG of the the FA $\mathcal{A}$ of the preceding example also had an edge from s_0 to s_1 labeled with the input $x = 0 \in X = \{0, 1\}$, then the corresponding transition function $\delta(s, x)$ would map s_0 into the set $\{s_0, s_1\}$. The resulting FA, illustrated in Figure 9.3, is nondeterministic. We call such an automaton an **NFA**. An NFA is just a nondeterministic FST with a string acceptance set A appended to the specification.

Definition 9.1.2 *For an NFA (or DFA) $\mathcal{A} = \langle \mathcal{T}, A \rangle$, a string $\mathbf{x}$ is accepted if and only if there exists a run $\mathbf{s} \in R(\mathcal{T}, \mathbf{x})$ whose last state is in A, that is, if $s_{|\mathbf{x}|} \in A$.*

Note that not all runs in the run set $R(\mathcal{T}, \mathbf{x})$ have to end in an accepting state. It is enough that a single run satisfies this requirement. For the example of Figure 9.3, the string $\mathbf{x} = (011)$ has the run set $R(\mathcal{T}, \mathbf{x}) = \{(s_0, s_1, s_2, s_3), (s_0, s_0, s_1, s_2)\}$. Whereas the first run ends in an accepting state, the second run does not. Nevertheless, the string is accepted.

As defined in Section 7.7.4, a string $\mathbf{x}$ of length n represents a finite sequence $(x_0, \cdots, x_{n-1})$, where $x_i \in X, \;\; 0 \leq i \leq n-1$, and a tape represents an infinite sequence $(x_0, x_1, \ldots)$. We will write either $|\mathbf{x}|$ or n as the length of the string. Physically, the index i of the sequence $\mathbf{x}$ can be thought of as an integral number of clock ticks, and thus measures something analogous to time. Symbols $\mathbf{w}, \mathbf{x}, \mathbf{y}$ will be preferred for denoting strings or tapes. ϵ is the unique string of length 0. We shall use upper case calligraphic symbols to denote **languages**, that is sets of strings or tapes. Some examples of languages are:

$$\mathcal{L}_1 = \{(0, 1, 0), (1, 1)\}, \;\; \{010, 11\}, \;\; \{11 \cdots 1, 00 \cdots 0\}.$$

Note that for compactness, we sometimes drop the comma separators between string letters.

For ease of reference we also define

$$\begin{aligned} X^* &= \{\mathbf{x} \mid |\mathbf{x}| \text{ is finite}\}, \\ X^+ &= X^* - \{\epsilon\}, \\ X^\omega &= \{\mathbf{x} \mid |\mathbf{x}| \text{ is infinite}\}, \end{aligned}$$

That is, $\mathcal{L}(X)^*$ is just the set of all strings that can be formed from a given input alphabet X, and $\mathcal{L}(X)^+$ is just the set of all such non-empty strings. For example, if $X = \{0, 1\}$, then $\mathcal{L}(X)^*$ is just the set all *finite* strings formed from (0)s and (1)s:

$$\{\epsilon, 0, 1, 00, 01, 10, 11, 000, \ldots, 111, \ldots, 00 \cdots 0, \ldots, 11 \cdots 1, \ldots\},$$

and $\mathcal{L}(X)^+$ is just the same set without the empty string ϵ.

Note that X^ω is just the set of all tapes on X. If $X = \{0,1\}$, then $\mathcal{L}(X)^\omega$ is just the set all *infinite* sequences formed from (0)s and (1)s:

$$\{(000\cdots), (100\cdots), (010\cdots), \ldots, (111\cdots)\}.$$

For $\mathbf{x}, \mathbf{y} \in X^*$ we define the *product* of $\mathbf{x}$ and $\mathbf{y}$ as the string $\mathbf{w} = \mathbf{xy} \in \mathcal{L}(X)^*$, obtained by **concatenating** $\mathbf{x}$ and $\mathbf{y}$. That is,

$$\begin{array}{rcll} w_i & = & x_i & , \quad 0 \le i < n = |\mathbf{x}|, \\ w_{n+i} & = & y_i & , \quad 0 \le i \le |\mathbf{y}|. \end{array}$$

Note such concatenation is defined for two strings, or for a string followed by a tape, but *not* for a tape followed by a tape (by definition, the first operand of a product must be finite).

We also define $\mathbf{x}\epsilon = \epsilon\mathbf{x} = \mathbf{x}$.

For a given $\mathbf{x}$ and i, j such that $0 \le i \le j < |\mathbf{x}|$, we define $\mathbf{x}(i,j)$ to be the sub-string, or **restriction** , of the sequence $\mathbf{x}$ to the time interval $\{k \mid i \le k < j\}$. Note that this interval includes x_i but not x_j. Cases of special interest are $\mathbf{x}(i,i) = \epsilon$, $\mathbf{x}(0, |\mathbf{x}| - 1) = \mathbf{x}$, and for $i \le j \le k$, $\mathbf{x}(i,k) = \mathbf{x}(i,j)\mathbf{x}(j,k)$. A restriction $\mathbf{y} = \mathbf{x}(0,i)$, for $i \le |\mathbf{x}|$, will also be called a **section**, or **prefix**, of the sequence $\mathbf{x}$. A prefix must be finite.

If $\mathbf{y}$ is a prefix (section) of $\mathbf{x}$, we say that $\mathbf{y} \le \mathbf{x}$. Thus the relation $\le$ induces a partial order on the set X^* (partial orders were discussed in detail in Chapter 3). For example, if $x = 011$, $y = 011110011$, and $z = 011110011110$, we would have $x \prec y \prec z$ in this partial order.

9.1.1 String Acceptance

In this section we describe Procedure DECIDE_FA of Figure 9.4, which decides whether a given string $\mathbf{x}$ is accepted by an NFA (or DFA). The procedure takes as its inputs the next state transition function $\delta(s,x)$ and the set of initial states S_0. It initializes a working set *From* to all states reachable by ϵ-moves from an initial state. Then, on each pass through the **for** loop, it computes *To* as the states which are x_i-successors of states in *From*. Each time this is done, Subprocedure EPSILON_UPDATE is called to update *From* to the set of states reachable by ϵ-moves from *To*. We illustrate this procedure by applying it to the NFA of Figure 7.35 of Page 295 to decide if string aab is accepted. Recall that in this NFA $S_0 = \{0\}$, $\delta(0,a) = 0$, $\delta(1,b) = \delta(0,\epsilon) = 1$, $\delta(2,c) = \delta(1,\epsilon) = 2$, and all the undefined transitions go to the unaccepting trap state (not shown). In the procedure, Line 2 computes *From* $= \{0,1,2\}$ since all three states are ϵ-reachable from the initial state 0. Then the first input symbol, $x_0 = a$ brings the NFA to the singleton state set *To* $= \{0\}$. Then (Line 4) *From* is again computed to be $\{0,1,2\}$. Thus the same sequence of events occur for the second symbol, $x_1 = a$. However, for $x_2 = b$, we get *To* $= \{1\}$ and *From* $= \{1,2\}$. In this case *From* contains the sole accepting state 2, so the procedure returns a 1 at Line 6, indicating acceptance.

```
Line Procedure DECIDE_FA(δ, S_0, x, A) {
1      From = S_0
2      From = EPSILON_UPDATE(δ, From)
       for(i = 0, ···, |x| − 1) {
3          To = δ(From, x_i)
4          From = EPSILON_UPDATE(δ, To)
5          if (From = ∅)return(0)
       }
6      if(From ∩ A ≠ ∅) return(1)
7      else return(0)
   }
```

Figure 9.4: Procedure for deciding string acceptance.

For the string *aba* we can summarize the behavior as follows.

Pass	Symbol	*From*(*Line*3)	*To*	*From*(*Line*4)
0	a	$\{0, 1, 2\}$	$\{0\}$	$\{0, 1, 2\}$
1	b	$\{0, 1, 2\}$	$\{1\}$	$\{1, 2\}$
2	a	$\{1, 2\}$	$\emptyset$	$\emptyset$

Thus the procedure returns 0, indicating that the string *aba* is not accepted.

9.1.2 Languages of Finite Automata

For a given DFA $\mathcal{A} = \langle X, S, \delta, S_0, A \rangle$, the set $\mathcal{L}(\mathcal{A})$ of strings accepted by $\mathcal{A}$ is called the language of $\mathcal{A}$. Note $\mathcal{L}(\mathcal{A}) \subseteq X^*$. For example, suppose the following possible input string of 15 input symbols was input to the DFA $\mathcal{A}$ of Figure 9.2:

$$\mathbf{x} = (000111011101111).$$

This string has the substring $\mathbf{y} = (11011101111)$ which is accepted by $\mathcal{A}$, and the section $\mathbf{w} = \mathbf{x}[0, 12] = (000111011101)$, which is not. The sets of strings accepted by a given FA $\mathcal{A}$ is called the **language**. Note that the products (catenation) $\mathbf{xy}$, $\mathbf{yx}$, and $\mathbf{wy}$ are all accepted.

In general, sets of strings are called **languages**, and will be denoted by calligraphic fonts, such as $\mathcal{L}$, with appropriate subscripts. $\mathcal{L}$ is a trivial set of strings if $\mathcal{L} = \emptyset$ or $\mathcal{L} = \{\epsilon\}$. The **product** of two sets of strings $\mathcal{L}_1$ and $\mathcal{L}_2$ will be defined as the usual Cartesian product:

$$\mathcal{L}_1\mathcal{L}_2 = \{\mathbf{x}_1\mathbf{x}_2 \mid \mathbf{x}_1 \in \mathcal{L}_1, \mathbf{x}_2 \in \mathcal{L}_2\}.$$

Since languages are sets, the **union** of two languages is just the set union of the two corresponding sets. It is customary to denote the union operation by "+", so

$$\mathcal{L}_1 \cup \mathcal{L}_2 = \mathcal{L}_1 + \mathcal{L}_2 = \{\mathbf{x} \mid \mathbf{x} \in \mathcal{L}_1 \text{ or } \mathbf{x} \in \mathcal{L}_2\}.$$

As an example, suppose $\mathcal{L}_1 = \{\mathbf{x}, \mathbf{y}\}$ and $\mathcal{L}_2 = \{\mathbf{w}, \mathbf{x}, \mathbf{y}\}$. Then

$$\mathcal{L}_1\mathcal{L}_2 = \{(\mathbf{x},\mathbf{w})(\mathbf{x},\mathbf{x}),(\mathbf{x},\mathbf{y}),(\mathbf{y},\mathbf{w}),(\mathbf{y},\mathbf{x}),(\mathbf{y},\mathbf{y})\}.$$

The following definition is the key to understanding **regular expressions**[2].

Definition 9.1.3 *For $\mathcal{X} \subseteq X^*$, the result $\mathcal{X}^*$ of the* **star operation** *is defined recursively as follows.*

$$\begin{aligned} \mathcal{X}^0 &= \{\epsilon\}, \\ \mathcal{X}^{n+1} &= \mathcal{X}^n\mathcal{X}, \\ \mathcal{X}^* &= \textstyle\bigcup_{n=0,1,\cdots} \mathcal{X}^n, \end{aligned}$$

A subtle point here is that although every member in the language $\mathcal{X}^*$ is finite (and therefore a string as opposed to a tape), $\mathcal{X}^*$ is an infinite set. If we delete the empty string, we get $\mathcal{X}^+ = \bigcup_{n=0,1,\cdots} \mathcal{X}^n$.

We call $\mathcal{X}^*$ the result of the **star operation** on the set $\mathcal{X}$, and note that $\emptyset^* = \{\epsilon\}$. Sets of sequences built from these constructions are quite significant.

Definition 9.1.4 Regular Sets (or Languages).
A set (or language) $\mathcal{X} \subseteq \mathcal{L}(X)^*$ is **regular**, *written $\mathcal{X} \in \mathcal{R}(X)$, if it can be obtained from the empty set and the singleton sets $\{x\}$ (for every $x \in X$) by a finite number of union, product, and star operations as defined above.*

Thus if $X = \{0, 1\}$, the following sets of strings can be identified as **regular expressions**, which we define informally here as a shorthand for regular languages[3]:

Regular Expression	Regular Language	*Description*
ϵ	$\{emptystr\}$	Set with one element — the empty string,
$\emptyset$	$\{\}$	The empty language — the set with no strings,
$(00 + 11)$	$\{00, 11\}$	Set with two elements — 00 and 11,
$0^+(00 + 11)$	$\{000, 011, 0000, 0011, \ldots\}$	Infinite set — any nonzero number of 0s, followed by 00 or 11,
$(0 + 1)^*001$	$\{001, 0001, 1001, \cdots\}$	Set of strings ending in 001.

Here parenthesization is determined by giving product precedence over union. Note that $\mathcal{L}(X)^*$ can be denoted $(0^*1^*)^* = (0 + 1)^*$.

Example 9.1.1 *Consider the alphabet $\{0, a, b\}$, and the regular expression $0^*(a + b)^*$. An NFA (top) and a DFA (bottom) accepting this regular language is given in Figure 9.5. The corresponding regular language is the set of all strings beginning with zero or more (0)s, followed by any empty or nonempty sequence of (a)s and/or (b)s. Note that the strings ba, 000, and $00aabbaab$ are in the language, whereas $0a0$, $000aaa000a$, and $00aabbaab0b$ are not, since from any state but the initial state, an*

[2]Regular expressions arise throughout text editing and programming, for example searching a text file for an expression with wild-cards, or listing a set of files with wild card specifiers.

[3]In [190] a careful distinction is made between regular expressions and regular languages

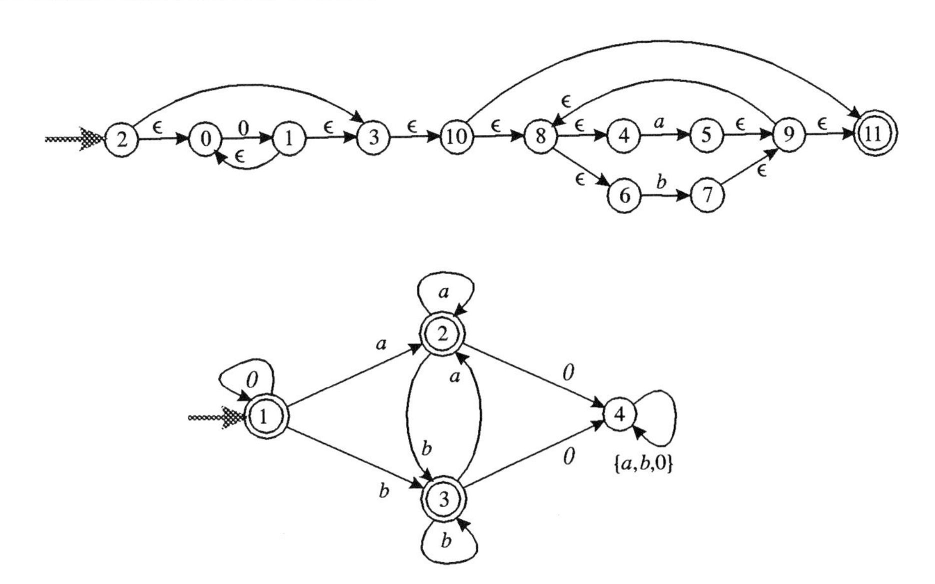

Figure 9.5: An NFA (top) and DFA (bottom) accepting the language of Example 9.1.1.

input symbol of 0 *takes the DFA to trap state* 4*. Similarly, an initial symbol other than* 0 *takes the automaton to either of the accepting states* 2 *or* 3*.*

Note in the DFA that the empty string is accepted, since the initial state 0 *is accepting, and that if a* 0 *is received after the first a or b,, the automaton goes to the trap state, wherein it remains forever.*

We shall treat the synthesis of an NFA from a given regular expression in Section 9.2. We explain there why the NFA of Figure 9.5 has, seemingly, too many states, and how its structure directly reflects the structure of the given regular expression.

The minimality of the associated DFA is explored further in Problem 2 of Section 9.8.

Note that sets of tapes (infinite sequences) are not regular languages, because each member of a regular language must be derivable from the alphabet X by a finite number of product, union, or star operations. Consequently, sets of strings $\mathcal{X} \subseteq \mathcal{L}(X)^*$ are sometimes called $*$-languages, and if also $\mathcal{X} \in \mathcal{R}(\mathbf{X})$, $\mathcal{X}$ will be called $*$-regular [163, 259] or just regular.

We conclude this section with a celebrated result [66] that shows that the set of regular languages are in one to one correspondence with the set of DFAs (and similarly for NFAs).

Theorem 9.1.1 *For every regular language $\mathcal{L}$, there is a DFA $\mathcal{A}(\mathcal{L})$, whose language is $\mathcal{L}$. Conversely, the language $\mathcal{L}(\mathcal{A})$ of every DFA $\mathcal{A}$ is regular.*

A simple algorithm that recursively computes the language of a given DFA and shows that this language is regular is given in [190, Page 319].

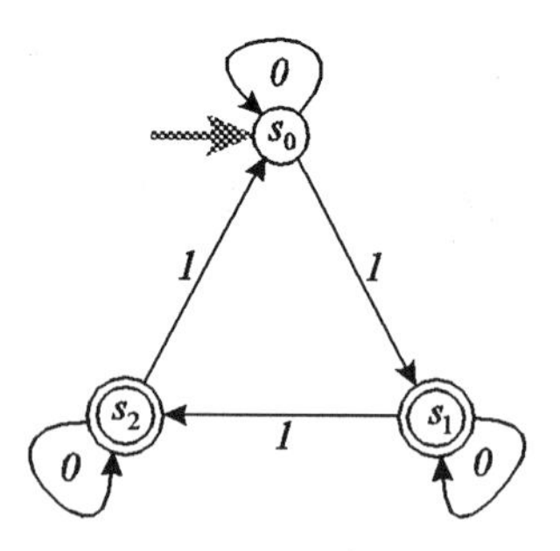

Figure 9.6: A DFA abstracted from the modulo 3 counter of Problem 5 of Page 314.

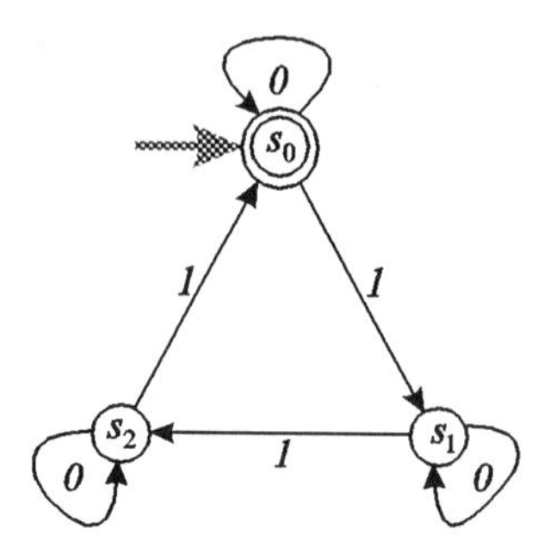

Figure 9.7: The complement of the DFA of Figure 9.6.

9.1.3 Complements of Languages

If $\mathcal{L}^1$ and $\mathcal{L}^2$ are languages (sets of strings), then we define

$$\mathcal{L}^1 - \mathcal{L}^2 = \mathcal{L}^1 \backslash \mathcal{L}^2 = \{x \in \mathcal{L}^1 \mid x \notin \mathcal{L}^2\}.$$

We include the second identity because set difference is sometimes denoted with a \ [163]. This allows us to define the **complement** of a language $\mathcal{L}$.

Definition 9.1.5 *The complement $\overline{\mathcal{L}}$ of a given language $\mathcal{L}$ defined on the alphabet X is*

$$\overline{\mathcal{L}} = X^* - \mathcal{L}.$$

For example if $\mathcal{L} = 1(1^*0^*)^*$, which is just the set of all strings defined on $X = \{0,1\}$ and beginning with a 1, then $\overline{\mathcal{L}} = 0(0^*1^*)^*$, which is just the set of all strings defined on $X = \{0,1\}$ and beginning with a 0.

This implies a similar definition for the complement of an automaton.

Definition 9.1.6 *Let $\mathcal{A} = \langle \mathcal{T}, A \rangle = \langle X, S, \delta, S_0, A \rangle$, be a DFA, then the* **complement**, *denoted $\overline{\mathcal{A}}$, is defined to be $\overline{\mathcal{A}} = \langle \mathcal{T}, \overline{A} \rangle = \langle X, S, \delta, S_0, (S - A) \rangle$.*

Note that the underlying FST is unchanged by the complementation, and that any string accepted by $\mathcal{A}$ will be rejected by $\overline{\mathcal{A}}$, and conversely. Further, $\mathcal{L}(\overline{\mathcal{A}}) = \overline{\mathcal{L}(\mathcal{A})}$.

For example, note that the complement of the FA of Figure 9.6 is that of Figure 9.7.

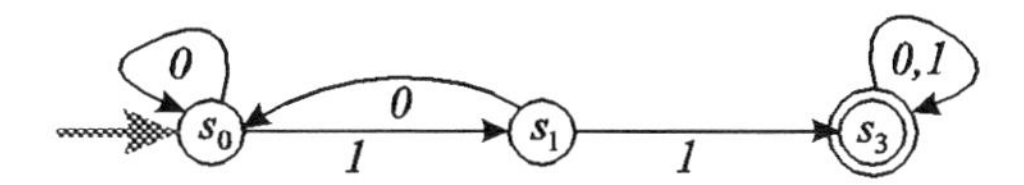

Figure 9.8: A simple DFA.

9.1.4 Examples

We first look at the language formed by the DFA specified in Figure 9.8. For this example, note that the following strings are accepted:

$$1010\cdots101101010\cdots11, 11, 00\cdots011.$$

However, the strings

$$1010\cdots10, 001001000, 00\cdots01,$$

are not accepted. In fact, it takes a sequence of 2 consecutive (1)s to get to the sole accepting state s_3 of this DFA, whose language may be expressed as

$$(0+1)^*11(0+1)^*.$$

To see this, note that no matter which of the three states the system is in, a sequence of two consecutive (1)s always take it to the accepting state s_3, from which it never exits, since it transitions back to itself for all inputs $x \in X$. On the other hand, any string which does not contain in a 11 sub-string will leave this DFA in the initial state s_0, since both states s_0 and state s_1 transition back to s_0 under a 0 input.

We next look at the language formed by the DFA $\mathcal{A} = \langle \mathcal{T}, A \rangle = \langle X, S, \delta, S_0, A \rangle$, specified in Figure 9.6. Here we have $X = \{0, 1\}$, $S = \{s_0, s_1, s_2\}$, $S_0 = \{s_0\}$, and $A = \{s_1, s_2\}$. This DFA is deceptively simple. For this example, note that these strings are accepted,

$$\{1010\}, \{01101010\}, \{0111011\},$$

which have two, four and 5 (1)s, respectively. However, the strings

$$\{101010\}, \{011011011\}, \{111111111\},$$

which have three, six, and nine (1)s, respectively are not accepted. The key observation in understanding how this DFA functions is to see that like a modulo 3 counter, this DFA stays in whatever state it is currently in, *until it receives a* 1 *input.* After it has received three (1)s, or six (1)s, or any number of (1)s which is 0 modulo 3 (that is, is a multiple of 3)[4], the DFA is always in its initial state, which is not accepting. Thus if a string has $183 = 3 * 61$ (1)s, it is rejected, but if has 184 (1)s, it is accepted. For any string, the index j of state s_j gives the residue modulo 3 of the number of (1)s in the string that brought the DFA from it initial state to the state s_j.

This is an illustration of the complement of a language. If we define $\mathcal{L}^3 = \{\mathbf{x} \in X^* \mid \mathbf{x} \text{ has a multiple of 3 1s}\}$, then we have $\mathcal{L}(A) = X^* - \mathcal{L}^3$.

[4]The residue modulo k of an integer n is defined as follows. First, let m be the largest integer such that $km < n$. Then the residue modulo k of n is just $n - (km)$. For $n = 26$, and $k = 3$, the residue modulo k of n is 2.

9.2 DFA Synthesis

A regular language, like any language, is a set of strings built from a basic underlying set symbol X, called the input alphabet. What makes a regular language special is that it is defined by a regular expression — that is, regular expressions represent regular languages. We will now show by construction that every regular language is accepted by some deterministic finite automaton. To prove this important fact, we:

1. Construct a binary parse tree, $T(\mathcal{R})$, for the given regular expression $\mathcal{R}$;

2. Construct from $T(\mathcal{R})$ an NFA $\mathcal{N}(\mathcal{R})$ which accepts the language represented by $\mathcal{R}$;

3. Construct from $\mathcal{N}(\mathcal{R})$ a DFA $\mathcal{D}(\mathcal{R})$ which accepts this same language.

This construction allows us to answer as follows two questions which are very important in text-editing, data retrieval, compilers, and symbol manipulation applications.

First, suppose we need to determine if $\mathcal{R}$ does indeed represent the language of some given DFA $\mathcal{D}_1$. After carrying out the above construction, we can check, using the methods for testing FSM equivalence given Section 7.8, if $\mathcal{D}_1 \equiv \mathcal{D}(\mathcal{R})$. If (and only if) equivalence is verified, we then know that $\mathcal{R}$ does represent the language accepted by DFA $\mathcal{D}_1$.

Second, suppose we wish to determine if two regular expressions $\mathcal{R}_1$ and $\mathcal{R}_2$ represent the same language. Using the above construction, we can conclude positively that $\mathcal{L}(\mathcal{R}_1) = \mathcal{L}(\mathcal{R}_2)$ if and only if $\mathcal{D}(\mathcal{R}_1) \equiv \mathcal{D}(\mathcal{R}_2)$.

Further, this construction also demonstrates that every set of strings accepted by an NFA can also be accepted by a DFA! In fact, as shown in [66] and in [190], the regular languages are exactly the same as the languages accepted by DFAs, which is one of the most important results in language theory.

The first step in our construction is to construct a binary parse tree for the given regular expression. This may be done by looking at the parsing rules for regular expressions, which are typically stated as follows.

```
# EXPR    ::= TERM { + TERM }
# TERM    ::= FACTOR { . FACTOR }
# FACTOR  ::= SUBEXPR { * }
# SUBEXPR ::= SYMBOL | ( EXPR )
# SYMBOL  ::= \w+
```

Stated in words, this means that an expression `EXPR` is either a `TERM` or a union of `TERM`s. Further a `TERM` is either a `FACTOR` or a product of `FACTOR`s, where a `FACTOR` is either a `SUBEXPR` or a closure symbol $*$. A `SUBEXPR` is recursively defined as a `SYMBOL` or an `EXPR` (`\w+` is a perl string matching any alphanumeric word).

The construction of a binary parse tree is illustrated, for the regular expression $(a^*b)^*$, in Figure 9.9. In this expression the parse rules identify two `FACTOR`s: the `SUBEXPR` (a^*b) and *, respectively. The top operator in the parse tree is therefore the closure operator *. Since the operand is a `SUBEXPR`, we recursively parse it into

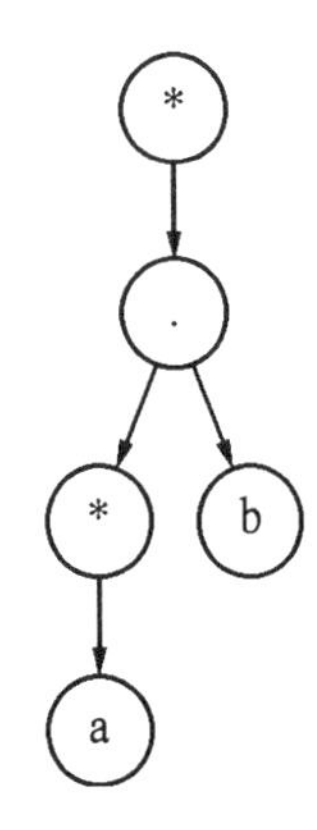

Figure 9.9: Binary Parse Tree for $(a^*b)^*$.

the product of FACTORs a^* (left operand) and b. After recursively parsing the left operand, we obtain the binary parse tree of Figure 9.9.

The next step is to construct an NFA from the given parse tree. Following conventional practice as described in [190], we choose to avoid cleverness. Instead, we construct a highly redundant NFA. Although extremely verbose, this NFA has the virtue that it can be trivially shown to accept the language of the regular expression embedded in the parse tree.

Since the parse tree has just three operands (product, union, and closure), the NFA can be constructed according to three mechanical rules, as illustrated in Figure 9.10, Figure 9.12, and Figure 9.13. The rules are applied by traversing the parse tree by DFS. As DFS completes for an operand node, the rule corresponding to the node's operator "fires". We illustrate the three rules in turn, and then give an example of the entire NFA construction for the regular expression $(a^*b)^*$.

The product rule is shown in Figure 9.10. If both left and right operands of a

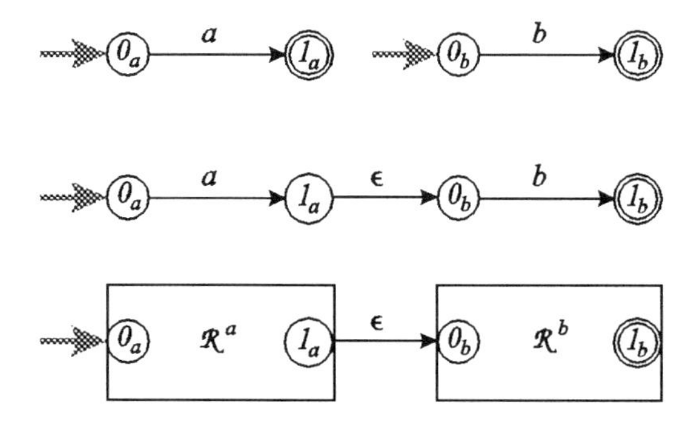

Figure 9.10: Rule for constructing an NFA which accepts the product of two regular languages.

product node were leaves of the tree, then we know that the corresponding FACTORs are atomic regular expressions (identified by the `\w+` case). In the figure, NFAs accepting these two simple languages are shown at the top. In such a case, it is clear by the definition of product that the NFA shown in the middle accepts the product language $\{ab\}$. Note here that the accepting state 1_a of the left NFA is demoted to ordinary internal node status, as is the initial state 0_b of the right NFA.

The key point of this construction rule is that the "ϵ-move" from state 1_a to state 0_b is unidirectional — there is no feedback or any other interaction between the two sub-NFAs. Note that this remains the case if the left and right NFAs were upgraded to automata representing arbitrarily complex NFAs representing arbitrary regular expressions $\mathcal{R}^a$ and $\mathcal{R}^b$, as shown at the bottom of the figure. That is, a string is accepted by the product NFA if and only if it consists of a prefix accepted by the left NFA, followed immediately by a suffix accepted by the right NFA.

To motivate the use of the seemingly redundant ϵ-moves in this context, we discuss an alternate rule which works correctly on some regular expressions, but produces incorrect results on others. To this end, consider the alternate, putatively "correct" composition rule of Figure 9.11. This candidate rule differs from the previous product

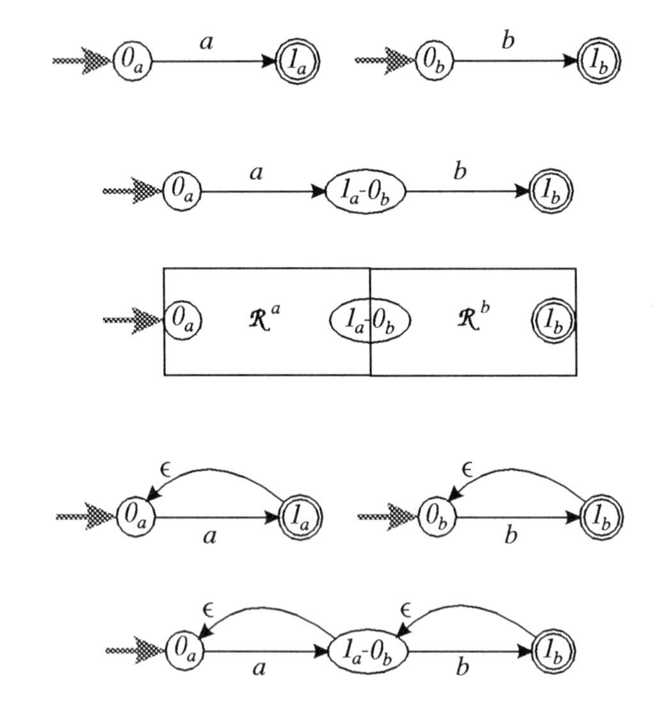

Figure 9.11: Incorrect rule for constructing an NFA which accepts the product of two regular languages.

rule by omitting the ϵ-move. As seen at the top of the figure, when the subNFAs accept the languages of atomic regular expressions, the candidate rule introduces no erroneous behavior. However, the generalization in the middle of the figure requires the rule to work for all left and right SUBEXPRs. Suppose, for the sake of contradiction, that the left and right SUBEXPRs were a^* and b^* respectively, as shown at the bottom of the figure. In this case the product automaton accepts the string *abab*, which produces the run

$$(0_a, 1_a - 0_b, 1_b, 1_a - 0_b, 0_a, 1_a - 0_b, 1_b).$$

Although this run ends in the accepting state, the product of a^*b^* is comprised only of strings of 0 or more contiguous (a)s, followed by 0 or more contiguous (b)s. Since this rule is not valid for all SUBEXPRs, we do not use it in our construction.

Of course, if efficiency became an issue, such rules could be applied conditionally to limit the size of the NFA.

Next we illustrate the union composition rule, using the example of Figure 9.12. At the top left of the figure, we again show NFAs for the case where the two operands

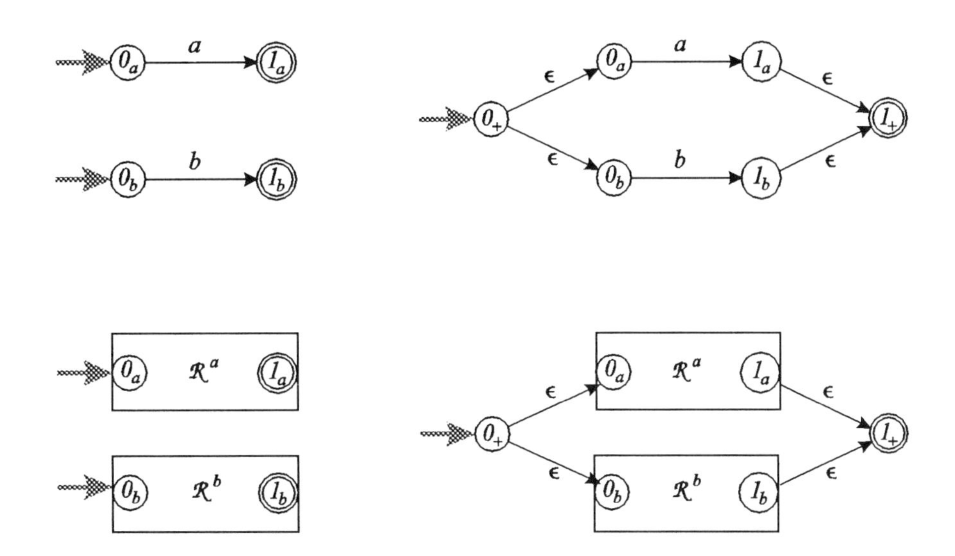

Figure 9.12: Rule for constructing an NFA which accepts the union of two regular languages.

of a union operation are atomic subexpressions. At the top right, we show a composed NFA which connects the two subNFAs "in parallel". Here external initial and accepting states are added, so that the original initial and accepting states of each subNFA are demoted.

Note again that due to the unidirectional nature of the ϵ-moves, there is no interaction between the two subNFAs — a string is accepted by the union NFA if and only if it is either a or b. Hence the language of the union NFA is just $a + b$. The situation is analogous if the left and right NFAs were upgraded to automata representing arbitrarily complex NFAs representing arbitrary regular expressions $\mathcal{R}^a$ and $\mathcal{R}^b$, as shown at the bottom of the figure. That is, a string is accepted by the union NFA if and only if it is a string accepted by the left NFA, or if it is a string accepted by the right NFA.

The closure rule is illustrated in Figure 9.13. At the top of the figure, we show the simple case of closure of an atomic subexpression. At the top right we add four ϵ-moves to construct a closure NFA. It is clear that the closure NFA accepts the language a^*. To see this, note that the empty string is accepted, since there is an ϵ-move from the initial state 0_ϵ to the accepting state 1_ϵ. Further, the input string a is accepted, since it produces, among other runs, the run $(0_\epsilon, 0_a, 1_a, 1_+)$. Similarly, the input string aa is accepted, since it produces, among other runs, the run $(0_\epsilon, 0_a, 1_a, 0_a, 1_a, 1_+)$. Note that this run cycles through the edge labeled a but cannot get back to the initial state 0_+.

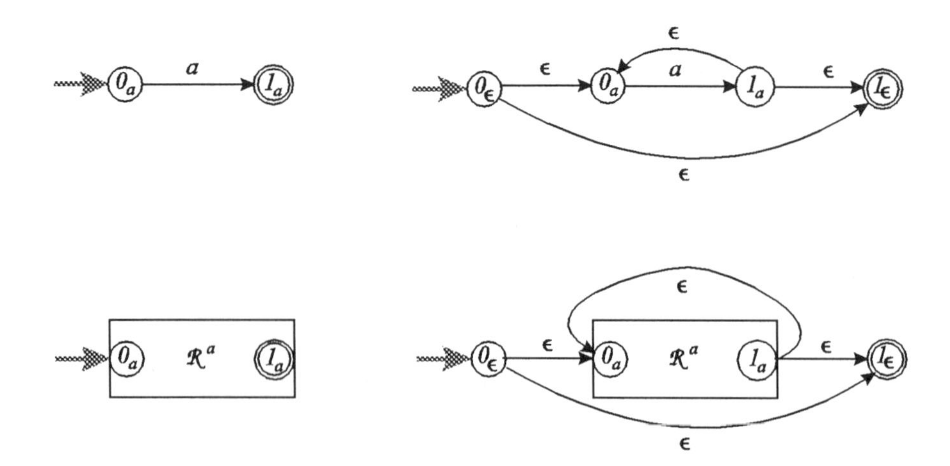

Figure 9.13: Rule for constructing an NFA which accepts the closure of two regular languages.

Again, the same behavior results if the atomic subexpression is replaced by an arbitrary regular expression, as shown at the bottom of the figure. Due to the directionality of the ϵ-moves, it is clear that only strings comprised of 0 or more repetitions of some string in the language $\mathcal{R}^a$ are accepted.

Now that the composition rules have been defined and illustrated by simple examples, we return to the construction of the DFA accepting the language represented by regular expression $(a^*b)^*$. The result is illustrated in Figure 9.14. Note on this an other computer generated graphics, the octagonal state indicates the initial state of the FST.

If we traverse the tree of Figure 9.9 by DFS, and apply the compositon rules in postorder, we first apply the closure rule (with a as operand), which produces the subgraph induced by states $\{0, 1, 2, 3\}$.

Then the product rule is applied (with a^* as left operand and b as right operand), which produces the subgraph induced by states $\{0, 1, 2, 3, 4, 5\}$.

Finally the closure rule is applied again (for the root node) which extend the previous subgraph into the entire NFA, with initial state 6 and accepting state 7.

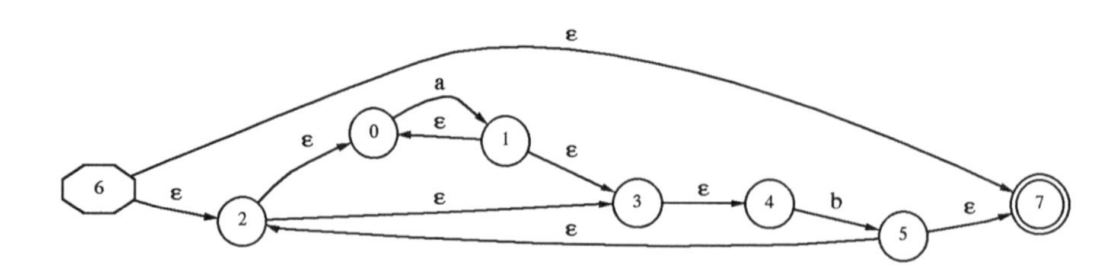

Figure 9.14: NFA whose language is $(a^*b)^*$.

9.2.1 Determinization of FSTs and FAs

It is well known that for every NFA (Nondeterministic Finite Automaton), there exists a DFA (Deterministic Finite Automaton) which accepts the same language. In this section we discuss how to find such a DFA. We have discussed above the specification of FAs in terms of both deterministic and nondeterministic transition functions, for which the range of δ is the power set 2^S (instead of just S, which is the range of deterministic transition functions). Since FAs are built from underlying FSTs, the same applies to FAs. In this section, we show how to map any transition function with range 2^S into a deterministic transition function [224, 66, 163]. The first method for doing this was given by Rabin and Scott [224] and called **subset construction** approach. This method has the disadvantage of potentially requiring $2^{|S|}$ states for the equivalent deterministic FST. However, this rarely occurs in practice.

The second, less frequently cited, method was given in [66] and called **deterministic image** approach. In this approach, the potential for generating a corresponding deterministic machine with exponentially more states is avoided.

9.2.2 The Subset Construction

The remaining step in our construction is due to Rabin and Scott [224], and is called the "subset construction". This simple procedure constructs a DFA that accepts the same language as a given NFA. The procedure is based on a modification of the FST simulation procedure DECIDE_FA of Figure 9.4 of Page 373. The modification is discussed in [190, Page 267]. The idea is that the *From* set of Line 2 is taken as the label of the initial state of the constructed DFA. Then for each symbol x in the alphabet X (instead of for each symbol in a given string) an edge is constructed, with label x, from the initial state to a new vertex labeled by *From* of Line 4. A new vertex is created only if no vertex already exists with this label. If such a vertex already exists, only a new edge is created.

Given an NFA $\mathcal{N} = \langle S^N, X, \delta^N, s_0\rangle$, the subset construction is thus based on traversing the NFA $\mathcal{N} = \langle S^N, X, \delta^N, s_0\rangle$ by DFS from its initial state s_0. The key idea is that in an NFA, if we are in state $s \in S^N$, we can think of ourselves as also being in any and all of the states in the set $S^N_\epsilon(s)$, which are the ϵ-successors of s. We initially define one state of the corresponding DFA for each of the $2^{|S^N|}$ subsets of the NFA state set S^N.

This process is repeated in turn for each new vertex thus created. The procedure terminates because there are only $2^{|S|}$ subsets of the original set of states, and the number of vertices in the created STG is bounded from above by $2^{|S|}$.

With this idea, we label the initial state, 0, of the corresponding DFA $\mathcal{D} = \langle S^D, X, \delta^D, \{0\}\rangle$, with the set of states $S_\epsilon(6) = \{0, 2, 3, 4, 6, 7\}$ which are ϵ-successors of the initial state 6 of the NFA. Note that the NFA and DFA have the same input alphabet, X, which is the set of atoms of the original regular expression. In our example, $X = \{a, b\}$.

Then we ask what DFA states are "reachable" from state 0, and define this property as follows. Let $S^N_i \subseteq S^N$ and $S^N_j \subseteq S^N$ be the two subsets of NFA states which label DFA states i and j. If there exists states $s_i \in S^N_i$ and $s_j \in S^N_j$ such that s_j is an x-successor of s_i for some $x \in X$, then we say that j is an x-successor of i in the DFA. For example, in the set $S_\epsilon(6) = \{0,2,3,4,6,7\}$ which labels initial state 0 of the DFA, there is a state 0 with a-successor state 1. since $S_\epsilon(1) = \{0,1,3,4\}$, we have state 1 in the DFA as an a-successor of state 0.

Similarly, from state $4 \in S_\epsilon(6) = \{0,2,3,4,6,7\}$, we have state 5 as a b-successor. Thus we have state $2 \in S^D$ with label $S_\epsilon(5) = \{0,2,3,4,5,7\}$.

Finally we define a DFA state as accepting if and only if its label set contains an accepting state. Under this definition, both $0 \in A^D$ and $2 \in A^D$, but $1 \notin A^D$.

Under the reachability rules just established for the subset construction, DFA states labeled with any other subsets of S^N are not reachable from the initial state 0, so these are not shown in the final result of Figure 9.15.

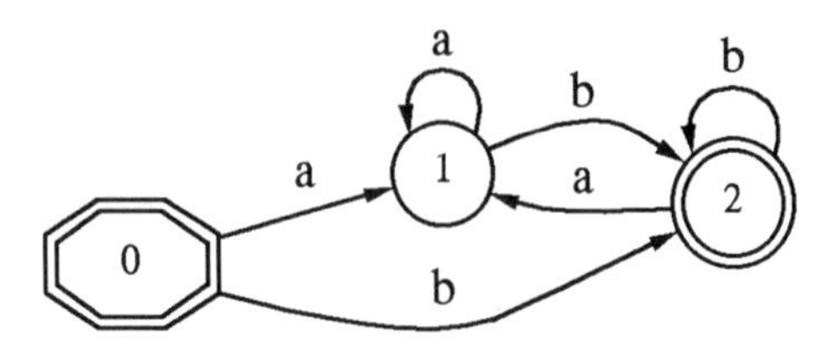

Figure 9.15: DFA whose language is $(a^*b)^*$.

Note that although the subset construction is conceptually simple, it may in the worst case generate a DFA with $2^{|S^N|}$ states. Since for practical regular expressions there may be a large number of states in the NFA, the number of DFA states could in principle become prohibitively large. However, the coalescing of states related by ϵ-moves into a single DFA state strongly controls this potentially explosive effect. Consequently, the disadvantageous worst case seldom arises in practice however, and the technique is widely used.

As the DFS completes at each operator node of the binary parse tree, the initial and accepting states may change identity according to the corresponding composition rule. For example, when using the product rule, the initial state of the composed NFA becomes the initial state of the left subNFA, while the accepting state becomes that of the right subNFA. Similarly, when the union or closure rules are applied, the initial state of the composed NFA is a new node "to the left of" the initial state of the left subNFA. It can thus be seen that the constructed NFA must end up with a single initial state and a single accepting state.

We now give an algorithm for the subset construction which codifies the foregoing informal discussion. The algorithm, shown in Figure 9.16, is based on traversing the NFA from its initial state using DFS, and simultaneously co-traversing the DFA being constructed. When one state s^N of the NFA is visited, we shall considered all states in the set $S_\epsilon^N(s^N)$ reachable by ϵ-moves from that state to be also visited in the NFA traversal.

A subprocedure, EPSILON_MOVES(s^N), is called to create $S_\epsilon^N(s^N)$. If a state with this same ϵ-move set has not previously been visited, we simultaneously create a state in the DFA, mark it as visited, and then recur as in ordinary DFS. In this way we create only the reachable states of the associated DFA.

Another call to X_EPSILON_MOVES(x, S_ϵ^N), is made to create the set $T_{x,\epsilon}^N(x, s^N)$ of states in the NFA reachable by a path whose first edge starts from a state in $S_\epsilon^N(s^N)$ and is labeled by an $x \in X$, $x \neq \epsilon$, and whose subsequent edges are all labeled by ϵ.

Notes: We initiate the co-DFS of the NFA and DFA from the single initial state of the NFA. This means that the argument of the initial recursive call is $S_\epsilon^N(s_0^N)$.

At Line 1, we accumulate all the states reachable by ϵ-moves from an x-successor of a state in S_ϵ^N.

At Line 2, we mark as visited a DFA state which has already been labeled.

At Line 3, the pair (num_dfa, TT) is added to the binary relation holding the correspondence between DFA state indices and NFA subsets.

At Line 4, we mark as accepting a new DFA state whose label contains the unique NFA accepting state , s_A^N.

At Line 5, we recur, passing as argument the label TT of the new DFA state, whose index is *num_dfa*

Note that two forms of nondeterminism give the NFA its expressive power. First, the ϵ-moves make it elementary to construct the DFA from the parse tree. Second, note that it is necessary to give only a partial specification of the NFAs, which are allowed to be complete. Since unspecified behavior is always regarded as leading to a non-accepting trap state (not shown in the NFA figure), such behavior cannot contribute to the language accepted by the NFA. However, note that the constructed DFA is completely specified as well as deterministic.

This construction was used by Kleene in the 50's to prove that for every NFA there is a DFA which accepts the same language.

9.2.3 The Deterministic Image

Definition 9.2.1 deterministic image of an FST:
Let $\mathcal{T} = \langle X, S, \delta, S_0 \rangle$ be a nondeterministic FST on X. The deterministic image of $\mathcal{T}$ is the deterministic FST $\mathcal{T}' = \langle X', S', \delta', S_0 \rangle$ defined as follows on a new input alphabet $X' \subseteq S \times X$. Here

1. *$S' = S$, that is, $\mathcal{T}'$ has the same states as $\mathcal{T}$;*

2. *$t = \delta'(s, (x, t)) : S \times X' \longmapsto S$. Thus δ' has domain $S \times (X \times S)$ and range S;*

3. *$\delta'(s, (x, t)) = t$ for all $t \in \delta(s, x)$.*

Note that if $\mathcal{T}$ is nondeterministic, $\delta : S \times X \mapsto 2^S$, there exists some pair (s, x) with $k > 1$ corresponding next states. That is, this pair has an image set $\delta(s, x) = \{t_1, \cdots, t_k\}$, $k > 1$, so $\delta'(s, (x, t_i)) = t_i$, $i = 1, \ldots, k$.

The key point of this simple construction is that each triple (s, x, t_j) such that $t_j \in \delta(s, x)$ is mapped into a single pair $(s, (x, t_j))$ with a unique image point t_j, $\forall 1 \leq j \leq k$. Thus $\mathcal{T}'$ is deterministic. Figure 9.17 shows a small FST and

```
Global Variables:
    NFA        𝒩 = ⟨S^N, X, δ^N, s_0^N, s_A^N⟩          STG of NFA
    Relation   Label : S^D ⟼ 2^{S^N}                   Initially 0
    Counter    num_dfa                                  Current DFA state count

Procedure DFS_SUBS_CONS(S_ε^N){
    TT = ∅
    foreach(x ∈ X) {
            T_{x,ε}^N = X_EPSILON_MOVES(x, S_ε^N)
1           TT = TT ⋃ T_{x,ε}^N
    }
    visited = 0
    foreach(k ≤ num_dfa) {
        if ((k, TT) ∈ Label) {
2           visited = 1
            break
        }
    }
    if (visited = 0) {
        num_dfa = num_dfa + 1
3       Label = Label ⋃{(num_dfa, TT)}
4       if(s_A^N ∈ TT)  A^D = A^D ⋃{num_dfa}
5       DFS_SUBS_CONS(TT)                               recur, passing new label.
    }
}
```

Figure 9.16: Algorithm SUBSET_CONSTRUCTION for determinizing a given NFA.

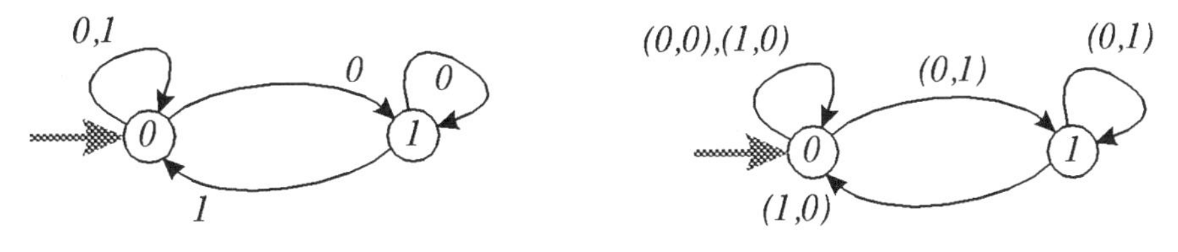

Figure 9.17: An FST and its deterministic image.

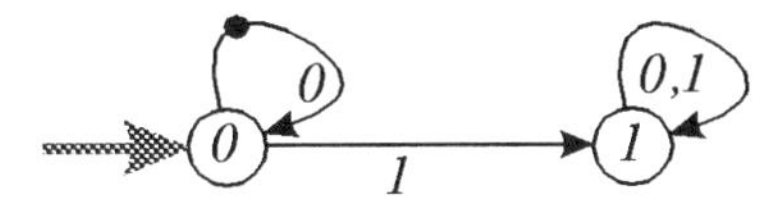

Figure 9.18: An L-automaton for expressing a safety property in formal verification.

its corresponding deterministic image. Here we have $X = \{0,1\} = S$, $S_0 = \{0\}$, and $X' = \{(0,0),(1,0),(0,1)\}$. Note that input symbol $(1,1)$ does not occur in X', since there is no transition to state 1 under input 1 in the original nondeterministic FST $\mathcal{T}$.

From an encoding viewpoint, one can see that if no state s has more that $k = 2(4,8,\ldots)$, x-successors for any $x \in X$, then one could encode all the nondeterminism with just $\log(k) = 1(2,3,\ldots)$ extra code bits. Then if there were n state variables and q primary input variables, and if $k << n+q$, encoding the nondeterminism into a deterministic next state function would come at very little additional expense.

The important property of the deterministic image is that for each input string or tape $x \in X$, the corresponding run set $R(\mathcal{T},\mathbf{x})$ is preserved. That is, for each run $\mathbf{s} \in R(\mathcal{T},\mathbf{x})$, there exists some string $\mathbf{x}' \in X^*$ such that $\mathbf{s} \in R(\mathcal{T}',\mathbf{x}')$. Thus the set of all possible runs for $\mathcal{T}$ is the same as the set of all possible runs of $\mathcal{T}'$.

Choueka used this fact to show that for every NFA there is a corresponding DFA which accepts the same language.

In the example FST of Fig 9.17, note that there are $3 \times 3 = 9$ length-2 strings of X'. Each of these produces a single run of $\mathcal{T}'$, which is deterministic. For example, the string $\mathbf{x}' = ((0,0),(0,1))$ in $\mathcal{T}'$, corresponding to the string $\mathbf{x} = (0,0)$ in $\mathcal{T}$, produces the run $\mathbf{s}' = (0,0,1)$. Since $R(\mathcal{T},(0,0)) = \{(0,0,0),(0,0,1)\}$, we have $\mathbf{s}' \in R(\mathcal{T},(0,0))$, as expected.

9.3 ω-Regular Automata

In this and the succeeding section, we offer the reader a taste of automata designed to accept input sequences of infinite length, called tapes. Our treatment is derived from the comprehensive work of Choueka[66] and Kurshan[163].

Tapes have corresponding infinite runs, so we shall assume that there STGs have no sink states. Consequently, though the STG has finite state, these infinite runs repeatedly traverse the cycles in the STG, which in this case must be a cyclic digraph. For example, the DFA of Figure 9.18 is cyclic and has two strongly connected components. In such a transition system, the response to a tape (infinite string) is well defined. If the DFA receives an arbitrarily large number of consecutive 0s, the execution path can stay in the first strong component (consisting of just the initial state) for an arbitrarily large number of cycles (clock ticks). However, once a 1-input is received, the execution path is bound to stay forever in the second strong component.

Kurshan [163] and others have shown that automata accepting such languages are extremely useful for the purpose of proving eventuality properties of sequential systems. Eventuality properties, such as "If I wait at this red light long enough, I am guaranteed that it will eventually turn green", are defined over infinite time. Tapes, ω-regular languages, and ω-regular automata are designed to deal with such properties.

We note here that the term ω-regular languages gets its name from the fact that ω is commonly used to denote the first infinite ordinal. While not strictly speaking a number, ω has the property that $n < \omega$ for all finite n.

One simple type of **ω-regular Finite Automaton**, called an **L-automaton**, is defined as follows. Since tapes are infinitely long, they have infinite runs, so one cannot refer to the final state of a run, as we do with DFAs. Consequently, the acceptance conditions for L-automata are expressed in terms of **recur edges** and **cycle sets**.

Qualitatively speaking, a tape (infinite sequence) is accepted by an L-automaton if there is a run of the sequence in the automaton such that:

- some recur edge is crossed an infinite number of times, or
- the set of states that are reached an infinite number of times are all contained in one cycle set.

We state this formally as follows.

Definition 9.3.1 *An L-automaton $\mathcal{A}$ is just an FST with an edge acceptance set E^A and node set C^A appended to the specification:*

$$\mathcal{A} = \langle \mathcal{T}, E^A, C^A \rangle = \langle X, S, \delta, S_0, E^A, C^A \rangle,$$

where

$$C^A = (C_1, \cdots, C_k).$$

Here E_A is just a simple set of STG edges, which impose edge-based acceptance conditions. However, C^A is composed of a set of k **cycle sets** *$(C_1, \cdots, C_k)$, which define vertex-based acceptance conditions. These sets work as follows. Consider an input string $\mathbf{x}$ and corresponding run $\mathbf{s}^{\mathbf{x}}$. Let $S^\infty(\mathbf{s}^{\mathbf{x}})$ be the set of states which occur infinitely often in the run $\mathbf{s}^{\mathbf{x}}$. Then the string $\mathbf{x}$ is* **accepted** *if and only if either,*

1. *the run $\mathbf{s}^{\mathbf{x}}$ has an infinite number of STG edges $(s_i^x, s_{i+1}^x) \in E^A$, or*
2. *there exists $C_i \in C^A$ such that $S^\infty(\mathbf{s}^{\mathbf{x}}) \subseteq C_i$.*

The L-automaton $\mathcal{A}$ is deterministic if the transition function δ of the underlying FST is, and complete if δ is.

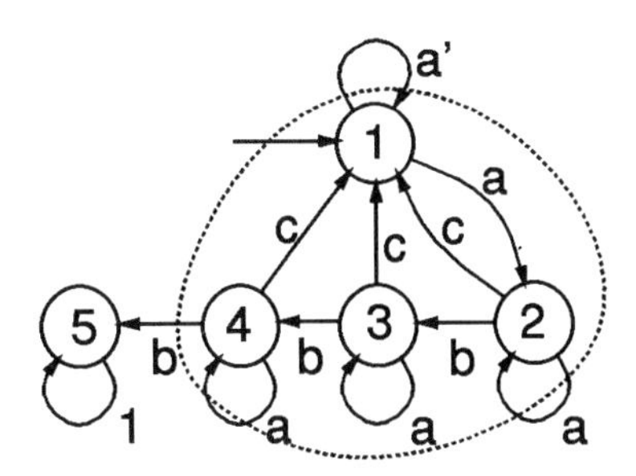

Figure 9.19: An L-automaton recognizing a class of tapes (infinite strings with at most two b inputs after an a, unless there is an intervening c).

An example of an L-Automaton is given in Figure 9.19. This deterministic L-automaton accepts the language over the alphabet $X = \{a, b, c\}$ composed of all the (infinite) sequences such that there are at most two b inputs after an a, unless there is an intervening c. Here, as in [163], we regard the input alphabet as the atoms of a Boolean algebra, so $a' = (b + c)$, or any input letter other than a. Similarly 1 means $a + b + c$, or any input letter. Note three consecutive b inputs between an a and a c lead to the trap state 5, whereas the first c after an a brings the automaton back to its initial state 1.

In this simple case, the acceptance condition is given by a single cycle set, so $C^A = \{\{1, 2, 3, 4\}\}$, as indicated by the set of states encircled by the dashed line in Figure 9.19.

Notice that DFA defined on the same FST, and with accepting set $A = \{1\}$, would accept *finite* strings such that there are at most two b inputs after an a, unless there is an intervening c. The language of this DFA would be

$$\mathcal{L} = (a')^*(a^*)(c + (b + bb)c).$$

An example of an L-automaton with only edge acceptance conditions is the automaton $\mathcal{A}$ given in Figure 9.18. The black dot on a given edge (s, t) indicates that $\mathbf{x}$ is accepted if any one of its corresponding runs passes through (s, t) an infinite number of times. In this case the complement $\overline{\mathcal{A}}$ is easy to express: every tape that has one or more 1 inputs is rejected by $\mathcal{A}$.

Such an acceptance condition can be efficiently checked by symbolically processing the cycles of a given STG. Such an automaton is frequently used in formal verification for checking safety properties such as the property that cars heading in orthogonal directions do not, ever, enter an intersection such as that controlled by the FSM of Figure 7.36 of Page 296.

Another example of an L-Automaton with acceptance defined in terms of recur edges only, is given in Figure 9.20. This deterministic L-automaton, over the alphabet

$$X = \{\text{a,...,z,A,...,Z,0,...,9}\},$$

accepts the language composed of sequences that contain an infinite number of "Bach" substrings. We again regard the input alphabet X as the atoms of a Boolean algebra, so

$$(B + a)' = \{\text{b,...,z,A,C,...,Z,0,...,9}\},$$

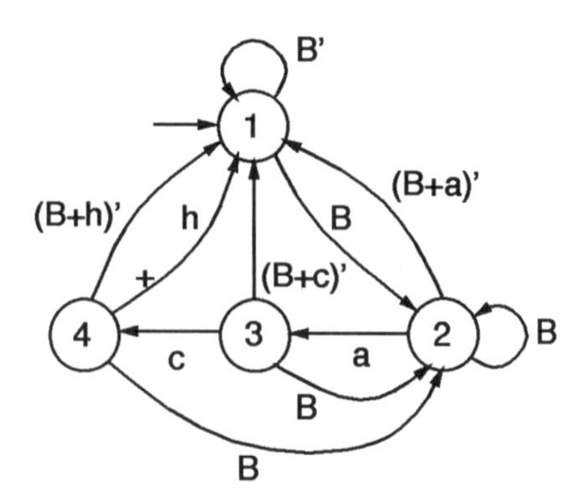

Figure 9.20: An L-automaton recognizing a tapes containing an infinite number of *Bach* substrings.

In this simple case, the acceptance condition is given by a single recur edge, indicated by the + sign on the $(4, 1)$ edge.

Unlike the previous example, in this case there is no immediate way to turn this automaton into one over finite strings.

9.4 Formal Verification with L-Automata

The approach to formal verification using this type of automaton typically involves defining a **property automaton**, and then forming the product of this automaton with the FSM being designed. The idea is to then check for the acceptance conditions on the edges and/or cycle subsets of the product. The details of such verification [263, 143] are discussed briefly in Section 9.5.3.

It is important to note that the star operation of Definition 9.1.3 of Page 374 applies, strictly speaking, only to finite sequences. Hence we need alternate notation to deal with sets of tapes like ω-regular languages. This situation is dealt with in the literature [66] by introducing two operations on tapes which are analogous to the $*$ operation on strings. We describe these two operations before turning to practical, BDD-based, formal verification.

9.4.1 ω-Regular Languages

A principal purpose of this chapter is to characterize automata which accept languages which are sets of tapes $\mathbf{x} \in X^\omega$. In this section we finally define ω-regular sets and present Choueka's basic lemma [66, Lemma 5.2, p129] [163, Chapter 6] for dealing with acceptance conditions for such tapes.

Definition 9.4.1 *For a given regular language $\mathcal{L} \subseteq X^*$, we define:*

$\lim \mathcal{L}$: *the* **limit** *of $\mathcal{L}$, written $\lim \mathcal{L}$, to be the set of all tapes $\mathbf{x} \in X^\omega$ which have an infinite number of sections (prefixes) in $\mathcal{L}$. That is,*

$$\begin{aligned} \lim(\mathcal{L}) &= \lim_{i\to\infty}\{\mathbf{x} \in \mathcal{L}(X)^\omega \mid \mathbf{x}(0, i_j) \in \mathcal{L}, j = 1, 2, \cdots\} \\ i_{j+1} &> i_j, \quad j = 1, 2, \cdots. \end{aligned}$$

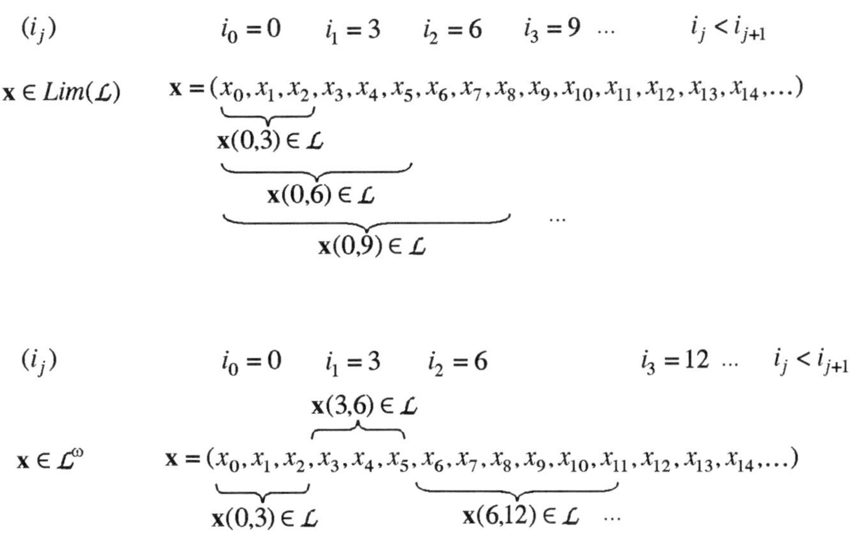

(i_j) $\quad i_0 = 0 \quad i_1 = 3 \quad i_2 = 6 \quad i_3 = 9 \; \ldots \quad i_j < i_{j+1}$

$\mathbf{x} \in Lim(\mathcal{L}) \quad \mathbf{x} = (x_0, x_1, x_2, x_3, x_4, x_5, x_6, x_7, x_8, x_9, x_{10}, x_{11}, x_{12}, x_{13}, x_{14}, \ldots)$

$\mathbf{x}(0,3) \in \mathcal{L}$

$\mathbf{x}(0,6) \in \mathcal{L}$

$\mathbf{x}(0,9) \in \mathcal{L}$ ...

(i_j) $\quad i_0 = 0 \quad i_1 = 3 \quad i_2 = 6 \quad i_3 = 12 \; \ldots \quad i_j < i_{j+1}$

$\mathbf{x}(3,6) \in \mathcal{L}$

$\mathbf{x} \in \mathcal{L}^{\omega} \quad \mathbf{x} = (x_0, x_1, x_2, x_3, x_4, x_5, x_6, x_7, x_8, x_9, x_{10}, x_{11}, x_{12}, x_{13}, x_{14}, \ldots)$

$\mathbf{x}(0,3) \in \mathcal{L}$ $\quad \mathbf{x}(6,12) \in \mathcal{L}$ ...

Figure 9.21: Illustration of $\lim(\mathcal{L})$ and $\mathcal{L}^{\omega}$.

Note that each i_j in the necessary **increasing infinite sequence** *points to the end of a finite string $\mathbf{x}(0, i_j) \in \mathcal{L}$, as shown at the top of Figure 9.21.*

$\mathcal{L}^{\omega}$: *the ω* **operation** *on $\mathcal{L}$, written $\mathcal{L}^{\omega}$, to be the set of all tapes $\mathbf{x} \in X^{\omega}$ which are composed of an infinite catenation of nonempty strings in $\mathcal{L}$. That is,*

$$\begin{aligned} \mathcal{L}^{\omega} &= \{\mathbf{x} \in X^{\omega} \mid \mathbf{x}(i_j, i_{j+1}) \in \mathcal{L}, \quad j = 1, 2, \cdots\} \\ i_{j+1} &> i_j, \quad j = 1, 2, \cdots. \end{aligned}$$

Note that consecutive indices i_j and i_{j+1} in the increasing infinite sequence point to the endpoints of finite strings $\mathbf{x}(i_j, i_{j+1}) \in \mathcal{L}$, which are catenated to form a tape in $\lim(\mathcal{L})$.

We illustrate this definition in Figure 9.21 by analyzing tapes in $\lim(\mathcal{L})$ and $\mathcal{L}^{\omega}$.

Some simple examples serve to illustrate these definitions. $0^{\omega} = \{(00\cdots)\}$, is a set containing a single all-0 tape. Similarly, $(10)^{\omega}$ is a single tape comprised of repeating (10) strings. Note whereas $(10)^{\omega}$ is well defined, $\lim(\mathcal{L})$ makes no sense if $\mathcal{L} = (10)$, since the lim operation requires an infinite sequence of prefixes.

However $\lim(10^*) = \{(100\cdots)\}$ which is a set containing a single tape containing one 1 followed by an infinite number of (0)s.

At this point a more substantial example is helpful to bring intuition to bear on these concepts. Suppose $\mathcal{L} = 10^*$. Then, noting that $\{1+0\}^{\omega}$ is the set of all tapes on $\{0, 1\}$, we can immediately characterize $\lim(\mathcal{L})$as follows.

$$\begin{aligned} \lim(\mathcal{L}) &= \{100\cdots\} && \text{(One tape with an initial 1).} \\ \mathcal{L}^+ &= 1(0+1)^* && \text{(Strings with an initial 1).} \\ \lim(\mathcal{L}^+) &= 1(0+1)^{\omega} && \text{(Tapes with an initial 1).} \\ \lim(\mathcal{L}^+) &\supseteq \{10^{\omega}, 1^{\omega}, (10)^{\omega}\} \end{aligned}$$

Note how the third identity derives from the second by simply replacing the $*$ with ω. It should now be evident that there is strong similarity between the $*$ operation

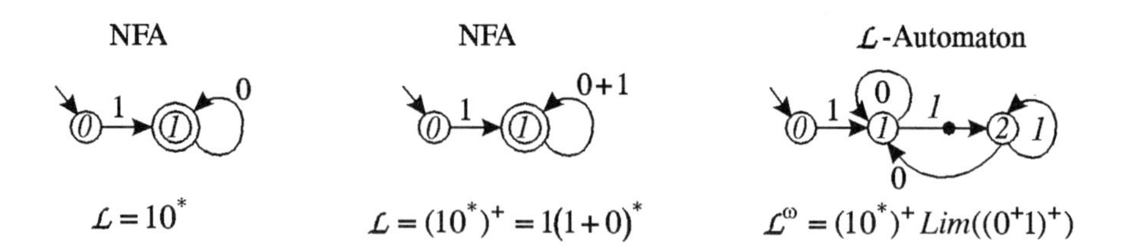

Figure 9.22: Automata Accepting $\lim(\mathcal{L})$ and $\mathcal{L}^\omega$.

and the lim operation. Also note why the inclusion of the fourth identity is obvious: $\lim(\mathcal{L}^+)$ is the set of all tapes beginning with a 1, and each member of the set on the right is such a tape.

The automata associated with some of these languages is illustrated in Figure 9.22 Note how $\mathcal{L}$ is the language of the NFA on the left, and $\mathcal{L}^+$ is the language of the NFA in the middle. We now show that $\mathcal{L}^\omega$ is the language of the ω-regular L-automaton on the right.

We can similarly characterize $\mathcal{L}^\omega$, by defining $\tilde{\mathcal{L}} = (0^+1) = (00^*1)$ as the set of strings on $\{0, 1\}$ with a block of one or more leading (0)s and a terminal 1. Thus $\tilde{\mathcal{L}}^+ = (0^*1)^+$, the set of strings formed by catenating substrings from $\tilde{\mathcal{L}}$. We have the following further identities.

$$
\begin{aligned}
\lim(\tilde{\mathcal{L}}) &= \lim((0^*1)^+) && \text{(Tapes formed by catenating strings from } \tilde{\mathcal{L}}\text{).} \\
\mathcal{L}^\omega &= 1(0^*1)^\omega && \text{(Tapes with leading 1 and infinitely many 10 substrings.).} \\
&= \mathcal{L}^+ \lim(\tilde{\mathcal{L}}).
\end{aligned}
$$

For any tape in $\mathcal{L}^\omega$, the initial 1 brings the NFA from its initial state to the middle state. Each subsequent substring consisting of an arbitrary number of (0)s followed by a single 1 which brings the automaton through the accepting recur edge (marked with a solid black circle). The recur edge is thus traversed and infinite number of times for such tapes.

The reader should note how the $\mathcal{L}^+$ prefix in the last identity is used to bring the associated runs to the 2-cycle containing the recur edge. In contrast, the $\lim(\tilde{V})$ suffix, with an infinite number of 10 substrings, is used to guide the run around the 2-cycle an infinite number of times.

Even though we picked $\tilde{\mathcal{L}}$ out of the air here, it can be shown that it is always possible to relate $\lim(\mathcal{L})$ and $\mathcal{L}^\omega$ by an appropriate choice of $\tilde{\mathcal{L}}$.

Lemma 9.4.1 *For every regular set $\mathcal{L} \subseteq X^*$, one can effectively find some corresponding set $\tilde{(\mathcal{L})} \subseteq X^*$ such that*

$$\mathcal{L}^\omega = \mathcal{L}^*(\lim \tilde{\mathcal{L}}).$$

An efficient constructive method for finding $\tilde{\mathcal{L}}$ is given in [66].

9.5 ω-regular Language Containment

ω-regular automata are typically employed to express desired temporal behavior such as eventuality or liveness properties. The aforementioned liveness property guarantees that some states of an FST will eventually be entered. The idea of eventuality is what gives importance to the notion of infinite sequences. That is, a state is said to be live, even though it is not entered for an arbitrarily long time, so long as it is eventually entered.

9.5.1 Lifting Acceptance Conditions to a Product L-Automaton

As stated above, the basic idea of formal verification using ω-regular automata is to form the product of the property automaton with the design FSM and to check for the acceptance conditions on the product.

The first step in checking language containment is to form an automaton $A_\times$, which is defined to be the product of some given FSM M_P (sometimes called a **Process** [163]) and a **task** L-automaton A_T (sometimes called the **property automaton**). We assume that M_P has deterministic transitions and A_T is deterministic.

Suppose that $v_i \in S_P$ is a generic state of M_P and $w_i \in S_T$ is a generic state of A_T. According to the definition of product of FSTs[5], there is an edge from state (v_1, w_1) to state (v_2, w_2) in $A_\times$ if there are edges from v_1 to v_2 in P and from w_1 to w_2 in A_T. The predicate on the edge of $A_\times$ is the product of the predicates of the two edges of M_P and A_T. If the product is empty, then the edge will never be traversed and can be dropped.

The initial state of the product automaton is the pair (s_P^0, s_T^0) of initial states of M_P and A_T.

The acceptance conditions are determined by the automaton A_T in the following way. If C_T is a cycle set of A_T, then the corresponding cycle set in $A_\times$ is given by:

$$C_\times = \{(v_i, w_j) \in S_P \times S_T : w_j \in C_T\}.$$

We say that the cycle set is **lifted** from A_T to $A_\times$. Recur edges are similarly lifted from A_T to $A_\times$. An edge from (v_1, w_1) to (v_2, w_2) is a recur edge of $A_\times$ if the edge from w_1 to w_2 is a recur edge in A_T. The resulting product machine is, therefore, also an L-automaton.

In many practical cases substantial computational savings may derive from specifying desired temporal properties as an automaton and the lifting them to the product with a practical FSM, so prospective VLSI CAD tool users/developers should note well this technique.

9.5.2 Example of Product L-Automaton

An example of a product L-automaton is given in Figure 9.23. The process M_P (top left) in this case is an FST that generates sequences of (a)s, (b)s, and (c)s. The task automaton A_T (top right) expresses the requirement that after any occurrence of a there must be an occurrence of b. Note how the recur edge and cycle set are lifted to the product automaton $A_\times$ (bottom).

Informally, the two conditions say that a sequence either eventually stays in state w_1 indefinitely (a occurs a finite number of times only), or it crosses the recur edge infinitely often (there is no infinite sequence of (a)s and c's without (b)'s). The

[5]See Definition 7.7.5 of Page 298.

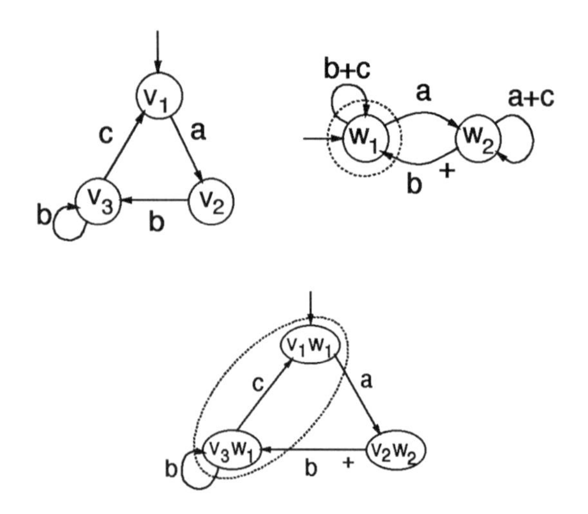

Figure 9.23: An example of a product automaton.

effect of creating the product automaton $A_{\times}$ (bottom) is to constrain the possible sequences that A_T receives to those produced by M_P. The language containment test then succeeds if and only if all possible sequences are accepted by $A_{\times}$.

As stated above, the key observation in testing ω-regular language containment is that the product automaton $A_{\times}$ has a finite number of states. Therefore, every infinite input sequence must produce a run that eventually repeatedly traverse cycles in $A_{\times}$. More specifically, each run must cycle within a single strongly connected component of $A_{\times}$[6].

We require that each reachable cycle in $A_{\times}$ satisfies one of the two following conditions.

- The cycle is entirely contained in a cycle set;
- The cycle contains a recur edge.

Tapes whose runs satisfy these conditions will be accepted.

9.5.3 BDD Representation of Cycle Sets and Recur Edges

It is customary in language containment checking to use BDD-based symbolic methods, and to represent the underlying FSTs with transition relations[7] rather than transition functions. Similarly, cycle sets are simply represented by characteristic functions, which are Boolean functions of binary encoding variables. For instance, if v_1w_1 is encoded as $x_1'x_2'$ and v_3w_1 is encoded as x_1x_2', then the characteristic function of the cycle set $\{v_1w_1, v_3w_1\}$ is $C_1 = x_2'$.

The representation of recur edges is obtained by augmenting the transition relation with one new variable r, so that the transition relation is a function of four sets of variables:

$$\widehat{T}(x, \sigma, r, y) : \{0,1\}^{m+2n+1} \longmapsto \{0,1\},$$

[6] See the definition of SCCs in Section 7.5.1.

[7] See Section 7.9.1 of Page 305 for a discussion of transition relations.

where the n-vectors x and y encode the present states and next states, respectively, σ encodes the primary input symbols which label the STG edges, and r is an extra code bit to identify recur edges. Thus $\widehat{T}(x,\sigma,1,y) = 1$ if and only if there is a recur edge, labeled by σ, from the state encoded as x to the state encoded as y. Similarly, $\widehat{T}(x,\sigma,0,y) = 1$ means that the edge is not a recur edge. We assume that the product automaton $A_\times$ is represented by the characteristic function of this augmented transition relation.

The language containment test is made practical by representing all these sets, and/or characteristic functions, and/or transition relations, by BDDs[8].

9.5.4 The Language Containment Algorithm

The following algorithm is adapted from [263]. As a preprocessing step, we transform the labeled digraph of the FST into an unlabeled digraph by a process known as **existential abstraction** of the primary input variables (σ). The abstracted[9] transition relation is $T(x,r,y) : \{0,1\}^{n+1} \longmapsto \{0,1\}$, where

$$T(x,r,y) = \exists_\sigma \widehat{T}(x,\sigma,r,y).$$

where $T(x,0,y) = 1$ if and only if there exists some non-recur edge (x,y) in the original STG (with any arbitrary input label). Similarly, $T(x,1,y) = 1$ if and only if there exists some recur edge (x,y) in the original STG (with an arbitrary input label).

The language containment check for L-automata can be reduced to checking that every reachable cycle in the STG of $A_\times$ either contains a recur edge or is entirely contained in one of the cycle sets. We present here an algorithm that can perform this check using symbolic operations on characteristic functions as discussed above. Neither states nor cycles nor recur edges are explicitly enumerated.

The containment check, given as Procedure LANGUAGE_CONTAINMENT in Figure 9.24, has five basic steps. In Steps 1 and 2, we compute the set of reachable states $R(x)$ and use it to simplify the computation. In Step 2, any cycles that remain are reachable from the initial state and do not use any recur edge. The language containment check gives an affirmative answer if and only if these cycles are contained in the cycle sets.

In Step 4, the expression

$$\exists_y(\tilde{T}^*(x,y) \cdot \tilde{T}^*(y,x) \cdot C_i'(y)),$$

for $NC_i(x)$, the expression before abstraction gives the pairs of states reachable from one another (i.e., on a cycle), such that one of the two elements is not in C_i (i.e., the cycle is not contained in C_i). This is illustrated in Figure 9.25. Here the existential abstraction (See Section 7.10.1 on Page 308) operator $\exists_y$ means that for each x in the resulting set $NC_i(x)$, "there exists some y such that the characteristic function in parenthesis evaluates to 1". There is a simple and efficient BDD implementation for this operation.

[8]See Section 6.1 of Page 220.
[9]See Definition 7.10.2 of Page 309.

Procedure LANGUAGE_CONTAINMENT$(T(x,r,y),\{C_i\})$ {

1. Compute the set $R(x)$ of states reachable from the initial state.
2. Remove the recur edges and restrict the relation to reachable states.
$$\widetilde{T}(x,y) = T(x,0,y) \cdot R(x).$$
3. Compute the transitive closure (See Section 1.4.6, on Page 24), $\widetilde{T}^*$, of $\widetilde{T}$.
4. For each cycle set C_i, compute the set of cycles of $\widetilde{T}$ not contained in C_i as follows:
$$NC_i(x) = \exists_y(\widetilde{T}^*(x,y) \cdot \widetilde{T}^*(y,x) \cdot C_i'(y)).$$
5. We then compute the intersection of the NC_i. The language containment check succeeds if and only if
$$\bigcap_{l \leq i \leq n} NC_{C_i} = \emptyset,$$
that is, if the intersection is empty.

}

Figure 9.24: ProcedureLANGUAGE_CONTAINMENT.

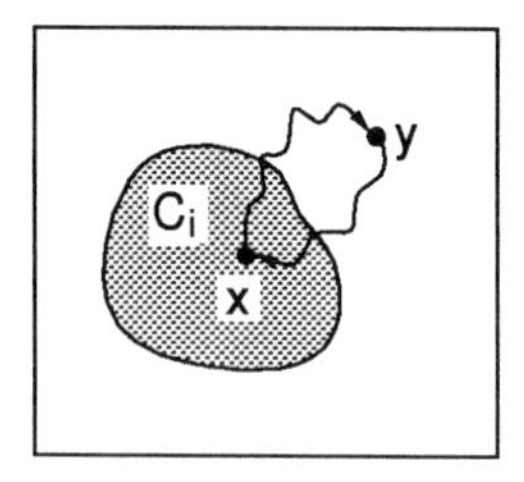

Figure 9.25: Illustration of non-containment in cycle set.

9.5.5 Example of Containment Check

We now demonstrate the language containment test on our previous example (Figure 9.23) of a product automaton. The test is illustrated in Figure 9.26. At the top we reproduce the previous product automaton with the lifted acceptance conditions. Note that there is one recur edge, $E^A = \{(v_2w_2, v_3w_1)\}$, and one cycle set $C_1 = \{v_1w_1, v_3w_1\}$. To emphasize that Procedure LANGUAGE_CONTAINMENT is designed for use in a BDD-based, symbolic approach, we again encode the states of the product automaton $A_\times$ as follows:

$$v_1w_1 \leftrightarrow x_1'x_2', v_3w_1 \leftrightarrow x_1x_2', v_2w_2 \leftrightarrow x_1'x_2.$$

After existentially abstracting the primary input vector σ, we identify the edges of the resulting STG with a 1 label if the edge is a lifted recur edge, or with a 0 label, otherwise. After this preprocessing, we call Procedure LANGUAGE_CONTAINMENT$(T(x,r,y),\{C$ which operates on the STG at the lower left of Figure 9.26.

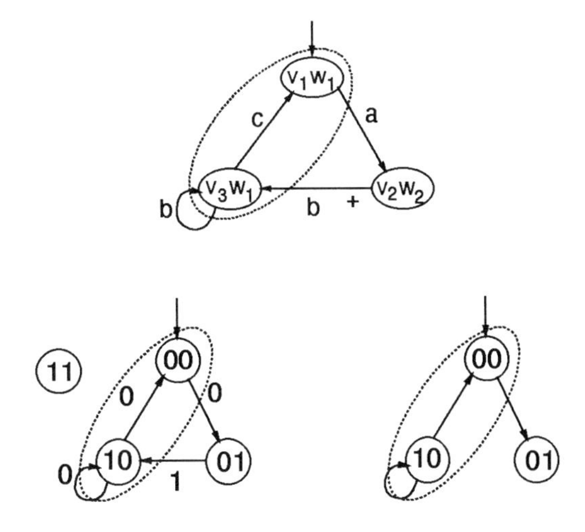

Figure 9.26: Language containment test on the product automaton of Figure 9.23.

The unreachable state encoded by 11 is eliminated from consideration in Steps 1 and 2 of Procedure LANGUAGE_CONTAINMENT. Step 2 also removes the recur edge (v_2w_2, v_3w_1), encoded as $(01, 10)$, thus leading to the STG at the lower right of the figure. Step 4 computes $NC_1 = \emptyset$, since the only cycle left is entirely contained in C_1. Thus

$$\bigcap_{l \leq i \leq n} NC_{C_i} \neq \emptyset,$$

is confirmed. Therefore, the language containment check is successful, since the language of the design FSM M_P is contained in the language of the task (property) automaton $\mathcal{A}_T$ if and only if the above intersection is empty.

Note that the other valid cycle $(00, 01, 10, 00)$ in the product STG at bottom left was eliminated from consideration when the recur edges were deleted. Thus we are assured that any cycles not contained in the intersection of the NC_i will be cycles closed by recur edges.

9.6 Notes

An excellent reference book for this chapter in particular is McEliece, Ash, and Ash [190]. Also relevant, although somewhat advanced, are Hopcroft's text [146] and the brilliant article by Choueka [66]. The best reference of all is the recent book by Kurshan, [163], which almost completely subsumes our treatment, but is too advanced for undergraduates, and has much greater scope as well. Along with the temporal logic CTL [98, 45, 51, 97, 52, 49, 50], ω-Regular Language containment [263, 59, 143, 142, 256, 10, 225] is one of the dominant vehicle being used for formal verification of sequential circuits.

9.7 Summary

In this chapter, we have covered three main topics:

1. Finite Automata and Regular Languages (Section 9.1);

2. Synthesis of DFAs from Regular Languages (Section 9.2);

3. ω-Regular Languages (Section 9.4.1).

In Section 9.1Finite Automata and Languages we covered the basic ideas of string acceptance by an FA. We showed that both DFAs and and NFAS were in one to one correspondence with regular languages. We concluded by defining the complements of a regular languages, which is sometimes easier to characterize. We discussed NFAs, which can have multiple runs for the same string input, We emphasized that a string is accepted by an NFA if and only if some run of the NFA for that string ends up in an accepting state. We noted that that this is true even in the case that another run of the same string is *not* accepted.

In Section 9.2 we described a systematic procedure for the synthesis of a DFA which Accepts a specified regular languages. The procedure was as follows. First, parse the regular language and build the corresponding NFA using the construction rules of Figure 9.10, Figure 9.12, and Figure 9.13. Second, use the subset construction to build a (non-minimal) DFA which accepts the same language, Finally, minimize this DFA using PARTITION_REFINEMENT of Section 7.4.2 of Page 269. We also described another method of determinization, called the deterministic image, which allows a non-deterministic FST to be represented by deterministic one (or, therefore, by a physical circuit), as in [225].

Finally, in Section 9.4.1, we defined ω-regular languages, which are to ω-regular automata as regular languages are to DFAs or NFAs. The difference is that instead of accepting states, ω-regular automata have infinitary acceptance conditions, based on states or edges that are traversed infinitely often. Then we showed how interesting temporal properties of sequential circuits or communication protocols could be phrased as an ω-regular language containment problem. We concluded with some examples and a BDD-based method for language containment checking.

9.8 Problems

1. Consider the DFA described by the STG

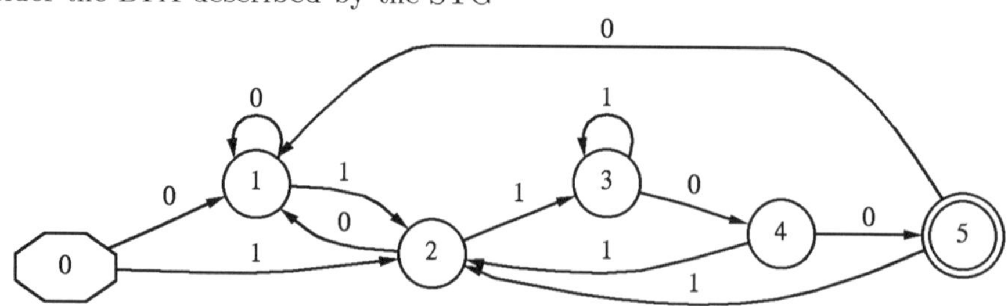

(a) Give the flow table description of a Moore FSM derived from this DFA according to Definition 7.4.5;

Present State	Next States		Output
	$x = 0$	$x = 1$	
0	1	2	0
1	1	2	0
2	1	3	0
3	4	3	0
4	5	2	0
5	1	2	1

Figure 9.27: Flow table equivalent Moore machine for Problem 1.

(b) Minimize the FSM $\mathcal{M}$ by applying Procedure STATE_EQUIVALENCE of Section 7.4.2;

(c) Now convert this minimized version of $\mathcal{M}$ back into a DFA $\mathcal{A}$;

(d) Show that $\mathcal{A}$ is equivalent to the the DFA of Figure 9.28 by showing that their STGs are isomorphic (identical given an appropriate node relabeling).

Solution. (a) The flow table of this STG is that of Problem 1 Since state 5 is the only accepting state, the outputs are 0 on all states except state 5 and a 1 is output on state 5.

(b) The 1-equivalence classes are $\{0, 1, 2, 3, 4\}, \{5\}$.
The 2-equivalence classes are $\{0, 1, 2, 3\}, \{4\}, \{5\}$.
The 3-equivalence classes are $\{0, 1, 2\}, \{3\}, \{4\}, \{5\}$.
The 4-equivalence classes are $\{0, 1\}, \{2\}, \{3\}, \{4\}, \{5\}$.
The 5-equivalence classes are the same as the 4-equivalence classes, so we conclude that $0 \equiv_1 1$.

(c) The STG of the minimized FA is the following

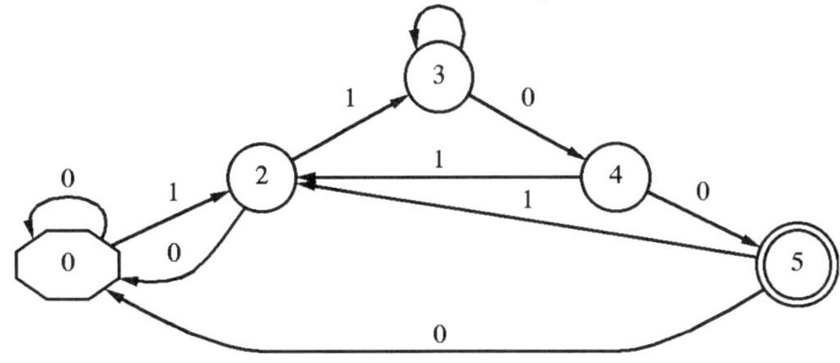

(d) The corresponding vertices in the isomorphic graphs are

$$(0, s_0), (2, s_1), (3, s_2), (4, s_3), (5, s_4).$$

$\square$

2. Consider the DFA $\mathcal{A} = \langle \mathcal{T}, A \rangle = \langle X, S, \delta, S_0, A \rangle$, where $A = \{5\}$, $S_0 = \{0\}$, where $\mathcal{T}$ is specified in the following tabular form.

Present State	Next States			Output
	$x = 0$	$x = a$	$x = b$	
0	1	2	3	1
1	1	2	3	1
2	4	2	3	1
3	4	2	3	1
4	4	4	4	0

(a) Draw the STG of this DFA, and transform the DFA into an equivalent Mealy machine $\mathcal{M}$ (according to Definition 7.4.5);

(b) Minimize the FSM $\mathcal{M}$ by applying Procedure STATE_EQUIVALENCE of Section 7.4.2;

(c) Now convert this minimized version of $\mathcal{M}$ back into a DFA $\mathcal{A}$;

(d) Show that $\mathcal{A}$ is equivalent to the the DFA of Figure 9.5 by minimizing the latter DFA and showing that the two minimized STGs are isomorphic (identical given an appropriate node relabeling).

3. Give the language accepted by the DFA of Figure 9.28

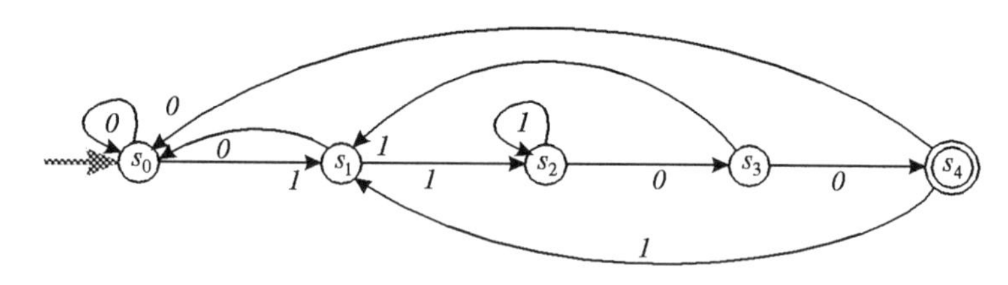

Figure 9.28: A DFA for recognizing a certain string.

Solution. The corrected automaton is a 1100-recognizer. That is, a DFA which accepted only strings ending in 1100. □

4. (a) Give the language accepted by the DFA $\mathcal{A}^5 = \langle \mathcal{T}^5, A^5 \rangle = \langle X, S, \delta, S_0, A^5 \rangle$, where $A^5 = \{1\}$, $S_0^5 = \{0\}$, and $\mathcal{T}^5$ is specified in the following tabular form.

δ	(s, x) (Present State,Input)							
	(0,0)	(0,1)	(1,0)	(1,1)	(2,0)	(2,1)	(3,0)	(3,1)
	2	3	1	1	2	2	2	1

(b) Give X and S for $\mathcal{T}^5$
(c) Draw the STG of $\mathcal{T}^5$.
(d) Give the SCCs of this STG

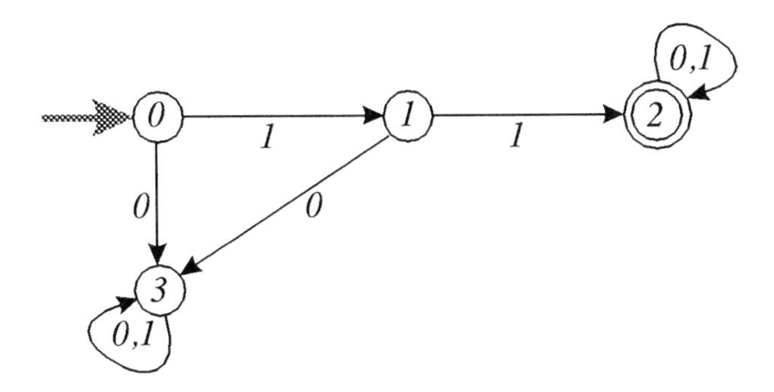

Figure 9.29: The DFA for Problem 5.

5. (a) Draw the STG of $\mathcal{T}$ underlying the DFA $\mathcal{A}^6 = \langle \mathcal{T}, A^6 \rangle = \langle X, S, \delta, S_0, A^6 \rangle$, where $A^6 = \{2\}$, $S_0^6 = \{0\}$, and $\mathcal{T}^6$ is specified in the following tabular form.

δ	(s,x) (Present State,Input)							
	(0,0)	(0,1)	(1,0)	(1,1)	(2,0)	(2,1)	(3,0)	(3,1)
	3	1	3	2	2	2	3	3

(b) Show why the language accepted by $\mathcal{A}^6$ is identical to that of $\mathcal{A}^5$.
Solution. (a) The STG is given in Figure 9.29. (b) Note the two STGs have the same X and S and are isomorphic (that is, they are identical except for the node labels). Hence the languages are identical. □

6. Show that the NFA $\mathcal{A}$ of Figure 9.3 of Page 371 is not a 111 recognizer, that is accepts strings ending in 111 and only such strings.

7. The problem of *Language Containment* is to show that the language $\mathcal{L}_1 = \mathcal{L}(\mathcal{A}_1)$ of some automaton $\mathcal{A}_1$ contains the language of another automaton $\mathcal{L}_2 = \mathcal{L}(\mathcal{A}_2)$. That is,

$$\mathcal{L}_2 \subseteq \mathcal{L}_1.$$

Show that the language $\mathcal{L}_1$ of the automaton of Figure 9.8 contains $\mathcal{L}_2$, the language of the automaton of Figure 9.2.
Solution. The language of the DFA of Figure 9.8 is just $\mathcal{L}_1 = (0+1)^*11(0+1)^*$. This regular language is just the set of all strings on the alphabet $\{0,1\}$ with a prefix ending in 11. Figure 9.2 is a 111-recognizer, that is, $\mathcal{L}_2 = (0+1)^*111$. Thus

$$\mathcal{L}_2 = (0+1)^*111 \subseteq (0+1)^*11 \subseteq (0+1)^*11(0+1)^* = \mathcal{L}_1.$$

□

8. (a) Consider the DFA obtained by removing the black dot on the $(0,0)$ edge of Figure 9.18 of Page 387. The acceptance set is then just the set $\{1\}$. Give the language of this DFA.

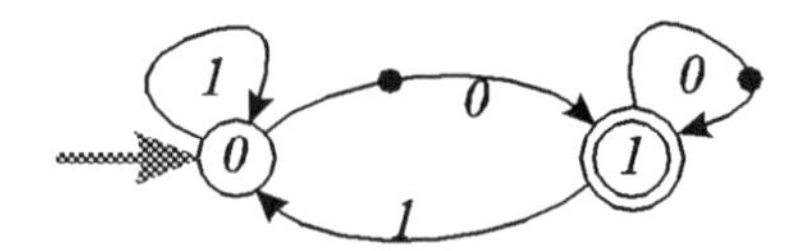

Figure 9.30: An L-automaton expressing a liveness property in formal verification.

(b) Now further modify this DFA by adding a 1-input to the $(0,0)$ edge label. Note the result is an NFA. Give the language of this NFA.

9. Show that the language $\mathcal{L}(\mathcal{A}_1)$ of the automaton of Figure 9.8 of Page 377 contains the language $\mathcal{L}(\mathcal{A}_3)$ of the automaton of Figure 9.3 of Page 371.

Solution. Since the automaton $\mathcal{A}_1$ of Figure 9.8 accepts any string with a prefix ending in 11, its language, $\mathcal{L}(\mathcal{A}_1)$, is just $(0^*(100^*)^*)11(0+1)^* = (0+1)^*11(0+1)^*$.

The automaton $\mathcal{A}_3$ of Figure 9.3 is nondeterministic, because of the extra 0-successor of the initial state s_0. Note that a string $\mathbf{x}$ is accepted by $\mathcal{A}_3$ only if its run $\mathbf{r}(\mathbf{x})$ ends in s_3. This implies that $\mathbf{r}$ must have a (s_0, s_1, s_2, s_3) prefix, which in turn implies that $\mathbf{x}$ must have either a 011 prefix or a 111 prefix. In either case, $\mathbf{x}$ contains a 11 prefix, and so is accepted by $\mathcal{A}_1$.

□

10. Suppose that $\mathcal{L} = (10)$. Show that in this case $\lim(\mathcal{L}^+) = \mathcal{L}^\omega$.

11. Consider the automaton $\mathcal{A}$ given in Figure 9.18. It was stated in Section 9.3 that $\mathcal{A}$ rejects every tape that has one or more 1.

(a) Give the language $\mathcal{L}^\omega$ of this automaton, expressing your result the ω operator.

(b) Express the complement of this language in terms of the lim operator.

Solution. Part (a)

Since $\mathcal{A}$ rejects every tape that has one or more 1only the single tape set, it must therefore accept the single tape $0^\omega = \{(00\cdots)\}$.

Part (b)

$\mathcal{L}(\overline{\mathcal{A}}) = \lim((0^*1^+0^*)^*)$. □

12. Another example of such an ω-regular finite automaton $\mathcal{A}$ is given in Figure 9.30. Note the similarity between this liveness automaton and the property automaton at the upper right of Figure 9.23. Characterize this language in terms of the input alphabet $X = \{0,1\}$ and the operations of Definition 9.4.1.

13. Show by counterexample that neither of the following statements are generally true.

$$\begin{aligned} \lim(\mathcal{L}) &= \mathcal{L}^\omega, \\ \lim(\mathcal{L}^*) &= \mathcal{L}^\omega, \end{aligned}$$

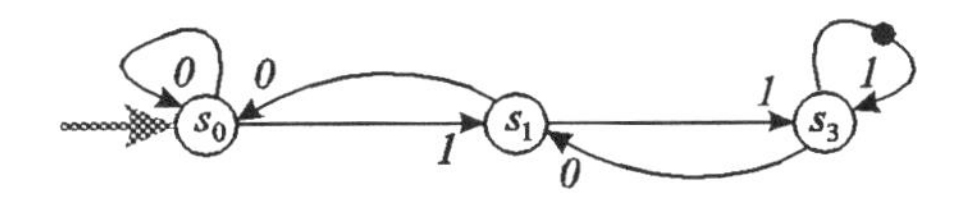

Figure 9.31: A simple L-automaton.

Hint: consider the case where $\mathcal{L} = (10*)$, as in Figure 9.22.

Solution. We have $10^\omega \in \lim(\mathcal{L})$, since this tape has an infinite number of prefixes which are in $\mathcal{L}$, and, therefore, in $\mathcal{L}$. However, $10^\omega \notin \mathcal{L}^\omega$, since the tape 10^ω has a single 1, whereas each tape in $\mathcal{L}^\omega$ has an infinite number of (1)s. This same argument also shows the second equation is not always true as well, since $\mathcal{L} \subseteq \mathcal{L}^*$. □

14. Consider the ω-regular automaton $\mathcal{A}$ of Figure 9.31.

 Which of the following languages are contained in the language $\mathcal{L}(\mathcal{A})$ of the ω-regular automaton $\mathcal{A}$?

 $$\begin{array}{ll} (a) & 11(0)^\omega \\ (b) & 1(1^+0^+111^+)^\omega \end{array}$$

 Explain.

15. Again consider the ω-regular automaton $\mathcal{A}$ of Figure 9.31. Which of the following languages are contained in the language $\mathcal{L}(\mathcal{A})$ of the ω-regular automaton $\mathcal{A}$?

 $$\begin{array}{ll} (a) & 10^*1(0^+1^+0^+)^\omega \\ (b) & (0+1)*(1^+0^+111^+)^\omega \end{array}$$

 Explain.

 Solution. (a): The language of (a) is *not* contained in $\mathcal{L}(\mathcal{A})$, although part of it is. For example, the language $(1\epsilon 1)(01110)^\omega$ is so contained. However, the language (b) also contains tapes like $101(010)^\omega$, which are not accepted because the associated runs get stuck in the $s_0 s_1 s_0$ cycle, and never pass through the recur edge.

 (b): The language of (b) is contained in $\mathcal{L}(\mathcal{A})$. This can be proved by first noting that the arbitrary prefix $(0+1)^*$ takes the automaton arbitrarily to either state s_0, s_1, or s_3. The containment was proved in Part (b) of Problem 14 for the s_1 case explicitly, and, implicitly, for the s_3 case as well.

 Thus it remains to consider the s_0 case. From state s_0, it is also true that any string in the language (1^+0^+111+) always brings the automaton to state s_3 and through the recur edge at least once. To see this, consider strings with a single leading 1. In this case, no matter how many (0)s come next, we are always left in state s_0, and the succeeding three or more (1)s always bring us through the recur edge. □

Part IV

Multilevel Logic Synthesis

Browning said "A man's reach should exceed his grasp, or what's a heaven for?". Compared to two-level logic synthesis, the problem of optimum multilevel logic synthesis is an impossible dream. Chip designers and tool designers, striving jointly and separately to scale this monstrous height, might well heed the words of Chaucer:

> A dronke man woot wel he hath an hous, But he noot which the righte wey is thider, And to a dronke man the wey is slider. And certes, in this world so faren we; We seken faste after felicitee, But we goon wrong ful often, trewely.
>
> (Knight's Tale, I. 1262-1267)

For the problem is an embarassment of riches. Every where we look, we see an unending sea of don't care conditions, impossible to fully enumerate, impossible to fully apply, yet fundamentally affecting the optimality of our results.

Roadmap of Part IV This part is comprised of four chapters, each dealing with a major subarea of multilevel. First, in Chapter 10, we treat the problem of "architecting", or determining a basic overall structure for a multilevel circuit. Often this operation is performed on a minimized two-level specification, and is done at the so-called technology independent level. However, it is also frequently performed on a previously optimized circuit, in order to change tradeoff strategies and optimize for some specific objective like speed or area. Our primary tool in this chapter will be the idea of algebraic factorization.

Algebraic factorization is fast, because it avoids exploration of exponentially (in the number of primary inputs) large search spaces. It is effective, because it can quickly identify large sets of common substructure in a set of Boolean formulae, and select specific subexpressions which can be most effective at reducing circuit size.

In addition to these salient properties, algebraic factorization enjoys a "conservation of literals" property [127, 128], which gives it remarkable powers in synthesis for testability. The key theorem is that if *only* algebraic optimization techniques are used to synthesize a multilevel circuit from a minimized two-level circuit, then the resulting multilevel circuit is 100% testable for all single and multiple stuck faults. Further, the tests for all these faults, which are provided automatically by the process of two-level minimization, remain valid for the synthesized multilevel circuit.

Second, in Chapter 11, we look at some proven techniques for locally optimizing a gicen multilevel circuit, given its initial topological structure. The key ideas are as follows.

1. Identify two-level subcircuits of the given multilevel circuit.

2. Compute the local don't care conditions which the multilevel circuit environment imposes on these subcircuits.

3. Minimize the subcircuits with respect to these don't care conditions, using the methods of Chapter 4 or Chapter 5.

In this chapter we show how to automatically compute the don't care conditions introduced briefly in Section 3.4.

Third, in Chapter 12, we review how tests for the stuck faults in the synthesized multilevel circuit may systematically and efficiently derived. Here we show how the test gerneration process can be made part of the multilevel optimization, and conversely — the two processes are profoundly inter-related.

We conclude in Chapter 13 with a discussion of how a multilevel circuit synthesized by a technology independent process like algebraic factorization may pseudo-optimally mapped into an arbitrary "library" of gates available in a specified technology. This is of profound importance since it provides intercourse between the mathematical world of automatic optimization and the practical world of VLSI technology.

Chapter 10

Multi-Level Logic Synthesis

10.1 Introduction

In Chapters 4 and 5 we have studied the synthesis of two-level circuits. These circuits are typically represented by SOP or POS forms. The most common implementation style for a two-level circuit is a PLA. Because of the simple structure of a two-level circuit, the optimization problems involved in their design are relatively well-understood. Indeed, we have seen efficient algorithms for both exact and heuristic minimization.

In this chapter and in the next two, we turn our attention to the synthesis of multi-level circuits, that is, circuits in which an arbitrary number of gates may lie on any path between a primary input and a primary output. Multi-level circuits are of great practical significance, because they represent the majority of the circuits designed in practice. This is especially true of circuits that are implemented with standard cells or gate arrays, either mask-programmable or field-programmable.

Multi-level circuits tend to be smaller and consume less power than their two-level counterparts. They are also faster in many cases. There exist functions, arithmetic functions provide several examples, for which two-level representations require exponentially many product terms, but multi-level representations only require linear or quadratic numbers of gates.

As mentioned in Section 4.2, the importance of two-level techniques is not seriously affected by the greater importance of multi-level implementations: Many ideas acquired in the study of two-level circuits will prove useful in the new context. One marked difference, however, is that the greater freedom we enjoy when designing multi-level circuits makes exact optimization algorithms of little practical use. We shall therefore concentrate on heuristic algorithms.

The greater freedom we enjoy in designing multi-level circuits manifests itself first in the many different structures that may be chosen to implement them. For instance, there are many different ways of designing an adder or a parallel multiplier; in general, we shall find implementations with different numbers of levels and different choices of gates. Given a circuit, we may decide to optimize it while retaining its initial structure, or we may find it advantageous to reduce or increase the number of levels, thereby **restructuring** the circuit. In practice, we normally end up combining the two approaches: We try different structures for the circuit and for each of them perform

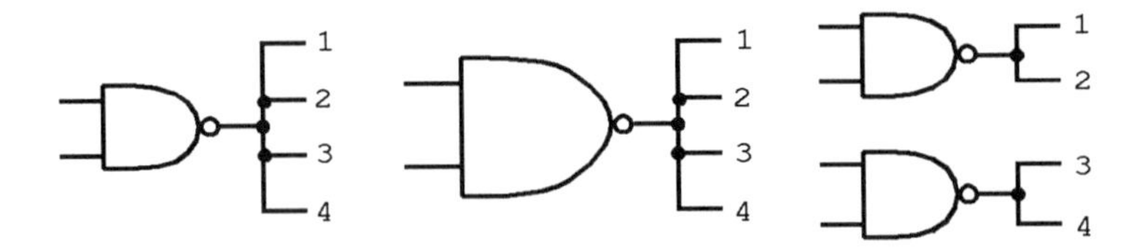

Figure 10.1: Example of Local Optimization.

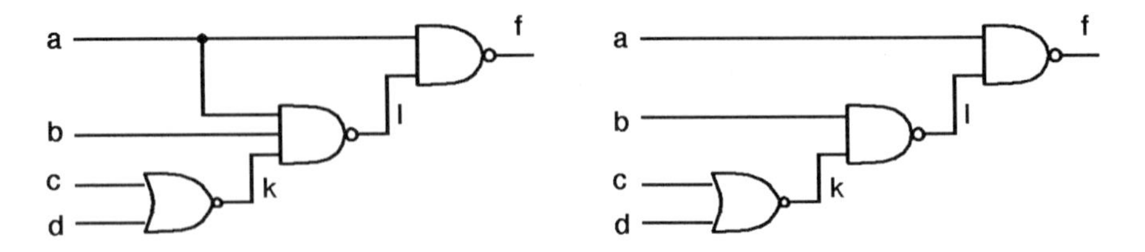

Figure 10.2: Another Example of Local Optimization.

local optimizations intended to reduce the area, delay, and power consumption, while leaving the structure of the circuit substantially unchanged.

We shall make the meaning of terms like "local optimization" and "structure" precise in the course of the next two chapters. For now, a couple of examples may suffice. A simple case of local optimization is the replacement of a two-input NAND gate by another two-input NAND gate with larger driving capability (lower output impedance), or by a pair of NAND gates connected in parallel, in order to make the circuit faster (and also, typically, larger). This is illustrated in Figure 10.1. Another local transformation is the removal of a redundant input connection from a gate, as the one illustrated in Figure 10.2. An instance of restructuring is the transformation of the circuit on the left in Figure 10.3 into the circuit on the right. These few examples should give an idea of the range of problems encompassed by multi-level synthesis; some tasks may be described as *technology dependent* (choice of a more powerful gate to increase speed), while other tasks, like the restructuring of Figure 10.3 may be carried out without specific knowledge of the implementation technology. (See the discussion in Section 2.5.) In this chapter we shall focus on restructuring, leaving technology-independent optimization to Chapter 11 and technology mapping to Chapter 13.

10.1.1 Networks and Algebraic Operations

Our formal model for a multi-level circuit is called **Boolean network**. We shall give a full definition of it in Chapter 11; an informal description will be adequate for the purposes of this chapter. A Boolean network is an acyclic graph.[1] Each node of the graph corresponds to a gate and an arc connecting two nodes corresponds to a connection between two gates. Gates are not restricted to be simple (e.g., NAND, NOR): In this chapter we shall admit arbitrarily complex functions for each node.

[1]The Boolean network is the basic data structure of SIS. The BLIF format is a simple format to describe Boolean networks. Every `.names` directive identifies a node of the network. All node terminals with the same name are connected to form a net.

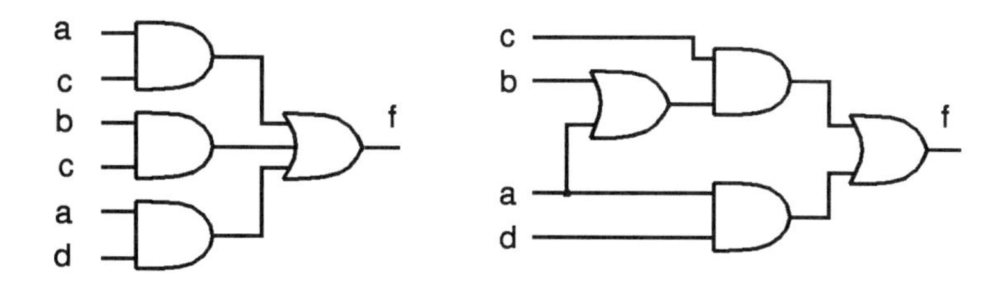

Figure 10.3: Example of Circuit Restructuring.

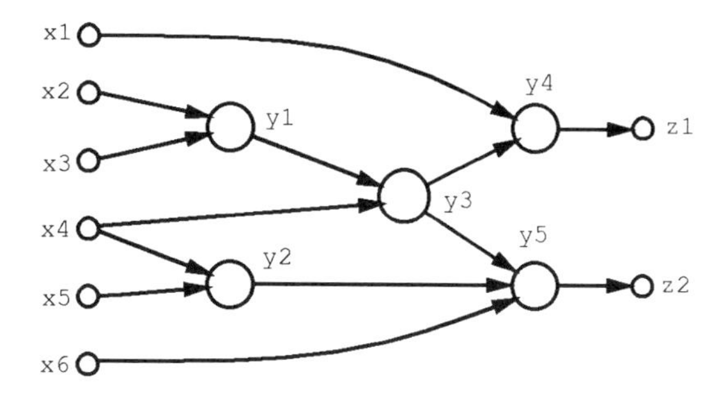

Figure 10.4: Example of Boolean Network.

An example is shown in Figure 10.4. For each node, a function must be given, for the description of the network to be complete. For example, we may have:

$$\begin{aligned}
f_1 &= f_1(x_2, x_3) &&= x_2' + x_3' \\
f_2 &= f_2(x_4, x_5) &&= x_4' + x_5' \\
f_3 &= f_3(x_4, y_1) &&= x_4' y_1' \\
f_4 &= f_4(x_1, y_3) &&= x_1 + y_3' \\
f_5 &= f_5(x_6, y_2, y_3) &&= x_6 y_2 + x_6' y_3'
\end{aligned}$$

Because the direction of the connections is generally understood, we shall henceforth omit the arrowheads. Also, we shall freely use the traditional symbols for gates instead of circles to indicate a node and its function, when the function is that of a simple gate.

Let us return to the example of Figure 10.3. The circuit on the left can be seen to correspond to the formula

$$f = ac + bc + ad,$$

while the circuit on the right corresponds to

$$f = c(a + b) + ad.$$

The latter form can be obtained from the former by factoring out c. Our treatment of restructuring will indeed be based on the factorization of switching formulae. There are other ways of factoring $ac + bc + ad$; for instance,

$$f = (a + bc)(c + d) \quad \text{or} \quad f = (c + ad)(a + b).$$

Although these factorizations are correct, it is not entirely obvious how to get the factorization starting from the SOP form. Indeed, there are so many different ways of factoring an expression, that it is common to apply restrictions on what types of factorizations are sought.

In particular, we shall mostly concentrate on a way of factoring expressions that is called—for historical reasons—**algebraic.** In this chapter, algebraic hints to elementary algebra—the algebra of expressions involving real numbers. It is used in contrast to the word "Boolean." Therefore, when in this chapter we refer to algebraic manipulation of formulae, we mean that we use those rules that are common to the algebra of real numbers and to Boolean algebras. When we refer to Boolean manipulation, we mean that we use all laws of Boolean algebras. Specifically, a few rules of Boolean algebras do not hold in the field of real numbers.

- Idempotence: $a \cdot a = a^2$ in the real numbers, but $a \cdot a = a$ in Boolean algebras.
- Complementation: There is no direct correspondent in the real field.
- Distributivity of '+' over '·': $a + bc = (a + b)(a + c)$ in Boolean algebras, but not in the real field.
- Absorption: $a + ab = a$ holds in Boolean algebras, but not in the real field.

The choice of the word "algebraic" to designate this restricted form of manipulation is a bit confusing, but its usage is so widespread in the literature[2], that replacing algebraic with another term would probably create more confusion that it would eliminate.

Another change in notational convention with respect to the previous chapters is important, but less disturbing: In the chapters on multi-level synthesis, we shall often consider a SOP form as a *set* of cubes (rather than a *sum* of cubes), and a cube as a *set* of literals (rather than a *product* of literals). For instance,

$$f = ac + bc + ad$$

is considered a set of three cubes:

$$f = \{c_1, c_2, c_3\},$$

with each cube considered a set of two literals:

$$c_1 = \{a, c\} \qquad c_2 = \{b, c\} \qquad c_3 = \{a, d\}.$$

This is done for notational convenience, and we always assume that the "meaning" of the SOP form is the join of the cubes, which in turn are the meet of their literals.

10.2 Representation Issues and Choices

As seen in Chapters 4 and 5, two-level minimization is a well developed science, with a solid body of theory and effective algorithms. In considering how multi-level logic theory could be similarly developed, it helps to recognize two things that are obscured when considering two-level implementations, but have been the subject to much discussion for multi-level logic. These are

[2]For instance, SIS has commands for **algebraic resubstitution** etc..

- how to represent the function of the network;
- how to represent the final implementation.

In two-level theory, these two issues are merged, in that the representation and the final implementation are the same, namely the SOP form. However, for multi-level we have a number of choices, and which of these is the best is not obvious:

- **Merged view.** Here, the network is represented so that each node is a valid "gate" chosen from a library of gates to be used in the final implementation. Thus representation and implementation are one.
- **Separated view.** In the separated view, two representations are allowed. One is technology independent, i.e. does not have any connection with the final building blocks to be used in the implementation. The other is the technology dependent view which uses only "valid" gates (i.e., found in a cell library or meeting some criterion). In the separated, technology independent view, there are also several choices:
- **General node.** This is the same as described for the Boolean network, namely that each node can be a representation of an arbitrary logic function. The advantage of this is that a theory can be more easily developed which is not handicapped by arbitrary restrictions which may change as the technology changes.
- **Generic node.** Here every node in the network is a generic node, like a 2-input NAND gate. The advantage is that each node is very simple; there is no need to store a general logic function at a node, since each node is the same function, only the inputs are different. Although there can be many more nodes than required for the general node description, experience indicates that some manipulations are much faster using this structure. The disadvantage is the network is finely decomposed in a particular way, and this may obscure some natural structures in the network.

For most of this sequel, we employ the general node representation since it includes all others as special cases, and a more complete theory and body of algorithms has been developed for this point of view. This choice motivates the definition of Boolean network that we shall encounter in Chapter 11.

10.2.1 Alternate Node Representations

Each node of a multi-level network has associated with it a representation of a switching function. The question of how this function is represented is important. Although any valid representation is allowed, some representations are preferred because they are:

- More efficient in memory;
- more indicative of the complexity of the final implementation;
- more efficient to manipulate.

In this section we survey some of the choices that can be made.

Sum-of-Products

The most obvious representation is the SOP form. This is very popular, possibly because of the historical influence of two-level, PLA, optimization problems. A sum-of-products is a set of cubes (product terms), e.g.,

$$abc' + a'bd + b'd' + b'e'f.$$

This is a natural choice in several ways. First, there are highly developed techniques for manipulating logic in this form, e.g., two-level minimizers, factoring, decomposition, tautology, AND, OR. Thus, even though we may prefer to have logic represented some other way, in order to manipulate it with present techniques, we may find it convenient to convert to sum-of-products, manipulate it, and then convert it back. Thus one can argue that this should be the nominal representation.

Factored Forms

Factored forms are probably a more natural representation for multi-level synthesis. A factored form is parenthesized expression like

$$(ab + b'c)(c + d'(e + ac')) + (d + e)(fg).$$

The argument for factored forms is that they are a natural multi-level representation. A factored form is isomorphic to a tree, where each internal node is an AND or OR operator, and each leaf is a literal. This provides a simple but efficient multi-level implementation of the function of the node. A representation which more accurately measures complexity is important in guiding the synthesis process, since synthesis can be seen as a sequence of transformations which may or may not be accepted depending on the quantity of complexity decrease obtained.

The count of the number of literals in a factored form is a good estimate of the complexity of the function, and can be translated directly into the number of transistors required for implementation. For instance,

$$f = ((a + bc)(c + d))'$$

can be implemented as a CMOS complex gate as in Figure 10.5. One can see that the number of transistors is exactly twice the number of literals. On the other hand, if f is implemented with simple gates only, as in Figure 10.6, the transistor count, though still related to the number of literals, is not so directly obtained from the factored form of f. (The circuit of Figure 10.6 requires 18 transistors, when implemented in CMOS.)

Of course, transistor count only indirectly measures area since wiring is also an important contributor to the total area. Thus, it has been suggested that a better area estimator would be the number of literals in the factored form plus a term proportional to the number of gates or nodes in the network or the number of terminals in the network.

Another argument for the factored form over the sum-of-products representation is that the factored form implicitly represents both the function and its complement. The complement factored form can be obtained by applying De Morgan's laws to the

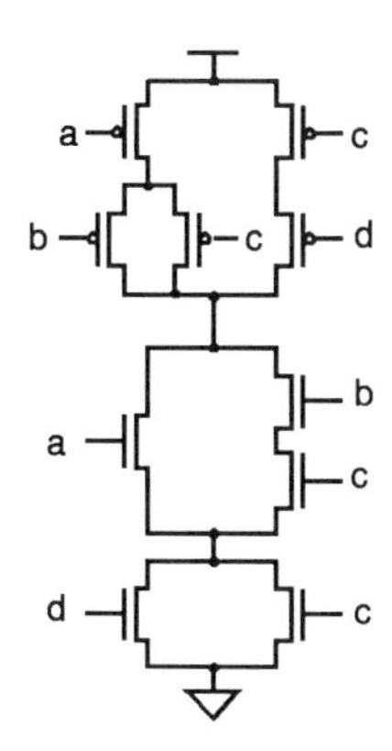

Figure 10.5: A CMOS Complex Gate Implementing $f = ((a + bc)(c + d))'$.

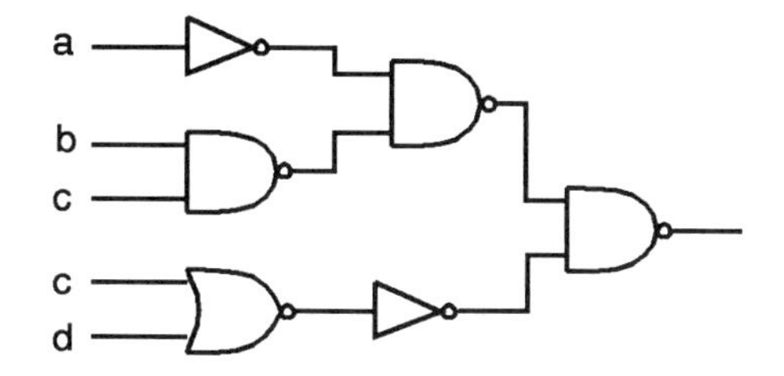

Figure 10.6: A Simple Gate Implementation of $f = ((a + bc)(c + d))'$.

factored form. Thus, ANDs are converted to ORs and vice versa, and literals are negated, producing a factored form for the complement which has the same literal count. For instance,

$$((a + bc)(c + d))' = a'(b' + c') + c'd'.$$

This result coincides with the notion that in a multi-level implementation a function and its complement are equally complex, only separated by an inverter. This is in contrast with the sum-of-products form where the number of cubes in a function can be exponentially larger than in its complement.[3] In this regard we may think of the factored form as representing both the function and its complement. Furthermore, factored forms subsume both SOP forms and POS forms as special cases.

The difficulty with factored forms is that they are more difficult to manipulate than two-level forms. This has stimulated recent work to extend methods for manipulation of sum-of-products to similar methods for factored forms. Three such results have appeared recently. The first [31] is motivated by the generic representation point of view, that, for speed of manipulation, each node in the Boolean network should be the same logic function, say, a NAND or NOR. The advantage of this is that the logic function of a node need not be stored, only the set of inputs. This leads to more efficient storage as well as possibly faster methods for manipulation. However, this representation, which is basically a form of factored form, suffers from the lack of developed algorithms. In [31] methods for kerneling and minimization were proposed. A second development [268] works from the standard factored form

[3]Consider, for instance, $f = x_1x_2 + x_3x_4 + x_5x_6 + \cdots$.

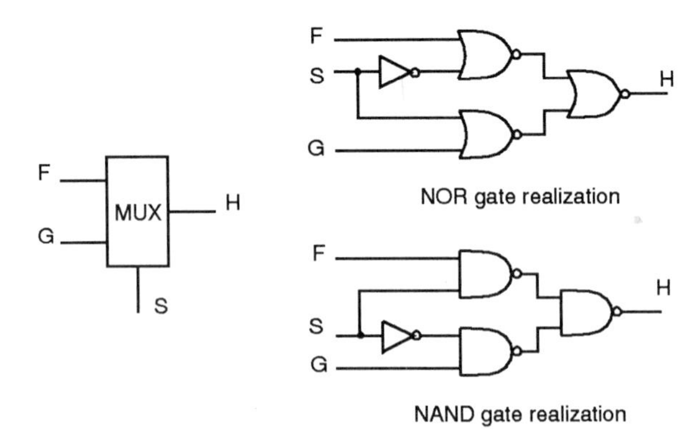

Figure 10.7: NAND and NOR Decompositions.

representation where a node is either an AND or an OR and most of the Boolean function manipulation methods are extended. However, one important but missing method is minimization. A third development [246] starts from a sum-of-products form and asks for a minimization procedure which has as its goal a minimal factored form. Here it is recognized that the normal goal of two-level minimizers (such as ESPRESSO), the minimal number of cubes, is inappropriate. A minimizer based on a minimal factored form has been developed.

Another lack in this area is some notion of optimality. Is a given factored form optimum? In the case of sum-of-products there is an effective answer via some form of Quine-McCluskey exact minimization which is quite efficient. (See Chapter 4.) However, for factored forms, the only known optimality result is the paper by [168] which is not practical for functions which depend on more than about 6-8 variables.

NAND or NOR Representations

An arbitrary Boolean function can be straightforwardly decomposed into NAND or NOR networks, that is, networks for which every node function is a NAND (NOR). This is illustrated in Figure 10.7, which shows the NAND and NOR decompositions of a 2-input MUX. Note

$$H = SF + S'G = ((SF)'(S'G)')' = \text{NAND}(\text{NAND}(S, F), \text{NAND}(S', G)).$$

There are three major advantages which are obtained when using this "generic." The first is fast simulation and simulation-like tasks connected with the internals of logic synthesis programs, and results from the inherently simpler data structures which result from this representation. The effect is large—from 1 to 2 orders of magnitude [68]. Third, rule-based or expert systems for logic optimization are easily formulated on NAND or NOR gates. These data structures also enable some efficient optimization strategies known as global flow methods [21].

The second main advantage stems from the fact that the NAND type representation is always complete with respect to inverter counts, whereas factored form representations ignore inverters, which must later be added at the technology mapping step. This can lead to a large discrepancy between the factored form literal counts and the transistor counts in the final implementation. The use of these generic node types also gives a better timing delay simulation model for use in the logic minimization package.

10.3 Representing Switching Functions in Factored Form

10.3.1 Factored Forms

An alternative representation (to sum-of-product form) of a logic function which is closer to the physical implementation of the logic is the factored form. It is the generalization of sum-of-product form allowing nested parenthesis. For example, each of the following is a factored form:

$$a$$
$$a'$$
$$ab'c$$
$$ab + c'd$$
$$(a+b)(c+a'+de)+f$$

where $a, a', b', c, \ldots$ are called literals of the factored form. Factored forms are often derived from given SOP forms; for example, the expression

$$ace + ade + bce + bde + e'$$

can be written in factored form as

$$e(a+b)(c+d)+e'.$$

In other words, a factored form is a sum of products of sums of products..., of arbitrary depth.

Factored forms are useful in estimating area and delay in a multi-level logic synthesis and optimization system. Factored forms in general are more compact than SOP or POS forms. For example, if the following factored form were expressed in SOP form, it would contain eight product terms for a total of 36 literals. In POS form would require six terms for a total of 12 literals.

$$(a+bc)(d+ef)(g+hi)$$

Furthermore, many commonly used logic operations can be easily performed on factored forms. De Morgan's laws applied to a factored form for a function f yield a factored form for f'; by contrast, application of De Morgan's laws to a SOP form for f yields a POS form for f' and vice versa. These properties make factored forms useful for the internal representation of logic functions in logic synthesis and optimization systems. The relative merits of the factored form representation were discussed in Section 10.2.1; now we give a formal definition of it.

Definition 10.3.1 *A factored form is defined recursively by the following rules:*

1. *A* product *is either a single literal or a product of factored forms.*

2. *A* sum *is either a single literal or a sum of factored forms.*

3. *A* factored form *is either a product or a sum.*

For example, each of the following is a factored form:

$$
\begin{array}{l}
x \\
y' \\
abc' \\
a + b'c \\
((a' + b)cd + e)(a + b') + e'
\end{array}
$$

where the first two are literals, the third is a product, the fourth is a sum, and the last one is a sum of products of sums of .... According to the definition, the following is not a factored form:

$$(a + b)'c$$

because it complements $(a+b)$ internally, which is not allowed by the definition. Like the two-level forms, factored forms are in general not unique, as illustrated by the following three equivalent factored forms:

$$
\begin{array}{l}
ab + c(a + b) \\
bc + a(b + c) \\
ac + b(a + c).
\end{array}
$$

10.3.2 Algebraic and Boolean Expressions

Definition 10.3.2 *An* **algebraic expression** $F = \{C_i\}$ *is one in which no cube contains another, that is* $C_i \not\subseteq C_j$, $i \neq j$. *An expression that is not algebraic is called* **Boolean**.[4]

For example, expression $a + bc$ is algebraic, and expression $a + ab$ is non-algebraic because $\{a\} \subset \{a, b\}$. (Remember that we consider a cube as a set of literals.) Note that an algebraic expression is not necessarily prime or irredundant: Some cubes may not be prime, as in $a + a'b$, and some cubes may be eliminated without changing the function, as in $xy + y'z + xz$.

Definition 10.3.3 *The* **support** *of an expression* F *is the set of variables,* $supp(F)$, *that* F *explicitly depends on, i.e.,* $supp(F) = \{x | \exists C \in F \text{ such that } x \in C\ x' \in C\}$. *Two expressions* F *and* G *are said to be* **orthogonal**, $F \perp G$, *if they have* **disjoint support**, *i.e.,* $supp(F) \cap supp(G) = \emptyset$.

For example, $\text{supp}(xy+x'z') = \{x, y, z\}$. Expressions $ab+c$ and $de'+f$ are orthogonal, whereas $ab + c$ and $c' + de$ are not.

[4] Brown [44] calls such an expression *absorptive*.

10.3.3 Algebraic and Boolean Factored Forms

As discussed above, there are many equivalent factored forms of a given function. The difference in the number of literals of these equivalent factored forms can be significant. For example, given the algebraic expression

$$abg + acg + adf + aef + afg + bd + ce + be + cd, \tag{10.1}$$

the following are three equivalent factored forms obtained by three different factorization algorithms:

$$\begin{array}{ll} (b+c)(d+e) + ((d+e+g)f + (b+c)g)a, & \text{(12 literals)} \\ (b+c)(d+e+ag) + (d+e+g)af, & \text{(11 literals)} \\ (af+b+c)(ag+d+e). & \text{(8 literals)} \end{array}$$

The first two of these equivalent factored forms are algebraic, whereas the third one is Boolean, as we shall see soon. Note also that there are 12 literals in the first factored form and only 8 literals in the last one.

The previous example also shows two different kinds of factored forms. Taking the first factored form and multiplying it out literal by literal to get a sum of cubes without using the Boolean identities $xx' = 0$ and $xx = x$, we can in fact recover the original sum-of-product form. But if we took the last factored form and multiplied it out in the same way, we would get a different, non-algebraic, SOP form because of the cube $afag$. This leads to the following definitions of algebraic and Boolean factored forms.

Definition 10.3.4 *A* **factored form** *F is said to be* **algebraic** *if the SOP expression obtained by multiplying F out directly (without using $xx' = 0$ and $xx = x$, and single-cube containment) is algebraic (or, equivalently, if all implied products in expanding to SOP form are algebraic). F is a* **Boolean** *factored form if it is not algebraic.*

For example, each of the following is an algebraic factored form:

$$\begin{array}{l} a + bc \\ (a+b)(c+d) \\ (b+c)(d+e+ag) + (d+e+g)af, \end{array}$$

and each of the following is a Boolean factored form:

$$\begin{array}{l} (a+b+c+d)(a'+b'+c'+d') \\ (af+b+c)(ag+d+e) \\ (a+b)((c+d)(e+f)+g) + b(e+h). \end{array}$$

None of the sum-of-products forms obtained by multiplying the Boolean factored forms out are algebraic. They either contain terms which are 0 ($xx'c = 0$), terms with redundant literals ($xxc = xc$), or redundant terms ($abc + ab = ab$).

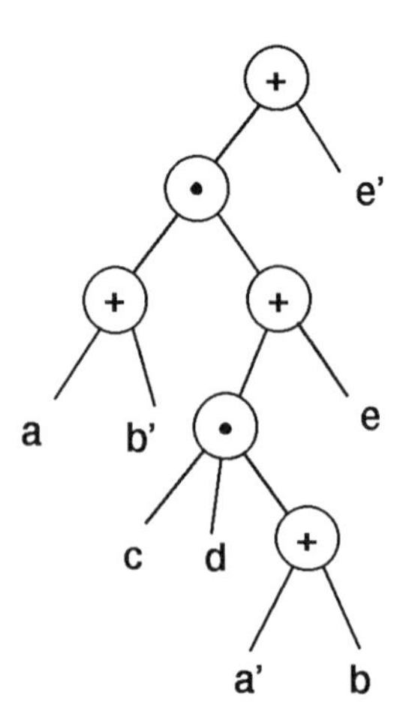

Figure 10.8: Factoring Tree for $((a' + b)cd + e)(a + b') + e'$.

10.3.4 Value of a Factorization

Definition 10.3.5 *The factorization value of an algebraic factorization $F = G_1 G_2 + R$ is given by*

$$\text{FACT\_VAL}(F, G_2) = \text{LITS}(f) - (\text{LITS}(G_1) + \text{LITS}(G_2) + \text{LITS}(R)) \qquad (10.2)$$

assuming G_1, G_2, and R are algebraic expressions.

Here, for any algebraic expression P, LITS(P) = Number of literals in SOP form of P. Another short example illustrates this definition. The algebraic expression

$$F = ae + af + ag + bce + bcf + bcg + bde + bdf + bdg$$

can be expressed in the factored form

$$F = (a + b(c + d))(e + f + g).$$

Note that only 7, rather than 24 literals are required.

If $G_1 = (a + bc + bd)$, and $G_2 = (e + f + g)$, then $R = 0$, and FACT_VAL(F, G_2) = (2)(3)+(2)(5) = 16. Note the factored form saved 17, rather than 16 literals. The extra literal saved comes from recursively applying the formula to the factored form of G_1.

10.3.5 Equivalent, Maximal, and Optimum Factorizations

Factored forms can be graphically represented as labeled trees, called factoring trees, in which each internal node including the root has a label of either "+" or "·", and each leaf has a label of either a variable or its complement. For example, the factoring tree of

$$((a' + b)cd + e)(a + b') + e',$$

is shown in Figure 10.8. Note that the product terms of a factoring tree need not be algebraic products.

Any sub-tree of a factoring tree is a *factor* of the factored form represented by the factoring tree. In other words, a factor of a factored form is any sum term or product term in the factored form. For example, $a(b+c)$ and $de+f$ are factors of the factored form $a(b+c)+de+f$.

Two factored forms are said to be *equivalent* if they represent the same logic function, and *syntactically equivalent* if their factoring trees, including the leaves, are isomorphic. For example, $a(b+c)+bc$ and $ab+c(a+b)$ are equivalent, and $(a+b)(c+d)e$ and $(c+d)e(a+b)$ are syntactically equivalent.

Some factored forms can be further factored. For example, $ab+ac$ can be further factored to $a(b+c)$, and $(a+b)c+(a+b)d$ can be further factored to $(a+b)(c+d)$.

Definition 10.3.6 *A factored form is* **maximally factored**, *if*

1. *For every sum of products, there are no two syntactically equivalent factors in the products.*

2. *For every product of sums, there are no two syntactically equivalent factors in the sums.*

For example, the following factored forms are not maximally factored

$$ab+ac$$
$$(a+b)(a+c),$$

because they contain trivial syntactically equivalent factors a in their products and sums, respectively. (Note that the product in the second factored form is not an algebraic product.) Hence, they can be further factored to

$$a(b+c)$$
$$a+bc.$$

Notice that in a Boolean algebra, "+" distributes over "·" and "·" distributes over "+". (See Chapter 3.) The above factorings are obtained by direct application of the distributive law. The transformation of $ab+ac$ to $a(b+c)$ may seem more obvious than that of $(a+b)(a+c)$ to $a+bc$, because $a(b+c)$ happens to be an algebraic product. This derives from the fact that we often think of a Boolean expression as a polynomial, in which multiplication distributes over sum. But, in Boolean algebras, the OR, "+", also distributes over the AND, "·". So, the second transformation should be just as simple. The fact that we are considering non-algebraic products like $(a+b)(a+c)$ emphasizes that the present discussion addresses underlying structure more specific than just the algebraic structure of expressions.

Given the factoring tree of an expression, the test for maximal factorization proceeds as follows: We visit the tree in post-order. At every non-terminal node we apply the definition.

10.3.6 Size, Unateness, and Cofactors of a Factored Form

The size of a factored form F is measured by the number of literals in the factored form and is denoted by $\rho(F)$. For example, $\rho((a+b)ca') = 4$ and $\rho((a+b+cd)(a'+b')) = 6$. A factored form is said to be *optimum* if no other equivalent factored form has fewer literals.

A factored form F is **positive unate** in x if literal x appears in F and literal x' does not appear in F, **negative unate** in x if literal x' appears in F and literal x does not appear in F. F is **unate** in x if it is either positive unate or negative unate in x. F is **binate** in x if it is not unate in x. For example, $(a + b')c + a'$ is positive unate in c, negative unate in b and binate in a.

The **cofactor** of a factored form F with respect to a literal l (i.e., x or x') is a factored form, denoted by F_l, obtained by replacing all occurrences of l with 1 and all occurrences of l' with 0, and simplifying the factored form using (only)the following identities of the Boolean algebra:

$$\begin{aligned} 1 \cdot x &= x \\ 1 + x &= 1 \\ 0 \cdot x &= 0 \\ 0 + x &= x. \end{aligned}$$

Notice that after this simplification (known as constant propagation), part of the factored form may appear as $x + x$. In this section, we take the viewpoint that x is not a literal, but another factored form. Later we introduce a generic factoring algorithm, Procedure Factor, in which the two x's may have different factored forms. Identifying these equivalent factored forms and obtaining the simplification $x + x = x$ is the subject of later sections. Here, we are only interested in obtaining a factored form which represents the cofactor of a function with respect to a literal.

The cofactor of a factored form F with respect to a cube c is a factored form, denoted by F_c, obtained by successively cofactoring F with each literal in c. For example, let $F = (x + y' + z)(x'u + z'y'(v + u'))$ and $c = vz'$. Then

$$\begin{aligned} F_{z'} &= (x + y')(x'u + y'(v + u')) \\ F_c &= (x + y')(x'u + y'). \end{aligned}$$

10.4 Division

A multiplicative inverse of x is a y such that $x \cdot y = 1$. Real numbers, with the exception of 0, have multiplicative inverses. Since a Boolean algebra does not have a multiplicative inverse, there can be no division operation. However, one can define operations which, when given functions f and p, find functions q and r such that $f = p \cdot q + r$. Every such operation is like the division operation and is therefore called *division* of f by p generating **quotient** q and **remainder** r. It is clear that such a division operation is not unique. Even for a given division operation, the resulting q and r may be dependent upon the particular representation of f and p. Consider, for instance, $f = x'z' + yz + xz$ and $p = x' + z$. It is easy to verify that:

$$\begin{aligned} f &= (x' + z)(x + z') + x'y \\ &= (x' + z)(y + z') + xz. \end{aligned}$$

Let f and g be Boolean functions satisfying $f \leq g$; then f can be written as a product, $f = g \cdot h$. It is sufficient to choose h such that

$$f \leq h \leq f + g'.$$

Similarly, if $f \cdot g \neq 0$, f can be written $f = gh + r$, with $r < f$. In this case, r must be chosen so that

$$f \cdot g' \leq r < f,$$

and, for a given r, h must be chosen so that

$$f \cdot r' \leq h \leq f + g'.$$

In the former case ($f \leq g$ and $r = 0$) we call g a **Boolean factor** of f and in the latter case we call g a **Boolean divisor** of f. For any logic function f there are *many* more Boolean divisors and factors than there are algebraic divisors and factor (discussed below). This poses the problem of a good choice in Boolean decomposition, since there are so many factors.

So far, we have considered Boolean functions. In practice, however, we have to deal with the formulae that represent those functions. Therefore, we now introduce two classes of division operations that work on SOP forms. Division for SOP forms is based on the concept of product; we now give a definition of product based on considering a cube as a set of literals and a SOP form as a set of cubes.

Definition 10.4.1 *The* product *of two cubes C and D is a cube defined by*

$$CD = \begin{cases} \emptyset & \text{if } \exists x (x \in C \cup D \ \text{ and } \ x' \in C \cup D) \\ C \cup D & \text{otherwise.} \end{cases} \tag{10.3}$$

The product *of two SOP expressions F and G is a SOP expression defined by*

$$FG = F \times G = \{CD | C \in F \ \text{ and } \ D \in G \ \text{ and } \ CD \neq \emptyset\}. \tag{10.4}$$

Notice that $CD = \emptyset$ if and only if $C \cup D$ contains both a literal and its complement.

Definition 10.4.2 *FG is an* **algebraic product** *if F and G have disjoint support otherwise FG is a* **Boolean product**.

For example,

$$(a + b)(c + d) = ac + ad + bc + bd$$

is an algebraic product and both

$$(a + b)(a + c) = a + ab + ac + bc$$

and

$$(a + b)(b' + c) = ab' + ac + bc$$

are Boolean products. As suggested by the Cartesian product symbol used in the definition, FG contains precisely $|F| \times |G|$ cubes if FG is an algebraic product.

Definition 10.4.3 *An operation* OP *is called* **division** *if, given two SOP expressions F and P, it generates SOP expressions Q and R ($\langle Q, R\rangle = \text{OP}(F, P)$) such that $F = PQ + R$.*

If PQ is an algebraic product, OP is called an **algebraic division**; otherwise PQ is a **Boolean product** and OP is therefore called a **Boolean division**. Suppose PQ is an algebraic product. If $R = \emptyset$, then P is an algebraic factor; otherwise, P is an **algebraic divisor**.

The algebraic divisors are a subset of the Boolean divisors, so the optimum choice of algebraic divisor may not be the best choice overall. Thus we are faced first with a choice of whether to look for Boolean or algebraic divisors, the former choice offering potentially higher quality at greater expense, and the latter offering low CPU expense and relative ease of evaluation. In general, we face two tasks in using either notion of division. First, to find a good candidate divisor, and the second to effect the division, i.e., to determine, given g and f, the coefficient h and remainder r so that $f = gh + r$. We now discuss briefly the concept of **weak division**, which is a specific example of algebraic division.

Definition 10.4.4 *Given two algebraic expressions F and P, a division is called* **weak division** *if:*

- *it generates Q and R such that PQ is an algebraic product;*
- *R has as few cubes as possible;*
- *$PQ + R$ and F are the same expression (have the same set of cubes).*

Given the expressions F and P, it can be shown that Q and R generated by weak division are unique. Often, F/P is used to denote the quotient of weak-dividing F by P.

We first examine informally how weak division works, with the help of an example. Let $F = ad+abc+bcd$ and $P = a+bc$. For each cube of P, p_i, we look for cubes of F, f_j, such that f_j has all the literals of p_i. In our case, for $p_1 = a$, we find $f_1 = ad = p_1 d$ and $f_2 = abc = p_1 bc$; for $p_2 = bc$, we find $f_2 = abc = ap_2$ and $f_3 = bcd = p_2 d$. We observe that d multiplies both $p_1 = a$ and $p_2 = bc$. Hence, we can factor it and write:

$$F = (a + bc)d + abc.$$

In general, weak division of Y by P is accomplished by considering each cube of P, p_i in turn. For each p_i, we consider every f_j; if f_j contains all the literals of p_i, we form the cube v_{ij} by deleting from f_j all the literals in p_i, and we add this cube to the set V^{p_i}. The cubes that appear in all the V^{p_i}'s form Q. The order in which the cubes are considered is immaterial. This is stated in form of pseudo-code in Figure 10.9.

We now use weak division to divide the function $F = ac + ad + bc + bd + e$ by $P = a + b$. $V^a = \{c, d\}$ is the set of elements that are multiplied by a in the original function F and $V^b = \{c, d\}$ is the set of elements that are multiplied by b in the original function. Hence, $V^a \cap V^b = \{c, d\}$, $Q = c + d$, $R = e$, and F factors into $(a + b)(c + d) + e$.

As a second example, consider $F = ad + aef + ab + b'cd + b'cef$ and $P = a + b'c$. We find $V^a = \{d, ef, b\}$ and $V^{b'c} = \{d, ef\}$. Hence, $Q = d + ef$ and $R = ab$; $F = (a + b'c)(d + ef) + ab$.

```
WEAK_DIV(F,P) {
    foreach (p_i) {
        V^{p_i} = ∅
        foreach (f_j) {
            if (f_j contains all the literals of p_i) {
                v_ij = f_j with the literals in p_i deleted
                V^{p_i} = V^{p_i} ∪ v_ij
            }
        }
    }
    Q = ∩_i V^{p_i}
    R = F − PQ
    return (Q, R)
}
```

Figure 10.9: Weak Division Algorithm.

10.5 Kernels and Co-Kernels

Having defined algebraic division, we can now introduce the concept of a **kernel** of an algebraic expression. The notion of a kernel was introduced in [38] to provide means for finding subexpressions common to two or more expressions. All operations used to find kernels are algebraic (i.e., algebraic product, algebraic division, etc.), but the word *algebraic* is omitted for brevity. In particular, algebraic division is done by WEAK_DIV.

Definition 10.5.1 *An expression is* **cube-free** *if no cube divides the expression evenly, that is,*

$$\neg\exists C \quad \textit{such that} \quad F = QC$$

(no remainder), and C is a cube.

For instance, $ab + c$ is cube-free; $ab + ac$ and abc are not cube-free. A cube-free expression must have more than one cube.

Definition 10.5.2 *The* **primary divisors** *of an algebraic expression F are the set of expressions*

$$D(F) = \{F/c \mid c \text{ is a cube}\}. \tag{10.5}$$

The **kernels** *of an expression F are the set of expressions*

$$K(F) = \{g \mid g \in D(F) \text{ and } g \text{ is cube-free}\}. \tag{10.6}$$

In other words, the kernels of an expression F are the cube-free primary divisors of F.

A cube c used to obtain the kernel $K = F/c$ is called a *co-kernel of* K, and $C(F)$ is used to denote the set of co-kernels of F. For example, the kernels and their corresponding co-kernels of the function

$$\begin{aligned} F &= adf + aef + bdf + bef + cdf + cef + bfg + h \\ &= (a+b+c)(d+e)f + bfg + h \end{aligned}$$

are listed below.

kernel	co-kernel	level
$d+e$	af, cf	0
$d+e+g$	bf	0
$a+b+c$	df, ef	0
$(a+b+c)(d+e)+bg$	f	1
$((a+b+c)(d+e)+bg)f+h$	1	2

$F/a = df + ef$ is a primary divisor, which is not cube-free.

Notice that a kernel may have more than one co-kernel and a co-kernel can be the trivial cube 1 if the original expression was cube-free.

For certain operations described in the following sections, it is nearly as effective and frequently more efficient to compute a certain subset of $K(F)$ rather than the full set. This leads to the following recursive definition for the **level of a kernel**.

Definition 10.5.3 *A kernel is of level 0 if it has no kernels except itself. A kernel is of level n if it has at least one kernel of level $n-1$, but no kernels (except itself) of level n or greater.*

Since each co-kernel is associated with a unique kernel, the above definition also serves to define the **level of a co-kernel**, which is just the level of the corresponding kernel.

The following theorem, [38], demonstrates the importance of kernels in the algebraic decomposition of sets of expressions.

Theorem 10.5.1 (Fundamental Theorem). *If two expressions F and G have the property that $K_F \in K(F)$ and $K_G \in K(G)$ imply that $|K_G \cap K_F| \leq 1$ (K_G and K_F have at most one term in common), then F and G have no common nontrivial algebraic divisors. (A nontrivial divisor has at least two terms.)*

This theorem is used to detect if two or more expressions have any common algebraic divisors other than single cubes. This can be done by computing the set of kernels for each logic expression, and forming nontrivial (more than one term) intersections among kernels from different functions. If this intersection set is empty, then we need only look for divisors consisting of single cubes (which is an easier task). In other words, we do not need to compute the set of all algebraic divisors for each expression to determine if there are common nontrivial algebraic divisors. This leads to great run time efficiency since the set of kernels is much smaller than the set of all algebraic divisors. Further, in the algorithm for computing kernels, the cube-free property of kernels leads to a very effective method for pruning the search tree for the kernels. On the other hand, if we do find a nontrivial intersection, then this is a candidate algebraic divisor common to two or more functions.

		1	2	3	4	5	6
		$abcd$	$abce$	$abde$	af	dpq	epq
2	$abce$	abc					
3	$abde$	abd	abe				
4	af	a	a	a			
5	dpq	d	0	d	0		
6	epq	0	e	e	0	pq	
7	g	0	0	0	0	0	0

Table 10.1: Cube Intersection Table.

10.5.1 Computation of Co-Kernels and Kernels

Let C^* be the set of cubes whose literals constitute the literal intersections of each subset of 2 cubes of an algebraic expression F. For each cube $c_i \in C^*$ define $K_i = F/c_i$ $=$ WEAK_DIV(F, c_i). Then c_i is a co-kernel of F, and K_i is its corresponding kernel. Because c_i is the literal intersection of cubes of F, no subcube may be factored out of each of the cubes of K_i. It may be shown that the set $C^*(F)$ contains all the level-0 co-kernels, and, perhaps, some co-kernels of higher level as well.

For example, let

$$F = abcd + abce + efg = abc(d+e) + efg = abcd + e(abc + fg),$$

then

$$C^* = \{abc, e\}.$$

The kernels of F are $d+e$ and $abc+fg$, with corresponding co-kernels abc and e, respectively.

This illustrates only one class of the co-kernels of F, known as level-0 co-kernels. Consider $F = abcd + abce + abde$. The pairwise intersections yield the level-0 co-kernels $\{abc, abd, abe\}$, and the 3-way intersection yields the level-1 co-kernel $\{ab\}$, whose kernel $c(d+e) + de$ has a level-0 subkernel $c+d$ and $d+e$.

To systematically derive C^*, we resort to a **cube intersection table** like the one for the algebraic expression

$$F = abcd + abce + abde + af + dpq + epq + g$$

shown in Table 10.1. Let n be the number of cubes in F. The table for F contains one column for each of the first $n-1$ cubes and a row for each of the last $n-1$ cubes. Since the intersection of c_i and c_j is the same as the intersection of c_j and c_i; and since we are not interested in the intersection of c_i and c_i, we can restrict ourselves to the part of the table below the main diagonal. The entry in position (i, j) is the intersection of c_i and c_j. (The rows are numbered from 2 to n.) Each non-zero entry is a co-kernel.

The cube intersection table can be extended to find all co-kernels, regardless of their level. This is shown in Table 10.2. The table contains four extra rows and columns corresponding to original pair-wise intersections with more than one literal.

		1	2	3	4	5	6	7	8	9	10
		$abcd$	$abce$	$abde$	af	dpq	epq	g	abc	abd	abe
2	$abce$	abc									
3	$abde$	abd	abe								
4	af	a	a	a							
5	dpq	d	0	d	0						
6	epq	0	e	e	0	pq					
7	g	0	0	0	0	0	0				
8	abc	abc	abc	ab	a	0	0	0			
9	abd	abd	ab	abd	a	d	0	0	ab		
10	abe	ab	abe	abe	a	0	e	0	ab	ab	
11	pq	0	0	0	0	pq	pq	0	0	0	0

Table 10.2: Extended Cube Intersection Table.

In this case the addition of the new rows and columns leads to the discovery of the level-1 co-kernel ab. One can easily verify that adding another row and column for ab to the table leads to no new co-kernels.

The complexity of this method, when computing level-0 co-kernels is $O(|F|^2 \log |F|)$, since all pairs of cubes of F need to be intersected, and then merged into a list of known co-kernels. This requires a membership test and a subsequent union operation, which can be done in $O(log|F|)$ by maintaining the list C^* in a priority queue (binary heap) with $O(log|F|)$ insertion cost. Of course, computing all the higher level co-kernels requires a recursion which examines all 3-way intersections, all 4-way intersections, and so on. This recursion leads to exponential worst case complexity because there are, potentially, the intersections of $2^{|F|}$ subsets of the cubes of F to investigate. There are other methods for the computation of kernels, but the one we have examined here, based on the intersection matrix, is adequate for our purposes.

10.6 Heuristic Factoring Algorithms

Factoring is the process of deriving a factored form of a given logic function represented in a sum-of-products form. The objective is to minimize the number of literals in the final factored form. Algorithms have been developed for solving exactly the problem of determining the optimum factored form [167]. However, in recent experiments using some modern extensions [271] for logic function manipulation, the complexity of these exact techniques still appears to be impractical for all but the smallest functions. The goal of this section is to develop fast heuristic factoring algorithms which rely on kernels [38] and find optimal factored forms.

Some factoring algorithms guarantee the results to be algebraic and others do not. One way of classifying factoring algorithms is by the following definition:

Definition 10.6.1 *A factoring algorithm is said to be* algebraic *if it is guaranteed to produce an algebraic factored form starting from an algebraic SOP expression. A factoring algorithm is* Boolean *if it is not algebraic.*

So far, the examples we have used are completely specified functions. In the process of multi-level logic optimization, we often encounter incompletely specified functions. Algorithms for factoring incompletely specified functions usually take very different approaches than those for factoring completely specified functions. Our treatment in Sections 10.6.1–10.6.5 is restricted to completely specified functions.

10.6.1 Generic Factoring Algorithm

A typical generic recursive factoring algorithm is given below.

```
FACTOR(F, DIVISOR, DIVIDE) {
    if (F has no factor) return (F)
    D = DIVISOR(F)
    (Q, R) = DIVIDE(F, D)
    return (FACTOR(Q) FACTOR(D) + FACTOR(R))
}
```

Like Procedure FACTOR, all heuristic factoring algorithms described in this section use the same top-down paradigm. Note the non-standard programming style, in which, subroutine names are passed as arguments.

Given an algebraic (or even Boolean) expression F in SOP form, Procedure DIVISOR(F) is used to find a candidate divisor, D, which, when substituted into F, can simplify the expression. Then, the quotient Q is found by dividing D into F using routine DIVIDE(F, D). Now, the function can be represented as a partial factored form $F = QD + R$ where R is the remainder. The algorithms then proceed to factor Q, D, and R separately using the same method.

If the expression F is algebraic, a single literal, co-kernel, or kernel of F can be returned by generic Procedure DIVISOR, and Procedure WEAK_DIV can be used for Procedure DIVIDE. But FACTOR is generic in the sense that any algebraic or Boolean procedures producing valid divisors and quotients can be used inside procedures DIVISOR and DIVIDE, respectively.

Certain refinements of FACTOR are needed to produce good results. We first use two examples to motivate the ideas behind the refinements. In each example, we list the original function F, the divisor D, the quotient Q, the initial remainder R, the partial factored form P, and the final factored form O given by FACTOR. It is important to point out that in the sequel, *discussions are restricted to algebraic operations only*.

Example:

$$\begin{aligned} F &= abc + abd + ae + af + g \\ D &= c + d \\ \text{FACTOR}(Q) &= ab \\ R &= ae + af + g \\ P &= ab(c + d) + ae + af + g \\ O &= ab(c + d) + a(e + f) + g \end{aligned}$$

■

Obviously, O is not optimal, because it is not maximally factored. It can be further factored to $a(b(c+d)+e+f)+g$.

The problem occurs when the quotient Q is a single cube, and some of the literals of Q also appear in the remainder R. To solve the problem we first check the quotient Q. If Q is a single cube, we pick a literal Q_1 of Q in the cube which occurs in the greatest number of cubes of F. We then divide F by Q_1, the chosen literal, to obtain a new divisor D_1. Now, F has a new partial factored form $Q_1D_1 + R_1$ and literal Q_1 does not appear in R_1. Notice that the new divisor D_1 has the original D as a divisor because Q_1 is a literal out of Q. When recursively factoring D_1, D will be discovered again.

Lemma 10.6.1 *If Q is a single cube, the procedure outlined above guarantees that the partial factored form $Q_1D_1 + R_1$ is maximally factored.*

Proof. Let $D_1 = cD_2$ where c is a common cube of D_1 and D_2 is cube-free. Now, the partial factored form is

$$F = Q_1cD_2 + R_1. \tag{10.7}$$

It is easy to see that every literal in c must be in the original cube Q (if the divisor D was cube-free). Since Q_1 is the literal of Q occurring in the greatest number of cubes of the original sum-of-products expression, no literal in c can occur in R_1. For otherwise, there would be more terms with that literal than terms with Q_1. The terms in D_2 come from terms of F that contain both Q_1 and that literal; $c \neq 1$ only if there are ties for best literal. In addition, D_1 was obtained by dividing F by literal Q_1, so Q_1 does not occur in R_1. Suppose D_2 is also a factor of R_1. Then the quotient of dividing F by D would have contained more than one cube since D_2 must contain D, which contradicts the fact that Q is a single cube. So, no factors of the product form Q_1cD_2 can be a divisor of R_1. Thus, the two sum terms have no common factor and the product terms Q_1, c, and D_2 have no common factor, since Q_1cD_2 is an algebraic product. F is by definition maximally factored at this level. □

Example:

$$\begin{aligned} F &= ace + ade + bce + bde + cf + df \\ D &= a + b \\ Q &= ce + de \\ P &= (ce+de)(a+b) + (c+d)f \\ O &= e(c+d)(a+b) + (c+d)f \end{aligned}$$

■

Again, the final factored form O is not maximally factored because $(c+d)$ is common to both products $e(c+d)(a+b)$ and $(c+d)f$. The final factored form should have been $(c+d)(e(a+b)+f)$.

The problem is that Q has a factor which is also a factor of R. The problem is solved by first making Q cube-free to get Q_1, then obtaining a new divisor D_1 by dividing F by Q_1. If D_1 is cube-free, we have obtained a partial factored form $F = Q_1D_1 + R_1$, and can recursively factor Q_1, D_1, and R_1. If D_1 is not cube-free, let $D_1 = cD_2$. Then, the partial factored form becomes $F = cQ_1D_2 + R_1$. Let $D_3 = Q_1D_2$, and we have a partial factoring $F = cD_3 + R_1$, which is the case illustrated by the previous example and can therefore be factored maximally. Therefore the solution is, if c exists, to take the most recurring literal in c and use that in recursively factoring the quotient and remainder.

Lemma 10.6.2 *If Q is not a single cube, the procedure outlined above maximally factors F at this level.*

Proof. Suppose D_1 is cube-free; then the partial factored form is $F = Q_1D_1 + R_1$. Here, Q_1 cannot be a factor of R_1 because Q_1 is used to obtain D_1 by dividing into F. D_1 cannot be a factor of R_1, because D_1 contains a factor D which, when dividing into F, gives quotient Q. So, the partial factored form is maximal at this level. If D_1 is not cube-free, then we have a partial factored form $F = cD_3 + R_1$ which is factored by our previous procedure. The previous lemma guarantees that the partial factored form obtained is maximal. □

Now, FACTOR can be improved by the procedures we have examined. The new routine is called GEN_FACTOR, which stands for **Generic Factoring**, and is shown in Figure 10.10. GEN_FACTOR takes as an input a SOP expression F, and two more parameters which are function names specifying how to find the initial divisor D and how to perform the division. As we shall see, by varying these two parameters, a spectrum of algorithms can be obtained with different run time versus quality trade-offs.

The function LF(F, C, DIVISOR, DIVIDE) is a variation of the literal factoring algorithm [32]: BEST_LITERAL selects a literal in C which occurs in the largest number of cubes of F. Instead of calling recursively LF on factors Q and R, we switch back to the generic factoring algorithm GEN_FACTOR. Then COMMON_CUBE(D) returns the largest common cube of D, and MAKE_CUBE_FREE(Q) eliminates the literals that appear in all the cubes of Q.

The following theorem shows that the results of GEN_FACTOR are always **maximally factored.**

Theorem 10.6.1 *If algebraic division is used for the* DIVIDE *operation, Algorithm* GEN_FACTOR *finds a maximally factored form.*

Proof. By induction on number of literals in the SOP form of F. If F has only one literal, it is obviously maximal. Suppose F has n literals, each of the factors passed to the next recursive call of GEN_FACTOR has no more than $n - 1$ literals, and, by the induction hypothesis GEN_FACTOR returns a maximally factored form. By the two previous lemmas, the factored form derived by GEN_FACTOR at this level is also maximal. So, the results of GEN_FACTOR are always maximally factored. □

Notice that this result uses algebraic division for DIVIDE everywhere. Whether the result extends to using Boolean division is still an open problem.

```
GEN_FACTOR(F, DIVISOR, DIVIDE) {
    D = DIVISOR(F)
    if (D = ∅) return (F)
    (Q, R) = DIVIDE(F, D)
    if (|Q| = 1) return (LF(F, Q, DIVISOR,DIVIDE))
    Q = MAKE_CUBE_FREE (Q)
    (D, R) = DIVIDE(F, Q)
    if (CUBE_FREE(D)) {
        Q = GEN_FACTOR(Q, DIVISOR , DIVIDE)
        D = GEN_FACTOR(D, DIVISOR, DIVIDE)
        R = GEN_FACTOR(R, DIVISOR, DIVIDE)
        return (Q)(D) + (R)
    }
    C = COMMON_CUBE(D);
    return (LF(F, C, DIVISOR, DIVIDE))
}

LF(F, C, DIVISOR, DIVIDE) {
    l = BEST_LITERAL(F, C)
    (Q, R) = DIVIDE(F, l)
    Q = GEN_FACTOR(Q, DIVISOR, DIVIDE)
    R = GEN_FACTOR(R, DIVISOR, DIVIDE)
    return (lQ + R)
}
```

Figure 10.10: Procedure GEN_FACTOR.

```
QUICK_FACTOR(F) {
    return (GEN_FACTOR(F), QUICK_DIVISOR, WEAK_DIV)
}

QUICK_DIVISOR(F) {
    if ( |F| ≤ 1 ) return (∅)
    if ( every literal of F appears exactly once ) return ∅
    return (ONE_LEVEL-0_KERNEL(F))
}

ONE_LEVEL-0_KERNEL(F) {
    if (every literal of F appears exactly once ) return (F)
    l = any literal appearing more than once in F
    F = MAKE_CUBE_FREE (F/l)
    return (ONE_LEVEL-0_KERNEL(F))
}
```

Figure 10.11: Procedures QUICK_FACTOR, QUICK_DIVISOR, and ONE_LEVEL-0_KERNEL.

Specific factoring algorithms are discussed in the succeeding sections, and are just instances of GEN_FACTOR with specific choices of DIVISOR and DIVIDE algorithms. Depending on a particular application of GEN_FACTOR, appropriate DIVISOR and DIVIDE algorithms are chosen to obtain desired run time versus quality trade-offs.

10.6.2 Quick Factor

The basic operations of GEN_FACTOR are DIVISOR and DIVIDE. In this section we introduce an algorithm which we call QUICK_FACTOR. QUICK_FACTOR is a version of GEN_FACTOR in which DIVISOR is replaced by QUICK_DIVISOR, which is a quick way of finding a useful divisor. Also, DIVIDE is replaced by WEAK_DIV, which is the algebraic division [38] operation. QUICK_FACTOR is defined in Figure 10.11. Also illustrated there is QUICK_DIVISOR, which is a simple modification of the algorithm used in [38] to find the kernels of a given algebraic expression. Also shown is the recursive subprocedure ONE_LEVEL-0_KERNEL , which quickly identifies just one level-0 kernel of an input algebraic expression. If F has only one term or is itself a level-0 kernel, then QUICK_DIVISOR returns $\emptyset$.[5] Otherwise, there is at least one multiple-cube divisor which is a level-0 kernel F and is not equal to F. This level-0 kernel is found by routine ONE_LEVEL-0_KERNEL, which arbitrarily picks a literal which appears more than once, divides F by the literal, and works recursively on the quotient. ONE_LEVEL-0_KERNEL terminates on the first level-0 kernel encountered, which occurs when every literal appears only once. This is a property of level-0 kernels, as one may easily verify from the definition.

[5] One may verify that it would also be correct to return F.

```
GOOD_FACTOR(F) {
    return (GEN_FACTOR(F, BEST_DIVISOR, WEAK_DIV))
}
```

BEST_DIVISOR returns the kernel with the highest factorization value.

Figure 10.12: Procedure for good factorization

10.6.3 Good Factor

Experiments have shown that QUICK_FACTOR is very fast and in many cases finds good factored forms. However, because QUICK_DIVISOR works only hard enough to find an arbitrary level-0 kernel, the results in some cases are not satisfactory. For example, in factoring the function

$$F = abg + acg + adf + aef + afg + bd + be + cd + ce,$$

QUICK_DIVISOR may have chosen a level-0 kernel $(d + e + g)$ which leads to the following factored form

$$F = a(g(b + c) + f(d + e + g)) + (d + e)(b + c)$$

which has 12 literals. However, if we spent more time to examine all the kernels and chose $(af + b + c)$ as the divisor, we would obtain a better factored form

$$F = (af + b + c)(d + e) + ag(f + c + b)$$

with 11 literals.

Good factoring (GOOD_FACTOR) tries to obtain a better result by working harder to find a good divisor to start with. In particular, GOOD_FACTOR(F) looks at all the kernels and picks one (K) which, when substituted into F, maximally reduces the total number of SOP literals of F and K. That is, the factorization value FACT_VAL is maximized (over the set of all kernels of F) by kernel K. This procedure is called BEST_KERNEL. GOOD_FACTOR is defined in Figure 10.12

10.6.4 Boolean Factor

Procedure GOOD_FACTOR can be further improved by replacing WEAK_DIV with **Boolean division**, which is performed in BOOL_DIV [39]. In the above example, GOOD_FACTOR obtained an 11 literal result. If BOOL_DIV were used to divide the divisor $(af+b+c)$ into F, the quotient and the remainder would have been $(ag+d+e)$ and 0, which would have lead to the following factored form

$$F = (af + b + c)(ag + d + e),$$

with only 8 literals. Although we do not show BOOL_DIV here, it is easily verified that this result is correct. Notice that this is no longer an algebraic factored form because of the term $afag$. This version of GEN_FACTOR is called **Boolean**

```
BOOL_FACTOR(F) {
    return (GEN_FACTOR(F, BEST_DIVISOR, BOOL_DIV))
}
```

Figure 10.13: Procedure BOOL_FACTOR.

factoring, and is given as BOOL_FACTOR in Figure 10.13. Since BOOL_DIV involves a two-level logic minimization step, it is a much more expensive operation than WEAK_DIV. Consequently, BOOL_FACTOR takes considerably more time than QUICK_FACTOR and GOOD_FACTOR. But, because it uses Boolean division, it is able in some cases to find some Boolean factored forms with significantly fewer literals than the alternative methods. This is shown by the following example, in which results for QUICK_FACTOR(F), GOOD_FACTOR(F), and BOOL_FACTOR(F) are given.

$$\begin{array}{rcll} F & = & abc'd' + abe'f' + a'b'cd + a'b'ef + cde'f' + c'd'ef & \text{24 literals} \\ \text{QUICK\_FACTOR}(F) & = & abc'd' + a'b'cd + ef(c'd' + a'b') + e'f'(cd + ab) & \text{20 literals} \\ \text{GOOD\_FACTOR}(F) & = & cde'f' + c'd'ef + a'b'(ef + cd) + ab(e'f' + c'd') & \text{20 literals} \\ \text{BF}(F) & = & (e'f' + c'd' + a'b')(ef + cd + ab) & \text{12 literals} \end{array}$$

It should be pointed out that this is a contrived example intended to show the potential power of Boolean factoring algorithms. In practice, most of the functions found in real circuits can be factored just as well by algebraic methods as by Boolean methods; algebraic methods are much faster.

10.6.5 Summary of Factoring Algorithms

To summarize, the factoring algorithms presented in this section are all based on a recursive paradigm: find first a divisor as the initial seed, and then use multiple divisions to try to improve the factors before recursively factoring them. All the algorithms use heuristics, i.e., at each step, the quality of the factoring is estimated by the literals in the sum-of-products form of each factor. Furthermore, because the initial divisors generated by DIVISOR are restricted to kernels only, the results of factoring largely depend on the initial sum-of-products forms. Most of these algorithms are implemented as part of the SIS program [249]. The experiments in Wang's thesis [268] show that the paradigm itself and the heuristics presented work well for most of the functions and the results are in general quite good.

This paradigm is evolved further by Wang to encompass "Complement Factoring," "Dual Factoring," and the factoring of incompletely specified function representations.

In multi-level logic optimization, the cost of a Boolean network has to be evaluated again and again. So, factoring algorithms are used constantly to estimate the number of literals in the factored forms of functions. In this application environment, the speed of the factoring algorithm is essential. QUICK_FACTOR seems to be particularly useful in this environment. In later stages of the optimization, a more accurate measure of network size and exact implementation of functions are needed. Therefore, more expensive algorithms such as GOOD_FACTOR or BOOL_FACTOR can be used. In fact, since all the algorithms presented are heuristic based, there is no theoretical guarantee that one can always outperform the others. So, when the quality of results is essential, one should try all the algorithms and keep the best result.

10.6.6 Rectangle Covering

The key problem in the algebraic operations presented above is the identification of a divisor. We have seen that kernels offer a good set of divisors both for factoring (or decomposition) and extraction. The problem of finding a kernel, and generally a common single- and multiple-cube divisor, can be reduced to a fairly general combinatorial optimization problem: Rectangle Covering [236, 33]. In addition to being elegant, this formulation favors the development of fast and effective algorithms, and is extensively used in the algebraic decomposition algorithms of SIS. Unfortunately, the rectangular covering problems is beyond the scope of this text. The interested reader is referred to the references.

10.7 Decomposition and Restructuring

10.7.1 Algebraic Resubstitution

The key operation in the process of algebraic substitution (or resubstitution) is the division of the cover F_i at node i in the network by the cover F_j or by F_j' at node j. During substitution, if F_j is an algebraic divisor of F_i, then F_i is transformed into

$$F_i = QF_j + R = QF_j + (Q_C F_j' + R_C). \tag{10.8}$$

Here the remainder R left after the division by F_j is, optionally, further decomposed through division by a representation, F_j', of the complement of F_j. Note that F_j' is different from the complement of the Boolean function F_j, which would require knowledge of the complete don't care set for node j, as will be seen in Chapter 11. This second division produces a quotient Q_C and remainder R_C.

For example, suppose a Boolean network contained nodes i and j with covers

$$\begin{aligned} F_i &= abd + cd + a'c'gh + b'c'gh + ef + ah, \\ F_j &= ab + c \\ F_j' &= a'c' + b'c'. \end{aligned}$$

Resubstitution of F_j into F_i would first obtain the quotient $Q = d$ and remainder $R = c'a'gh + c'b'gh + ef + ah$ and then $Q_C = gh$ and $R_C = ef + ah$, leading to the reduced expression

$$F_i = dF_j + (ghF_j' + ef + ah).$$

Note that this process results in a cover for node i with only 9 literals, whereas it formerly had 17. If the second division were omitted, F_i would have ended up with 14 literals. Note that this operation may increase the delay of the critical path in the network, so that potential area (literal) savings need to be weighed against possible performance degradation.

As typically implemented, the resubstitution process attempts this for each pair F_i, F_j in the Boolean network, seemingly implying as many as $2n(n-1)$ algebraic divisions, if there are n nodes in the network. However, observe that since the operation is algebraic, there is no point in attempting to divide F_i by any F_j which is the cover of a node j which is not in the direct fanout of one of the direct fanins of node i. Such considerations lead to a set of filters which are inexpensive to apply and yet are very successful in circumventing most of the useless divisions.

The filters may be summarized as follows. The function F_j is not an algebraic divisor of F_i if:

1. j is not a fanout of a fanin of i;

2. F_j contains a literal not in F_i;

3. F_j has more terms than F_i;

4. for any literal, the number of times it occurs in F_j exceeds that in F_i;

5. F_i is in the transitive fanin of F_j.

Notice that Filters 2 and 4 must be applied to F_j' as well. If F_i contains x but not x' and F_j contains x', F_j will not divide F_i, but F_j' still may. In some cases, we are not interested in the result of division if the quotient F_i/F_j is only a single cube. This can be detected by another useful filter: if for any literal, the count for F_j equals the count for F_i, then (F_i/F_j) is at most a single cube. In the above example literals b and c occur exactly once in both F_i and F_j, so $Q = F_i/F_j = d$ is a single cube. Resubstitution in SIS is performed by the `resub` command.

10.7.2 Selective Node Elimination

The key operation in the process of node elimination is the replacement of a literal y_j which appears in a cover F_i by the algebraic SOP expression F_j, and similarly for the literal y_j' and the cover complement F_j'. The elimination process is driven by a quantity known as the elimination value, e_value$_j$, of node j. This is defined as the increase in the total number of literals in the network if F_j were substituted into its fanout to replace literal y_j, and the ensuing product expanded to SOP form (and similarly for F_j'). An approximation to this, valid when all products are algebraic, may be given as follows.

We consider a function F_i in the fanout of F_j to be represented by

$$F_i = \tilde{F}_i y_j + \hat{F}_i y_j' + R_i. \tag{10.9}$$

We are interested in the increase in the literal count when we substitute the SOP representation of F_j into this representation of F_i. If we do this for every node in the direct fanout of F_j we will have eliminated node j from the network. Because a small savings of literals can come at the cost of increased levels of logic and thus increased delay, we need to be able to eliminate nodes with small value. Also, because the process of factoring a two-level network into a multi-level network is heuristic, it is useful to be able to iterate. In order to iterate we will have to partially collapse the function in order to try a different factorization/decomposition strategy. This partial collapsing is done by node elimination.

Thus we now consider a single node in the fanout of node j and determine the increase in the number of literals associated with node elimination. In order to proceed with this computation we must agree as to the measure or value of a node in a Boolean network. There are two possible choices. The number of literals in the SOP representation of the node or the number of literals in the factored form representation of the node. The factored form literal count is useful for two reasons. First, it more closely approximates the actual transistor count for the final implementation of the node. Second, the literal count for the factored form is the same as the literal count for the complement of the function. Since we use both the function and its complement in node elimination having both values readily available is useful.

Therefore we first let n_i be the number of times that either y_j or its complement appears in the factored form representation of F_i. Further, we let L_i be the number of literals in the factored form for F_i. Now we see that an approximation to the number of literals in the factored form of F_i after elimination is $n_i L_j - n_i$, where the first term approximates the increase by substituting in F_j and the second term accounts for the elimination of y_j and its complement from the expression for F_i. Now, to find the complete e_value (cost) of the elimination for the entire network we need to sum over the fanout of node j and then subtract out the literal count of node j which will be eliminated. This leads to the following expression

$$\text{e\_value}_j = \left(\sum_{i \in \text{ fanout of } j} n_i (L_j - 1) \right) - L_j \tag{10.10}$$

In the above resubstitution example, if we consider reversing the process and eliminating node j into node i the elimination value of node j is computed as follows.

$$\begin{aligned} L_j &= 3 \\ n_i &= 2 \end{aligned}$$

so, if i is the only node in the fanout of j, we have

$$\text{e\_value}_j = 2(3 - 1) - 3 = 1.$$

Note that the first equation accurately portrays the literal savings in the Boolean network accountable to node j, only if all the indicated products are algebraic. While this is indeed the case in the example given, this will not necessarily be the case in the context of multilevel logic optimization, in which Boolean operations may be freely intermixed with algebraic ones. In the more general case, it may be regarded as an approximation to the true elimination value.

As typically implemented, the elimination operation is an iterative process, carried out with respect to a specified value threshold, v. Thus the SIS command "`eliminate` v", applied to a given Boolean network, would consider all nodes in the network in a certain order. For each node j encountered, if $\text{e\_value}_j < v$, F_j would be substituted into its fanout, and then node j eliminated from the network. On each pass through the network, a flag is initially lowered and then raised if any node is thus eliminated.

```
QUICK_EXTRACTION(N) {
    apply QUICK_DECOMPOSITION to each node of the network
    perform all profitable pairwise algebraic substitutions
    eliminate all single literal functions
    eliminate all functions with small value
    return resulting modified Boolean network N+;
}
```

Figure 10.14: Procedure QUICK_EXTRACTION.

If the flag is still lowered at the end of any pass, then we know that no further elimination is possible, and the whole process terminates.

Iteration is necessary because subsequent node eliminations may affect the e_value of nodes already processed in the current pass. In the SIS program, the processing of nodes is done in depth first search (postorder), where the search is initialized at the primary inputs, and the elimination value of node j is attempted only after the conditional elimination of all nodes in the fanout of j has already been completed.

In the above example, node j would not be affected if $v \leq 0$, but if $v > 0$, the elimination operation on the Boolean network would put F_i back into its original form and eliminate node j, assuming node j has no other fanout.

10.7.3 Extraction

Extraction is the operation that identifies common subexpressions and manipulates the Boolean network accordingly. Algebraic decomposition and algebraic resubstitution and node elimination can be combined to provide an effective extraction algorithm.

In particular, suppose that we define a procedure QUICK_DECOMPOSITION , which applies QUICK_FACTOR to a given node and then creates a new node in the network for each new factor of this node. This provides a very fast method for breaking down a Boolean network. It may be combined with algebraic substitution and elimination to form a fast QUICK_EXTRACTION procedure as shown in Figure 10.14.

At the end of the QUICK_DECOMPOSITION step, each node of the network cannot be factored; each literal appears only once. Substitution identifies identical nodes and one is substituted into the other leaving a node whose logic function is a cube with a single literal. These are eliminated along with the nodes that have small value, typically those which do not fanout.

Example:

Consider a Boolean network with 2 nodes whose (algebraic) expressions are:

$$\begin{aligned} F_1 &= abc + abd + ae + af + g \\ F_2 &= ace + ade + bce + bde + cf + df \end{aligned}$$

Assuming that QUICK_DECOMPOSITION obtained the same factorization that these two expressions had in a previous example (see Section 10.6.1), the result of QUICK_DECOMPOSITION would be

$$\begin{aligned} O_1 &= aY_1 + g \\ Y_1 &= bX_1 + e + f \\ X_1 &= c + d \\ O_2 &= X_2 Z_2 \\ Z_2 &= eY_2 + f \\ Y_2 &= a + b \\ X_2 &= c + d, \end{aligned}$$

which would require 18 SOP form literals. Then algebraic resubstitution would recognize that $X_1 = X_2$, leading to

$$\begin{aligned} O_1 &= aY_1 + g \\ Y_1 &= bX_1 + e + f \\ X_1 &= c + d \\ O_2 &= X_1 Z_2 \\ Z_2 &= eY_2 + f \\ Y_2 &= a + b, \end{aligned}$$

which would require 16 SOP form literals. Finally, the elimination operation might eliminate the node producing Y_2, resulting in a final SOP form of

$$\begin{aligned} O_1 &= aY_1 + g \\ Y_1 &= bX_1 + e + f \\ X_1 &= c + d \\ O_2 &= X_1 Z_2 \\ Z_2 &= e(a + b) + f, \end{aligned}$$

which costs 16 SOP form literals (or 15 factored form literals). ■

The motivation behind this is that QUICK_FACTOR is very fast but still identifies good kernels to factor each single function well. The kernels become nodes of the Boolean network and algebraic substitution operation identifies common ones. Thus common divisors identified in this way are ones that are also near best for factoring. Of course, this is not always the best choice, and not all common divisors are found, but the method is very fast and the results quite good.

10.8 Notes

The paper by Brayton and McMullen [38] formed the basis for this chapter, and is of historical importance in VLSI CAD. Not only is it one of the most widely cited papers in the synthesis literature, but it described algorithms that perhaps implemented in the greatest number of practical tools, including the currently dominant tools worldwide: SIS and the SYNOPSYS design compiler.

Algebraic synthesis is surprisingly effective at area minimization for multilevel circuits (although far from perfect). An added bonus of algebraic minimization tools is their impact on **testability**. It was first shown by Morrison and Jacoby (their work on a 1986 conjecture by Karen Bartlett is reported in [129, 127, 128]) that two-level logic minimization, followed by algebraic factorization, provides and preserves an existing complete single stuck-fault test set. That is, the test vectors provided as by-products of the two-level minimization test for all stuck-faults. Further, these tests also test all multiple stuck faults. The algebraic factorization algorithms not only reduce circuit area and provide a good overall "structuring" of a given circuit, but maintain the applicability of the two-level test set. This led to the surprising capability of designing compact multilevel circuits that were 100% testable for all multiple stuck faults by (provided) test vectors.

10.9 Summary

In this chapter we have examined how both algebraic and Boolean factored forms can be used to compactly represent multilevel logic. The algebraic methods discussed are anomalous in the sense that they do not obey the laws of Boolean algebra of Chapter 3. Along the way, we have characterized Syntactically equivalent, maximally factored, and optimum factored forms.

We have presented the ideas and given illustrative examples for the elegant theory of kernels and co-kernels of [38], as expressed in Theorem 10.5.1 of Page 426. We have shown how kernels and co-kernels can be computed and used to root out any and all common subexpressions in the algebraic subexpressions implicitly present in two-level or multilevel logic. Methods have been given for computing all or part of the sets of kernels and co-kernels.

We have presented in detail the "recursive generic factoring" approach, which is the key idea in the SIS suite of algebraic synthesis tools. We have shown how the idea of "weak (algebraic) division" plays a key role in this approach, and have given many examples and problems which demonstrate and motivate these concepts.

10.10 Problems

1. For each of the following expressions, say whether they are algebraic, prime, irredundant.

   (a) $abc + a'b + bcd$;

   (b) $abc + abd + a'b'c$;

   (c) $abc + bcd + bd$.

   Solution.

   (a) $abc + a'b + bcd$ is algebraic because no term is included in another term. It is not prime, because $bcd \leq bc$ and bc is an implicant of the function.

It is not irredundant, because bcd can be dropped; it is contained in the consensus term of abc and $a'b$.

(b) $abc+abd+a'b'c$ is algebraic. It is prime, because no consensus terms exist among the product terms (hence, all terms are prime). It is irredundant, because all primes are essential.

(c) $abc + bcd + bd$. It is not algebraic, because $bcd \leq bd$. It is neither prime nor irredundant, because bcd is not prime and can be dropped.

□

2. Is the following formula a factored form?

$$((a+c)b' + (c+d)')ef'.$$

3. For each of the following factored forms, say whether they are algebraic.

(a) $ab + (c+d)(a+b')$;

(b) $ab'c + a'bc$;

(c) $c(a' + b + c') + bd$;

(d) $(a+bc)(d+ae)$.

4. Compute the factorization value for each of the following pairs of SOP and factored form.

$$\begin{array}{ll} ab + ac + ad + b'c + b'd & ab + (c+d)(a+b') \\ ad + ae + bcd & (a+bc)(d+ae) \end{array}$$

Solution. The factorization value for the first pair is given by:

$$10 - (2+2+2) = 4.$$

For the second pair, we have:

$$7 - (3+3+0) = 1.$$

□

5. Draw the factoring tree for $ab + (c+d)(a+b')$.
Solution. The tree is shown in Figure 10.15. □

6. Are the following pairs of factored forms equivalent? Are they syntactically equivalent?

$$\begin{array}{ll} a(b'+c') + a'(b+c) + a'c + b'c & (a+b+c)(a'+b'+c') \\ w(xy+z) + x'y' & x'y' + w(z+xy) \end{array}$$

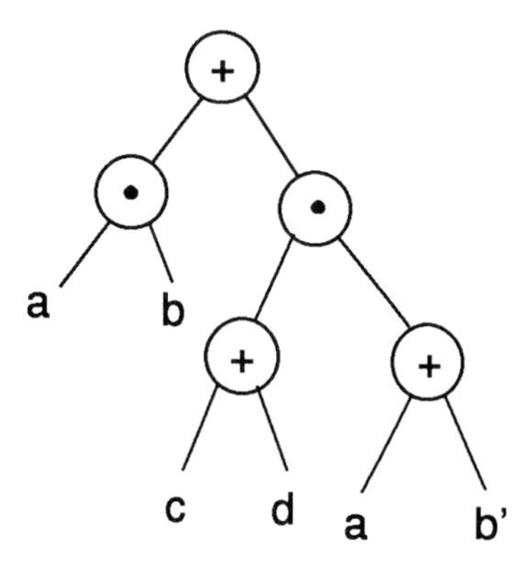

Figure 10.15: Factoring Tree for Problem 5.

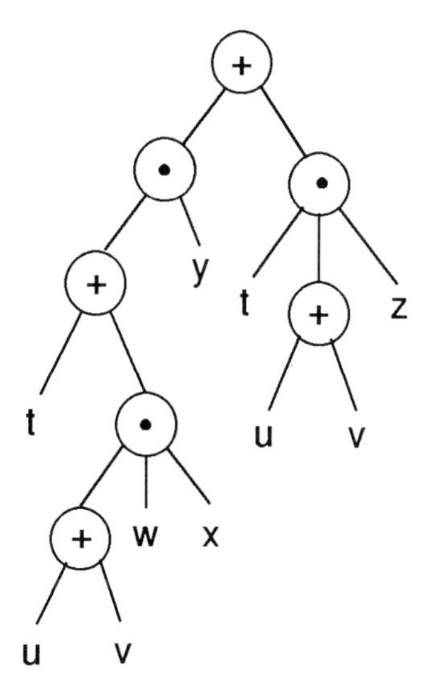

Figure 10.16: Factoring Tree for Problem 8.

7. For the SOP $F = abc+abe+ade+af+g$ and the divisor $G_2 = b(c+e)+de+f$

   (a) compute the factorization value;

   (b) show the factoring tree.

8. For the following factored form, draw the factoring tree and say whether the formula is maximally factored by applying Definition 10.3.6 to the tree.

$$(t + (u + v)wx)y + t(u + v)z.$$

Solution. The factoring tree is shown in Figure 10.16. The factored form is maximally factored, because there are syntactically equivalent factors ($u+v$ and t), but they do not appear as factors of two sums belonging to the same product of sums, or as factors of two products belonging to the same sum of products. □

9. Compute the positive and negative cofactors of $(t + (u + v)wx)y + t(u + v)z$ with respect to u.
   Solution. The positive cofactor is obtained by setting $u = 1$:

$$(t + wx)y + tz.$$

   The positive cofactor is obtained by setting $u = 0$:

$$(t + vwx)y + tvz.$$

□

10. Are the following two forms *syntactically* equivalent?

$$F_1 = (a + bc)(b + c) + d$$

$$F_2 = d + (a + b)(ab + c)$$

11. Use SIS to obtain a factored form for $ab' + cd + ad + ac$. Use the command `print_factor`, abbreviated `pf`. You may choose either BLIF or EQN as input format. Include your input file and the output from SIS.
    Solution. The input file in EQN format looks like this:

```
g = a*b' + c*d + a*d + a*c;
```

    The following is the output of SIS. when given the commands `ps`, `p`, and `pf`.

```
eqn2            pi= 4   po= 1   nodes=  1       latches= 0
lits(sop)=   8  lits(fac)=   6
     {g} = a b' + a c + a d + c d
     {g} = a (d + c + b') + c d
```

□

12. In this problem you are to run the multi-level synthesis tool SIS to find a good factored form. You may input your data in either EQN or BLIF format, but your output must be BLIF. Try to obtain the 23 literal result

$$(ab + gh)(rs + u + d + e) + r(t(ab + uws) + ghw) + dp + eq$$

    for the function of Problem 15. Give the SIS optimization directives you used, and the order in which you used them.

13. Divide

$$F = ab + ac + ad' + bc + bd'$$

    by $G = a + b$.

Solution.

f_j	p_i	V^a	V^b
ab	a	b	
ab	b		a
$\surd$ ac	a	c	
$\surd$ ad'	a	d'	
$\surd$ bc	b		c
$\surd$ bd'	b		d'

$$Q = V^a \cap V^b = c + d'$$
$$R = F - QD = ab$$

Cube ab, which is not marked, forms the remainder. We can therefore write $F = (c + d')(a + b) + ab$. □

14. Divide

$$F = aef + agh + ai + bef + bgh + bi + cdef + cdgh + cdi + abh + cei$$

by $G = a + b + cd$.

15. Apply the WEAK_DIV procedure to the algebraic expression F

$$F = abrs+abrt+abd+abe+abu+ghrs+ghrw+ghd+ghe+ghu+dp+eq+rstuw$$

and the divisor $D = ab + gh$. Obtain $Q = F/D$ and R.

Solution.

F	p_i	V^{ab}	V^{gh}
$\surd$ $abrs$	ab	rs	
$abrt$	ab	rt	
$\surd$ abd	ab	d	
$\surd$ abe	ab	e	
$\surd$ abu	ab	u	
$\surd$ $ghrs$	gh		rs
$ghrw$	gh		rw
$\surd$ ghd	gh		d
$\surd$ ghe	gh		e
$\surd$ ghu	gh		u
dp	–		
eq	–		
$rstuw$	–		

$$Q = V^{ab} \cap V^{gh} = rs + d + e + u$$
$$R = F - QD = abrt + ghrw + dp + eq + rstuw$$

In the table, the cubes that are not marked are those in the remainder. If we were to recur on R we would eventually reach the following factored form

$$(ab+gh)(rs+u+d+e)+r(t(ab+uws)+ghw)+dp+eq.$$

□

16. For $F = abc + abd' + a'bd + abef$ find the primary divisor corresponding to the cube ab. Is the result a kernel?

Solution. The result of dividing F by ab is $c + d' + ef$. Since this result is cube-free, it is a kernel. Notice that it is a kernel of level 0, because each literal appears exactly once. □

17. For the function $F = uwxy + uvxy + tx + tz + rsw + rsv + vwxy$

(a) compute the level-0 kernels;

(b) for each kernel of part (a), express F in $ck + r$ form, where c is the co-kernel, k is the kernel and r is the remainder.

Solution.

(a) In the following table, the null intersections are indicated by empty entries.

	$uwxy$	$uvxy$	tx	tz	rsw	rsv
$uvxy$	uxy					
tx	x	x				
tz			t			
rsw	w					
rsv		v			rs	
$vwxy$	wxy	vxy	x		w	v

After trying each co-kernel, we find that the level-0 kernels are $(v + w)$, $(u + v)$, $(u + w)$, and $(x + z)$.

(b)

$$\begin{aligned} F &= uxy(w+v)+tx+tz+rsw+rsv+vwxy \\ &= rs(w+v)+uwxy+uvxy+tx+tz+vwxy \\ &= wxy(u+v)+uvxy+tx+tz+rsw+rsv \\ &= vxy(u+w)+uwxy+tx+tz+rsw+rsv \\ &= t(x+z)+uwxy+uvxy+rsw+rsv+vwxy \end{aligned}$$

□

18. Compute all the level-0 co-kernels and kernels of the following function, by drawing the cube intersection table.

$$abc + abd' + bcd' + abe + bce + ade.$$

19. Apply procedure QUICK_FACTOR to the expression

$$F = bcd + abd + acd + de + bdf.$$

Solution. First, QUICK_FACTOR finds a divisor by calling QUICK_DIVISOR. QUICK_DIVISOR then calls ONE_LEVEL-0_KERNEL, which selects one literal that appears more than once in F. Suppose this literal is, in our case, b. It is then simple to verify that the chosen divisor is $c + a + f$. The choice of b is the best that QUICK_DIVISOR may make. Indeed, the following is the complete list of the kernels and co-kernels of F. (It can be verified by building the cube intersection table.)

$$\begin{aligned} F/d &= bc + ab + ac + e + bf \\ F/bd &= c + a + f \\ F/ad &= b + c \\ F/cd &= b + a \end{aligned}$$

Note that of the above kernels, the latter three are level-0. Of those, $a + c + f$ is the one with the most literals and hence the best factorization value.

Let us now summarize the recursive calls made by QUICK_FACTOR on our function. In the following,

$$\begin{aligned} F' &= bc + ab + ac + e + bf \\ F'' &= a + c + f \\ F''' &= ac + e \end{aligned}$$

QUICK_FACTOR(F) :	$D = \text{QUICK\_DIVISOR}(F) = a + c + f$
	$(Q, R) = \text{WEAK\_DIV}(F, D) = (bd, acd + de)$
	Q is a single cube so call literal factor routine
LF(F, Q) :	$l = \text{BEST\_LITERAL}(F, bd) = d$
	$(Q, R) = \text{WEAK\_DIV}(F, d) = (bc + ab + ac + e + bf, \emptyset)$
	now recur on Q, returning lQ
QUICK_FACTOR(F') :	$D' = \text{QUICK\_DIVISOR}(F') = a + c + f$
	$(Q', R') = \text{WEAK\_DIV}(F', D') = (b, ac + e)$
	Q' is a single cube so call literal factor routine
LF(F', Q') :	$l' = \text{BEST\_LITERAL}(F', b) = b$
	$(Q', R') = \text{WEAK\_DIV}(F', b) = (a + c + f, ac + e)$
	now recur on Q' and R', returning $l'Q' + R'$
QUICK_FACTOR(F'') :	$D'' = \text{QUICK\_DIVISOR}(F'') = \emptyset$ so return F''
QUICK_FACTOR(F''') :	$D''' = \text{QUICK\_DIVISOR}(F''') = \emptyset$ so return F'''

Returning through the entire recursion stack we get:

$$F = lQ = l(l'Q' + R') = d(b(a + c + f) + ac + e)$$

□

20. Apply Procedure QUICK_FACTOR to the expression

$$F = uwy + vwy + xy + uz + vz.$$

Show the calls made by QUICK_FACTOR. Assume that ONE_LEVEL-0_KERNEL returns w as the chosen literal.

QUICK_DIVISOR(F)=u+v

21. Repeat Problem 20, assuming this time that ONE_LEVEL-0_KERNEL returns u as the chosen literal.

22. For the following expressions, perform algebraic substitution of H into F and G. Don't forget to consider the complement of H.

$$\begin{aligned} F &= x'y'z + xyz + x'yz' + xy'z' \\ G &= x'yz + xy'z + xyz + xyz' \\ H &= x'y + xy' \end{aligned}$$

Solution.

$$\begin{aligned} F/H &= z' \\ F &= H(F/H) + x'y'z + xyz \quad = \quad H(F/H) + R_F \\ H' &= xy + x'y' \\ R_F/H' &= z \\ F &= z'H + zH' \qquad \leftarrow \\ G/H &= z \\ G &= H(G/H) + xyz + xyz' \quad = \quad H(G/H) + R_G \\ R_G/H' &= 0 \\ G &= zH + xyz + xyz' \qquad \leftarrow \end{aligned}$$

□

23. Perform resubstitution of

$$G = ab' + c$$

into

$$F = ab'd + cd + ab'fg + cfg + a'c'e + bc'e + aef.$$

24. For the Boolean network of Figure 10.17, say how many resubstitutions should be attempted, if the filters of Section 10.7.1 are applied. Perform the possible resubstitutions and draw the resulting Boolean network.

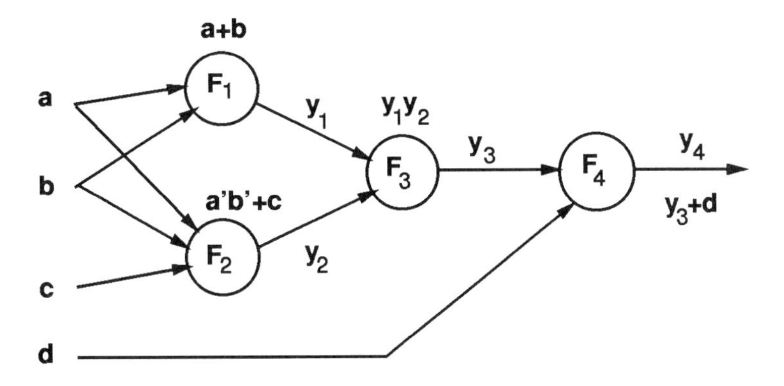

Figure 10.17: Boolean Network for Problem 24.

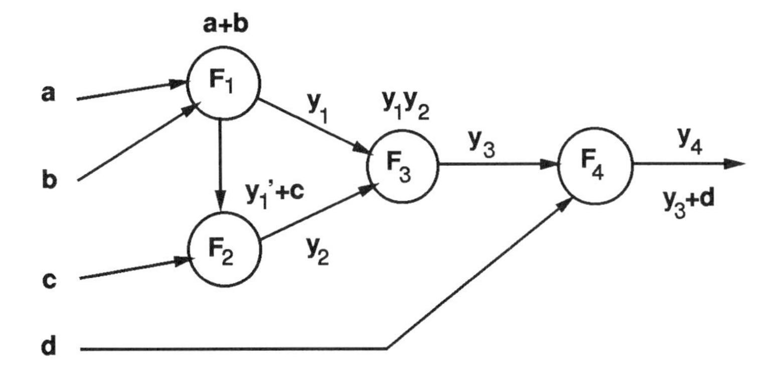

Figure 10.18: Boolean Network for Problem 24 after Resubstitution.

Solution. In solving this problem, let us keep in mind that resubstitution works with the local functions at the nodes—not the global functions. For instance, the function of F_3 is $y_1 y_2$, rather than $(a + b)c$.

There is a total of 12 pairs of functions to be considered: Each node with the other three. The first filter alone, however, eliminates all cases except those involving F_1 and F_2. For instance, we cannot resubstitute F_2 into F_3, because they have no inputs in common. It is then clear that neither F_2 nor F_2' divide F_1, because F_2 contains c, which does not appear in F_1. Similarly, F_1 does not divide F_2, because a and b do not appear in F_2. On the other hand, $F_1' = a'b'$ passes all the filters. Therefore, we attempt the division of F_2 by F_1'. The result is:

$$F_2 = y_1' + c.$$

The resulting Boolean network is shown in Figure 10.18. In summary, one substitution was possible. Notice that the network of Figure 10.18 can be further simplified. For instance, applying elimination or minimization would result in a smaller literal count. Also notice that resubstitution has in this case increased the number of levels in the network from three to four. □

25. Given

$$
\begin{aligned}
F_1 &= ab + c \\
F_2 &= y_1(d + ef) + ade \\
F_3 &= y_1 e + y_1'(de' + f)
\end{aligned}
$$

find the elimination value of F_1.

Solution. We have in this case $L_1 = 3$. The elimination value is given by:

$$\text{e\_value}_1 = 1 \cdot 2 + 2 \cdot 2 - 3 = 3.$$

This says that eliminating F_1 will increase the literal count by 3, if all products are algebraic. This is easily verified to be the case in this problem. □

26. Apply the extraction algorithm to:

$$
\begin{aligned}
F_1 &= abde + abfg + cde + cfg + aef + bgh \\
F_2 &= abdf + abeh + cdf + ceh + abd'h' + cd'h'
\end{aligned}
$$

Specifically, show the decomposition of F_1 and F_2. (Assume that ONE_LEVEL-0_KERNEL selects literals appearing more than once in alphabetic order: For instance, if both a and b appear more than once, a is chosen.) Indicate what resubstitution are made. Use 0 as threshold for elimination. Finally show the resulting Boolean network, indicating for each node its function.

Solution. We begin by factoring F_1. The level-0 kernel chosen by QUICK_DIVISOR is $de + fg$. Division yields:

$$F_1 = (ab + c)(de + fg) + aef + bgh.$$

This is maximally factored and is the result returned by QUICK_FACTOR. Proceeding similarly for F_2, we get $df + eh$ as divisor and

$$F_2 = (ab + c)(df + eh + d'h')$$

as maximal factorization.

Decomposition creates one node for each factor of F_1 and F_2.

$$
\begin{aligned}
F_1 &= X_1 X_2 + aef + bgh \\
X_1 &= ab + c \\
X_2 &= de + fg \\
F_2 &= Y_1 Y_2 \\
Y_1 &= ab + c \\
Y_2 &= df + eh + d'h'
\end{aligned}
$$

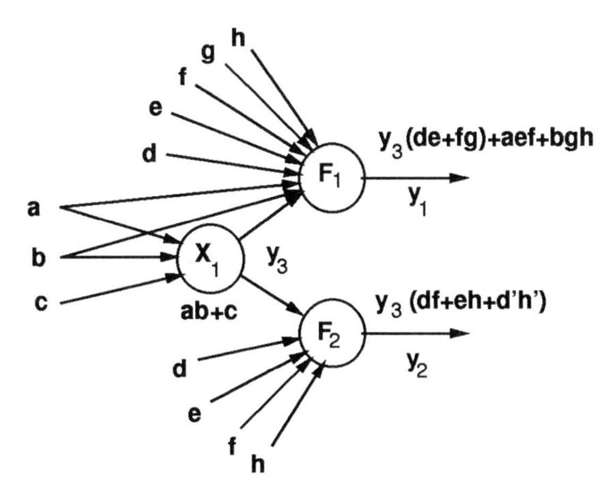

Figure 10.19: Boolean Network for Problem 26 after Extraction.

Resubstitution now identifies that $X_1 = Y_1$. Let us assume that X_1 is resubstituted into Y_1. Y_1 is then reduced to a single literal function which is eliminated.

$$\begin{aligned} F_1 &= X_1X_2 + aef + bgh \\ X_1 &= ab + c \\ X_2 &= de + fg \\ F_2 &= X_1Y_2 \\ Y_2 &= df + eh + d'h' \end{aligned}$$

Finally, we proceed to elimination. The nodes X_2 and Y_2 have elimination value -1; therefore, they are eliminated. The resulting equations are:

$$\begin{aligned} F_1 &= X_1(de + fg) + aef + bgh \\ X_1 &= ab + c \\ F_2 &= X_1(df + eh + d'h') \end{aligned}$$

The Boolean network is shown in Figure 10.19. As a result of extraction, the number of literals in factored form goes from 22 to 21. □

27. In this problem you will use SIS to optimize `rd53.pla`, a benchmark circuit which you can find in

```
.../sis/ex/comb/mcnc91/tlex/rd53.pla
```

Try the scripts that you will find in

```
.../sis/sis_lib
```

(You do not need to specify the path to use these scripts from SIS, because they are in the standard library.) Find the script that gives the least number of literals in factored form. You can try to find a literal count in factored form smaller than the best literal count given by any standard script. You can do that by combining scripts and plain SIS commands. Document your work.

Solution. The scripts we need to consider are only those concerned with technology-independent optimization. If we run them, we get the following results.

```
sis> rp rd53.pla
sis> ps
rd53.pla        pi= 5   po= 3   nodes=  3       latches= 0
lits(sop)= 144  lits(fac)=  71
sis> so script.algebraic
sis> ps
rd53.pla        pi= 5   po= 3   nodes= 17       latches= 0
lits(sop)=  65  lits(fac)=  65

sis> rp rd53.pla
sis> so script.boolean
sis> ps
rd53.pla        pi= 5   po= 3   nodes=  7       latches= 0
lits(sop)=  47  lits(fac)=  37

sis> rp rd53.pla
sis> so script.rugged
sis> ps
rd53.pla        pi= 5   po= 3   nodes=  8       latches= 0
lits(sop)=  42  lits(fac)=  37

sis> rp rd53.pla
sis> so script
sis> ps
rd53.pla        pi= 5   po= 3   nodes=  7       latches= 0
lits(sop)=  47  lits(fac)=  37

sis> rp rd53.pla
sis> so script.espresso
sis> ps
rd53.pla        pi= 5   po= 3   nodes= 36       latches= 0
lits(sop)= 207  lits(fac)= 123
```

The best result for the scripts is 37 literals in factored form. There are several ways of improving on that result. One consists of repeatedly applying the same script:

```
sis> rp rd53.pla
sis> ps
rd53.pla        pi= 5   po= 3   nodes=  3       latches= 0
```

```
lits(sop)= 144  lits(fac)=  71
sis> so script.rugged
sis> ps
rd53.pla        pi= 5   po= 3   nodes=  8       latches= 0
lits(sop)=  42  lits(fac)=  37
sis> so script.rugged
sis> ps
rd53.pla        pi= 5   po= 3   nodes=  7       latches= 0
lits(sop)=  43  lits(fac)=  36
sis> so script.rugged
sis> ps
rd53.pla        pi= 5   po= 3   nodes=  8       latches= 0
lits(sop)=  42  lits(fac)=  36
sis> so script.rugged
sis> ps
rd53.pla        pi= 5   po= 3   nodes=  8       latches= 0
lits(sop)=  42  lits(fac)=  36
```

After the fourth application there is no change in either literals in SOP form or literals in factored form. Hence, we decide to stop. □

Chapter 11

Multi-Level Minimization

11.1 Introduction

In Chapter 10 we have examined the restructuring of a multi-level circuit. Sometimes restructuring is applied to an initial two-level representation of the circuit to extract a multi-level network. More often, especially in the case of large circuits, the initial circuit is also multi-level and is derived by a description in some HDL. (See Section 2.2.) In both cases, after restructuring, a designer normally tries to improve the resulting multi-level circuit by local optimization. It may be the case, for instance, that the function attached to a given node may be simplified, leading to a smaller and more testable circuit. The transformation may also reduce the delay of the circuit, by reducing the capacitive loads (though there are exceptions to this). The changes to the network caused by local transformations may set the stage for further profitable restructuring, leading to an iterative process.

Technology-independent local optimization is often referred to as *multi-level minimization*, because of its close relationship with two-level minimization techniques. We shall see how the notions of **primality**, **irredundancy**, and **don't care conditions** extend to the multi-level world. Primality and irredundancy will allow us to define (local) optimality for a multi-level circuit; they also provide the link between the optimality of a network and its **testability**.

The study of don't cares in multi-level networks is central to multi-level minimization. Unlike two-level circuits, don't care conditions are both external (i.e., provided as part of the problem) and internal (due to the structure of the circuit). We shall examine the origin, formulation, and computation of these don't care conditions, thus filling out the details missing from the brief description of don't care sets in terms of intervals in a Boolean Function algebra (Section 3.4). We shall also see the limits of the idea of don't care conditions that are exposed when we deal with multi-level networks.

To deal with local transformations, a view of the circuit structure is needed that is more refined than that required in the case of algebraic restructuring. This is because the computation of the internal don't cares relies on the network structure. Therefore, we start this chapter with a detailed definition of what has come to be called the **Boolean network**.

11.2 Boolean Networks

In the following definition and in the sequel, we reserve boldface type for vectors of Boolean functions (lower case) for representations of functions (upper case).

Definition 11.2.1 *A* Boolean network *is an interconnection of n Boolean functions defined by a five-tuple* $(\mathbf{f}, \mathbf{y}, \mathbf{I}, \mathbf{O}, \mathbf{d^X})$, *consisting of*

1. $\mathbf{f} = (f_1, f_2, \ldots, f_n)$ – *a vector of completely-specified logic functions ("gates" of the network).*

2. $\mathbf{y} = (y_1, y_2, \ldots, y_n)$ – *a vector of logic variables (signals of the network) that are in one-to-one correspondence with* $\mathbf{f}$. *(y_i corresponds to f_i.)*

3. $\mathbf{I} = (I_1, I_2, \ldots, I_p)$ – *a vector (ordered set) of y-indices that identify the corresponding p externally controllable signals as primary inputs.*

4. $\mathbf{O} = (O_1, O_2, \ldots, O_q)$ – *a vector (ordered set) of y-indices that identify the corresponding q externally observable signals as primary outputs.*

5. $\mathbf{d^X} = (d_1^X, d_2^X, \ldots, d_q^X)$ – *a vector of completely-specified logic functions that specify the set of don't care minterms on the outputs of* η.

In order to avoid subscripted subscripts, it is convenient to also think of the ordered sets $\mathbf{I}$ and $\mathbf{O}$ as functions. For example, $I(i) = I_i$ is the ith element of the set $\mathbf{I}$, $i = 1, 2, \ldots, p$. We shall use the vector $\mathbf{x}$ as a synonym for the I-components of y, that is, $x_i = y_{I_i} = y_{I(i)}$, $i = 1, 2 \ldots, p$ or, more compactly $\mathbf{x} = \mathbf{y_I}$. Similarly, we shall use the vector $\mathbf{z}$ as a synonym for the O-components of $\mathbf{y}$, that is, $z_i = y_{O_i} = y_{O(i)}$, $i = 1, 2, \ldots, q$ or, more compactly $\mathbf{z} = \mathbf{y_O}$. We shall reserve the symbols n , p , q for the lengths of the vector $\mathbf{y}$, $\mathbf{x}$, and $\mathbf{z}$, respectively.

The supports of the functions f implicitly define a **digraph**, or directed graph $G^\eta = (V^\eta, E^\eta)$ for which there is a one-to-one correspondence between the vertices (nodes) $v_i \in V^\eta$ and both the component function f_i of $\mathbf{f}$ and the component signal y_i of $\mathbf{y}$. In the graph, a directed edge (or arc) connects node i to node j if $y_i \in$ SUPP(f_j), where SUPP(f_j) is the set of variables on which f_j depends. Since there is one node for each function in the network $|V^\eta| = n$; we shall reserve m for the number of edges in the network, that is, $|E^\eta| = m$.

For combinational logic, the digraph G^η is usually a DAG or directed acyclic graph, but in theory this is not necessarily the case. Our primary interest is in combinational digital circuits. A digital circuit is combinational if its output response depends only on the current input pattern, and not on any previous input patterns. A sufficient condition for a Boolean network to represent a combinational digital circuit is that the Boolean network graph G^η be acyclic. Note that this condition is not necessary; there exist combinational circuits that are represented by cyclic Boolean network graphs; furthermore, there exist Boolean functions for which the cyclic Boolean network is optimum, with an appropriate definition of optimality [55]. However, because of electrical considerations such as noise-margin, switching time, and settling time cyclic combinational circuits are only of academic interest for modern technologies (e.g., ECL and CMOS). Therefore, we assume from now on that the Boolean network graph is an acyclic graph.

Adjacency and path relations are defined in terms of the digraph G^η. A node i is a *predecessor* or *input* or *fanin* of a node j if $(i,j) \in E^\eta$. The set of all predecessors of node j is denoted by the set P_j; accordingly, $\text{SUPP}(f_j) = \{y_i | i \in P_j\}$. A node j is a *successor* or *output* or *fanout* of a node i if $(i,j) \in E^\eta$. The set of all successors of node j is denoted by S_j. If there is a path from node i to node j in G^η, i is called *transitive predecessor* or **transitive fanin** of node j, and the set of all transitive predecessors of node j is denoted by P_i^*. If there is a path from node i to node j in G^η, j is called *transitive successor* or **transitive fanout** of node i, and the set of all transitive successors of node i is denoted by S_i^*. By convention, i is not a member of any of the set P_i^*, S_i^*, P_i, or S_i. It is convenient to think of S_j, S_j^*, P_j, and P_j^* as functions. For example, $S_j(i)$ is the i-th element of S_j, $i = 1, 2, \ldots, |S_j|$.

Note that $f_{I(i)}$ is not defined because by definition, $f_{I(i)}$ is an external logic function which makes $x_i = y_i(i)$ externally controllable. Thus $I(i)$ is a source node of G^η, that is, a node for which $P(i) = \emptyset$. Similarly, the variables z_i in the vector $\mathbf{z}$ are externally observable. By definition, the functions $f_{O(i)}$, $i = 1, \ldots, m$ must be literal-functions i.e., $f_{O(i)} = y_j$ for some j. Further the variables $y_{O(i)}, i = 1, \ldots, m$ must not be used by any other function. That is, $y_i \notin \text{SUPP}(f_j)$, $\forall i \in O, j \in 1, 2, \ldots, n$. If $i \in O$, i is by definition a sink node of G^η, that is, a node for which $S(i) = \emptyset$.

Each signal in a Boolean network represents the voltage on a segment of interconnect in the circuit implementing the Boolean network. This wire segment is called a *net*. A net has exactly one logical signal, which corresponds to a specific node signal y_i in the Boolean network. This signal is called the *source terminal*, which determines the logic value on the net. The rest of the net provides interconnect to the *sink terminals*, which are the inputs to the nodes in the fanout S_i. The set of all source and sink terminals for all nets is also called the *pins* of the net. Thus, the network η has one *output pin* for each node in G^η, and one *input pin* for each edge of G^η (the pin corresponds to the head of the directed edge).

We shall also refer to each of the m edges i, j as a *connection*, denoted c_{ij}. Each connection has an associated logic variable y_{ij}, a single fanout $S_{ij} = j$, and a single fanin $P_{ij} = i$. Clearly $S_{ij}^* = S_j^* \cup \{j\}$, and $P_{ij}^* = P_j^* \cup \{i\}$. Although in a fault-free network $y_{ij}(x) = y_i(x)$, $\forall x \in B^p$, it is useful in network optimization and testing to maintain this distinction. Note that strictly speaking, the connections $c_{ij}, j \in O$ are sufficient to represent connections to the primary outputs, that is, the buffer functions $f_j, j \in O$ are not strictly needed in the definition.

We note that the network functions and connections give rise to a set of don't care conditions represented by the implicitly defined vector of don't care functions $\mathbf{d} = (d_1, d_2, \ldots, d_n)$ whose components are in one to one correspondence with those of $\mathbf{f}$. These don't care sets are an implicit property of the specified Boolean network, and are characterized below. Together, the specified function vector $\mathbf{f}$ and the implicit don't care function vector $\mathbf{d}$ combine to define a vector $\mathbf{ff}$ of incompletely specified functions

$$\mathit{ff}_i = (f_i d_i', d_i, r_i), \quad i = 1, \ldots, n, \tag{11.1}$$

where

$$r_i = (f_i + d_i)'. \tag{11.2}$$

We shall see that each connection $c_{ij} \in \eta$ also has its own unique associated don't care set d_{ij}, giving rise to the incompletely specified connection functions

$$ff_{ij} = (f_i d'_{ij}, d_{ij}, r_{ij}), \quad i = 1, \ldots, n, \quad j \in S(i) \tag{11.3}$$

where

$$r_{ij} = (f_i + d_{ij})'. \tag{11.4}$$

Note that the functions f_i are *local functions* in that they are the local function attribute of node i given in the specification of η. We also associate with node i a global function $f_i^*(x)$ whose support is a subset of the primary inputs of η, and which satisfies $f_i^*(x) = \gamma(I, f_i(y))$. Here γ is a *composition operator*, which satisfies the recursive definition

$$\gamma(A, f_j(Y)) = \begin{cases} y_j & \text{if} \quad j \in A, \\ f_j & \text{if} \quad P_j \subseteq A, \\ f_j(\gamma(A, f_{P_j(1)})), \gamma(A, f_{P_j(2)}), \ldots, \gamma(A, f_{P_j(|P_j|)}) & \text{else.} \end{cases} \tag{11.5}$$

When $A = I$, the result is the global function f_i^*.

Note $f^*(x) = f(y(x))$. This reflects the fact that for each possible input vector $\mathbf{x}$, there is a unique corresponding signal vector $\mathbf{y}$. Note that for $i = 1, \ldots, n$, f_i^* is not part of the specification of the Boolean network but can be derived by **flattening** (or *composing* or *eliminating*) all function nodes in the transitive fanin P_i^*.

In the sequel we shall use the five-tuple $\mathbf{N} = (\mathbf{F}, \mathbf{y}, \mathbf{I}, \mathbf{O}, \mathbf{D^X})$, to denote a representation of a Boolean network, where $\mathbf{F}$ and $\mathbf{D^X}$ are representations of $\mathbf{f}$ and $\mathbf{d^X}$.

Figure 11.1 shows a 2-bit full adder circuit. The first input is the carry-in, the next two are the least significant bits of the input words, and the last two the most significant bits. The 3 outputs are the least and most significant output bits, and the carry-out, respectively. This is a so-called *iterative network*, since the subnetwork for the second bit addend and carry-out is identical to subnetwork for the first. For this network, we have $\mathbf{I} = \{1, 2, 3, 4, 5\}$, $\mathbf{x} = (y_1, y_2, y_3, y_4, y_5)$, $\mathbf{O} = \{18, 19, 20\}$, $\mathbf{z} = (y_{18}, y_{19}, y_{20})$, and $V^\eta = \{1, 2, \ldots, 20\}$. Note that XOR gate 12 produces logical signal y_{12}. The output of this gate is the source terminal of the corresponding net, which has a single sink terminal on the input to output buffer 18. Note that the variables associated with the node indices marked by prime (in the figure only) are "external" to the Boolean network under consideration, and cannot be referred to, that is, cannot be in the support of any of the functions f_i. Nevertheless, the literal-functions $f_{O(i)} = y_{P(O(i))}$ (the output buffers) are part of the network, and these nodes are properly in the fanout of the nodes $P(O(i))$. The successor and predecessor sets include

$$\begin{array}{llllllll}
P_{12} & = \{1,6\}, & P_{12}^* & = \{1,2,6,3\}, & S_{18} & = \emptyset, & S_{18}^* & = \emptyset \\
S_{12} & = \{18\}, & S_{12}^* & = \{18\}, & P_{18} & = \{12\}, & P_{18}^* & = \{1\text{–}3,6,12\} \\
P_{14} & = \{8,13\}, & P_{14}^* & = \{1\text{–}3,7,8,13\}, & P_3 & = \emptyset, & P_3^* & = \emptyset \\
S_{14} & = \{15,16\}, & P_{14}^* & = \{15\text{–}17,19,20\}, & S_3 & = \{6\text{–}8\}, & S_3^* & = \{6\text{–}8,12\text{–}20\}
\end{array}$$

The global functions include

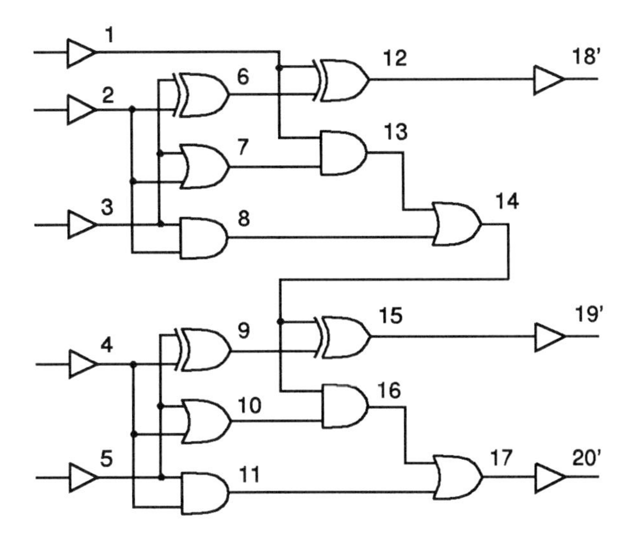

Figure 11.1: Example for Boolean Network. (Input and Output Elements are Buffers and are Considered Part of the Network.)

$$\begin{array}{rclcl} f_3^* & = & \gamma(I, f_3) & = & y_3 \\ f_{12}^* & = & \mathrm{XOR}(\gamma(I, f_6), \gamma(I, f_1)) & = & XOR(XOR(y_2, y_3), y_1) \\ f_{14}^* & = & \gamma(I, f_8) + \gamma(I, f_{13}) & = & x_2 x_3 + x_1(x_2 + x_3) \end{array}$$

Note that f_{12}^* gives the least significant sum bit (3-bit XOR), and f_{14}^* gives the internal carry-bit (majority function). The implicit don't care sets for the nodes and connections include

$$\begin{array}{rcl} d_{10} & = & x_4 x_5 + x_1' x_2' + x_2' x_3' + x_1' x_3' \\ d_{14,16} & = & x_4 x_5 + x_4' x_5' \\ d_{17} & = & \emptyset \end{array}$$

Derivation of these don't care sets will be given later. However, it is easy to verify that the indicated minterms are don't care. Consider, for example, d_{10}. If $x_4 = x_5 = 1, y_{11} = 1$, so $y_{17} = 1$, *independent* of the value of y_{10}. Similarly, if any two of the set $\{y_1, y_2, y_3\}$ are 0 the carry input from the first bit will be 0, so $y_{16} = 0$ *independent* of y_{10}.

Note that in this simple example, the fanins of the primary output buffers have only a single fanout. Exceptions to this are numerous in practice.

11.2.1 Network Cost

The logic functions in the Boolean network can be represented in many ways. In the sequel we use both the sum-of-product form and the factored form. The sum of products is the standard representation for a two-level logic function, typically a *minimal* (i.e., prime and irredundant) two-level representation is used. As discussed in Chapter 10, a factored form is a tree representation of a logic function using the operators AND (denoted by a dot or by concatenation), OR (denoted by +), and NOT (denoted by either a prime or an overbar and used only to produce complemented literals). More precisely, a factored form was defined in Section 10.3.1 as either a literal, or sum or product of factored forms.

The *literals-in-sum-of-products-form* cost function for a Boolean network is the sum over all nodes of the number of literals in the SOP representation for the function at each node.

The *literals-in-factored-form* cost function for a network is the number of literals in an optimal factored form for each expression in the network. The optimization problem of deriving an optimal factored form for a Boolean function is called *factoring* and was discussed in Section 10.6.

An important contrast between two-level and multi-level minimization is that in two-level minimization, the SOP form was the natural method of representation because it directly described how the function could be implemented as a PLA. This is no longer the case for multi-level logic. As an example, consider the cover

$$G = ad + ae + bd + be + cd + ce.$$

This has 12 literals, but it can also be represented as

$$G = (a + b + c)(d + e),$$

which has only 5 literals. Multi-level implementation of this in terms of NOR or NAND gates would be

$$\begin{array}{rcl} y_1 &=& \text{NOR}(a, b, c) \\ y_2 &=& \text{NOR}(d, e) \\ y_3 &=& \text{NOR}(y_1, y_2) \end{array} \qquad \text{or,} \qquad \begin{array}{rcl} y_1 &=& \text{NAND}(a, d) \\ y_2 &=& \text{NAND}(a, e) \\ y_3 &=& \text{NAND}(b, d) \\ y_4 &=& \text{NAND}(b, e) \\ y_5 &=& \text{NAND}(c, d) \\ y_6 &=& \text{NAND}(c, e) \\ y_7 &=& \text{NAND}(y_1, y_2, y_3, y_4, y_5, y_6), \end{array}$$

The NOR representation requires 7 literals, whereas the NAND representation, which basically reproduces the structure of the original SOP form, requires 12+6=18 literals. Thus the initial SOP representation for G is not a good indication of the ultimate complexity of its implementation. This can also be seen by considering the complement of G in SOP form

$$G' = a'b'c' + d'e'$$

and noting that the only difference in implementing G or G' is the cost of one inverter. Rudell has shown that for multi-level logic the factored (e.g., $G = (a + b + c)(d + e)$) form literal count correlates very well with final placed and routed chip area [156].

Because of the specification of external don't care sets, a Boolean network is a representation (or an implementation) of a set of incompletely specified Boolean functions, in the same way that a PLA or SOP form with output don't cares is. For multi-level minimization, we have several objectives. One is to minimize area, and a measure that seems to be well correlated with this is the total number of literals in all the function representations at the nodes. However, this measure takes no account of potentially large numbers of inverters which appear in the circuit during techmapping, so another measure of interest is the number of edges in a *NOR decomposition* of the network (or, equivalently, *NAND*), where the NOR gates have appropriately bounded fanin and fanout counts. Because this cost measure fully accounts for inverters in the Boolean network, and the factored form measure ignores inverters, it suffers less of a discontinuity between the pre- and post-techmapping cost.

Another measure of great interest is the delay through the network. In general, one is interested in implementing a set of functions that meet certain delay constraints while minimizing area. Note that the primary objective for two-level minimization, the number of cubes in the representation, is of interest for multi-level only in so far as it tracks the total number of literals.

11.3 Don't Cares in Multi-Level Networks

Don't cares conditions that arise in the design of digital systems are classified according to their origin into two classes.

11.3.1 Satisfiability Don't Cares

Satisfiability don't cares occur because there are input combinations to a circuit that can never occur. These may arise because of the way the digital system is specified, e.g., in a microprocessor design certain instruction codes may not be used and therefore will never occur in a valid input. Another example occurs when one block of combinational logic is the input to another. The first block may have output bit patterns that will never occur because of the type of logic function being implemented. Since these outputs are inputs to the next block, then the bit patterns that do not occur are don't cares for the second block of logic. In either case, we can interpret the bit patterns that never occur as Boolean vectors corresponding to "states" of the network that can never be reached. In the testing literature, these states are referred to as *non-controllable states* or as *non-justifiable states*.

Because of the fundamental connection to the satisfiability question, don't care conditions of this type will be referred to generically by labels that include the acronym "SDC," for Satisfiability Don't Care set, and similarly the acronym "ODC" will be used for Observability Don't Care set. However, the actual don't care sets themselves will be denoted by lower case function names, $\bar{s}$, and $\bar{o}$, respectively. Note that we use the overbar indicating complementation, because the don't care conditions correspond to non-satisfiability or non-observability.

Refining our treatment, we notice that signal vectors $\mathbf{y} \in B^n$ exist that do not satisfy all the Boolean equations $y_i = f_i(y)$. In fact, there are 2^n possible signal vectors, but of these only 2^p can occur for a specified set of Boolean equations. This is true because the response of a network to *primary input minterm* (one in which every primary input has a specified binary value) is unique. By an input vector we mean the vector $\mathbf{x}$, with $x_{I(i)} \in B, \quad i = 1, \ldots, p$, that is, with each component set to some specific value in $B = \{0, 1\}$. Recall that each node of a Boolean network is associated with two functions—a local function $f_j : B^n \to B$ and a global function $f_j^* : B^p \to B$. We shall refer to the space B^p as the *primary input space* and B^n as the *extended space* $(n \geq p)$. Sometimes it is convenient discuss local functions in terms of their local support, and still assume that all Boolean equations in the network are satisfied. The signal vectors that cannot occur are called *satisfiability don't care conditions*. Since we can think of these also as non-controllable states, there is a clear connection to the concept of *controllability*, which plays a key role in the testing literature.

11.3.2 Observability Don't Cares

The second class of don't care conditions arises because of filtering effects that prevent local changes to the network from being observable at the primary outputs of the network. These conditions are called *observability don't cares*. Again there is a profound connection to testing, since observability measures play a key role in ATPG (Automatic Test Pattern Generation) algorithms.

This close interaction between testing and logic synthesis areas will be repeatedly observed in the sequel. (See Chapter 12.) As we shall see, it is impossible to declare a network area-optimal without first implicitly deriving tests for all single stuck-at faults.

Observability don't cares occur because of the way an output is used. It may happen that because of the circuitry that fans out from a set of signals, perturbations in the signal values in this set cannot be observed (that is, they have no effect) on signals at the pre-specified observation points (true primary outputs). In this case the conditions under which the signals cannot be observed are don't cares for the functions producing the perturbed signals. In the example of one combinational logic block feeding another, the second block serves as a filter for the first and can cause non-observability of some of the outputs under certain input conditions. For example, suppose that we have two blocks of logic, the first computing a data flow function, and the second implementing an enable signal that controls whether the data flow result is latched at the outputs. Clearly the output of the data flow function is not observable under the conditions that disable the latch. Thus these are observability don't care conditions for the data flow logic block. In the parlance of the testing literature, one says that under these conditions, perturbations (for example, due to stuck-at faults) of the data flow logic are not able to *propagate*.

We shall show that both primary output and internal nodes of a Boolean network are, in the most general case, incompletely specified functions, with SDC and ODC contributions to their don't care sets.

11.3.3 Use of Don't Cares in Minimization

When we simplify an intermediate node v_i of a multi-level Boolean network using a two-level minimizer, we sometimes use only the satisfiability don't care sets for a subset of the nodes that can be substituted with "high probability" into the node being optimized. An optimal two-level minimizer substitutes some set of variables, corresponding to the nodes of the network, into f_i that results in a minimal literal count for f_i. For example, let

$$\begin{aligned} t &= sk' + s'abcd + s'a'b'cd \\ k &= ab + a'b' \end{aligned}$$

$$
\begin{aligned}
s &= ef + e'f' \\
r &= cd
\end{aligned}
$$

where a, b, c, d, e, f are primary inputs. If we simplify t using the satisfiability don't cares, given, as we shall see, by

$$(k \oplus (ab + a'b')) + (s \oplus (ef + e'f')) + (r \oplus (cd)),$$

we find $t = sk' + s'kr$. The Boolean function at t in the Boolean space B^n does not change, since only the satisfiability don't cares are used for simplification.

11.3.4 Internal and External Don't Cares

Don't cares may be further classified as being either *External Don't Cares*, arising from the external environment in which the network is embedded, or *Internal Don't Cares*, arising from the structure of the network itself. We can view the external don't cares as deriving from these same conditions applied to the larger network in which the Boolean network is hierarchically embedded. This will become clearer after some detailed discussions.

Although our definition uses functions in defining a Boolean network, synthesis programs must read data files containing representations of these functions. Thus, in minimizing multi-level logic represented by a Boolean network, we assume that we are given an initial representation of each function f_i, and also a representation of each primary outputs don't care function d_i^X. The don't care conditions common to all outputs are the external satisfiability don't cares, $\bar{s}^X$, which are the primary input patterns that will never occur. The external observability don't cares, $\bar{o}_i^X$, are those that are specific to the separate output functions, usually arising from the way each output is used.

The i-th completely specified "don't care" function d_i^X must come from the specification of the environment in which the Boolean network η lives. We refer to d_i^X as the external don't care set of primary output i, which arises from the two phenomena discussed below.

11.3.5 External Satisfiability Don't Care Conditions

First, for a particular design the designer may decree that a particular primary input vector $x \in B^p$ *will never occur*. The vector $x \equiv y_I$ constitutes a don't care minterm, and such minterms are don't care for all primary outputs. The set of all such minterms is denoted by $\bar{s}^X$.

11.3.6 External Observability Don't Care Conditions

Second, the designer may state that for any primary output $i \in O$, the value of y_i *will not be used* for a set of primary input vectors (minterms) in the set $\bar{o}_i^X$. (Keep in mind that external don't care sets $\bar{s}^X$ and $\bar{o}_i^X$ derive from satisfiability and observability don't care conditions, respectively, that occur in the larger external network in which the subject Boolean network is to be embedded.) Thus for each primary output the total external don't care set can be written

$$d_i^X = \overline{s}^X + \overline{o}_i^X, \ i = 1, \ldots, q. \tag{11.6}$$

This equation gives the completely specified functions d_i^X, $i = 1, \ldots, q$ (don't care sets) associated with the primary outputs of a Boolean network. A principal objective of the sequel is to identify representations of the analogous don't care sets for each of the incompletely specified functions associated with the intermediate variables of a given Boolean network (and their corresponding internal nodes).

For the external don't care set, often there is no need to distinguish between these two types of don't cares, and in these cases we simply refer to d_i^X, rather than to specific terms of the external don't cares. However, we note that if $A \subseteq O$ is an arbitrary subset of O, then

$$\overline{s}^X \leq \prod_{\{i \in A\}} d_i^X, \tag{11.7}$$

which is true because $\overline{s}^X$ is lumped into each term on the right hand side of the previous equation.

We will also see that the use of a don't care set to specify how the output will be used is, in general, insufficient to capture this information completely. This generalization will enable us to handle Boolean networks with nodes having multiple output Boolean functions, rather than just single output Boolean functions, as well as networks that are hierarchically embedded in a larger network. It has been proposed that the notion of equivalence relations be used for this [40, 35] and that leads to the concept of **Boolean relations**. However, we continue here the tradition, coming from PLA synthesis, of using external don't cares to capture some of this information.

11.4 Internal Satisfiability Don't Cares

One of our main objectives is to iteratively perform single-node optimizations on all nodes of a Boolean network. To identify how this synthesis task may be done optimally, we now characterize formally the internal (implicit) don't care conditions applicable at internal nodes of a Boolean network.

Satisfiability Don't Cares are extremely important, as they are easy to compute and approximate, and are the basis of some powerful invariance relations for Boolean networks. These don't cares are a result of the existence of the additional intermediate variables, y_j, introduced at the intermediate nodes of a Boolean network. As an example, consider the network described by the Boolean equations

$$\begin{aligned} y_1 = F_1 &= a'b' \\ y_2 = F_2 &= c'd' \\ y_3 = F_3 &= y_1'y_2', \end{aligned}$$

which implement $F_3^* = (a+b)(c+d)$. For any node that uses the intermediate variables, y_1 and y_2, we have the option of eliminating y_1 and y_2 or expanding the Boolean space to include these variables. If we do the latter, there are combinations of variables that will never occur. For example, the combination $y_1 = 0, \quad a = 0, \quad b = 0$

will never occur, and in general, since $y_1 \neq a'b'$, then will never occur. The don't care conditions are thus the min-terms of the function

$$y_1(a+b) + y_1'(a'b').$$

A general Boolean network introduces many intermediate nodes and variables with the relations

$$y_j = f_j(y), \tag{11.8}$$

where y is the set of all logic signals in the network. Since we require that the Boolean network be acyclic, f_j must really depend on only the subset P_j^* of the y variables. The set of all satisfying truth assignments of the above relation is

$$s_j = (y_j \equiv f_j) = (y_j \oplus f_j)'.$$

In the extended space $B^{|y|} = B^n$, the satisfiability don't care set is given by

$$\overline{s}^* = \sum_{j \notin I} \overline{s}_j = \sum_{j \notin I} y_j \oplus f_j = \sum_{(j \notin I)} (y_j f_j' + y_j' f_j).$$

This is called the *overall satisfiability don't care set*, and we again use the superscript $*$ to denote a transitive property. Sometimes it is appropriate to consider only the don't care relations associated with the transitive fanin of node i. This is denoted as

$$\overline{s}_i^* = \sum_{j \in P_i^*} (y_j f_j' + y_j' f_j). \tag{11.9}$$

This don't care function gives all the input vectors that will never occur, due to the network structure, and is called the satisfiability don't care set because each of the relations

$$y_j = f_j(y)$$

must be satisfied during the correct operation of the network. The part of $\overline{s}^*$ contributed by the f_j, namely, $\overline{s}_j = (y_j f_j' + y_j' f_j)$, is called the *local satisfiability don't care set* of node j. For the above example, $\overline{s}_2 = y_2(c+d) + y_2' c' d'$.

Similarly, an SDC may be defined for the connections c_{ij} of η:

$$\overline{s}_{ij} = \sum_{k \in P_{ij}^*} (y_j f_k' + y_k' f_k). \tag{11.10}$$

Primary output nodes $i \in O$, which have no fanout specified in η, (11.10) gives the entire internal don't care set.

11.5 Observability Don't Cares

These don't care conditions are subtler in origin, and harder to compute and approximate than those of the satisfiability don't care set. In fact, it can be shown that the present definitions of Observability Don't Cares in the literature are insufficient to completely characterize multiple output circuits. For the multiple output case the concept of Boolean relations must be used.

Because of the complexity of ODCs, both in their computation and in their definition we will content ourselves with a superficial study. As we stated earlier, the ODCs arise when a particular input assumes a controlling value on a gate thus rendering the other inputs don't care. More specifically, we are concerned with the effect on the output signal and only indirectly on the effect on intermediate signals.

For example, when the first input to a 2-input OR gate is 1, the output is 1 regardless of the value of the second input. In a dual manner, a 0 input on an AND gate forces the output to be 0 regardless of other input values. These forcing inputs are called *controlling* inputs.

Thus, we can imagine a situation in which a Boolean network has an output node, j, which is an OR gate with two intermediate variables y_1 and y_2 as inputs. If we are trying to minimize y_2, in addition to the SDCs, we must include the fact that the output of y_2 is a don't care when $y_1 = 1$. When working in the expanded space of variables B^n, these ODCs can add important minimization potential. For example, consider the following three node Boolean network.

$$\begin{aligned} y_1 &= xw \\ y_2 &= x' + y \\ F &= y_1 + y_2 \end{aligned}$$

If we attempt to minimize y_1, we must calculate ODCs and SDCs. The SDCs for y_1 are given by

$$SDC_{y_1} = y_2 \oplus (x' + y) = y_2 x y' + y_2' x' + y_2' y.$$

Given our previous discussion we can write the ODCs for y_1 as

$$ODC_{y_1} = y_2 = 1.$$

The cover for this overall two-level function is given by

w	x	y	y_2	y_1
1	1	–	–	1
–	–	–	1	–
–	1	0	1	–
–	0	–	0	–
–	–	1	0	–

Through proper use of the don't cares we can minimize this to yield

$$\begin{aligned} y_1 &= w \\ y_2 &= x' + y \\ F &= y_1 + y_2 \end{aligned}$$

or, after "sweep"

$$\begin{aligned} y_2 &= x' + y \\ F &= w + y_2 \end{aligned}$$

This minimization requires both the ODCs and the SDCs; either alone is not sufficient. From a circuit point of view there is a redundancy which is removed by the minimization.

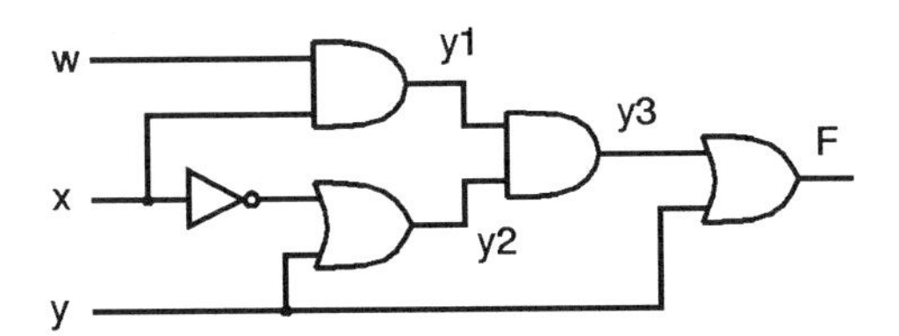

Figure 11.2: An Example Network for the Computation of Observability Don't Cares.

Let us consider a slightly different example in which the output gate is replaced by a XOR gate. This yields the following Boolean network.

$$
\begin{aligned}
F &= y_1 \oplus y_2 \\
y_1 &= wx \\
y_2 &= x' + y
\end{aligned}
$$

Examining this network we see that y_1 has the same SDCs as in our previous network. However, if we ask under what conditions on y_2 the output is unaffected by the value of y_1, we are led to the conclusion that there are no ODCs for this node. Thus, under this new set of don't cares we cannot minimize y_1.

Two more examples will increase our understanding of observability don't cares. First we shall consider a network with a depth of more than two. Consider the Boolean network defined by the following equations and shown in Figure 11.2.

$$
\begin{aligned}
F &= y_3 + y \\
y_3 &= y_1 y_2 \\
y_2 &= x' + y \\
y_1 &= wx.
\end{aligned}
$$

We can easily write the satisfiability don't cares for y_1. The observability don't cares, however, will involve two terms. It is still true that when $y_2 = 0$ the output F is unaffected by y_1, but in addition, when $y = 1$ the output is also unaffected by y_1. Thus we see that in constructing the ODCs we must consider all ways in which the value of the node under consideration will not affect the primary output. From our analysis, $ODC_{y_1} = y_2' + y = x + y$. Therefore y_1 is redundant, because $f_1 = wx \leq ODC_{y_1}$.

Since a Boolean network can have a complex SOP expression at every node, we can construct a two level Boolean network that will still present us with a challenge in calculating the ODCs. Consider the following Boolean network.

$$
\begin{aligned}
F &= y_1 w + y_2 w' + y_1' y_2' \\
y_1 &= xw \\
y_2 &= x' + y
\end{aligned}
$$

If we examine this network, we can find that the ODCs for y_1 are given by

$$ODC_{y_1} = w' y_2 + w y_2'.$$

11.5.1 Computing ODCs with the Boolean Difference

We can compute ODCs for individual nodes using the Boolean difference formula. The Boolean difference of a function with respect to a variable states the conditions under which the function is sensitive to that variable. That is, when the functions output will depend on the value of that variable. In formula, it is given by:

$$\frac{\partial f}{\partial x} = f_x \oplus f_{x'}. \tag{11.11}$$

If we consider F to be the output of our network, then F is sensitive to y_i exactly when $\frac{\partial F}{\partial y_i} = 1$. The complement of the Boolean difference gives the conditions under which the function is not sensitive, which is precisely the definition of the observability don't cares. Thus although there are some subtleties that we shall omit, we can simply write

$$ODC_{y_i} = \overline{\frac{\partial F}{\partial y_i}}. \tag{11.12}$$

This can be very complicated to compute for a variable that is deeply nested in a circuit, requiring a Boolean difference chain rule; it is useful as a conceptual framework for thinking about ODCs, and it shows how intimately ODCs are related to testing and automatic test pattern generation.

It should be clear by now that the calculation of the ODCs for a given node in a Boolean network can be a very complicated task. In the multi-level minimization program SIS a subset of these don't cares are chosen. It should also be clear that not all the SDCs are going to be useful in minimizing a given node; SIS also chooses a subset of the SDCs to use in the minimization of a node.

11.6 Prime and Irredundant Networks

A key concept in logic optimization is that of Boolean equivalence. In the multilevel context, we wish to establish when a given Boolean network, η, can be replaced by another one, $\tilde{\eta}$. This is true when the two sets of corresponding primary output global functions are equivalent modulo the external don't cares; that is, when η and $\tilde{\eta}$ represent the same set of incompletely specified functions $\mathbf{f}_i = (f_i, d_i, r_i), \quad \forall i \in O$.

Definition 11.6.1 *Boolean networks* $\eta = (\mathbf{f}, \mathbf{y}, I, O, \mathbf{d}^X)$, *and* $\tilde{\eta} = (\tilde{\mathbf{f}}, \tilde{\mathbf{y}}, \tilde{I}, \tilde{O}, \mathbf{d}^X)$, *are* equivalent modulo the don't care sets $\mathbf{d}^X$ *(written* $N \overset{\mathbf{d}^X}{\equiv} \tilde{N}$*), if there exists a permutation* Q *of* $\{1, 2, \ldots, q\}$ *such that for each Primary Output* $\tilde{z}_i(\mathbf{x})$ *in* $\tilde{O}$, $\tilde{z}_i(\mathbf{x}) = z_Q(i)(\mathbf{x})$ *for all* $x \notin \mathbf{d}^X$.

The permutation, Q, in Definition 11.6.1 is needed to identify the proper correspondence between the primary outputs of the two Boolean networks which may be very different structurally. For simplicity, we assume, without loss of generality, that Q is the identity permutation.

Actually, this definition is restrictive, and assumes that the *only* don't care conditions are those described above, involving single outputs only. However, as discussed above, there exist in nature don't care conditions defined over *groups of outputs*, and these are not included in the above definitions. A general definition of equivalence can be stated, but this requires more information about the external environment than just the external single output don't care set d_i^X for $i \in O$.

For example, the external environment may have outputs i and j connected only to the input pins of an exclusive-or gate, in which case the environment would be unable to distinguish between outputs $y_i = 1$, $y_j = 0$ and $y_i = 0$, $y_j = 1$.

Note that our definition of equivalence requires only that the primary outputs of two Boolean networks match for each care input vector. In particular, it is not necessary to have identity or even correspondence between the intermediate variables of the two networks. For example, a 4-level network could be equivalent to a 2-level network. The task of minimizing a representation N of a Boolean network η consists of iteratively transforming N into an equivalent representation $\tilde{N}$ where $\tilde{N}$ is smaller than N in some sense. Two properties of minimality, similar to those for the classical 2-level case, are especially relevant to the multilevel case (Boolean networks having these properties are 100% testable for stuck-at faults). These properties are defined next.

Definition 11.6.2 (Prime and Irredundant Boolean Networks) *Given a Boolean network representation* $N = (\mathbf{F}, \mathbf{y}, I, O, \mathbf{D}^X)$, *a cube* C *of the 2-level representation of* F_i *is* prime, *if no literal of* C *can be removed without causing the resulting network representation* $\tilde{N}$ *to be* not *equivalent to* N. *In more formal terms, let* $\tilde{N} = (\tilde{\mathbf{F}}, \mathbf{y}, I, O, \mathbf{D}^X)$, *be a Boolean network representation for which* $\tilde{F}_j = F_j$, $\forall j \neq i$ *and* $\tilde{F}_i = (F_i - \{C\}) \cup \tilde{C}$, *where* $\tilde{C}$ *is* C *with one of its literals removed. Then* C *is prime if* $N \overset{\mathbf{D}^X}{\not\equiv} \tilde{N}$. *Similarly, a cube* C *of* F_i *is* irredundant *if* C *cannot be removed from the representation of* F_i *without causing the resulting network* $\tilde{N}$ *to satisfy* $N \overset{\mathbf{D}^X}{\not\equiv} \tilde{N}$. *A Boolean network* $N = (\mathbf{F}, \mathbf{y}, I, O, \mathbf{D}^X)$, *is said to be prime, if all the cubes in each of the representations* F_i *of* N *are prime, and irredundant if all of these cubes are irredundant.*

Note that these two concepts are associated with local minima of a cost function which is nondecreasing in the total number of cubes and literals required to represent the incompletely specified logic functions, realized by the given Boolean network. These local minima represent networks that are 100% testable for conventional single stuck-at faults—otherwise redundant logic could be removed to obtain smaller networks. Hence to prove even local optimality, all faults must be proved irredundant. This is almost equivalent to generating tests for all faults. Thus one can see that area optimization subsumes automatic test pattern generation (ATPG)—a fact that may cause the distinctions between the testing and synthesis fields to blur increasingly as time goes on.

11.7 Two-Level Minimization with Multi-Level Don't Cares

The complete don't care set for a given node in a Boolean network is given by:

$$CDC_i = ExtSDC + ExtODC + SDC_i + ODC_i. \tag{11.13}$$

The first two terms are the external don't cares which will be the same for ever node in the Boolean network. The last two terms are the internal satisfiability and observability don't cares.

With these don't cares we can apply ESPRESSO and find a minimal cover for the node. Unfortunately, once we change the function of a given node, the don't cares for other nodes change. Thus, if we minimize node i and then we modify any node j that was important in the calculation of the don't cares for node i, we may still be able to minimize node i further because of changes in the don't care sets.

This poses a difficulty for minimization programs like SIS. SIS proceeds from the input nodes to the output nodes in breadth-first order and then iterates until no further improvement is possible. Normal practice is to use a *script*, which is a sequence of commands involving factorization, resubstitution, node minimization iterations, followed by node elimination, further network restructuring and minimization either a fixed number of times or until no cost improvement is found.

Clearly we have very weak notions of optimality involved in multiple level synthesis. However, if the complete don't care set is calculated, we can guarantee that at convergence of any of these sequences of iterations the multi-level network will be prime and irredundant. If an approximation to the complete don't care set were used, we would not be able to make this statement.

11.8 Notes

The various methods for exploiting don't cares in multilevel logic minimization grew out of three key approaches. First, the transduction approach was pioneered by Muroga [204][1]. However, his the full generality of his method for quasi-optimally selecting gate representations from the set of "permissible functions[2]" was not clarified until the publication of [204].

Second, the "multilevel don't care approach" was presented in [16, 15]. In this approach, the essential mechanisms of ESPRESSO were brought to bear on multilevel networks through the definition and exploitation of satisfiability and observability don't cares. The third influential method was the "Global Flow" approach [22, 21, 40]. While this approach could be interpreted in terms of the other two, it showed how techniques originally directed toward problems in compiler optimization could be brought to bear on logic minimization.

All of these approaches are directly or indirectly incorporated into state of the art multilevel logic synthesis programs like SIS or the SYNOPSYS design compiler, which grew out of the early efforts by DeGeus and co-workers, [16] and by Rudell [39], whose elegant architecture for the MIS package was profoundly influential.

11.9 Summary

In this chapter, we have presented a more detailed discussion of the internal structure of a circuit. We have formally defined the Boolean network as the graph model of a

[1] Muroga's ideas on multilevel minimization were the first to appear [148].
[2] The set of permissible functions is directly related to intervals in the Boolean function algebra.

combinational logic circuit (Section 11.2).

We then used the notion of the Boolean network to formally characterize the don't care conditions that arise naturally in multilevel Boolean networks (Section 11.3). This completes and expands upon the introductory treatment given in Section 3.4 in terms of intervals in Boolean function algebras.

Then we characterized the "Satisfiability Don't Cares" that arise naturally in Boolean networks in terms of "can't occur" vectors of signal values. Supposing that a Boolean network has n nodes/gates and $p \leq n$ primary inputs, we have observed that of the possible 2^n possible configurations of the logic signal vector $\mathbf{y} = (y_1, \ldots, y_n)$ only 2^p of them can actually occur in any given network. The other configurations of the $\mathbf{y}$-vector can't occur, and are therefore "don't care".

A constructive method was given for computing the satisfiability don't care set for any gate in the Boolean network. Fortunately, most practical circuits typically result in substantial don't sets for gates involved in reconvergent fanout. This is the rule, rather than the exception, because reconvergent fanout is the natural concomitant of the practice of re-using logic to save area whenever possible. Unfortunately, however, there are usually so many don't care conditions, that it is impractical to compute them all for use in multilevel logic minimization.

In Section 11.5, "Observability Don't Cares" were defined formally in terms of Boolean differences. We thus showed that whereas the satisfiability don't cares were defined by the nature of the gates and how they were connected, the observability don't care computation required the presence of primary outputs — if there are no observable outputs, there can be no observability don't cares.

Having thus completed the don't care "picture", we then showed some simple techniques for exploiting Don't Cares in logic minimization. We concluded by showing how to make a given Boolean network Prime and Irredundant. In this respect, we demonstrated that these two kinds of don't care conditions is complete. That is, once we have used these don't cares to make the network Prime and Irredundant[3], the resulting optimized network is locally optimal in the sense that no literal or gate can be removed without changing the behavior of the network

11.10 Problems

1. Prove that $(a + b) \oplus (a + c) = a'(b \oplus c)$.

   Solution.

$$
\begin{aligned}
(a+b) \oplus (a+c) &= (a+b)'(a+c) + (a+b)(a+c)' \\
&= a'b'(a+c) + (a+b)a'c' \\
&= a'b'c + ba'c' \\
&= a'(b \oplus c)
\end{aligned}
$$

   Note that this simple lemma can save time when calculating Boolean differences. □

[3]Note that a prime and irredundant network is 100% testable for stuck faults.

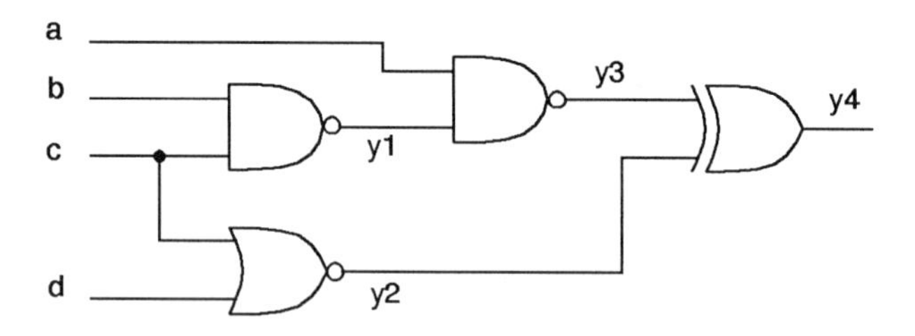

Figure 11.3: Network for Problem 3.

2. Compute $\frac{\partial f}{\partial x}$ and $\frac{\partial g}{\partial x}$ for:

$$f(x, y, z) = x + yz \qquad \text{and} \qquad g(x, y, z) = xy + z'.$$

Solution.

$$\frac{\partial f}{\partial x} = f_x \oplus f_{x'} = 1 \oplus yz = y' + z'$$

$$\frac{\partial g}{\partial x} = g_x \oplus g_{x'} = (y + z') \oplus z' = z(y \oplus 0) = yz$$

□

3. Find the complete satisfiability don't care set of the network of Figure 11.3 and write it in tabular form.

Solution.

a	b	c	d	y_1	y_2	y_3	y_4	
–	1	1	–	1	–	–	–	
–	0	–	–	0	–	–	–	$y_1 \oplus F_1 = y_1bc + y_1'b' + y_1'c'$
–	–	0	–	0	–	–	–	
–	–	1	–	–	1	–	–	
–	–	–	1	–	1	–	–	$y_2 \oplus F_2 = y_2c + y_2d + y_2'c'd'$
–	–	0	0	–	0	–	–	
1	–	–	–	1	–	1	–	
–	–	–	–	0	–	0	–	$y_3 \oplus F_3 = y_3y_1a + y_3'y_1' + y_3'a'$
0	–	–	–	–	–	0	–	
–	–	–	–	–	1	1	1	
–	–	–	–	–	0	0	1	
–	–	–	–	–	1	0	0	$y_4 \oplus F_4 = y_4y_2y_3 + y_4y_2'y_3' + y_4'y_2y_3' + y_4'y_2'y_3$
–	–	–	–	–	0	1	0	

□

4. Find the observability don't care set of y_1 in the network of Figure 11.4. Use the observability don't care to simplify F_1. Draw the resulting network.

Solution. The observability don't cares for y_1 are given by $b' + c$. If we simplify $F_1 = ab' + a'b$ with this don't care set, we get $F_1 = a'$. The resulting network is shown in Figure 11.5. □

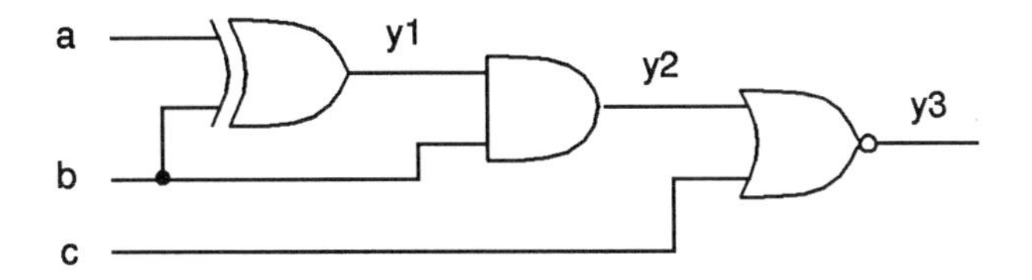

Figure 11.4: Boolean Network for Problem 4.

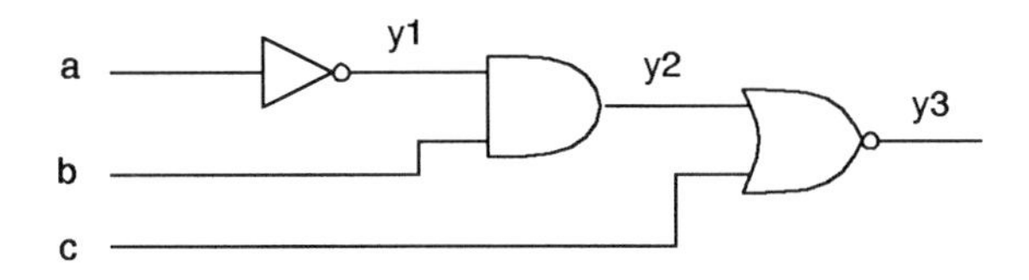

Figure 11.5: Simplified Boolean Network for Problem 4.

5. For

$$f = x_1 x_2 (x_3 + x_4') + x_1' x_3 x_4,$$

compute $\frac{\partial f}{\partial x_1}$.

6. Simplify F_1 in the network of Figure 11.6. Use both satisfiability and observability don't cares. Apply the heuristic minimization method to two-level minimization. Draw the simplified network.

Solution. The observability don't care set for y_1 is $ODC_{y_1} = a' + y_2'$. We see that in order to relate these conditions to the inputs of F_1, we need the satisfiability don't cares due to F_2. These are given by

$$y_2 c + y_2 d + y_2' c' d'.$$

We now put our on-set and don't care set in cubical form:

a	b	c	d	y_2	y_1
–	0	0	–	–	1
0	–	–	–	–	–
–	–	–	–	0	–
–	–	1	–	1	–
–	–	–	1	1	–
–	–	0	0	0	–

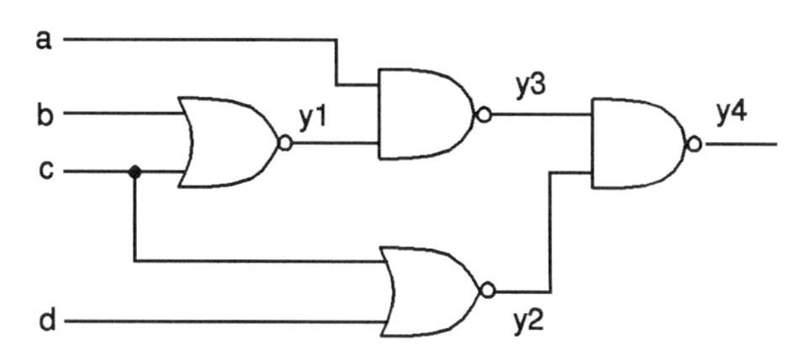

Figure 11.6: Boolean Network for Problem 6.

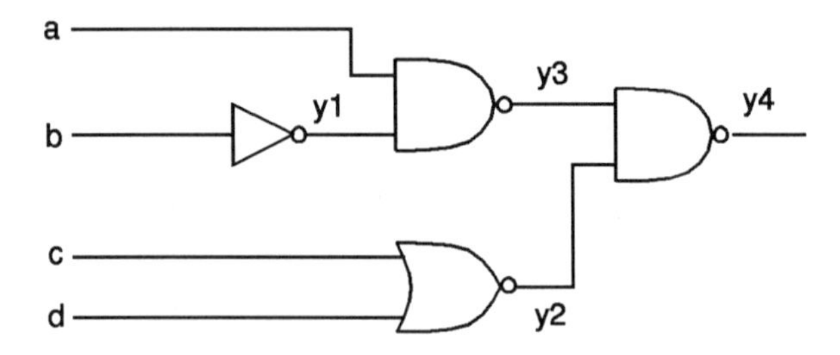

Figure 11.7: Simplified Boolean Network for Problem 6.

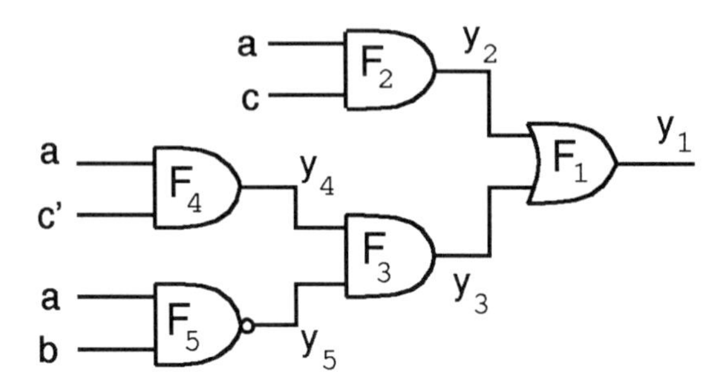

Figure 11.8: Circuit for Problem 7.

We can see that the only cube of the on-set can be expanded in the c direction, yielding $F_1 = b'$. The resulting modified network is shown in Figure 11.7. Notice that we could have immediately recognized that a' is not useful in the simplification of F_1 and we could have dropped it before attempting simplification. □

7. For the circuit shown in Figure 11.8

   (a) compute the entire SDC set and present it in tableau form;

   (b) compute ODC_{y_4} and ODC_{y_5} using the method of Boolean differences;

   (c) show how F_4 and F_5 can each be reduced to single literal covers using their respective observability don't cares;

   (d) give the expressions for y_1 in terms of primary inputs before and after the minimizations you performed in part (c).

Chapter 12

Automatic Test Generation for Combinational Circuits

12.1 Introduction

The yield of a manufacturing process is the fraction of fault-free products. The yield of IC fabrication processes varies widely and is sometimes lower than 50%. In similar cases, less than one in two fabricated circuits is functioning properly. If DL is the defect level, i.e., the fraction of defective parts after testing, Y is the yield, and T is the fault coverage of the test (1 for a perfect test and 0 for a totally ineffective test), then it can be shown that, under some simplifying assumptions,

$$DL = 1 - Y^{(1-T)}.$$

For a yield of 50% and a coverage of 90%, this formula gives a defect level of 6.7%, which is much larger than what is normally acceptable (hundreds of parts per million). Testing is therefore essential to insure the quality of a product and tests must be of high quality.

There are two fundamental methods of testing a circuit:

- Functional testing;
- structural testing.

In functional testing, few or no assumptions are made on the failure mechanisms and the possible faults. Moreover, in pure functional testing, no assumptions are made on the structure of the circuit under test (CUT). Functional testing, as the name implies, is concerned with verifying that the CUT performs as expected in most (ideally all) situations. Structural testing, on the other hand, is based on an assumed fault set. It consists of verifying whether any of the faults in the set is actually present in the CUT. The faults that are considered are alterations of the structure of a fault-free circuit, hence the knowledge of the structure of the CUT is essential in this form of test.

In the following we concentrate on structural testing of combinational circuits. We choose structural testing over functional testing, because it is better at generating

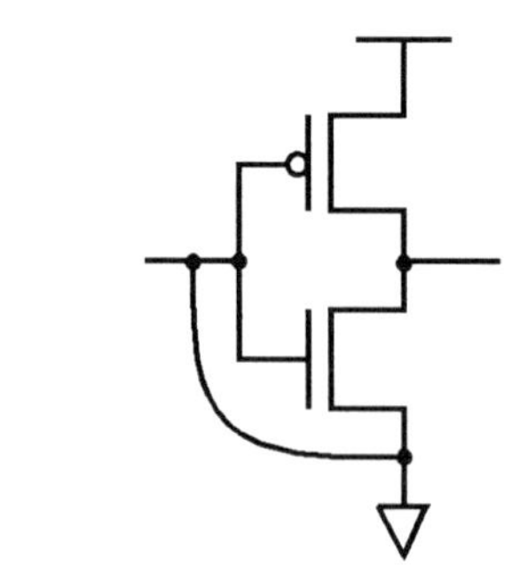

Figure 12.1: A short-circuit in a CMOS inverter.

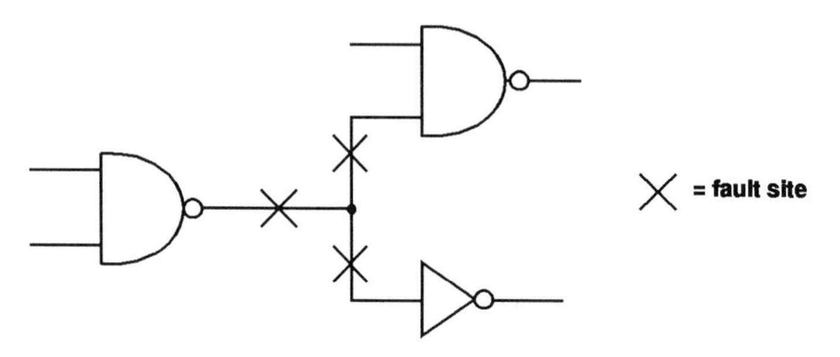

Figure 12.2: Stuck-at faults.

high-quality tests. We limit ourselves to combinational circuits to keep our presentation simple. However, in so doing we shall be able to examine many important concepts that apply to sequential testing as well. Finally, we shall consider circuits composed of simple gates (NAND, NOR, AND, etc.). This assumption is also made for the sake of simplicity. We shall see how to generate a test for a given fault and we shall see how test generation relates to logic optimization.

12.2 Faults and Fault Models

Faults in electronic circuits may be due to many different failure mechanisms, e.g., shorts, defective soldering, wrong value of the transistor threshold voltage. At the logic level one tries to give an abstract representation of the effects of a fault on the behavior of a circuit. In the case represented in Figure 12.1, the short between the input lead of the inverter and ground causes the input to be stuck at the value 0. The short-circuit is thus represented at the logic level as a *stuck-at-0* fault at the input of the inverter.

There are many possible fault models, of which the *single stuck-at-0/1* fault model is the most commonly used. Not all faults can be modeled as stuck-at faults. However, a good test for stuck-at faults usually detects many faults of other types too. When the single stuck-at fault model is used, the faults are single gate terminals stuck at either 0 or 1. Notice that more than two faults are defined for a single wire, if the wire drives multiple gates. This is illustrated in Figure 12.2. The faults on the *stem* are distinguished from the faults on the *branches* of the fan-out tree. Because of this, the total number of faults is $2N$, where N is the number of gate terminals. However,

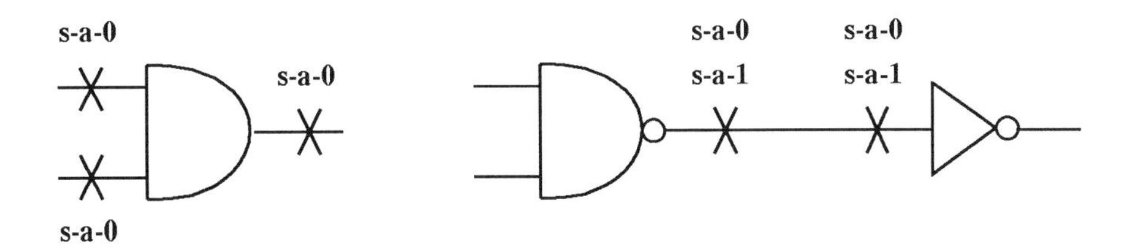

Figure 12.3: Equivalent faults.

the number of faults to be actually considered can be reduced thanks to the following definition.

Definition 12.2.1 *Let f_1 and f_2 be two faults of a circuit C. Let F be the function performed by C when no fault is present. Let F_1 and F_2 be the functions performed by C in the presence of f_1 and f_2, respectively. Then faults f_1 and f_2 are* equivalent *if and only if $F_1 = F_2$.*

If n faults are equivalent, it is sufficient to generate a test for one of them, in order to cover all of them. The two typical examples of equivalent faults are given in Figure 12.3. In the example to the left, we see that all stuck-at-0 faults at the terminals of an AND gate are equivalent. Suppose that a stuck-at-0 fault is present at one input of the AND gate and that we cannot observe the faulty input directly (it is not a primary output). As a result of the fault, the output is forced to 0. Since we do not observe the faulty input, we cannot tell whether the fault is on the input or the output of the gate. In the example to the right, we see that, if there is no fanout, the faults at the output of the NAND gate are equivalent to the faults at the input of the inverter.

In the case of a two-input OR gate, the three stuck-at-1 faults are equivalent. The three stuck-at-0 faults, on the other hand, are not equivalent. In general, finding all pairs of equivalent faults is difficult. However, applying the two criteria illustrated in the previous examples is straightforward and identifies most equivalent faults. We call the process of identifying equivalent faults *fault collapsing*, and we shall assume in the sequel that fault collapsing is performed prior to test generation.

It is obviously possible to consider *multiple* stuck-at faults. A multiple stuck-at fault consists of the simultaneous presence of several single stuck-at faults. The main problem with multiple faults is their number. Given M possible fault sites, there are $3^M - 1$ multiple stuck-at faults. In the following we only consider single faults.

We conclude this section with two additional remarks on fault equivalence. First, the faults on the stem of a fanout tree are not equivalent, in general, to the faults on the branches. This is why we consider them separately. Second, there is a special case that deserves consideration. Suppose that fault f_1 is such that it does not alter the behavior of the circuit; that is, $F_1 = F$. Then such a fault is *untestable* or *undetectable*. We sometimes call an untestable stuck-at fault *redundant*, because it is always associated with a redundancy (redundant gate or connection) in the circuit[1].

[1]Note that a connection may be redundant with respect to the logical operation of the Boolean network, and yet be critical to the performance of the gate (see Figure 12.23 on Page 491)

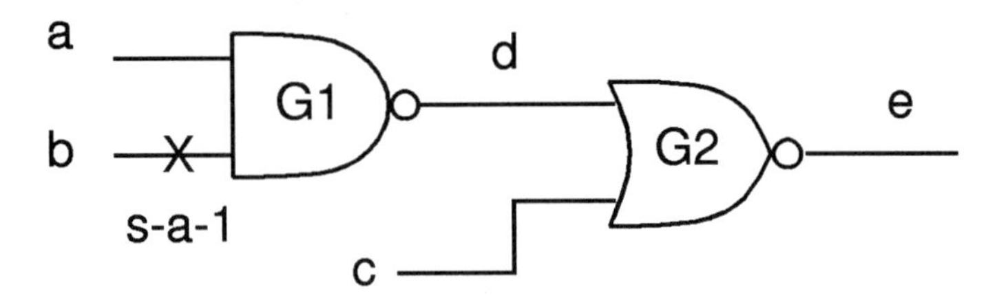

Figure 12.4: A simple combinational circuit.

12.3 Automatic Test Generation

In this section we present a procedure to generate a test for a given stuck-at fault in a combinational circuit. A test for a stuck-at fault in a combinational circuit is simply an assignment of zeroes and ones to the primary inputs of the circuit that causes different outputs in the good and in the faulty circuits. Such an assignment is called an *input vector* or vector, for short. There are many methods to generate tests for combinational circuits. Some of them are called *algebraic*, because are based on the algebraic manipulation of the expressions representing the functions of the circuits. Other methods, including the one that we shall examine, are called *topological* because they deal with the gates and their interconnections, i.e., with the topology of the circuit. Topological methods are based on intuitive concepts and can be very efficient. All methods are referred to as ATPG methods, where ATPG stands for *Automatic Test Pattern Generation*.

12.3.1 Excitation and Sensitization

Consider the simple circuit if Figure 12.4. Suppose we want a test for the stuck-at-1 fault on input b. It is clear that b must be 0 in our test. If not, the output of $G1$ would not depend on the presence of the fault. The general form of this elementary observation is the following: A test must *excite* the fault, that is, must cause the value complementary to the faulty value to appear at the fault site.

Proceeding in our example, let us consider what would happen if we set a to 0: The output of $G1$ would be 1 regardless of the other input. Once again, we would not be able to detect the fault. Hence, a must be set to 1. By a similar argument, c must be set to 0. The only test for our fault is thus 101. The assignment $a = 1$ sensitized the output of $G1$ to the value on b. Similarly, the assignment $c = 0$ sensitizes the output of $G2$ to the value on d. Together, the two assignments create a *sensitized path* that connects the fault site to the output of the circuit. A sensitized path is a sequence of gates such that their outputs are sensitized to the presence of the fault.

In general, a test for a fault must sensitize at least one path from the fault site to one of the primary outputs of the circuit. We can therefore divide test generation in two tasks: Excite the fault and propagate it to the outputs by sensitizing one or more paths.

Consider now the circuit of Figure 12.5 and suppose that we want to generate a test for the stuck-at-0 fault on e. This time the excitation condition requires that e be 1 in the fault-free circuit. However, e is not a primary input and we must work

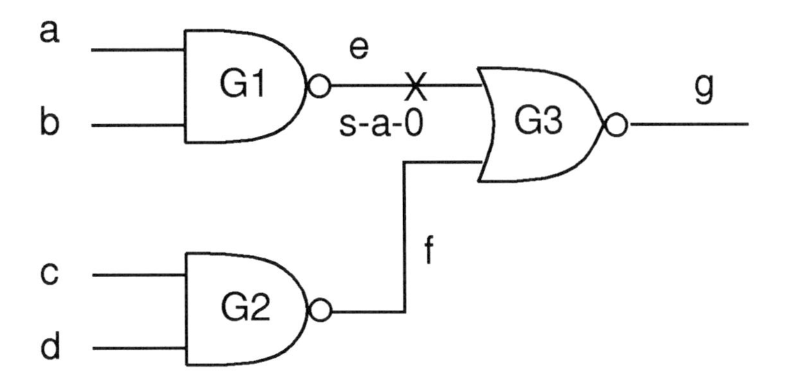

Figure 12.5: Another simple combinational circuit.

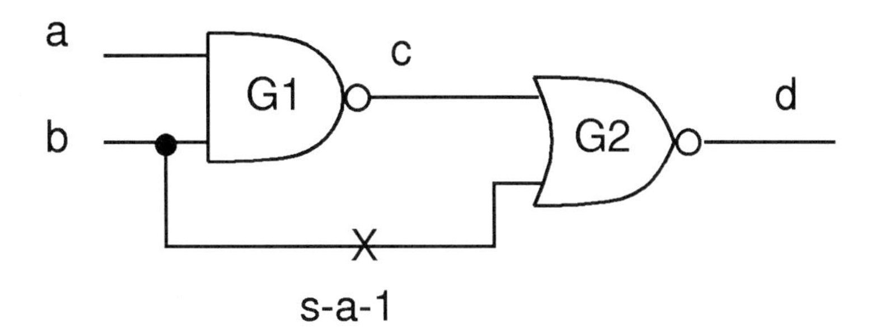

Figure 12.6: A redundant combinational circuit.

our way backwards. In this simple example, we easily see that our goal of setting e to 1 can be achieved in several ways. We can set either a or b to 0, or both.

The sensitization of a path to the output requires $f = 0$. Also in this case, we have to proceed backwards until we find primary inputs. We find that $c = d = 1$ is the only assignment that produces $f = 0$. In conclusion, we have found three possible tests for our fault: 0011, 0111, and 1011.

In general we shall be content of finding one test only, but in this example we want to emphasize that there may be several tests for one fault. In this case, we found multiple excitation conditions and a unique sensitization condition. It is possible to have multiple sensitization conditions as well.

The other important remark on this example is that in general, the excitation and sensitization conditions give us assignments to internal nodes of the circuits and we have to derive suitable primary input assignments that will produce those internal values. This may not be always possible, as illustrated by the next example. Let us consider the stuck-at-1 fault on the input of $G2$ driven by b in Figure 12.6. The fault is excited if and only if $b = 0$. However, the sensitization of $G2$ requires $c = 0$, which in turn implies $a = b = 1$. Hence the two requirements are contradicting. We conclude that there is no test for our fault.

Two important remarks are in order here. First, the circuit of Figure 12.6 is redundant. One can easily verify that d equals 0 identically. It is a general truth that an untestable stuck-at fault corresponds to a redundancy in the circuit. If a stuck-at-i

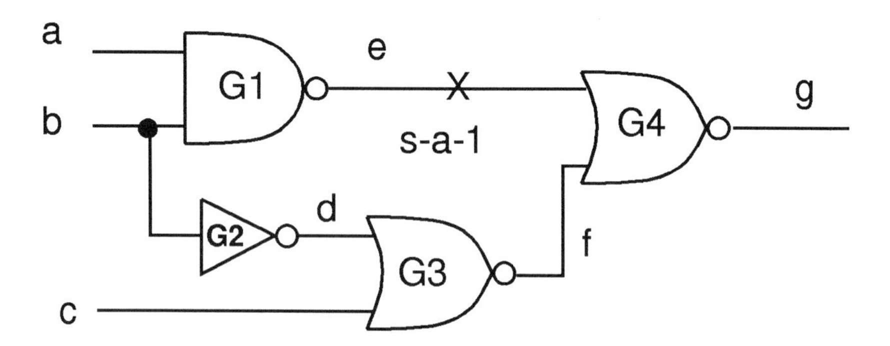

Figure 12.7: A combinational circuit.

fault is untestable at node a, then replacing a by the constant value i will not change the behavior of the circuit. Otherwise there would be a test. The circuit after the replacement is simpler than the original one. The connection between redundancy and untestable faults will be discussed in Section 12.4. Notice that the stuck-at-1 fault on the stem of b is also untestable, while the stuck-at-1 on the input of $G1$ connected to b is testable.

The second remark is that the conflicting requirements on the value of b in our example derived from the fact that we reached b through two different paths. It is true in general that such a condition may occur only in the presence of nodes with multiple fanout (in this case b itself). More specifically, the trouble is caused by paths that have a common source (e.g., b) and a common sink (in this case gate $G2$). This situation is given the name of *reconvergent fanout*. It is safe to say that reconvergent fanout is the root of all evil, as far as test generation is concerned.

Not all conflicts indicate that a fault is untestable. We have seen that sometimes excitation and sensitization conditions may be satisfied in different ways. If at some point we make an arbitrary decision between two choices and later come to a conflict, we have to return to the point where we made the choice and try the alternative. Consider, for instance, the circuit of Figure 12.7. The excitation conditions give $a = b = 1$. Sensitization requires $f = 0$. Looking at $G3$, we see that there are two ways of achieving that. Suppose we initially choose to set d to 1. In order to get a 1 on d, we need a 0 on b. This conflicts with the previous requirement that b be 1. This conflict, however, does not indicate that the fault is untestable. It just says that we cannot achieve sensitization by setting d to 1. We have to reverse our decision and try $c = 1$. We then succeed and 111 is our test. Returning on one's step and reversing a previous choice is called **backtracking**.

Even though in this case it was apparent that c was a better choice, on larger circuits it may be difficult to avoid wrong choices that have to be reversed. Indeed, the difference between various algorithms for test generation often lies in their abilities at guessing the right choice and avoiding as much backtracking as possible.

We conclude this section by recalling some useful facts and introducing some notation. AND, OR, NAND, and NOR gates have a *controlling* value. The controlling value is the value that, when present on at least one input, forces the output to a known value (the *controlled* value). For instance, the controlling value for AND and

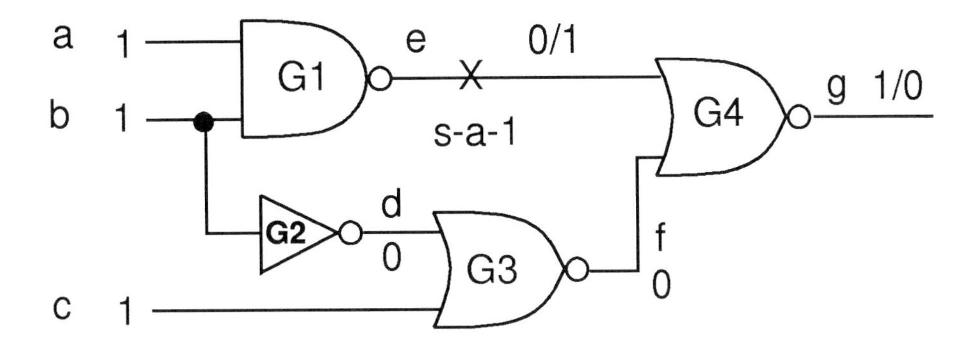

Figure 12.8: Use of compound values.

NAND gates is 0. For OR and NOR gates is 1. For negative gates (NAND and NOR) the controlled value is the complement of the controlling value. For AND and OR, it is the same.

The complement of the controlling value of a gate is the *non-controlling value.* At this point of our discussion, it should be clear that if we want to sensitize a path through given inputs of a gate, we must set all the other inputs—called the *side inputs*—to non controlling values. For instance, if we want to sensitize a path through one input of a three-input NAND gate, we must set the remaining two inputs to one. If we want to sensitize two paths through a three-input NOR gate, we must set the remaining input to 0.

Notice that XOR and XNOR gates do not have controlling values. Instead, it is important to realize that we can only have an odd number of inputs sensitized, if we want the output to be sensitized. This is because the output of a XOR or XNOR will always change in response to a single change in the inputs. However, two changes 'cancel out.' Familiarity with these simple facts will help in the next section.

12.3.2 A Simple Test Generation Algorithm

We have seen that the key idea in generating test is to generate a difference between the good and the faulty circuit and then propagate it to the primary outputs by creating one or more sensitized paths. The values of the good circuit and the faulty circuit at the nodes along a sensitized path are complementary. This will be indicated by 1/0 or 0/1, where the first value is the one of the fault-free circuit. The use of these *compound values* is illustrated in Figure 12.8. The values 0/0 and 1/1 are simply indicated by 0 and 1, respectively. We can see that the only sensitized path in this circuit (the same as in Figure 12.7) originates at node e and reaches the output g through gate $G4$. We also see that, when 111 is applied to the inputs, the fault-free circuit outputs a 1, whereas the faulty circuit produces a 0.

In the literature, the symbols D and $\overline{D}$ are used to indicate 1/0 and 0/1, respectively. Indeed, the first complete algorithm for test generation, the **D-algorithm**, owes its name to the use of these symbols. D stands for defect.

We are now ready to delineate a simple algorithm for test generation. The inputs to the algorithm are a description of the circuit and a single stuck-at fault. The output is a test if one exists, or the indication that the fault is untestable.

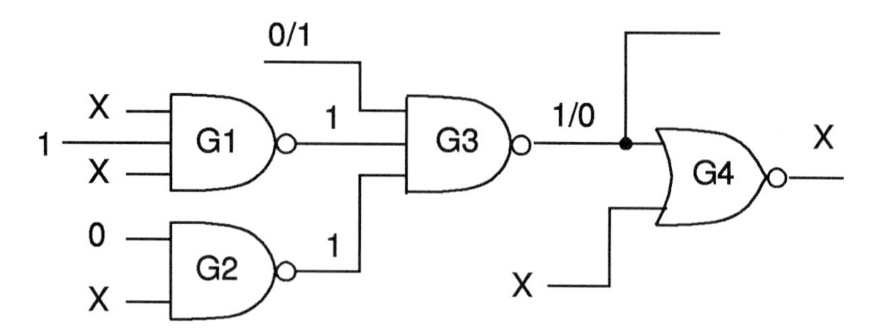

Figure 12.9: Frontier element ($G4$) and unjustified element ($G1$).

The algorithm initializes all the lines in the circuit to *unassigned* and then builds the test by assigning values to the lines that are required for excitation and sensitization and to the lines that drive them, if they are not primary inputs. We shall use X to indicate *unassigned*, and 0, 1, 1/0, and 0/1 for the possible assigned values. An element whose output is unassigned and such that one of its inputs carries a compound value 1/0 or 0/1 is said to belong to the **frontier**. The frontier tells us how far the symptoms of the fault have been propagated. An element whose output has been assigned, but whose assigned inputs do not imply the output is said to be *unjustified*. Examples of frontier and unjustified elements are given in Figure 12.9, where a fragment of a circuit is shown. The output of $G2$ is implied by the 0 on one of its inputs; however, the output of $G1$ is not implied by its inputs: Hence, $G1$ is unjustified. Before we proceed to detail the algorithm, we need to discuss several preliminaries.

From the examples, we have seen that we may have to choose among several options and possibly retract from some choices because of conflicts. When dealing with large circuits we may often make choices that lead to other choices, which in turn lead to other choices. We need a data structure that will allow us to keep track of these multiple, cascaded choices and will guide us in an orderly examination of all options.

Such a data structure is the so-called **decision tree**, a binary tree, similar to the search tree of the covering problem. Every node corresponds to a signal for which a choice is made. Along the arcs, we annotate the signals whose values are implied by the choices made. As a first example, let us revisit the example of Figure 12.7. The decision tree has a single non-terminal node and is shown in Figure 12.10. The values along the arc going into the node are uniquely determined by the excitation and sensitization conditions. As discussed in the previous section, we are faced with a choice[2] when we try to find an input assignment that will cause f to be 0. If we decide to try $d = 1$ first, we create a node labeled d and an arc out of it labeled 1.

When we try to propagate backwards the implications of this choice, we immediately find that there is a conflict for b. We then abandon this path in the decision tree and backtrack to the last decision node—in our case the only one. This time we set $d = 0$, because we know that there is no test for $d = 1$. This assignment forces c

[2]When we discuss *implications* later on, we shall see that in this example we do not really need to make choices.

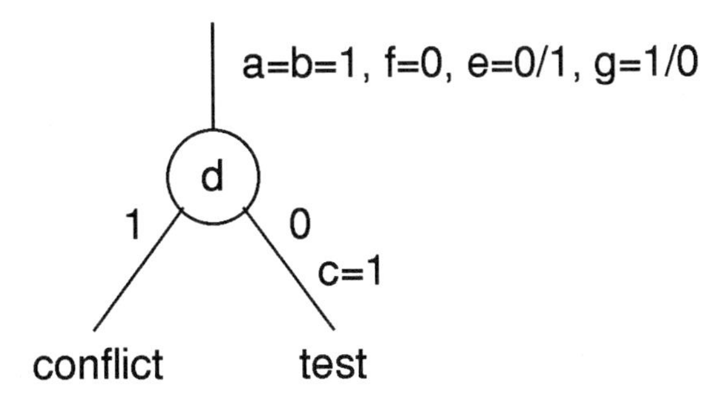

Figure 12.10: Decision tree for the example of Figure 12.7.

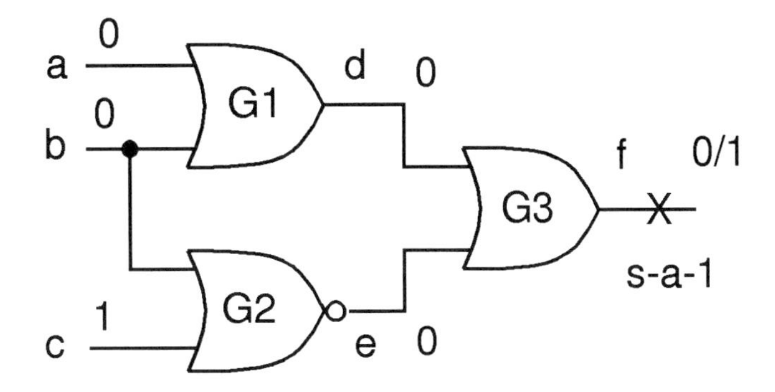

Figure 12.11: Example of implications.

to 1, because otherwise $f = 1$. We realize at this point that we have achieved all our objectives: The frontier has been propagated to a primary output and no unjustified lines are left. Hence, we have a test and we terminate.

Notice that in this case we may have claimed termination because we had propagated the fault symptoms to one output and all primary inputs had been assigned values. However, in general, the requirement that no unjustified lines are left is sufficient. This less stringent requirement will sometimes generate tests where some primary inputs and possibly some internal lines are left unassigned.

12.3.3 Implications and Backtracking

The parallel between the search tree of the covering problem of Section 4.7 and the decision tree of the ATPG algorithm is not just superficial. In the covering problem, it is important that at each node the matrix be maximally simplified by applying all reduction techniques. When generating a test, it is important to find as many possible implications of the choices made along the path leading to the current node. In both cases, the objective is to minimize the number of nodes of the tree that are actually visited, or, almost equivalently, the number of *backtracks*. Indeed, in the ATPG literature, the number of backtracks performed by a given algorithm on a given example is one of the most important figures of merit.

The importance of carrying out as many implications as possible is illustrated in Figure 12.11. By just using the implications of the excitation condition ($f = 0$ in the fault-free circuit), we derive a complete test. Indeed, $f = 0$ implies $d = e = 0$ and

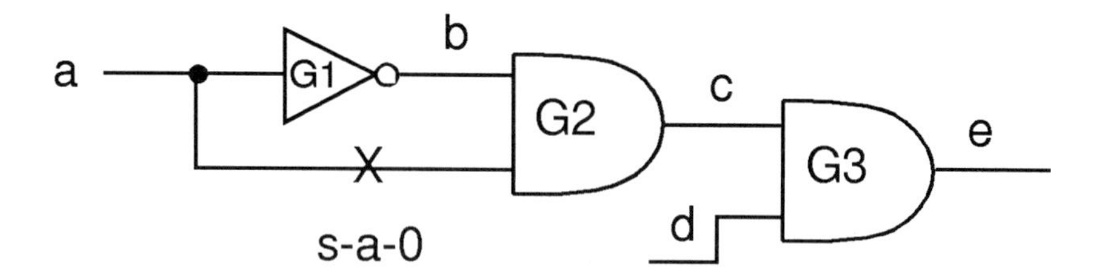

Figure 12.12: Another example of implications.

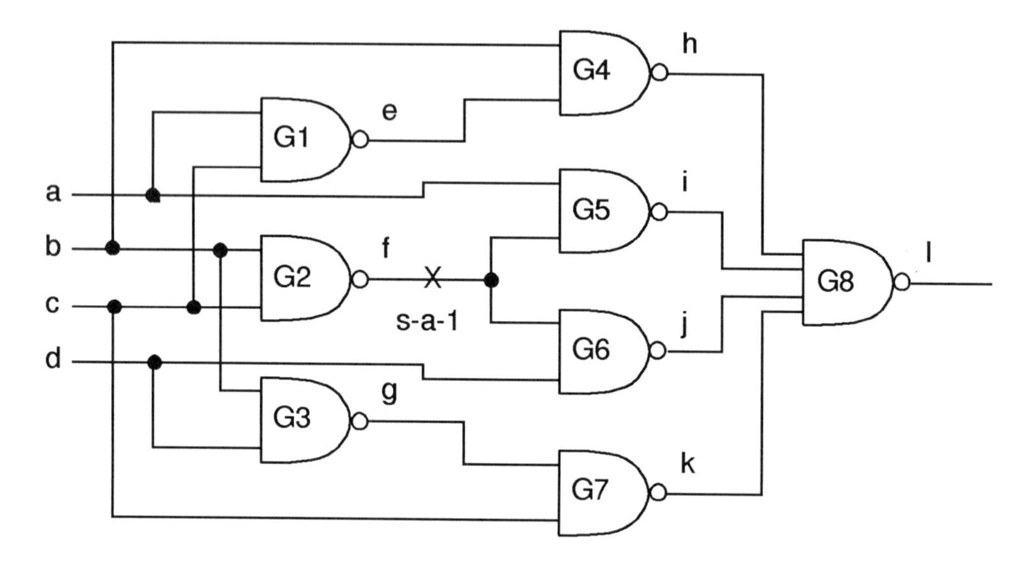

Figure 12.13: Schneider's example.

$d = 0$ implies $a = b = 0$. Finally, $e = 0$ and $b = 0$ jointly imply $c = 1$.

Another interesting example of implications is shown in Figure 12.12. The fault excitation condition implies $a = 1$. This in turn implies $b = c = 0$. As a result, the frontier is empty—there are no frontier elements—and we conclude that there is no test for the given fault.

A third, more complex, example of implications is illustrated in Figure 12.13. There we can see that all possible sensitized paths must go through gate $G8$. Those inputs to $G8$ that are not reachable from the fault site f (h and k) must have non-controlling values for sensitization to occur. This dictates in this case $h = k = 1$. Combined with $b = c = 1$, these implications further give $e = g = 0$ and $a = d = 1$. Finally, i and j are implied to be $1/0$ and l is seen to be $0/1$. We have found a test without any backtrack (actually, without even a choice)[3]. The condition we have exploited at gate $G8$ is called *unique sensitization.*

A further analogy between ATPG and the covering problem is given by the choice of the splitting variables. In both cases a judicious choice may substantially reduce

[3]The test 1111 creates two sensitized paths in this case. Since this test is the only one for the fault (it was obtained without any choice), there is no test with a single sensitized path for this example. Indeed the circuit of Figure 12.13 was used to show that an algorithm that tries to sensitize only one path at the time may not work.

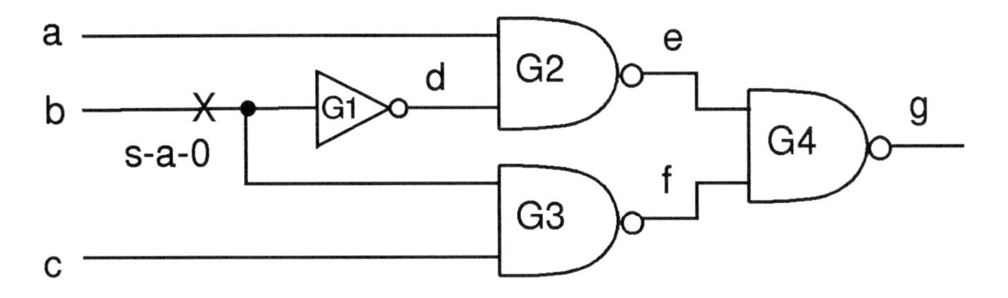

Figure 12.14: ATPG example.

the amount of work. For instance, in the case of the example of Figure 12.7, splitting on c rather than d would have saved one backtrack. We shall return to this issue in more detail, after presenting the outline of our test generation algorithm.

1. Apply the fault excitation condition.

2. Perform the implications of the last assignment.

3. If the fault symptoms have reached at least one primary output, justify the remaining unjustified lines. If justification fails, backtrack and go to Step 2. Otherwise, exit: A test has been found.

4. If the frontier is empty, backtrack and go to Step 2.

5. If the frontier consists of one gate only, perform the resulting implications (this is discussed later) and go to Step 2.

6. Choose one signal that is not reachable from the fault site and assign to it either 1 or 0. Create a corresponding node in the decision tree. Go to Step 2.

In Step 6, the restriction on what signals may be chosen is imposed so that we can restrict the chosen values to 0 and 1. If the selected line were reachable from the fault site, it might have a 0/1 or 1/0 value. We want to avoid this possibility, to keep the algorithm simple.

In Step 5, the case of a frontier composed of a single element is considered. In this case, all unassigned inputs are set to non-controlling values. Clearly, a controlling value on one of those inputs would stop the propagation. Furthermore, a compound value on one of those inputs is not possible, because there the frontier contains only one gate. Indeed, there should be an input to the only frontier element that is X and is reachable from the fault site. However, along the path there should be a frontier element for this to happen.

Let us now see how the algorithm is executed for the circuit of Figure 12.14. Initially, all lines are set to X.

1. (Step 1) The excitation condition causes b to be 1/0.

2. (Step 2) Performing the implications of $b = 1/0$, we find $d = 0/1$.

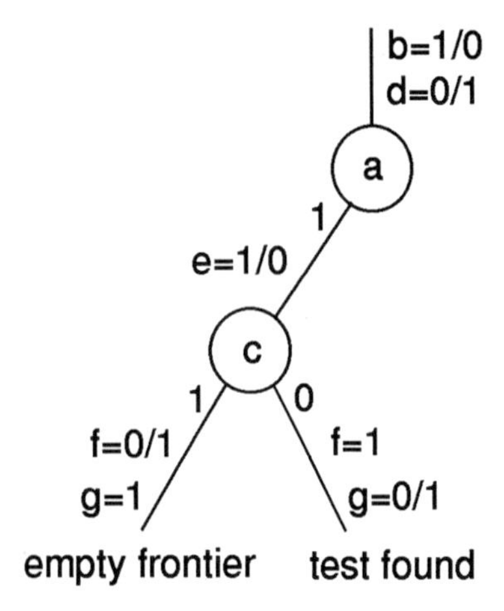

Figure 12.15: Decision tree for the example of Figure 12.14.

3. (Step 6) The frontier consists of two elements: $G2$ and $G3$. We choose $a = 1$ and create a node in the decision tree for this choice. Our goal, in selecting $a = 1$, is to move the frontier forward (to $G4$) by allowing propagation of the fault symptoms through $G2$.

4. (Step 2) The implication of $a = 1$ is $e = 0/1$.

5. (Step 6) The frontier now contains $G3$ and $G4$. We choose $c = 1$ and create a node in the decision tree.

6. (Step 2) The implications of $c = 1$ are $f = 1/0$ and $g = 1$.

7. (Step 4) The frontier is now empty. We need to backtrack. This is done by reversing the last choice, i.e., by returning f and g to the X value and by setting $c = 0$.

8. (Step 2) The implications of $c = 0$ are $f = 1$ and $g = 1/0$.

9. (Step 3) The fault symptoms have reached the output and no unjustified lines remain. A test has been generated (110) and the algorithm returns.

The decision tree generated for this example is shown in Figure 12.15.

12.3.4 Choice of the Decision Variables

Let us now return to the discussion of the choice of a line at Step 6 of the algorithm. The way we formulated the algorithm, we need only comply with the restriction that the line we choose is not reachable from the fault site. This, of course, leaves several strategies possible. We briefly review some of them. In the problems, we shall rely on our intuition to select an appropriate line to be assigned.

We first notice that our formulation of the algorithm allows us to select both internal lines and primary inputs. We can also select both unjustified and unassigned elements. General strategies can be obtained by restricting our choices according to

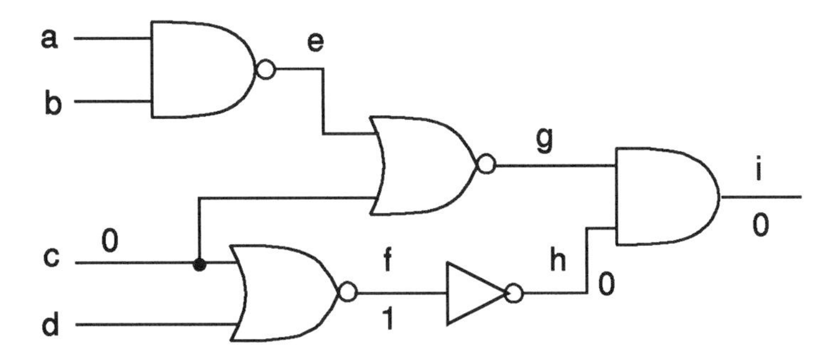

Figure 12.16: Example of backtrace.

these two dichotomies. For instance, in the original D-algorithm we always select an unassigned input to a frontier gate, until we reach a primary output, or until the frontier disappear. Then we always select an unassigned input to an unjustified gate. This strategy is simple, but sometimes inefficient.

In the PODEM algorithm, we always select a primary input. The process whereby we identify a suitable primary input (one that is likely to advance our cause), and, the value to assign to it, is called *backtrace*[4]. The backtrace procedure is given an *objective*—an internal line and a desired value for it—and it traces a path backwards in the circuit (whence the name) until an input is found. The initial objective of backtrace is chosen so as to drive the frontier forward.

Figure 12.16 illustrates an example of backtrace. The initial objective is to set $i = 0$. When the procedure goes through an AND gate having 0 as objective, it selects one input of the gate and a value of 0 as new objective. Suppose that h is chosen. The new current objective is $h = 0$ and the next step is to backtrace through the inverter. There is no choice involved in this case and the resulting current objective is $f = 1$. When the procedure goes though a NOR gate having 1 as objective, it selects one input and a value of 0 as new objective. In our case, let us suppose $c = 0$ is chosen. Since c is a primary input, the procedure terminates. Notice that $c = 0$ does not suffice to guarantee $i = 0$ and that the choices made by the backtrace procedure are not entered in the decision tree. Backtrace is just a heuristic that identifies a primary input that may help reaching the objective. At every step of backtrace, the choice of which input to follow is again heuristic.

In the FAN algorithm, we select either fanout points or *head-lines*. A head-line is a line such that all the gates preceding it do not fan out. The selection process is based on a procedure called *multiple backtrace* that is an enhancement of the backtrace procedure used by PODEM. We do not go into the details of how multiple backtrace works; we just give a rationale for FAN's approach. We said that reconvergent fan-out is what makes testing difficult. We choose fan-out points for assignments in the hope to expose possible conflicting assignments that may occur at those lines early. On the other hand, we can always justify a head-line to either 0 or 1, because it is the output of a sub-circuit without fan-out. Hence we want to determine what value the head-line should have first and postpone the actual justification to when we know we have a test.

[4]Not to be confused with *backtracking*.

12.3.5 Putting the Pieces Together

Now that we have an algorithm to generate a test for a given fault, we can address the issue of how this algorithm is used to generate a complete test for a circuit. Given a circuit, we generate all possible stuck at faults and then identify equivalent faults. Only one fault, called representative, is chosen from each set of equivalent faults. The circuit and the list of representative faults are then passed to the ATPG program. The ATPG program works in conjunction with a **fault simulator**. A fault simulator is a program that determines which faults from a given list are detected by a given set of tests.

The ATPG program picks one fault from the list and tries to generate a test for it. If no tests exists, the fault is marked untestable and removed from the list. If a test is found, it is passed to the fault simulator. The point of doing that is that the test generated for one fault may actually detect many other faults: The fault simulator is run and all detected faults are removed from the list. Then the ATPG program picks one of the remaining faults and iterates the process until no faults are found.

There are two advantages in coupling a test generator to a fault simulator. First, time can be saved, because simulating faults is faster than generating tests. Second, the number of tests generated is kept small, by not adding tests devised for faults that are already covered.

In practice, several techniques complement the basic scheme we just outlined. For instance, it is common practice to apply a set of randomly generated tests to a circuit. The cost of generating (pseudo) random tests is negligible, and they cover many faults. It is also common practice to set limits on the number of backtracks performed for a given fault. Faults that require too many backtracks are aborted. This is done to prevent a few faults from degrading performance substantially. Finally, we can mention that the number of tests can be reduced, without affecting the fault coverage, by applying *reverse* fault simulation. Once a set of tests has been created, it is simulated in reverse order of generation. Those tests that provide no additional coverage are dropped[5].

As a final remark to this section, it is important to emphasize that test generation is computationally expensive. For large circuits, the requirements of test must be taken into account at the design stage. Various techniques have been developed that go under the collective name of *Design for Testability.*

12.4 Redundancy Removal

Our ATPG algorithm can be used to simplify circuits by a method called *redundancy removal*. The method is based on the observation we made earlier that an untestable stuck-at fault signals the redundancy of the circuit. If, for instance, the stuck-at-1 fault at a connection is untestable, it means that that connection can be replaced by a constant 1, without changing the function performed by the circuit. Redundancy removal can be applied to the circuit either before or after technology mapping. In the

[5]It is also possible to minimize the number of tests by setting up a covering problem. The columns correspond to the tests and the rows to the faults.

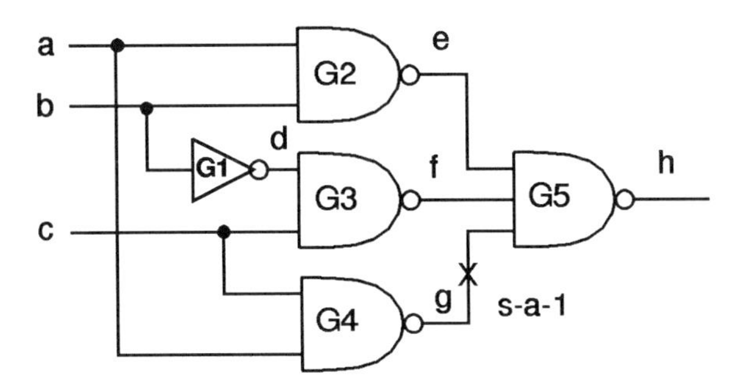

Figure 12.17: A redundant circuit.

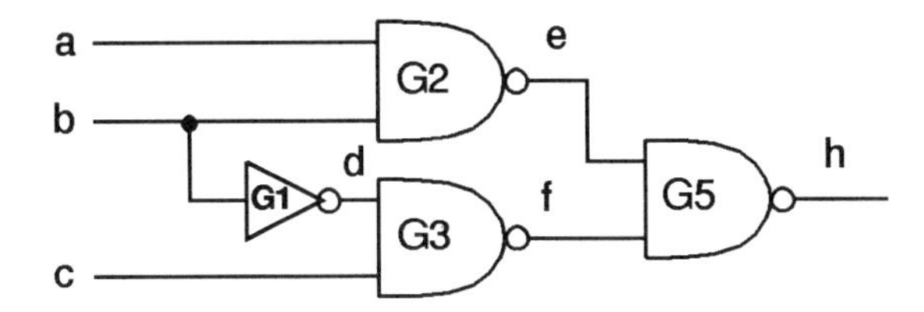

Figure 12.18: Irredundant circuit derived from the one of Figure 12.17.

latter case, removing redundancies may produce a circuit that is no longer mapped. Rework is then required.

Let us consider the circuit of Figure 12.17. If we try to generate a test for g stuck-at-1, we find that $a = c = 1$ are necessary to excite the fault. Also, unique sensitization dictates $e = f = 1$. One easily sees that $e = 1$ and $a = 1$ imply $b = 0$, but $f = 1$ and $c = 1$ imply $b = 1$. Hence, the fault is untestable[6]. Therefore, the lowest input to $G5$ is removed (which is equivalent to fixing it to 1) and $G4$ is removed from the circuit, because it feeds no gates. The result is illustrated in Figure 12.18.

In general, if an input line of a gate is fixed to a non-controlling value, the input is removed. If the input line is fixed to a controlling value, the gate is removed and its output is fixed to the controlled value of the gate. This constant value is then propagated. Suppose a constant 0 is applied to one input of a two-input NOR gate. The gate simplifies to an inverter. A similar consideration applies to NAND gates. If after redundancy removal, the circuit being simplified contains pairs of cascaded inverters, those pairs can be removed.

The next example illustrates an important point. In general, multiple redundancies cannot be removed simultaneously. One can easily verify that both input stuck-at-1 faults for the circuit of Figure 12.19 are untestable. It is equally clear that we cannot replace both inputs by constant values. In general, some redundancies can be eliminated simultaneously because they do not interfere, but the removal process remains intrinsically serial. This is illustrated also by the following example. There is only one untestable fault in the circuit of Figure 12.20, namely the input of $G1$ connected to b stuck-at-0. However, when the corresponding redundancy is removed, the circuit of Figure 12.21 is obtained, which contains an untestable fault (e stuck-at-1). The result of removing also this redundancy is shown in Figure 12.22.

[6]Notice that $G4$ implements the consensus term of the terms implemented by $G2$ and $G3$.

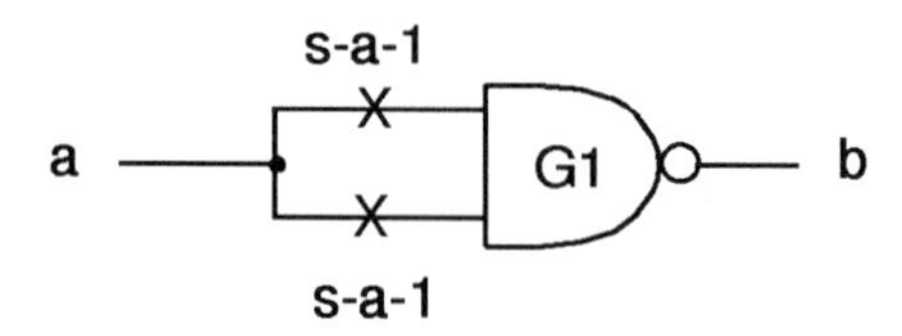

Figure 12.19: Circuit with multiple redundancies that cannot be simultaneously removed.

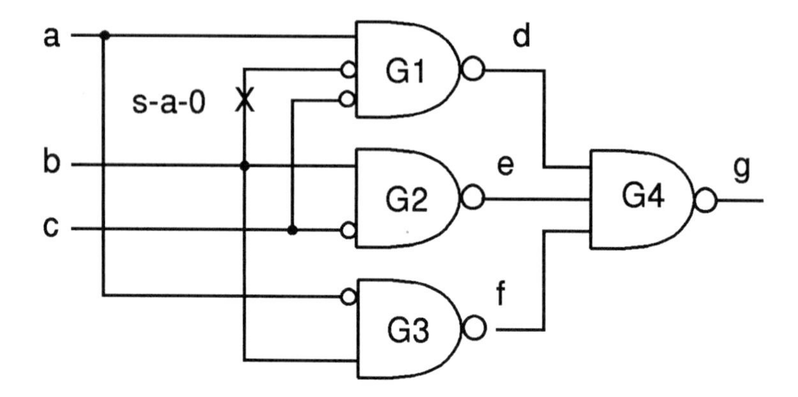

Figure 12.20: Circuit where the removal of one redundancy exposes another redundancy.

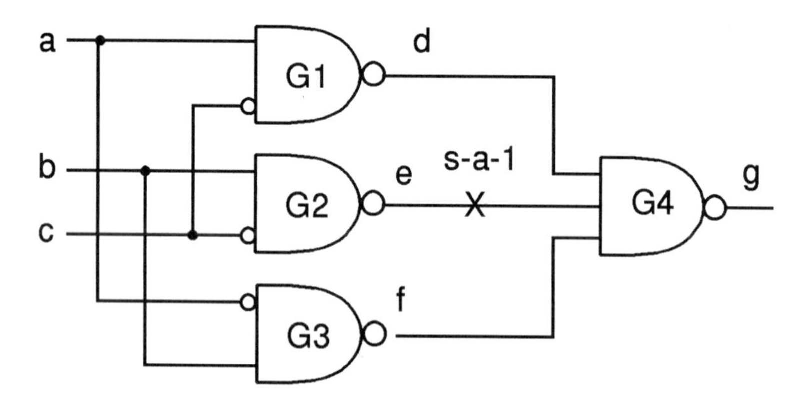

Figure 12.21: Circuit of Figure 12.20 after the removal of the only redundancy.

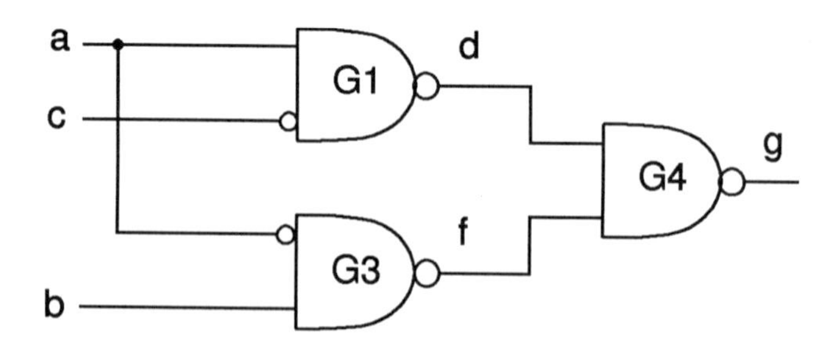

Figure 12.22: Circuit of Figure 12.21 after the removal of the remaining redundancy.

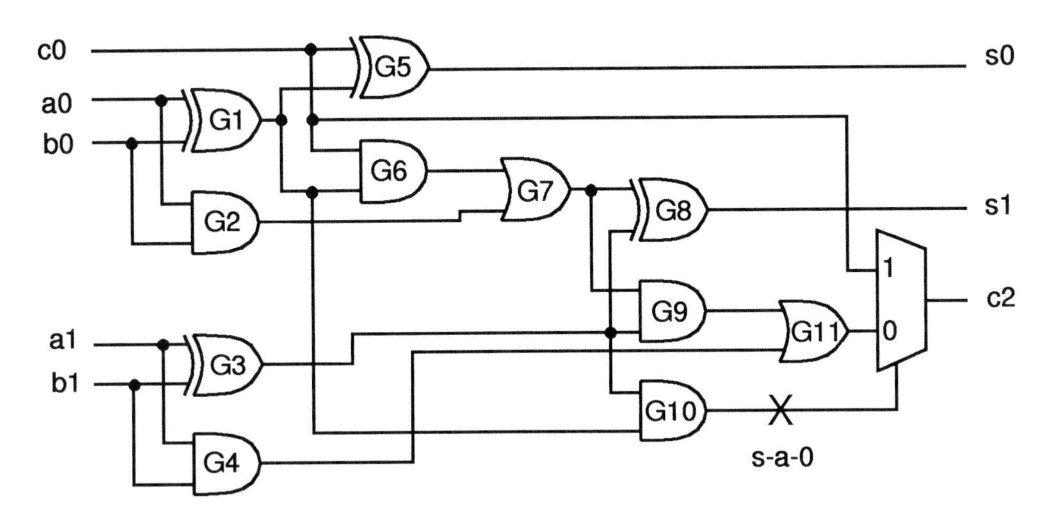

Figure 12.23: A 2-bit carry-skip adder.

It should also be noticed that the result of redundancy removal depends in general on the order in which redundancies are removed. To see why this is true, consider a two-level circuit implementing the complete sum for a function. Initially, all gates corresponding to non-essential primes will be (individually) redundant. However, as soon as some redundancies are removed, other redundancies will disappear, as primes become relatively essential.

Redundancy removal has the desirable property that it increases the testability of the circuit, and it reduces the area. In general, it also improves the performance of the circuit, by reducing the capacitive loads and the number of series transistors in the gates. There are, however, counterexamples to this behavior, the most famous of which is the carry-skip (or carry-bypass) adder. A 2-bit carry-skip adder is shown in Figure 12.23. The stuck-at-0 fault on the control input of the multiplexer is untestable. However, if the redundancy is removed, the circuit is transformed into a 2-bit ripple-carry adder—a slower circuit. Specialized techniques exist to address this kind of problems.

We conclude this section with a short comparison of redundancy removal to another optimization technique for multi-level circuits that we have examined. If we consider a gate as a node of a Boolean network, we see that eliminating a redundancy corresponds to expanding a cube of the node cover in the input part or dropping a cube from the cover. Therefore, there is a strong similarity between what the don't care based method does and what redundancy removal does. A moment's thought should suffice to see that the implications used in redundancy removal and the satisfiability don't cares are strictly related. If, for instance, $a = 1$ implies $b = 0$ in a circuit, then ab is a cube of the satisfiability don't care for that circuit and vice versa. Redundancy removal actually works with an implicit representation of the don't cares.

The other observation that stems from seeing redundancy removal as a series of expansions and reductions is that it is possible to generalize redundancy removal to a method that allows moves that temporarily increase the cost of the network. This idea is analogous to the one used in heuristic minimization and the algorithm that is based on it has been aptly called *multi-level tautology*.

12.5 Notes

General books on the subject of testing digital circuits are [56, 42, 233, 107, 199, 1, 96]. The more recent books also cover design for testability. Seminal papers on testing combinational circuits are [219, 232, 248, 116, 108, 247]. The test generation procedure that we have presented in Section 12.3 freely combines features of the D-algorithm [232], PODEM [116], and FAN [108]. For a sophisticated use of implications, the reader is referred to [247]. An interesting method, proposed in [166]; it is now implemented in SIS. The connection between redundancy, don't cares, and implications has been studied in [30, 15, 34, 127]. Multi-level tautology is discussed in [126]. T he issue of redundancy versus delay is considered in [160, 192, 242]. The subject of testing two-level combinational circuits, specifically PLAs, has received considerable attention *circa* 1980. A review can be found in [254]. The reader interested in sequential testing and redundancy removal is referred to [139, 185, 58, 110, 205, 61, 62, 57, 65, 64].

12.6 Summary

In this chapter, we have completed the brief, introductory treatment that was given in Section 1.4.4. Three main topics were covered. First, we gave, in Section 12.2, a summary of the various faults models in widespread usage at the time of writing.

Second, our main focus was the treatment of automatic test pattern generation, in Section 12.3. After treating issues of excitation and sensitization (especially the method of unique sensitization). This led us naturally to the subject of logical implications. (Section 12.3.3).

Third, we discussed at some length how the basic tools of the ATPG trade could be used to perform redundancy removal (Section 12.4) on a given multilevel logic circuit. This operation leaves the Boolean network in a locally optimum "prime and irredundant" state.

12.7 Problems

1. Divide the single stuck-at faults defined for the circuit of Figure 12.24 into sets of equivalent fault, by applying the two simple criteria described in Section 12.2.

   Solution. We have five inputs, six two-input gates and one one-input gate. This gives us a total of $5 \cdot 2 + 6 \cdot 6 + 1 \cdot 4 = 50$ single stuck-at faults. Here is the list divided in sets of equivalent faults. There are 18 such sets.

```
Class  1 s_a_0 on gate G7, output
Class  2 s_a_1 on gate G7, output       s_a_0 on gate G7, input j
         s_a_0 on gate G7, input k      s_a_0 on gate G5, output
         s_a_0 on gate G6, output
```

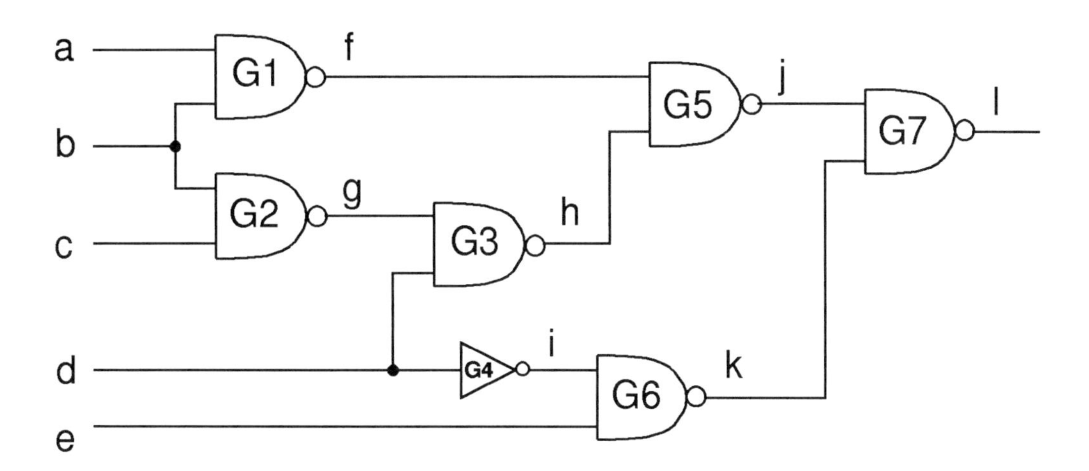

Figure 12.24: Combinational circuit for Problems 1–4.

```
Class  3 s_a_1 on gate G7, input j    s_a_1 on gate G5, output
         s_a_0 on gate G5, input f    s_a_0 on gate G5, input h
         s_a_0 on gate G1, output     s_a_0 on gate G3, output
Class  4 s_a_1 on gate G7, input k    s_a_1 on gate G6, output
         s_a_0 on gate G6, input i    s_a_0 on gate G6, input e
         s_a_0 on gate G4, output     s_a_1 on gate G4, input d
         s_a_0 on input e
Class  5 s_a_1 on gate G5, input f    s_a_1 on gate G1, output
         s_a_0 on gate G1, input a    s_a_0 on gate G1, input b
         s_a_0 on input a
Class  6 s_a_1 on gate G5, input h    s_a_1 on gate G3, output
         s_a_0 on gate G3, input g    s_a_0 on gate G3, input d
         s_a_0 on gate G2, output
Class  7 s_a_1 on gate G1, input a    s_a_1 on input a
Class  8 s_a_1 on gate G1, input b
Class  9 s_a_1 on gate G3, input g    s_a_1 on gate G2, output
         s_a_0 on gate G2, input b    s_a_0 on gate G2, input c
         s_a_0 on input c
Class 10 s_a_1 on gate G3, input d
Class 11 s_a_1 on gate G2, input b
Class 12 s_a_1 on gate G2, input c    s_a_1 on input c
Class 13 s_a_1 on gate G6, input i    s_a_1 on gate G4, output
         s_a_0 on gate G4, input d
Class 14 s_a_1 on gate G6, input e    s_a_1 on input e
Class 15 s_a_0 on input b
Class 16 s_a_1 on input b
Class 17 s_a_0 on input d
Class 18 s_a_1 on input d
```

□

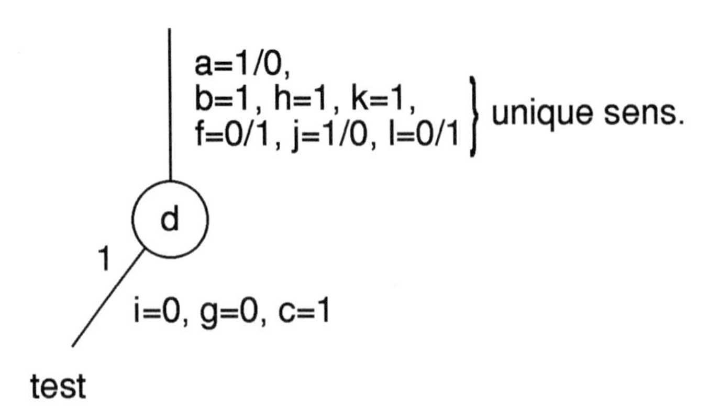

Figure 12.25: A decision tree for Problem 2.

2. For the circuit of Figure 12.24, find a test for the fault a stuck-at-0. Draw the decision tree. Use all possible implications at every step.

   Solution. A possible decision tree is shown in Figure 12.25. Since e is unassigned when we find a test, we have actually found two tests: 11110 and 11111. The entire set of faults is:

abcde
1111–
11–00

   This entire set could be obtained by continuing the exploration of the decision tree after the first successful leaf is reached. This is not done often in practice, because of the extra cost. Notice the important role of unique sensitization in this problem. □

3. Repeat the previous problem for the fault g stuck-at-0.

4. Repeat the previous problem for the fault h stuck-at-1.

5. For the circuit of Figure 12.26, find a test for the fault f stuck-at-1. Draw the decision tree. Use all possible implications at every step.

   Solution. A possible decision tree is shown in Figure 12.27. In this case the set of all tests is given by:

abcd
––11

   □

6. For the circuit of Figure 12.28 find a test for the fault h stuck-at-0. Draw the decision tree. Use all possible implications at every step.

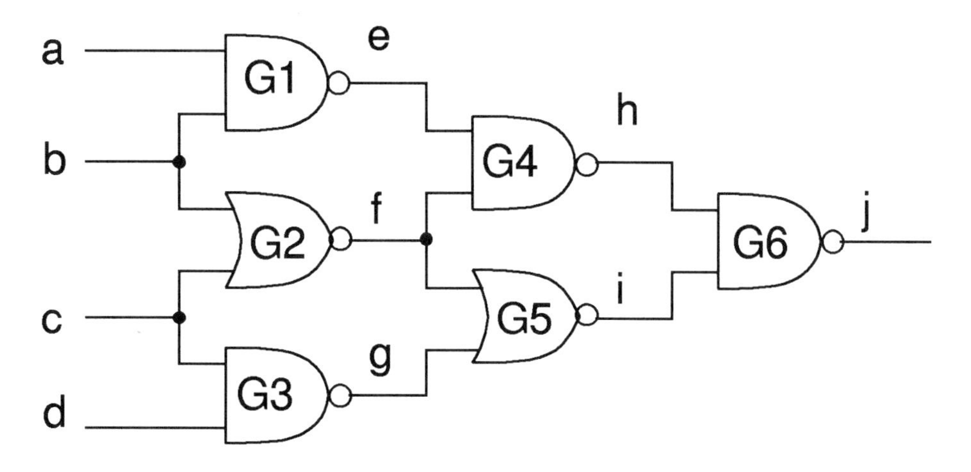

Figure 12.26: Combinational circuit for Problems 5–9.

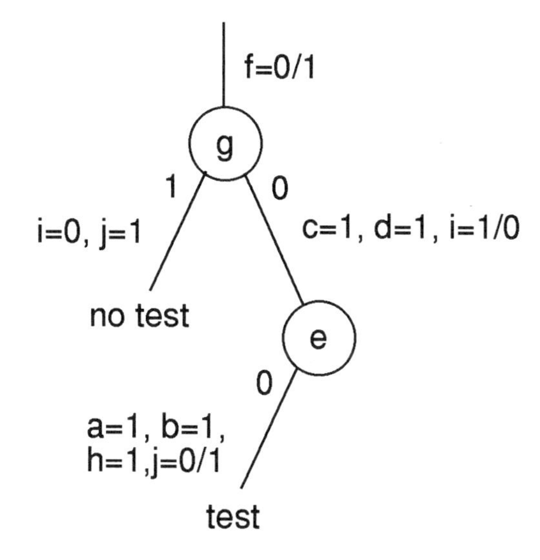

Figure 12.27: A decision tree for Problem 5.

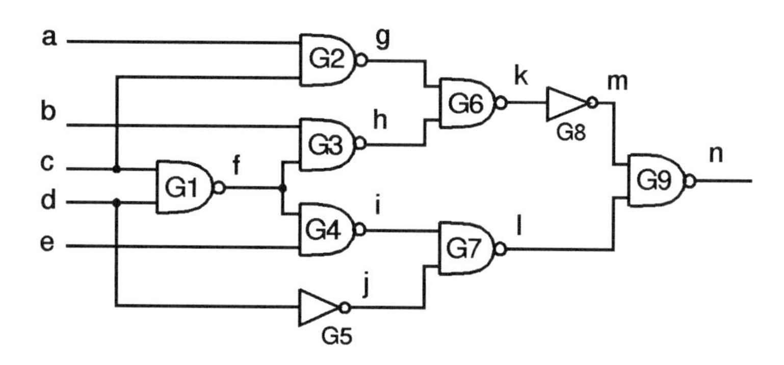

Figure 12.28: Circuit for Problem 6.

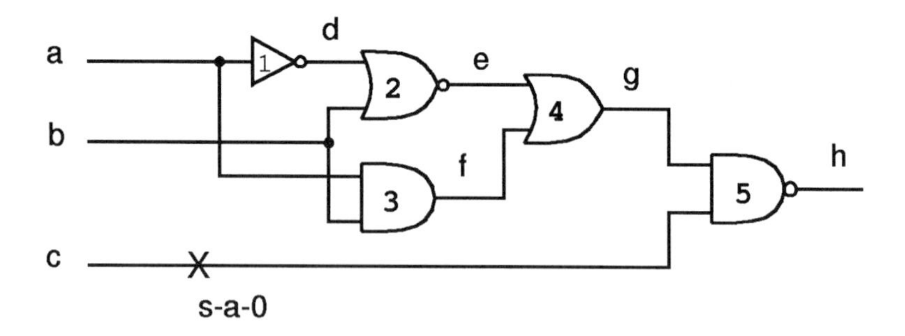

Figure 12.29: Circuit for Problem 8.

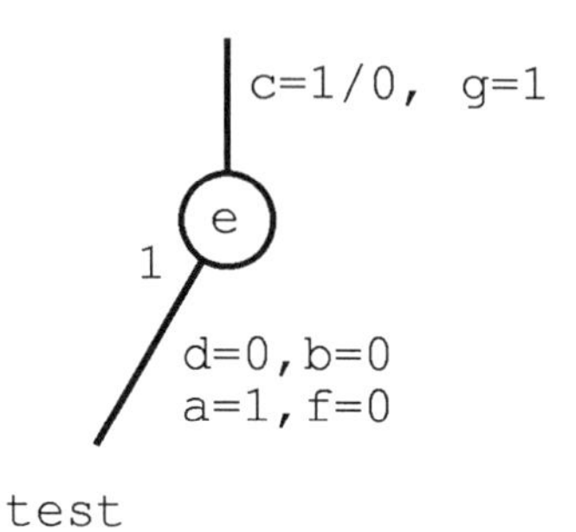

Figure 12.30: Decision tree for Problem 8.

7. Repeat Problem 6 for the stuck-at-1 fault on the input of Gate $G4$ connected to f.

8. For the circuit of Figure 12.29, generate a test for the fault "c stuck-at-0." Show the decision tree.

   Solution. A decision tree is shown in Figure 12.30. The resulting test is 101. One can verify that 111 is the only other test for "c stuck-at-0." □

9. Apply redundancy removal to the simplification of the circuit of Figure 12.26. Give the faults that you choose as targets and draw the intermediate circuits that you obtain. For every run of the test generation algorithm, draw the decision tree. [Hint: the function implemented by the circuit is $c' + d'$.]

   Solution. Since the function does not depend on a and b, as suggested by the hint, we first direct our attention to faults that explicitly remove that dependency. Since a feeds a NAND gate, we choose a stuck-at-0, to maximize the impact. When we try to generate a test for a stuck-at-0, we get the following assignments by excitation and unique sensitization:

$$a = 1/0, b = 1, f = 1, i = 1.$$

   However, $i = 1$ implies $f = 0$, a contradiction. Hence, a stuck-at-0 is untestable as expected. The simplified circuit is shown in Figure 12.31. We now consider b stuck-at-1 in the simplified circuit. Imposing $b = 0/1$, we get $c = 0$, $f = 1/0$,

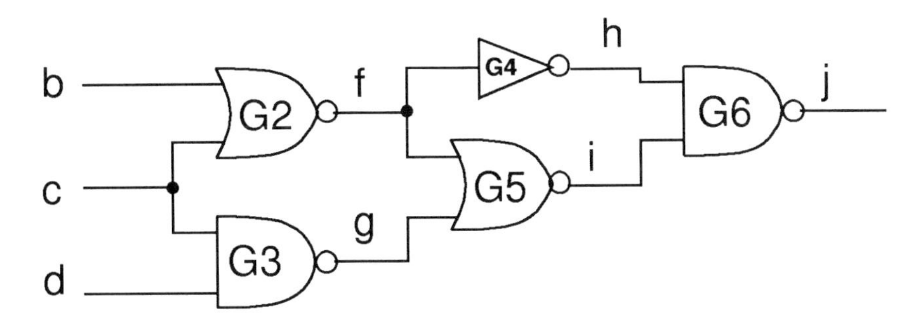

Figure 12.31: Circuit of Figure 12.26 after removal of one redundancy.

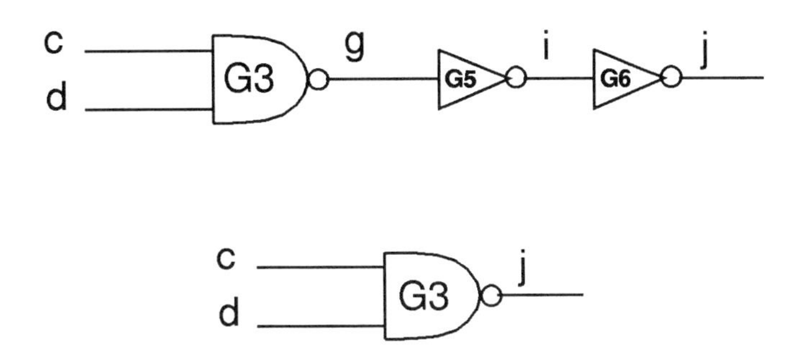

Figure 12.32: Circuit of Figure 12.31 after removal of one redundancy (top) and after further removal of the inverter pair (bottom).

$g = 1$, $i = 0$, $j = 1$. Hence, the frontier is empty and the fault is untestable. After removal of the redundancy, the top circuit of Figure 12.32 is obtained. Finally, when the string of two inverters is removed, the lower circuit in Figure 12.32 is obtained. As anticipated, it implements $c' + d'$. □

10. Apply redundancy removal to the simplification of the circuit of Figure 12.33. Give the faults that you choose as targets and draw the intermediate circuits that you obtain. For every run of the test generation algorithm, draw the decision tree.

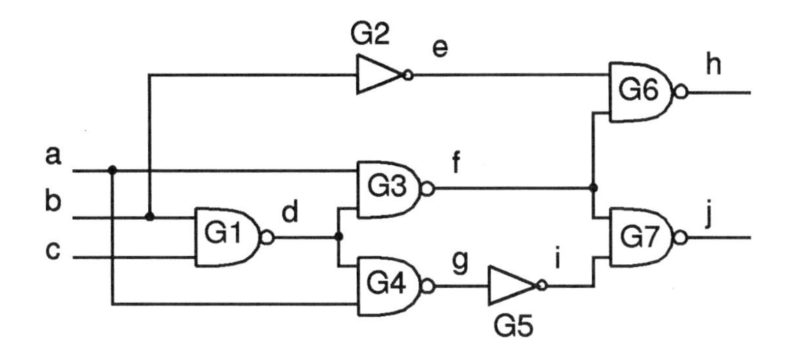

Figure 12.33: Circuit for Problem 10.

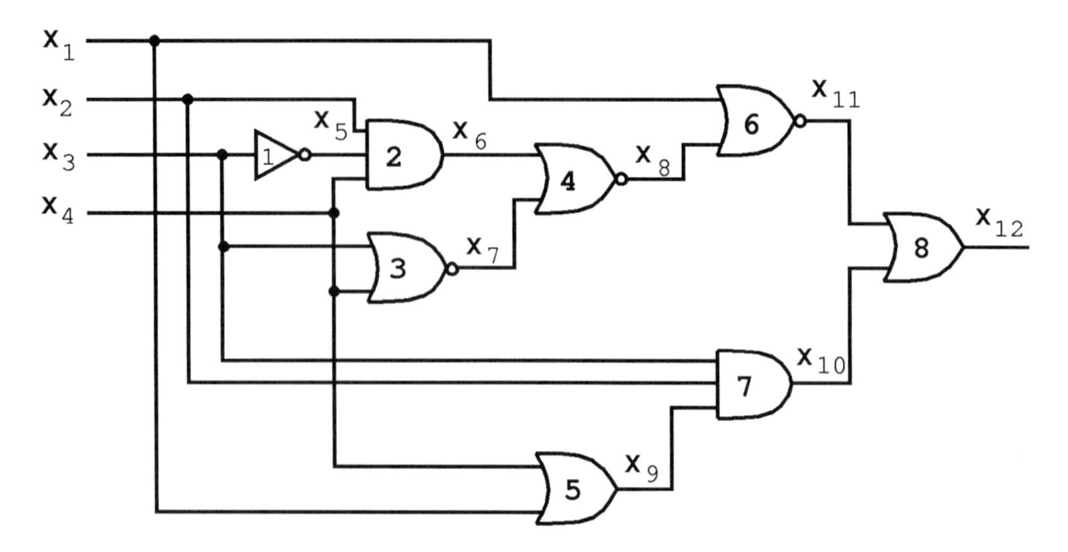

Figure 12.34: Circuit for Problem 11.

11. This problem is on redundancy removal and specifically on the dependence of the final result on the order in which redundancies are removed. Verify, by trying to generate tests for them, that, in the circuit of Figure 12.34, the faults:

    (a) x_5 stuck-at-1;

    (b) input of Gate 2 connected to x_4 stuck-at-1;

    are untestable. Prove also that the fault "input of Gate 5 connected to x_4 stuck-at-0" is testable, by generating a test for it. Verify that after removing the redundancy corresponding to "input of Gate 2 connected to x_4 stuck-at-1," the fault "input of Gate 5 connected to x_4 stuck-at-0" remains testable. Finally, verify that the removal of the redundancy corresponding to the fault "x_5 stuck-at-1" makes the fault "input of Gate 5 connected to x_4 stuck-at-0" untestable. Draw the circuit that results from removing that redundancy as well. For all test generation attempts, show the decision tree.

    Solution. We try to generate a test for "x_5 stuck-at-1." We get by implication:

$$x_5 = 0/1, x_3 = 1, x_7 = 0, x_2 = 1, x_4 = 1, x_9 = 1, x_{10} = 1, x_{12} = 1.$$

    Hence, the implications cause the frontier to disappear and no test exists. The decision tree has no decision nodes in this case.

    We now try to generate a test for "input of Gate 2 connected to x_4 stuck-at-1." We get by implication:

$$x_4 = 0, x_5 = x_2 = 1, x_3 = 0, x_7 = 1, x_8 = 0.$$

    Also on this case the frontier disappears. Hence, there is no test and the decision tree has no decision nodes.

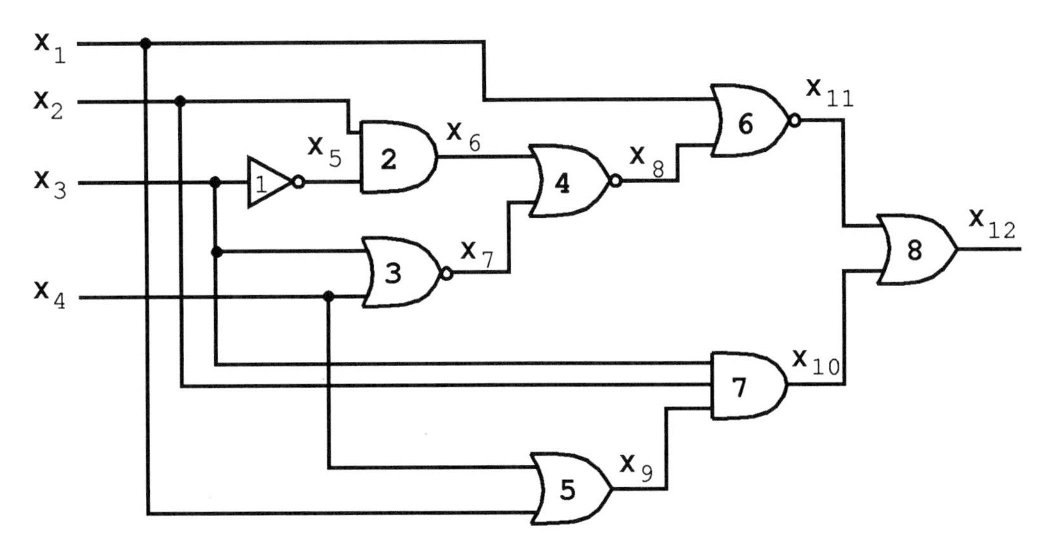

Figure 12.35: Circuit of Figure 12.34 after removal of "input of Gate 2 connected to x_4 stuck-at-1."

Let us now consider "input of Gate 5 connected to x_4 stuck-at-0." We get by implication:

$$x_4 = 1, x_1 = 0, x_2 = x_3 = 1, x_{11} = 0, x_8 = 1, x_6 = x_7 = 0, x_5 = 0,$$

$$x_9 = x_{10} = x_{12} = 1/0.$$

The test 0111 is found and the fault is therefore testable. The circuit obtained by removing the redundancy associated with "input of Gate 2 connected to x_4 stuck-at-1" is shown in Figure 12.35. We can verify that "input of Gate 5 connected to x_4 stuck-at-0" is still testable. Indeed, if we try to generate a test, we get:

$$x_4 = 1, x_1 = 0, x_2 = x_3 = 1, x_1 1 = 0, x_5 = 0, x_6 = 0, x_7 = 0, x_8 = 1,$$

$$x_9 = x_{10} = x_{12} = 0/1.$$

Hence, 0111 is still a test. One can also verify that "x_5 stuck-at-1" is now testable. Indeed, by implication we obtain:

$$x_5 = 0/1, x_3 = 1, x_2 = 1, x_1 = 0, x_{10} = 0, x_9 = 0, x_4 = 0, x_7 = 0,$$

$$x_6 = 0/1, x_7 = 1/0, x_{11} = x_{12} = 0/1.$$

Let us now consider the effect of removing "x_5 stuck-at-1." The resulting circuit is shown in Figure 12.36. If we try to generate a test for "input of Gate 5 connected to x_4 stuck-at-0," we now get:

$$x_4 = 1, x_1 = 0, x_2 = x_3 = 1, x_9 = x_{10} = 0/1, x_6 = 1, x_8 = 0,$$

$$x_{11} = 1, x_{12} = 1.$$

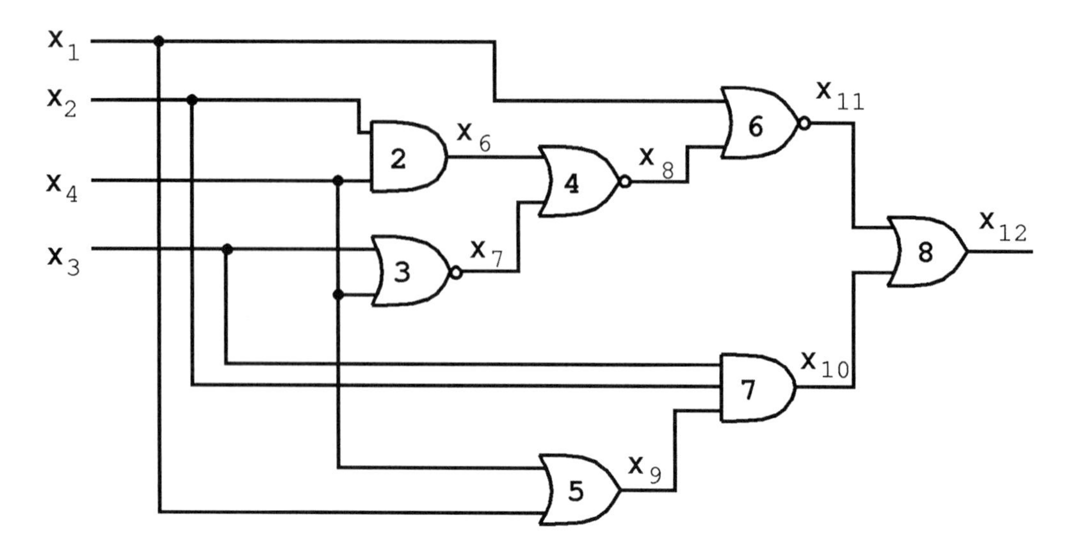

Figure 12.36: Circuit of Figure 12.34 after removal of "x_5 stuck-at-1."

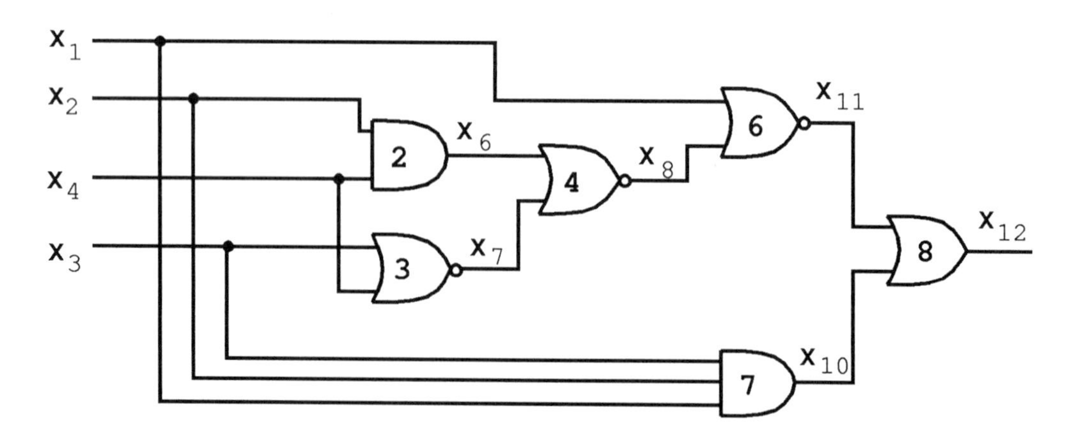

Figure 12.37: Circuit of Figure 12.36 after removal of "input of Gate 5 connected to x_4 stuck-at-0."

Hence the frontier disappears and there is no test. The circuit is redundant and can be simplified as in Figure 12.37. □

12. Run the **atpg** command of SIS on the circuit of Figure 12.28. Use the **-d** option and write the test patterns to a file. Include your BLIF file, the output of the atpg -d command, and the tests generated.

Solution. The BLIF file looks like the following:

```
.model pb12.4
.inputs a b c d e
.outputs n
.names c d f
11 0
.names a c g
11 0
.names b f h
11 0
.names e f i
11 0
.names d j
1 0
.names g h k
11 0
.names i j l
11 0
.names k m
1 0
.names l m n
11 0
.end
```

The session with SIS goes like this:

```
sis-3>atpg -d pb12.4.pat
18 total faults
RTG: covered 17 remaining 1
17 faults covered by RTG
S_A_1: NODE: i  INPUT: f
Redundant
faults: 18        tested: 17        aborted: 0        redundant: 1
```

Finally, these are the patterns generated by the **atpg** command.

```
# atpg test patterns for pb12.4
.inputs  a b c d e
00100
```

```
11011
01110
00101
10010
10110
01100
```

What tests are generated depends on the order of the gates in the BLIF file. □

13. Run the `red_removal` command of SIS on the circuit of Figure 12.33. Use the `-d` option. Include your BLIF file and the output of the `red_removal -d` command. Also include the output produced by `p` and `ps` before and after redundancy removal. Compare your result with that obtained in Problem 10.

Solution. This is the BLIF file for the circuit of Figure 12.28.

```
.model pb12.5
.inputs a b c
.outputs h j
.names b c d
11 0
.names b e
1 0
.names a d f
11 0
.names a d g
11 0
.names g i
1 0
.names e f h
11 0
.names f i j
11 0
.end
```

The session with SIS goes like this:

```
sis-1>rl pb12.5.blif
sis-2>ps
pb12.5          pi= 3   po= 2   nodes=  7        latches= 0
lits(sop)=  12  lits(fac)=  12
sis-3>p
     {h} = e' + f'
     {j} = f' + i'
     d = b' + c'
     e = b'
     f = a' + d'
```

```
    g = a' + d'
    i = g'
sis-4>red_removal -d
18 total faults
RTG: covered 13 remaining 5
13 faults covered by RTG
S_A_0: NODE: {j}        INPUT: f
Redundant
9 total faults
S_A_0: NODE: d  INPUT: b
Redundant
5 total faults
S_A_0: NODE: c  OUTPUT
Redundant
S_A_1: NODE: c  OUTPUT
Redundant
0 faults remaining after using previous tests
sis-5>ps
pb12.5          pi= 3   po= 2   nodes=  2       latches= 0
lits(sop)=   2  lits(fac)=   2
sis-6>p
    {h} = a + b
    {j} = -1-
```

We see that the result is consistent with the one obtained in Problem 10. The only difference is that SIS has performed a `sweep` command. □

Chapter 13

Technology Mapping

In the most used paradigm for logic synthesis, after a technology independent optimization of a set of logic equations, the result has to be mapped into a feasible circuit which is optimal with respect to area and satisfies a maximum critical-path delay. In this paradigm, the role of technology mapping is to finish the synthesis of the circuit by performing the final gate selection from a particular library. The algorithms chosen for technology mapping are simplified because they can be constrained by the structure of the equations produced by the technology-independent optimizations. It is not the role of technology mapping to change the structure of the circuit radically, for example, by finding common subexpressions between two or more parts of the circuit. Likewise, it is not the role of technology mapping to reduce the number of levels of logic along the critical path. The role of technology mapping is the actual gate choice to implement the equations—for example, choosing the fastest gates along the critical path, and using the most area-efficient combination of gates off the critical path.

There are several characteristics which are desirable for a technology mapping algorithm. These are:

1. Adapt easily to different libraries.
2. Support irregular collections of logic functions.
3. Handle detailed technology-dependent cost functions.
4. Efficient execution time.

First, it is desirable that the technology mapping algorithm be able to adapt to a variety of different libraries with minimal effort. This is difficult because many libraries have an irregular collection of logic functions available as primitives. An algorithm which depends on characteristics of a particular library (for example, availability of a complete set of CMOS and-or-invert gates) is of limited use. Also, an algorithm which is geared to a subset of the gates in a library is limited in its optimization potential. To achieve the goal of library adaptability, an approach to technology mapping should be *user-programmable*. The user should be able to provide new gates to the technology mapper without understanding its detailed operation, and these gates should be used effectively.

During technology mapping, simple cost functions such as transistor count or levels of logic will not provide high-quality circuits. Instead, it is necessary to consider more detailed models for the cost of a gate in the actual target technology. This detailed level of modeling, coupled with gates which have irregular area and delay cost functions, complicates the technology mapping process.

Therefore, to provide high-quality results for different libraries and circuits, a technology mapping algorithm must make few assumptions about the relative cost and performance of the gates in a library, and must be prepared to model accurately the cost functions which are optimized.

While it is always desirable to have an efficient algorithm, generally the execution performance of the technology mapping algorithm is less important than the quality of the final result. This is true for the last optimization of a circuit before fabrication. However, the steps of technology-independent optimization and technology mapping are iterated by a logic synthesis system if the performance goals are not initially met. Technology mapping in this case operates as an accurate predictor for the quality of a technology-independent representation and these results are fed back to the technology-independent optimization to improve the final implementation. Therefore, it is desirable that a technology mapping algorithm support a fast execution mode as well as a slower mode which provides higher-quality optimization.

The two basic approaches followed for technology mapping are:

1. rule-based techniques [14, 149, 83];

2. graph covering techniques [39, 158].

Rule-based techniques have the same structure as rule-based techniques for technology independent optimization. It is important to mention that a rule-based system can combine the technology independent and technology mapping stages providing, in principle, a more global view of logic optimization. However, the nature of rule-based systems is to perform local optimization, thus yielding an interesting trade-off with the paradigm that separates the two stages but offers a more global view of each of the stages.

Even though they suffer from flexibility and problems of large execution time, local transformation techniques have demonstrated the ability to produce high-quality results.

In this chapter, we focus on the graph covering based techniques. These techniques match well the requirements listed above.

13.1 Graph Covering and Technology Mapping

The approach of using Directed-Acyclic-Graph(DAG)-covering for technology mapping in logic synthesis was first proposed by K. Keutzer in the program DAGON [158]. His thesis was that technology mapping for logic synthesis is closely related to the problem of code generation for programming language compilers, and hence the advanced techniques that have been developed for code generation should be applicable to technology mapping.

The problem of code generation in a compiler is to map a set of expressions onto a set of machine instructions for the target machine. Extensive research into compilers has led to efficient ways of formulating and solving this problem [4]. Each machine instruction is decomposed into a directed acyclic graph (DAG) of atomic operations, called a pattern. Each instruction has a cost associated with it which represents the relative cost, in execution time, of choosing that instruction. The sequence of high-level expressions is also represented by a DAG of atomic operations. The optimal code generation problem is equivalent to finding an optimum cost cover of the subject DAG by the pattern DAGs.

A similar approach is taken for the technology mapping problem. A set of base functions is chosen such as a two-input NAND-gate and an inverter. The logic equations are optimized in a technology-independent manner and are then converted into a graph where each node is restricted to one of the base functions. This graph is called the **subject graph**. The logic function for each library gate is also represented by a graph where each node is restricted to one of the base functions. Each graph for a library gate is called a **pattern graph**. For any given logic function there are many different representations of the function using the base function set. Therefore, each library gate is represented by many different pattern graphs.

The technology mapping problem is viewed as the optimization problem of finding a minimum cost covering of the subject graph by choosing from the collection of pattern graphs for all gates in the library. A *cover* is a collection of pattern graphs such that every node of the subject graph is contained in one (or more) of the pattern graphs. The cover is further constrained so that each input required by a pattern graph is actually an output of some other pattern graph. For area optimization, the cost of the cover is defined as the sum of the areas of the individual gates. For minimum delay optimization, the cost of a cover is defined as the critical path delay of the resulting circuit using an appropriate delay model. For the more typical optimization problem of optimizing for minimum area under a given timing constraint, any cover which results in a circuit with critical path delay greater than that allowed for any output is considered an illegal cover; thus, the minimum-area legal cover is the optimization goal. If there are no legal covers, the cover of minimum delay is considered the desired solution.

The critical parts of the procedure are the selection of the set of base functions and the optimization technique used to solve the covering problem.

13.2 Choice of Base Functions

The choice of a set of base functions is arbitrary as long as the base function set is functionally complete.[1] However, this decision does influence the number of patterns needed to represent the gates in a library and the quality of the solution provided by DAG-covering. The goal is to find the base-function set which provides the highest level of optimization and produces a small set of patterns. In SIS [249, 236], a base-function set of a two-input NAND-gate and an inverter is used. This choice is motiv-

[1] A functionally complete set is a set of functions that can express any other function. For instance, two-input ANDs, ORs, plus inverters allow one to represent any switching function. Another example is provided by two-input NANDs (or NORs) and inverters.

ated by the fact that this set can be proved [236] to be as good in terms of optimization potential as any other set containing two-input NOR-, AND-, OR-gates, and inverters.

When both a NAND-gate and NOR-gate are used in the base-function set, the number of patterns to represent some functions increases. For example, using both a two-input NAND-gate and a two-input NOR-gate, a large number of pattern graphs are required for all representations of the gate $f = (ab + cd)'$. Variations such as three NAND-gates (with inverters), three NOR-gates (with inverters), and other representations using both NAND-gates and NOR-gates are possible patterns for this gate. Using only the two-input NAND-gate reduces the number of patterns to one.

The covering paradigm implies that each node of the subject graph is covered by a pattern, but cannot be split and partially covered by two patterns. Therefore, the granularity of the base function set affects the optimization potential. Thus, a fine resolution base-function set allows for more covers, and hence better quality solutions. However, this has a price—more patterns are required to represent the logic function for some gates. In DAGON, two-input, three-input, and four-input NAND-gates are used as the base-function set. With this approach, the logic function

$$f = (abcd + efgh + ijkl + mnop)'$$

requires only one pattern—a tree of five four-input NAND-gates. Representing all patterns for this same function using two-input NAND-gates and inverters requires eighteen patterns. However, given the possibility for improved optimization, the finer resolution base function appears to be the better approach.

13.3 Creating the Subject Graph

A logic network has many representations as graphs of components from the base-function set and each representation is a potential subject graph for DAG-covering. Each starting point leads to a graph cover of possibly different cost. Even if the covering problem is solved exactly, every one of these starting points should be considered for an optimum solution.

Therefore, heuristics are used to find an optimal form for the subject graph. As mentioned in the introduction, these optimizations include algebraic decomposition and Boolean simplification techniques using technology-independent cost functions. The number of nodes in the subject graph is used as a technology-independent estimate of the area of the circuit. The total number of literals in SOP form is effectively the same area estimator. The longest path from an input to an output in the subject graph is used to estimate the delay of the circuit.

The goal of technology-independent optimization should be to find a representation for the circuit which provides a good starting point for DAG-covering. The optimized equations are then transformed into two-input NAND-gate and inverter form in a straightforward manner. The sum of products of each node is translated into an AND-OR circuit. Each gate is realized as a balanced tree of two-input NAND gates and inverters. Different decompositions give different results, in general.

13.4 The DAG-Covering Problem

DAG-covering-by-DAGs is NP-hard even with only three pattern graphs (inverter, two-input NAND, two-input NOR) and if each subject graph node has no more than two incoming and outgoing edges [159].

An exact covering algorithm has been proposed in [236] based on a branch-and-bound procedure. However, the complexity of the algorithm is so large that only trivial problems could be solved. Anyway, it is debatable whether we need to solve this problem exactly, since the subject graph is already the result of a heuristic mapping and hence does not reflect the most general optimization problem that needs to be solved. Hence, a more effective approach would be to develop a heuristic DAG-covering algorithm. However, this is still an open problem (L. Lavagno at Berkeley has experimented with a number of heuristic with some degree of success).

An alternative approach to the DAG-covering problem is to simplify it so that the simplified problem could be solved effectively (for example in linear time). Of course, the quality of the final solution will depend on the reduction of the search space.

Keutzer in DAGON [158] has proposed reducing the DAG-covering problem to a set of tree-covering-by-trees problems. His procedure is based on the following steps:

1. Partition the subject graph into trees;
2. Cover each tree optimally;
3. Piece the tree-covers into a cover for the subject graph.

This approach has been proven quite effective. In particular, it can be proven that, if the cost function is additive such as area, the tree-covering problem can be solved with a linear complexity algorithm based on dynamic programming. DAGON is a technology mapping program written by Keutzer on top of the tree manipulation tool TWIG [260], which was originally developed to provide a flexible framework for building efficient algorithms for tree matching and for solving the tree-covering problem. TWIG uses the Aho-Corasick [2] string-matching algorithm for matching and the Aho-Johnson [4] dynamic programming algorithm for optimal tree covering.

Its weak points are in the loss of global view due to the step of partitioning into trees. Covers across partition boundaries are not allowed. It is interesting to see whether different partitioning algorithms can substantially improve the results obtained with this procedure.

The approach followed in MIS [39, 239] is patterned after DAGON. To improve the quality of the solution, additional covers are exposed by replacing any straight interconnection between gates with a pair of inverters. The search space is augmented substantially at little cost. The recent work of Lehman and Watanabe takes this idea one (or more) steps further and expands the search space to include logic reachable by local algebraic resynthesis [172].

13.5 Tree Covering by Dynamic Programming

In this section we address the problem of optimally covering a tree circuit by means of tree patterns. A tree circuit is a single output circuit in which each gate, except the

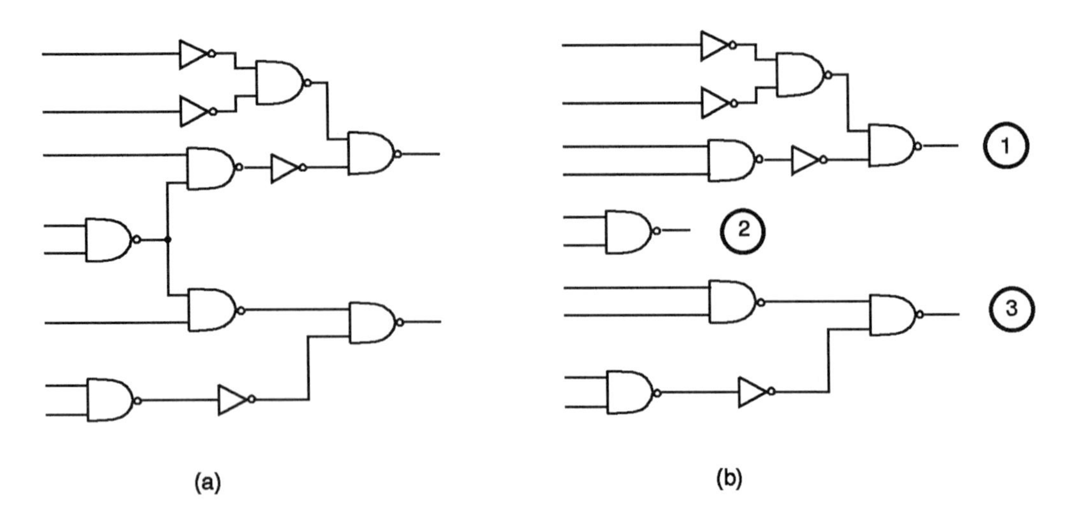

Figure 13.1: Splitting a DAG into a Forest of Trees.

output, feeds exactly one other gate. We assume that the circuit, called the *subject graph*, is made of two-input NAND gates and inverters. The tree patterns are also small circuits made of two-input NAND gates and inverters; they represent the cells of the library.

Not all circuits of interest are trees; indeed, most are not. However, we can easily decompose a DAG into trees, by splitting it at the fanout points. This is illustrated in Figure 13.1. The circuit on the left is not a tree, because there is one gate that feeds two other gates. By splitting at the fanout point, we get the three trees on the right of Figure 13.1. Each tree is mapped individually and the solution for the original circuit is obtained by connecting the solutions for the three trees. From this point on, therefore, we concentrate on mapping a tree.

The mapping procedure consists of two phases:

- pattern matching; and

- tree covering.

In the matching phases we find all possible ways in which a library pattern may cover some nodes of the subject tree. To fix ideas, let us consider the library of Figure 13.6 and the third tree from Figure 13.1(b). In the library of Figure 13.6, every gate is represented by a single pattern. (In general this is not true.) Also, for each gate, the cost is given. For instance, the cost of the three-input NAND gate is 3.

For each node, we see if we can match any of the library patterns. Clearly, all NAND gates in the subject tree are matched by the pattern for the two-input NAND gate, and all inverters in the subject tree are matched by the pattern for the inverter. For the output node, however, there we can also match the three-input NAND gate.

In general, after the matching phase, each node will have a (non-empty) set of matches. The set is not empty, because the library includes the inverter and the two-input NAND gate. The second phase—covering—selects one optimum matching for each node.

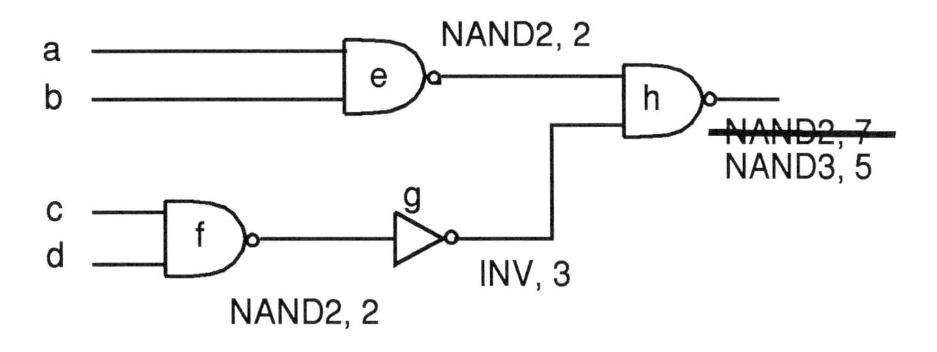

Figure 13.2: A Subject Tree and its Matches.

Let us consider our example, which is reproduced, for convenience, in Figure 13.2. If we select the three-input NAND gate to match the output node, then we cover gates f and g, besides h; hence, we do not need matches for f and g. If, on the other hand, we choose the two-input NAND gate, then we need to select gates to cover f and g. We see that some nodes that appear in the subject tree may become, in the final mapped circuit, internal to a gate.

We now present a systematic approach to selecting the matchings that follows a strategy called *dynamic programming*[2]. This approach guarantees a cover of minimum cost and is very efficient. Its run time grows linearly with the size of the subject tree and with the number of library patterns.

We start from the primary inputs and we consider the gates in some *topological order*. A topological order is any order in which no gate precedes one of its fanouts. For instance, (e, f, g, h) and (f, e, g, h) are two topological orders for the example of Figure 13.2; we shall follow (e, f, g, h).

For each gate, we determine the optimum cover of the subtree rooted at the gate output. in our example, gates e, f, and g pose no problem, because they have only one possible match. The cost of a cover is determined as the cost of the match at the node, plus the optimum costs of the covers of the nodes that are inputs to the match. The optimum covers of the primary inputs have cost 0, of course. Applying this rule, the optimum cost for e and f is 2. For g, we have to sum the cost of the inverter matching g to the optimum cost of covering f, which has been determined to be 2. Hence, we get the value of 3 shown in Figure 13.2. Notice the importance of proceeding in topological order: It guarantees that the optimum cost of the inputs is known when a match is considered.

Let us consider now the choice of the match for the primary output of the circuit. Since there are two possible matches, we shall compute two solutions and choose the best. If we choose the two-input NAND to cover h, then we need to implement covers for the outputs of e and g. We already know that an optimum cover for e has a cost of 2 and an optimum cover for g has a cost of 3. Since the two-input NAND gate has a cost of 2, we get a minimum cost of 7, if this match is chosen.

If, on the other hand, we select the three-input NAND gate as match, we only

[2]Dynamic Programming is a strategy for optimization that can be applied to many problems [18].

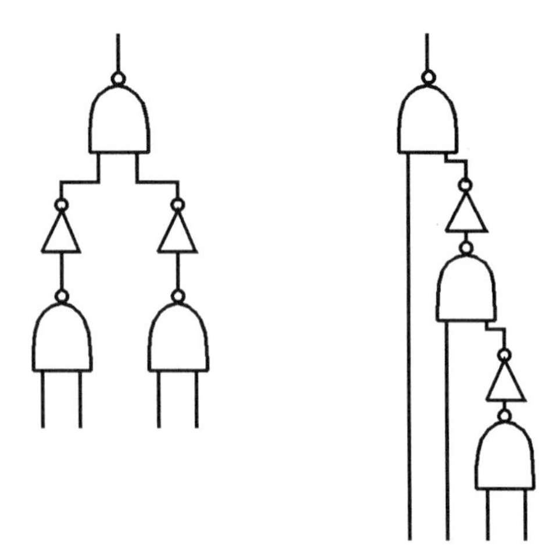

Figure 13.3: The Two Possible Patterns for a Four-Input NAND Gate.

need to implement a cover for e, at an optimum cost of 2. Since the three-input gate costs 3, we get a total cost of 5. This latter solution is obviously better than the previous one and is therefore chosen as the optimum cover for the primary output.

In summary, for each node we consider all possible matches. A given match requires some nodes as inputs. Because we proceed in topological order, we know the optimum cost of covering those nodes; hence we can compute the optimum cost, under the assumption of choosing the given match. If we do the same for all possible match of a node, we compute an optimum cover for that node, by simply selecting one of the matches that give minimum cost. We finally proceed backwards from the primary output, and collect all the optimum matches for the nodes that need to be implemented in the optimum cover of the output.

This strategy does not work for DAGs in general, because there may be conflicting requirements on how to map a node that fans out to several gates.

13.6 Decomposition

In this section we discuss, with the help of examples, issues related to the decomposition of the Boolean network and the library gates into two-input NAND gates and inverters.

First of all, we notice that the decomposition is not unique. In Figure 13.3, we show that a four-input NAND gate may be described in two different ways. The pattern on the left is a *balanced* tree: The four inputs are equally split between the two gates at the first level. The pattern on the right, on the other hand, corresponds to an *unbalanced* tree: One input comes from the left and three inputs come from the right. The pattern where one input comes from the right and three inputs come from the left is isomorphic to this one and is not considered a distinct pattern for the purpose of matching. It is the matching procedure that takes care of trying all possible permutations of the inputs.

The non uniqueness of the decomposition forces us to consider all distinct (up to isomorphism) pattern for the library gates. For the circuit to be mapped, however, we

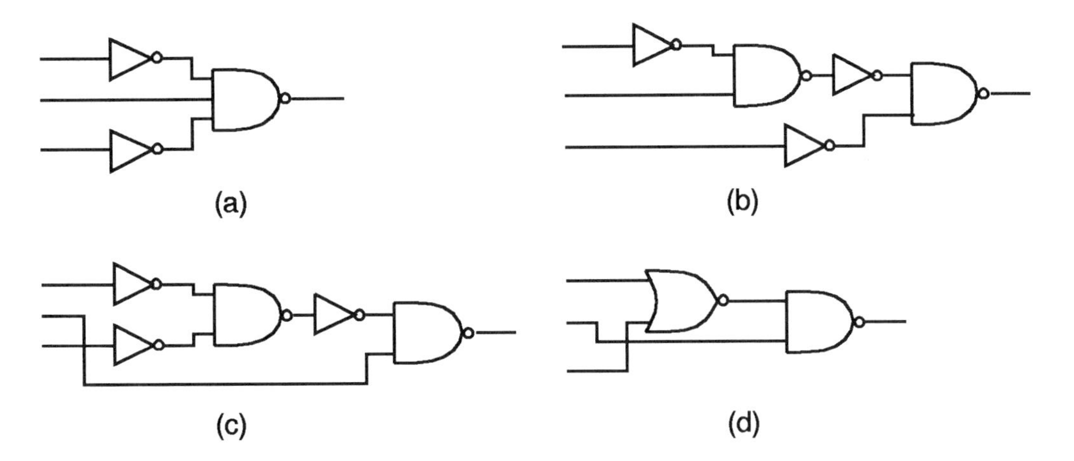

Figure 13.4: Two Possible Decompositions of the Same Circuit.

only consider one decomposition. There are, in general, too many decompositions for a large circuit to make it possible for one to examine all of them in a reasonable time. We should be aware, however, that this restriction to one specific decomposition may lead to suboptimal mappings[3].

Consider Figure 13.4. The circuit to be mapped is given in Part (a). A first decomposition is shown in Part (b). When we apply tree covering with the library of Figure 13.6, we get back the same circuit of Part (a), which costs 5. If we use the decomposition of Part (c), however, we get the mapped circuit of Part (d), which has a cost of 4.

The example of Figure 13.4 also illustrates another aspect of decomposition. If we allow three-input NAND gates in the subject trees, besides two-input NAND gates and inverters, then the circuit of Part (a) is already in decomposed form. Therefore, the cheaper mapping cannot be found.

13.7 Delay Optimization and Graph Covering

Synthesis for performance is increasingly important due to the push towards electronic systems with more optimal performance. Thus, a solution for technology mapping must consider timing in a direct way. If the delay were independent of the gate driven (i.e., a constant load model is used), then a dynamic programming algorithm of linear complexity could be applied as well. Thus far, even though this model is not accurate by any means, technology mapping for delay was carried out under this assumption. The results obtained were reasonable but by no means optimal. In fact, for a general delay cost function, the optimal cover depends of the forward part of the tree and hence the dynamic programming algorithm cannot be applied as is.

Rudell [236] has suggested a method to solve the minimum delay optimization problem for trees and the constrained-by-timing area optimization problem. His idea is based on a binning technique for the pin-loads as follows.

[3]It is important here to keep in mind that the optimality that we have claimed for the tree covering procedure applies only with respect to a given decomposition of the subject graph and to the covering of the trees.

1. The unique set of pin-loads is determined and binning functions are constructed;

2. Obtain an array of solutions at each node of the subject tree, one per bin;

3. The arrival time for each cover for each load value is computed;

4. At each input of the cover, the optimal solution for driving the corresponding pin-load is selected.

5. The final cover is chosen based on the external load at the root of the tree.

The cover obtained by this technique is a minimum delay cover. Note that this approach subsumes all technology mapping related problems such as phase assignment and discrete sizing. It can also be generalized to solve the problem of technology mapping for optimum area cover with delay-constraints.

The complexity of the algorithm is still linear but it depends on the number of load-pins and arrival-time bins. For a reasonable library, we can have as many as 100 different pin-loads and 10,000 arrival bins (.01 ns for 100ns) yielding 1,000,000 solutions per node! Hence to make this algorithm practical, Rudell devised an approximate technique that uses only a fixed number of bins. A clustering algorithm provides a good value for the bins so that the approximation due to the insufficient number of bins is minimized. A straight-forward implementation of the algorithm runs only four times slower than the standard algorithm.

13.8 Notes

The ruled base approach to technology mapping was pioneered by [83]. The tree covering approach was first proposed in [158]. Rudell [236] formulated technology mapping as a binate covering problem. The usual approach to matching disregards don't care information. A more accurate approach is provided by *Boolean matching* [180]. An interesting recent development combines the decomposition and mapping phases [172]. Technology mapping for performance is addressed in [261]. After technology mapping, gate resizing may reduce area, delay, and power [100, 13]. Technology mapping for Field Programmable Gate Arrays (FPGAs) is covered in [203, 103].

13.9 Summary

Technology mapping transforms an abstract representation of a multilevel logic circuit into an interconnection of gates from a library. We have examined the approach to technology mapping that uses tree covering. The Boolean network is first decomposed in simple gates—typically NANDs and inverters: The result is called the subject graph. Matching then identifies all possible ways in which a gate of the subject graph can be implemented by a gate in the library. The best combination of matches is then chosen by a dynamic programming approach. Dynamic programming solves exactly only the covering part of the problem. Therefore, the results are of good quality, but no global optimality is guaranteed.

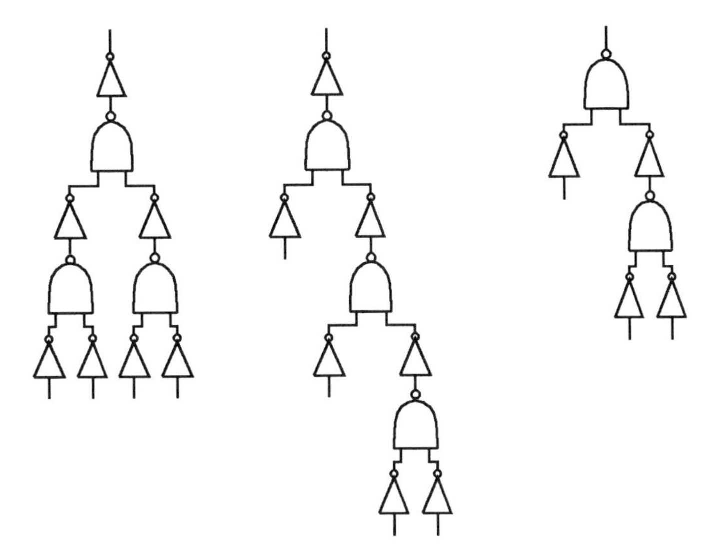

Figure 13.5: Library Patterns for Four-Input NOR and Three-Input OR.

13.10 Problems

1. Draw the patterns, in terms of two-input NAND gates and inverters, for a four-input NOR gate and for a three-input OR gate.

   Solution. There are two patterns for the four-input NOR gate and one pattern for the three-input OR gate. They are shown in Figure 13.5. □

2. In this problem you will technology map the circuit shown in Figure 13.7, minimizing for area. The library to use is

function	cost
$INV(x_1) = x_1'$	1
$NAND2(x_1, x_2) = (x_1 x_2)'$	2
$NAND3(x_1, x_2, x_3) = (x_1 x_2 x_3)'$	3
$NOR2(x_1, x_2) = (x_1 + x_2)'$	2
$OAI21(x_1, x_2, x_3) = ((x_1 + x_2) x_3)'$	3

   In both parts you should illustrate the way you decomposed the library cells into pattern trees, show the best cost solution trace through the subject tree and show your final cover for the circuit.

   (a) Use the dynamic programming method we used in class with no modifications. Your total cost should be no more than 14.

   (b) Modify the basic method we used by inserting a pair of series inverters into every wire that does not already have an inverter at one end (including the root and leaves of the subject tree). You will need to modify your library slightly by

      i. adding a dummy *FEEDTHRU* element, which matches two series inverters and has a cost of 0;

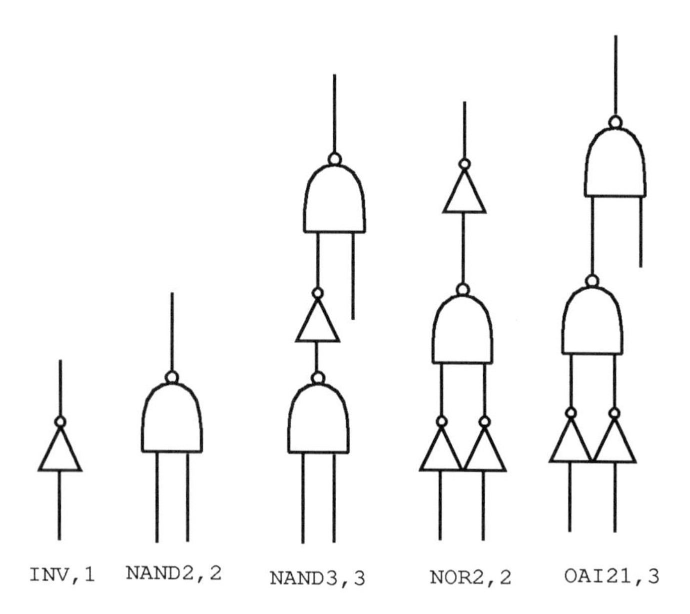

Figure 13.6: Library of Pattern Trees.

ii. inserting an inverter pair into each *internal* wire of your pattern trees that does not already touch an inverter.

Perform dynamic programming with these modifications and you should obtain a better solution. Since the modified method requires that you synthesize some nodes (i.e., between inverter pairs) which were not in the original circuit, it is desirable that you use as few of these as possible. This can be achieved by choosing the solution which uses the fewest synthesized nodes when comparing solutions of equal cost.

(c) The *depth* of a tree mapping is the length of the longest path, measured in terms of library cells, from a leaf to the root of the subject tree. What is the depth of each of your solutions in parts (a) and (b)? (Observe that the *FEEDTHRU* has a depth of 0, since it is not a physical cell).

(d) In a few sentences, explain how you could modify the costing scheme of (a) to minimize the depth of the solution.

Solution.

(a) See Figures 13.6–13.8. In particular, in Figure 13.7, all possible matches are shown for all gates. For each match, the optimal cost is given. Those matches that give sub-optimal costs are struck out.

(b) See Figures 13.9–13.11.

(c) We can see by inspection of the final covers that the solutions to parts (a) and (b) both have depth 4.

(d) To calculate depth instead of area, we look at all the fanin nodes to the current match and use the *longest*. To this value we add the depth of the current match (usually 1, although the *FEEDTHRU* has depth 0). This

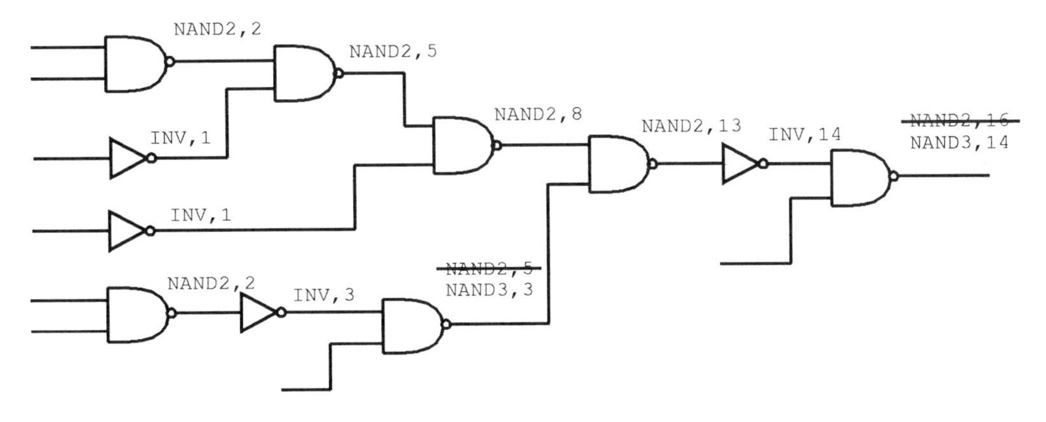

Figure 13.7: Best Solution Trace.

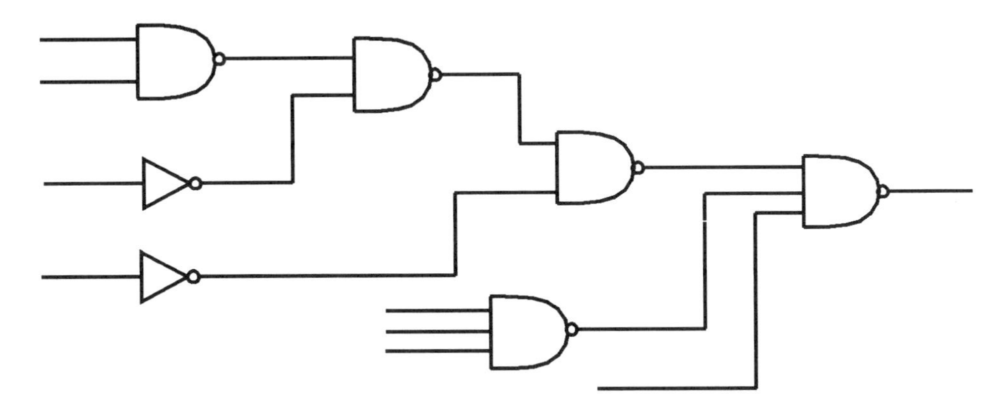

Figure 13.8: Final Cover.

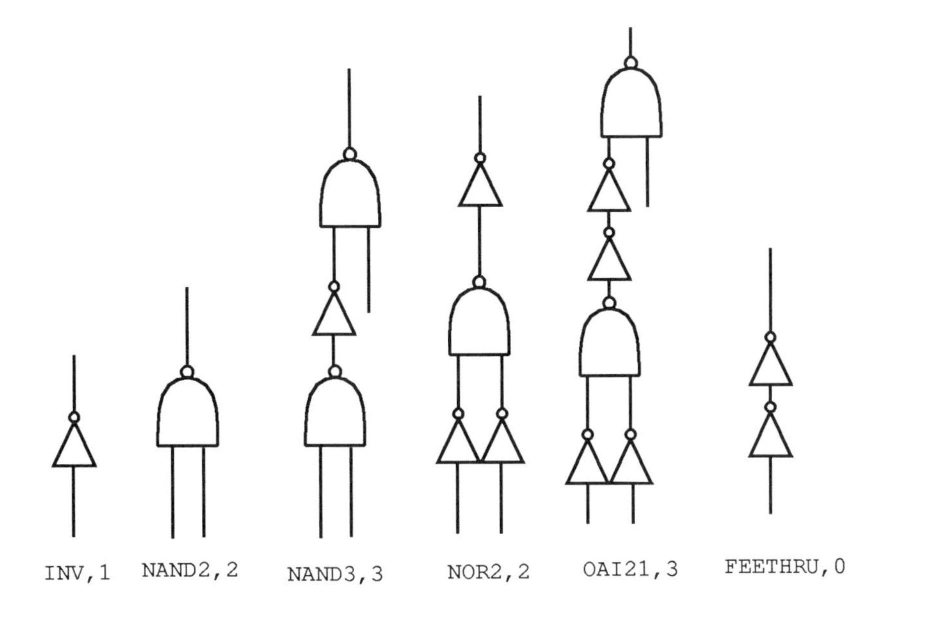

Figure 13.9: Modified Library of Pattern Trees.

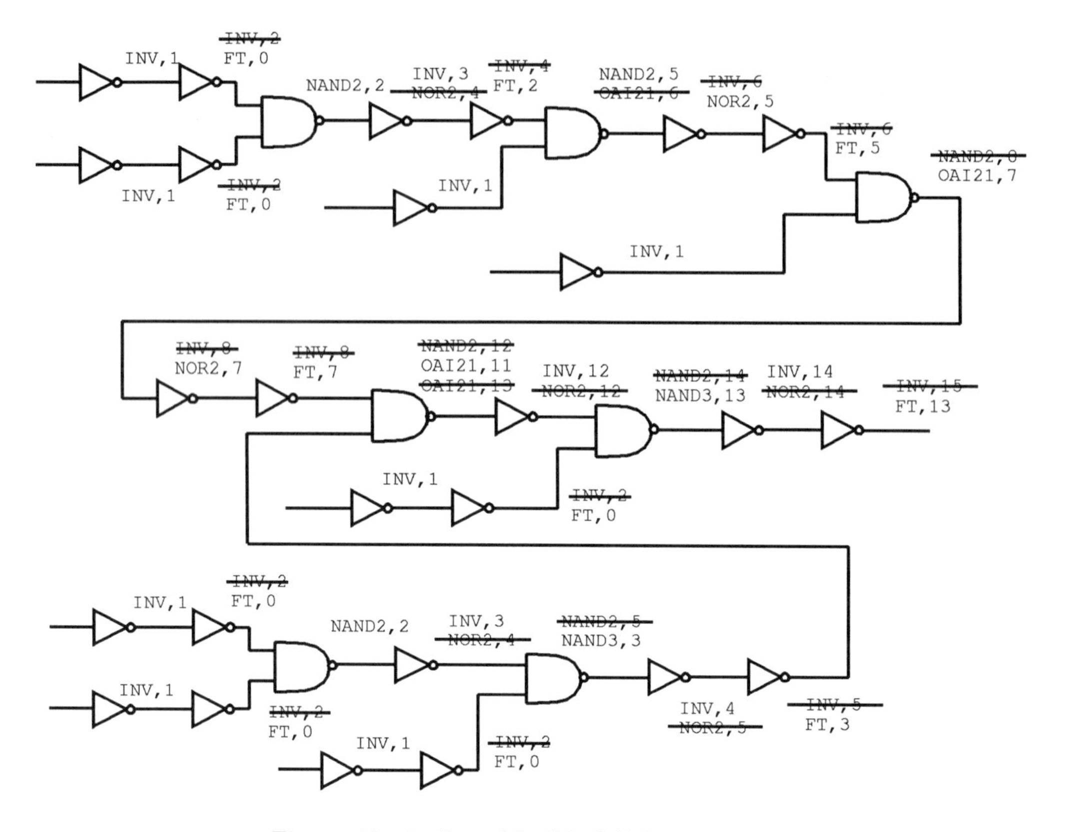

Figure 13.10: Best Modified Solution Trace.

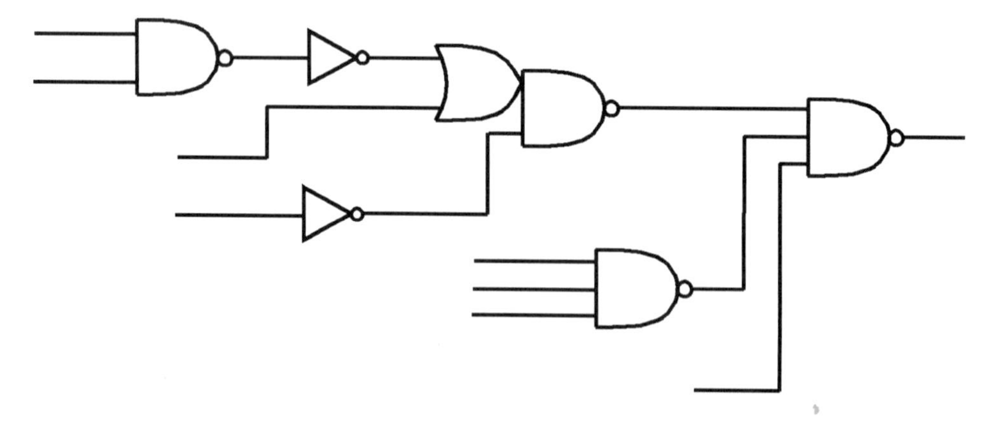

Figure 13.11: Modified Final Cover.

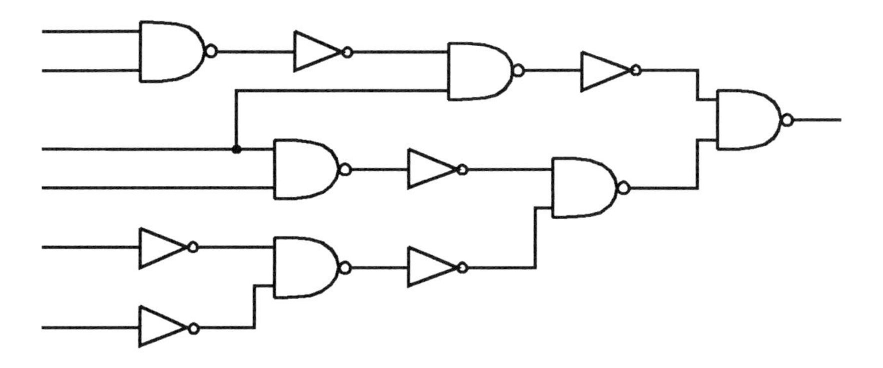

Figure 13.12: Boolean Network for Problem 3.

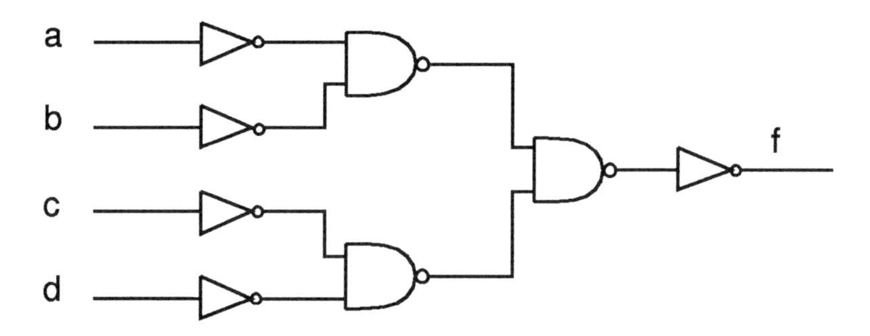

Figure 13.13: Boolean Network for Problems 4 and 5.

becomes the cost of the current solution. Of all the possible solutions at the node under computation, we keep the best one. This guarantees that every tree has minimum depth.

□

3. Apply the dynamic programming algorithm to the network of Figure 13.12. Do not insert inverter pairs. Use the library of Figure 13.6.

4. Apply the dynamic programming algorithm to the network of Figure 13.13. First, solve the problem without inserting inverter pairs and using the library of Figure 13.6. Then insert inverter pairs and use the library of Figure 13.9. Draw the two solutions and say which one has lower cost.

5. This problem is on using SIS for technology mapping. You have to describe the library of Figure 13.6 in the GENLIB format and then use this library to map the circuit of Figure 13.13. For documentation on the GENLIB format, read `~fabio/sis1.1/sis/doc/genlib.doc`. There are also several examples of libraries in the standard SIS library (In a typical installation, the path for user fabio would be `~fabio/sis1.1/sis/sis_lib`).

   Use 1 for `<input-load>` and 999 for `<max-load>`. Use 1.0 for all block delays and 0.0 for all fanout delays.

Describe the circuit of Figure 13.13 in BLIF format, using one `.names` directive for each gate. Check the solution obtained by SIS with the `map` command against your solution of Problem 4. Include your library and circuit descriptions, as well as the output from SIS.

Solution. The library is:

```
GATE zero    0   O=CONST0;
GATE one     0   O=CONST1;
GATE inv     1   O=!a;           PIN * INV 1 999 1.0 0.0 1.0 0.0
GATE nand2   2   O=!(a*b);       PIN * INV 1 999 1.0 0.0 1.0 0.0
GATE nand3   3   O=!(a*b*c);     PIN * INV 1 999 1.0 0.0 1.0 0.0
GATE nor2    2   O=!(a+b);       PIN * INV 1 999 1.0 0.0 1.0 0.0
GATE oai21   3   O=!((a+b)*c);   PIN * INV 1 999 1.0 0.0 1.0 0.0
```

The `blif` description of the circuit is:

```
.model pb11.4
.inputs a b c d
.outputs f
.names a abar
0 1
.names b bbar
0 1
.names c cbar
0 1
.names d dbar
0 1
.names abar bbar n1
11 0
.names cbar dbar n2
11 0
.names n1 n2 n3
11 0
.names n3 f
0 1
.end
```

The output of SIS should look like this:

```
sis-1>rl pb11.4.blif
sis-2>rlib pb11-4.genlib
sis-3>map
WARNING: uses as primary input drive the value (0.00,0.00)
WARNING: uses as primary input arrival the value (0.00,0.00)
WARNING: uses as primary input max load limit the value (999.00)
WARNING: uses as primary output required the value (0.00,0.00)
WARNING: uses as primary output load the value 1.00
```

```
sis-4>p
     n1 = a' b'
     n2 = c' d'
     {f} = n1' n2'
sis-5>print_gate
n1          nor2             2.00
n2          nor2             2.00
{f}         nor2             2.00
```

Note that SIS automatically inserts the inverter pairs. □

Appendix A

ASCII Codes

<table>
<tr><td>00</td><td>nul</td><td>01</td><td>soh</td><td>02</td><td>stx</td><td>03</td><td>etx</td><td>04</td><td>eot</td><td>05</td><td>enq</td><td>06</td><td>ack</td><td>07</td><td>bel</td></tr>
<tr><td>08</td><td>bs</td><td>09</td><td>ht</td><td>0a</td><td>nl</td><td>0b</td><td>vt</td><td>0c</td><td>np</td><td>0d</td><td>cr</td><td>0e</td><td>so</td><td>0f</td><td>si</td></tr>
<tr><td>10</td><td>dle</td><td>11</td><td>dc1</td><td>12</td><td>dc2</td><td>13</td><td>dc3</td><td>14</td><td>dc4</td><td>15</td><td>nak</td><td>16</td><td>syn</td><td>17</td><td>etb</td></tr>
<tr><td>18</td><td>can</td><td>19</td><td>em</td><td>1a</td><td>sub</td><td>1b</td><td>esc</td><td>1c</td><td>fs</td><td>1d</td><td>gs</td><td>1e</td><td>rs</td><td>1f</td><td>us</td></tr>
<tr><td>20</td><td>sp</td><td>21</td><td>!</td><td>22</td><td>”</td><td>23</td><td>#</td><td>24</td><td>$</td><td>25</td><td>%</td><td>26</td><td>&</td><td>27</td><td>’</td></tr>
<tr><td>28</td><td>(</td><td>29</td><td>)</td><td>2a</td><td>*</td><td>2b</td><td>+</td><td>2c</td><td>,</td><td>2d</td><td>-</td><td>2e</td><td>.</td><td>2f</td><td>/</td></tr>
<tr><td>30</td><td>0</td><td>31</td><td>1</td><td>32</td><td>2</td><td>33</td><td>3</td><td>34</td><td>4</td><td>35</td><td>5</td><td>36</td><td>6</td><td>37</td><td>7</td></tr>
<tr><td>38</td><td>8</td><td>39</td><td>9</td><td>3a</td><td>:</td><td>3b</td><td>;</td><td>3c</td><td><</td><td>3d</td><td>=</td><td>3e</td><td>></td><td>3f</td><td>?</td></tr>
<tr><td>40</td><td>@</td><td>41</td><td>A</td><td>42</td><td>B</td><td>43</td><td>C</td><td>44</td><td>D</td><td>45</td><td>E</td><td>46</td><td>F</td><td>47</td><td>G</td></tr>
<tr><td>48</td><td>H</td><td>49</td><td>I</td><td>4a</td><td>J</td><td>4b</td><td>K</td><td>4c</td><td>L</td><td>4d</td><td>M</td><td>4e</td><td>N</td><td>4f</td><td>O</td></tr>
<tr><td>50</td><td>P</td><td>51</td><td>Q</td><td>52</td><td>R</td><td>53</td><td>S</td><td>54</td><td>T</td><td>55</td><td>U</td><td>56</td><td>V</td><td>57</td><td>W</td></tr>
<tr><td>58</td><td>X</td><td>59</td><td>Y</td><td>5a</td><td>Z</td><td>5b</td><td>[</td><td>5c</td><td>\</td><td>5d</td><td>]</td><td>5e</td><td>^</td><td>5f</td><td>_</td></tr>
<tr><td>60</td><td>‘</td><td>61</td><td>a</td><td>62</td><td>b</td><td>63</td><td>c</td><td>64</td><td>d</td><td>65</td><td>e</td><td>66</td><td>f</td><td>67</td><td>g</td></tr>
<tr><td>68</td><td>h</td><td>69</td><td>i</td><td>6a</td><td>j</td><td>6b</td><td>k</td><td>6c</td><td>l</td><td>6d</td><td>m</td><td>6e</td><td>n</td><td>6f</td><td>o</td></tr>
<tr><td>70</td><td>p</td><td>71</td><td>q</td><td>72</td><td>r</td><td>73</td><td>s</td><td>74</td><td>t</td><td>75</td><td>u</td><td>76</td><td>v</td><td>77</td><td>w</td></tr>
<tr><td>78</td><td>x</td><td>79</td><td>y</td><td>7a</td><td>z</td><td>7b</td><td>{</td><td>7c</td><td>|</td><td>7d</td><td>}</td><td>7e</td><td>~</td><td>7f</td><td>del</td></tr>
</table>

Figure A.1: Table of ASCII Codes.

Appendix B

Supplementary Problems

1. Prove that for a Boolean function f:

   (a) $f = xf_x \oplus x'f_{x'}$;

   (b) $x' = 1 \oplus x$;

   (c) $f = x\frac{\partial f}{\partial x} \oplus f_{x'}$.

   [Hint: Use the first two results to prove the third.]

2. Find two simplified formulae for the Boolean difference $\frac{\partial f}{\partial x} = f_x \oplus f_{x'}$ that are valid when f is positive and negative unate in x, respectively.

3. Find the complete sum for the following function:

$$f = w'xy' + wy'z + w'yz + xyz'.$$

   Apply the recursive algorithm based on

$$CS(f) = ABS([x_1 + CS(f(0, x_2, \ldots, x_n))] \cdot [x_1' + CS(f(1, x_2, \ldots, x_n))]).$$

4. For the following completely specified function,

$$f = w'y'z' + w'xz + xyz' + wyz' + wx'z + x'y'z',$$

   find a SOP with the minimum number of product terms, by applying the recursive algorithm based on

$$CS(f) = ABS([x_1 + CS(f(0, x_2, \ldots, x_n))] \cdot [x_1' + CS(f(1, x_2, \ldots, x_n))]),$$

   to find the complete sum and then solving the covering problem.

5. For the following cover, verify the validity of the expansion from 1——01|1 to 1——0–|1 by setting up and solving a tautology problem.

tuvwxyz	f
1----01	1
-01-1-1	1
-101-00	1
0--10-0	1
1--000-	1
11--0--	1
----1-0	1
1--1--0	1
-11--0-	1
-00--1-	1

In solving the tautology problem, apply the properties of unate functions whenever possible.

6. Check whether the following function is the tautology.

wxyz	f
110-	1
1-01	1
1---	1
-1-1	1
011-	1
00--	1
0010	1

Use the properties of unate functions whenever possible. When splitting, choose the variables in alphabetic order, and test the positive cofactor first.

7. Find all kernels and co-kernels for the following function.

$$F = abcg + adeg + fg + cdf + bce.$$

For each kernel, indicate its level.

8. Apply the *QUICK_FACTOR* algorithm to:

$$abf + acdef + fg + dg.$$

Assume that *ONE_LEVEL-0_KERNEL* selects literals appearing more than once in alphabetic order: For instance, if both a and b appear more than once, a is chosen. Between a and a', a is chosen.

9. Factor the following function with the *QUICK_FACTOR* algorithm.

$$F = abcg + adeg + fg + cdf + bce.$$

Assume that *ONE_LEVEL-0_KERNEL* always chooses among the literals appearing more than one cube the first in alphabetic order.

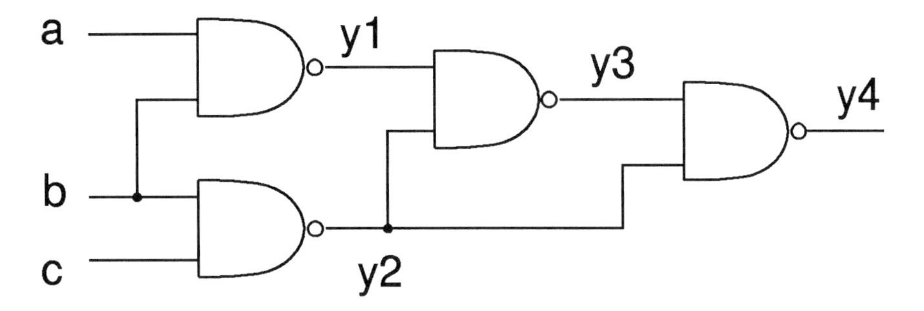

Figure B.1: Boolean Network for Problem 13.

10. Apply the extraction algorithm to:

$$\begin{aligned} F_1 &= abde + abf + a'cde + a'cf + g \\ F_2 &= ab'de + ab'f + ab'g + acde + acf + acg \end{aligned}$$

Specifically, show the decomposition of F_1 and F_2. (Assume that *ONE_LEVEL-0_KERNEL* selects literals appearing more than once in alphabetic order: For instance, if both a and b appear more than once, a is chosen. Between a and a', a is chosen.)

Indicate what resubstitution are made. Use 0 as threshold for elimination. Finally show the resulting Boolean network, indicating for each node its function.

11. Compute all kernels and co-kernels for the following function,

$$f = acd + abc + abe + bde,$$

using the cube intersection matrix. For each pair kernel/co-kernel, write its level.

12. Apply the extraction algorithm to:

$$\begin{aligned} F_1 &= abc + a'b' + a'c' \\ F_2 &= bc + d \end{aligned}$$

Specifically, show the decomposition of F_1 and F_2. (Assume that *ONE_LEVEL-0_KERNEL* selects literals appearing more than once in alphabetic order: For instance, if both a and b appear more than once, a is chosen. Between a and a', a is chosen.)

Indicate what resubstitution are made. Use 0 as threshold for elimination. Finally show the resulting Boolean network, indicating for each node its function.

13. Find the ODCs of y_2 in the circuit shown in figure B.1. Use the Boolean difference method. Notice that the standard procedure cannot be applied, because y_3 depends on y_2.

14. Simplify G_4 in the network of Figure B.2. Use both satisfiability and observability don't cares. Apply the heuristic minimization method to two-level minimization. Draw the simplified network.

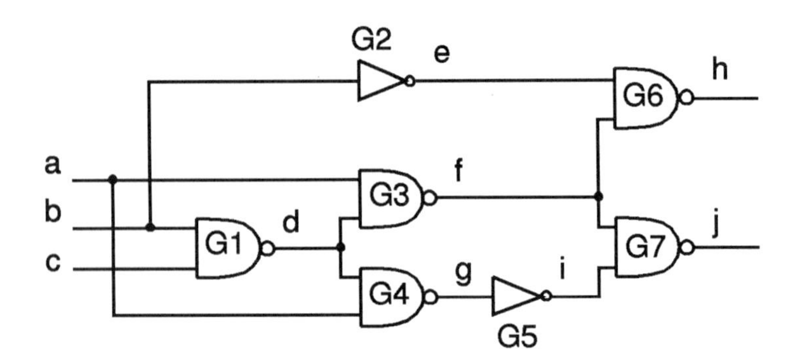

Figure B.2: Boolean Network for Problem 14.

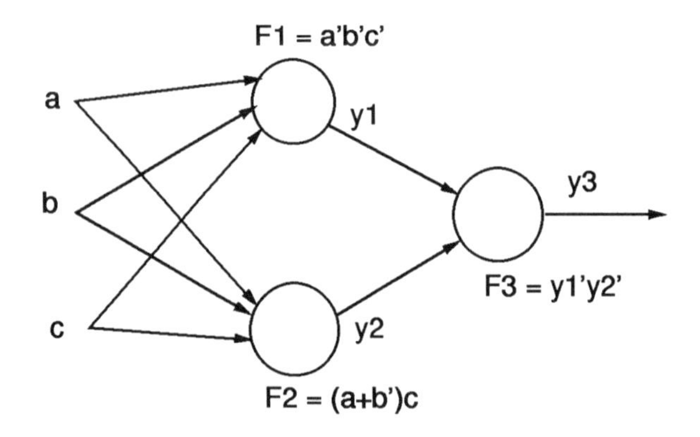

Figure B.3: Boolean Network for Problem 15.

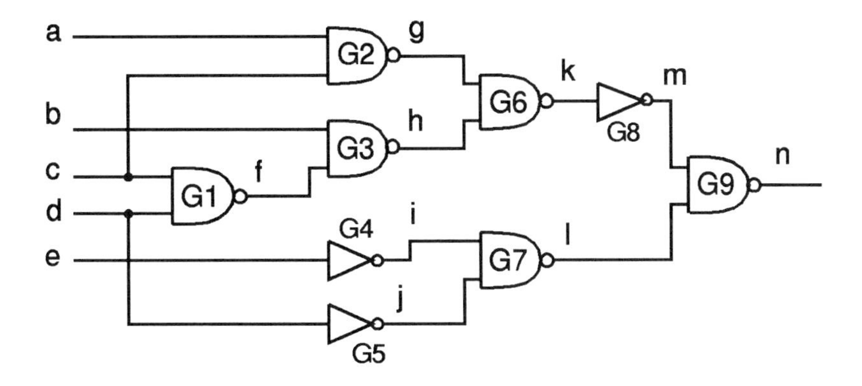

Figure B.4: Boolean Network for Problem 16.

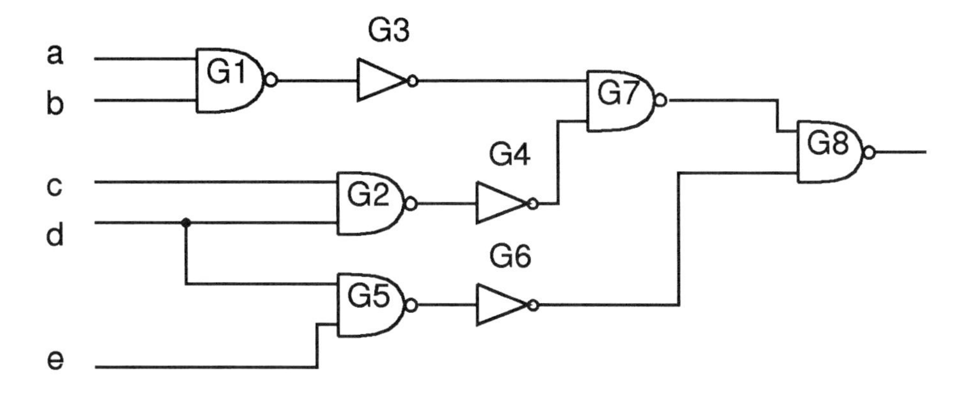

Figure B.5: Circuit for Problem 17.

15. For the Boolean network of Figure B.3, simplify F_1 to $a'b'$. Use observability and satisfiability don't cares. Solve the simplification problem by the heuristic two-level minimization procedure.

16. Apply the dynamic programming algorithm to the network of Figure B.4. First, solve the problem without inserting inverter pairs and using the library of Figure 13.6. Then insert inverter pairs and use the library of Figure 13.9. Draw the two solutions and say which one has lower cost.

17. Apply the dynamic programming algorithm to the network of Figure B.5.

    Use the library of Figure 13.6. Do not insert inverter pairs. Show the trace of the algorithm and draw your final solution.

18. Apply the dynamic programming algorithm to the technology mapping of the network of Figure B.6. Use the library of Figure 13.6. Do not insert inverter pairs. Show the trace of the algorithm and draw your final solution.

19. For the circuit of Figure B.7,

    consider the two faults:

    - Stuck-at-1 on input d of Gate 3;

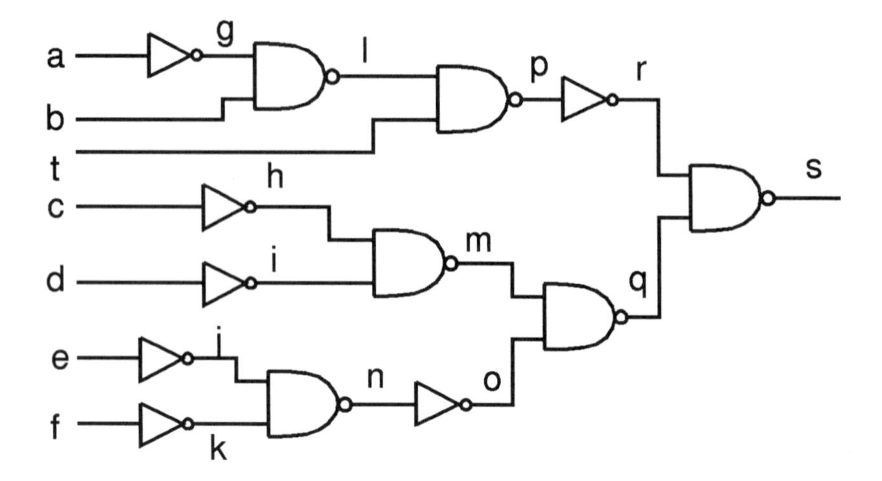

Figure B.6: Circuit for Problem 18.

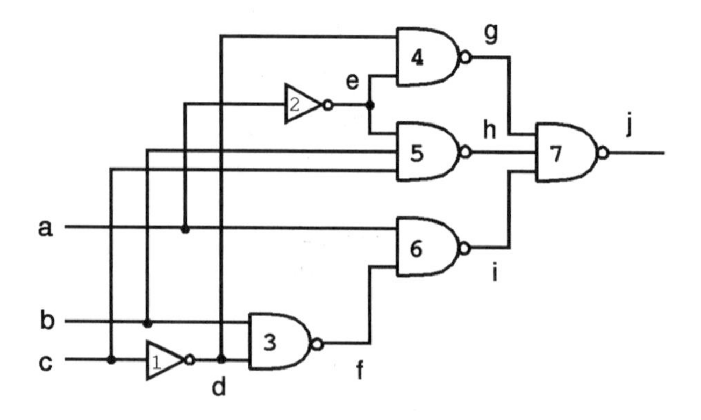

Figure B.7: Circuit for Problem 19.

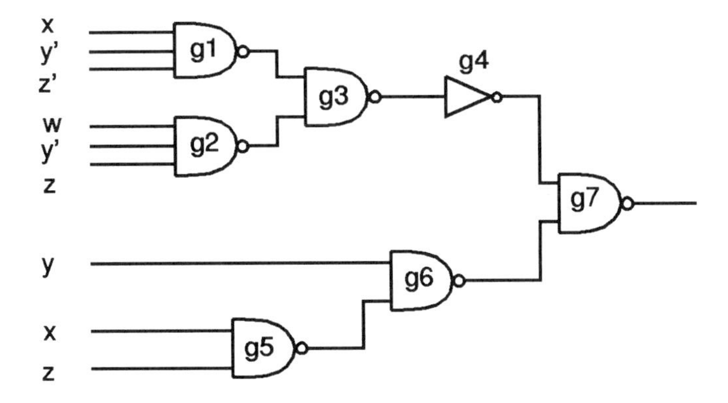

Figure B.8: Circuit for Problem 20.

- Stuck-at-1 on input e of Gate 5.

Do the following:

(a) Prove that exactly one of the two faults is untestable, by trying to generate tests for both.

(b) Remove the redundancy associated with the redundant fault. Draw the resulting circuit.

(c) Prove the the fault that was testable has now become untestable.

(d) Draw the final simplified circuit.

In all test generation attempts, detail your work, by giving all the implications and choices (if any) that you need to reach your result.

20. For the circuit of Figure B.8. consider the two stuck-at-1 faults:

- On the input of g_1 connected to y';
- on the input of g_5 connected to x.

Which of the two faults is untestable? Explain. After the redundancy associated to the untestable fault is removed, is it possible to test for the other fault? If not, simplify the circuit accordingly.

21. Apply redundancy removal to the circuit of Figure B.9. Target the following faults:

(a) stuck-at-0 on the input to gate G_4 connected to c;

(b) stuck-at-0 on the output of G_1.

For each fault show the decision tree. (It may have no nodes.) Draw the simplified circuit.

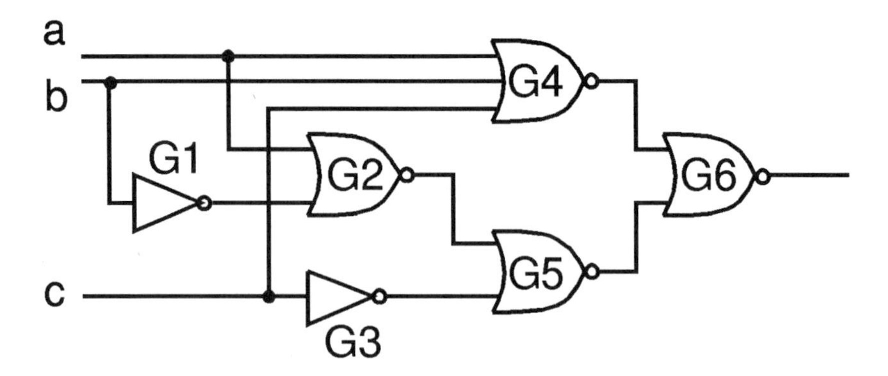

Figure B.9: Circuit for Problem 21.

22. Solve the following covering problem by the branch-and-bound algorithm. Assume unit costs for all the columns.

$$\begin{array}{c} \\ 1 \\ 2 \\ 3 \\ 4 \\ 5 \end{array} \begin{array}{c} \begin{array}{cccc} 1 & 2 & 3 & 4 \end{array} \\ \left[\begin{array}{cccc} 1 & 1 & - & 1 \\ - & 1 & 0 & 1 \\ 1 & 0 & - & - \\ 0 & 0 & 1 & 0 \\ - & 1 & - & 0 \end{array} \right] \end{array}$$

When splitting, choose the longest column and, in case of tie, choose the column of lowest index. Also, when two columns dominate each other, retain the one with the lower index.

Draw the search tree, and indicate for each node the lower bound.

23. Solve the following binate covering problem. Assume unit costs for all columns.

$$F = \begin{array}{c} \begin{array}{ccccc} x_1 & x_2 & x_3 & x_4 & x_5 \end{array} \\ \left[\begin{array}{ccccc} 1 & 1 & 1 & - & - \\ 1 & - & - & 1 & - \\ - & - & 0 & - & 1 \\ 0 & - & 1 & - & - \end{array} \right] \end{array} \begin{array}{c} \\ 1 \\ 2 \\ 3 \\ 4 \end{array}$$

Say whether the given matrix is cyclic. Split initially on x_1. Draw the search tree and detail your work.

24. Solve the following binate covering problem by the branch-and-bound algorithm. Assume unit costs for all the columns.

	x_1	x_2	x_3	x_4	x_5	x_6
1	1	–	–	–	–	–
2	0	–	–	–	1	1
3	–	0	1	1	–	–
4	–	–	1	1	–	–
5	–	–	–	1	1	–
6	0	–	1	–	1	–
7	–	1	–	–	–	0
8	–	0	–	1	0	–
9	–	–	1	–	0	1

When splitting, choose the longest column and, in case of tie, choose the column of lowest index. Also, when two columns dominate each other, retain the one with the lower index.

Draw the search tree, and indicate for each node the lower bound.

25. Suppose that a binate covering problem P is given, defined by:

$$\min \sum_{j=1}^{n} w_j x_j$$

$$F(x_1, \ldots, x_n) = 1,$$

with $F = \prod_{i=1}^{m} C_i$. (Each C_i is a sum of literals.)

Consider a *relaxed* problem $\hat{P}$ defined by:

$$\min \sum_{j=1}^{n} w_j x_j$$

$$\hat{F}(x_1, \ldots, x_n) = 1,$$

with $\hat{F} = \prod_{i=1}^{p} C_i$, $p < m$. In words, $\hat{F}$ is composed of the first p clauses of F. Prove that:

(a) The cost of an optimum solution to $\hat{P}$ is a lower bound to the cost of the optimum solution to P.

(b) If an optimum solution to $\hat{P}$ is also a solution to P, then it is optimum for P.

[Hint: $F \leq \hat{F}$.]

26. For the following flow table, draw the compatibility table.

	00	01	11	10
1	3,0	1,–	—	—
2	6,–	2,0	1,–	—
3	–,1	—	4,0	—
4	1,0	—	—	5,1
5	—	5,–	2,1	1,1
6	—	2,1	6,–	4,1

27. For the flow table of Problem 26, compute the maximal compatibles. [Hint: You should find five maximal compatibles.]

28. For the flow table of Problem 26, compute the prime compatibles and their class sets. [Hint: You should find thirteen prime compatibles.]

29. Find a minimum cost solution for the flow table of Problem 26 that is composed of maximal compatibles only. To do so, set up and solve a binate covering problem. In this problem, you are *not* required to build the reduced table. Give the details of your solution of the binate covering problem. [Hint: The matrix is 13×5.]

30. For the flow table of Problem 26, here replicated, find the reduced flow table corresponding to the solution:

$$a = \{1, 2, 5\}, b = \{3, 6\}, c = \{4, 5\}, d = \{4, 6\}.$$

	00	01	11	10
1	3,0	1,–	—	—
2	6,–	2,0	1,–	—
3	–,1	—	4,0	—
4	1,0	—	—	5,1
5	—	5,–	2,1	1,1
6	—	2,1	6,–	4,1

31. For the following flow table, find a cover with the minimum number of states. Specifically, draw the compatibility table, compute the maximal and the prime compatibles, set up and solve the binate covering problem, and build the reduced table.

	00	01	10
a	a,1	b,1	b,–
b	d,–	a,–	c,0
c	d,0	a,1	—
d	b,–	a,0	b,–

32. For the following flow table, find a cover with the minimum number of states. Specifically, draw the compatibility table, compute the maximal and the prime compatibles, set up and solve the binate covering problem, and build the reduced table.

	00	01	11	10
1	1,–	2,1	—	—
2	3,–	—	2,1	4,1
3	2,–	2,1	3,–	—
4	—	1,0	3,–	4,–

33. Apply the fan-out oriented algorithm of *Mustang* to the state encoding of the FSM described by the following flow table. Use minimum-length codes.

	0	1
1	1,00	2,01
2	3,10	2,01
3	3,00	4,11
4	1,10	1,10

Show the matrices S and Z, the attraction graph, the ranking of the states for the embedding algorithm, and the codes derived by the embedding algorithm.

Bibliography

[1] Abramovici, M., Breuer, M. A., and Friedman, A. D. *Digital Systems Testing and Testable Design.* Computer Science Press, New York, 1990.

[2] Aho, A., and Corasick, M. Efficient string matching: an aid to bibliographic search. *Communications of the ACM* (June 1975), 333–340.

[3] Aho, A. V., Hopcroft, J. E., and Ullman, J. D. *Data Structures and Algorithms.* Addison-Wesley, Reading, MA, 1983.

[4] Aho, A. V., and Johnson, S. C. Optimal code generation for expression trees. *Journal of the Association for Computing Machinery* (July 1976), 488–501.

[5] Akers, S. B. Binary decision diagrams. *IEEE Transactions on Computers C-27*, 6 (June 1978), 509–516.

[6] Armstrong, D. B. A programmed algorithm for assigning internal codes to sequential machines. *IEEE Transactions on Electronic Computers EC-11* (Aug. 1962), 466–472.

[7] Ashar, P., Ghosh, A., Devadas, S., and Newton, A. R. Implicit state transition graphs: Applications to sequential logic synthesis and test. In *Proceedings of the IEEE International Conference on Computer Aided Design* (Santa Clara, CA, Nov. 1990), pp. 84–87.

[8] Avedillo, M. J., Quintana, J. M., and Huertas, J. L. New approach to the state reduction in incompletely specified sequential machines. In *IEEE International Symposium on Circuits and Systems* (New Orleans, LA, May 1990), pp. 440–443.

[9] Avedillo, M. J., Quintana, J. M., and Huertas, J. L. A new method for the state reduction of incompletely specified sequential machines. In *Proceedings of the European Design Automation Conference* (Glasgow, UK, Mar. 1990), pp. 552–556.

[10] Aziz, A., Balarin, F., Brayton, R. K., Cheng, S.-T., Hojati, R., Krishnan, S. C., Ranjan, R. K., Sangiovanni-Vincentelli, A., and Shiple, T. HSIS: A BDD-based environment for formal verification. In *Proceedings of the Design Automation Conference* (San Diego, CA, June 1994).

[11] Bahar, R. I., Frohm, E. A., Gaona, C. M., Hachtel, G. D., Macii, E., Pardo, A., and Somenzi, F. Algebraic decision diagrams and their applications. In *Proceedings of the International Conference on Computer-Aided Design* (Santa Clara, CA, Nov. 1993), pp. 188–191.

[12] Bahar, R. I., Frohm, E. A., Gaona, C. M., Hachtel, G. D., Macii, E., Pardo, A., and Somenzi, F. Algebraic decision diagrams and their applications. Internal report, University of Colorado, Boulder, Apr. 1993.

[13] BAHAR, R. I., HACHTEL, G. D., MACII, E., AND SOMENZI, F. A symbolic method to reduce power consumption of circuits containing false paths. In *Proceedings of the International Conference on Computer-Aided Design* (San Jose, CA, Nov. 1994), pp. 368–371.

[14] BARTLETT, K. A weak division approach to multilevel synthesis. Master's thesis, University of Colorado, 1986.

[15] BARTLETT, K. A., BRAYTON, R., HACHTEL, G., JACOBY, R., MORRISON, C., RUDELL, R., SANGIOVANNI-VINCENTELLI, A., AND WANG, A. Multi-level logic minimization using implicit don't cares. *IEEE Transactions on Computer-Aided Design of Integrated Circuits and Systems CAD-7*, 6 (June 1988), 723–740.

[16] BARTLETT, K. A., COHEN, W., DE GEUS, A., AND HACHTEL, G. Synthesis and optimization of multilevel logic under timing constraints. *IEEE Transactions on Computer-Aided Design of Integrated Circuits and Systems CAD-7*, 6 (June 1988), 723–740.

[17] BEARDSLEE, M., KRING, C., MURGAI, R., SAVOJ, H., BRAYTON, R., AND NEWTON, A. SLIP: An environment for system level interactive partitioning. In *Proceedings of the IEEE International Conference on Computer Aided Design* (Nov. 1989), pp. 280–283.

[18] BELLMAN, R. *Dynamic Programming.* Princeton University Press, Princeton, NJ, 1957.

[19] BENINI, L., AND DE MICHELI, G. State assignment for low power dissipation. In *Proceedings of the Custom Integrated Circuits Conference* (San Diego, CA, May 1994), pp. 136–140.

[20] BENINI, L., AND DE MICHELI, G. State assignment for low power dissipation. *IEEE Jour. Solid State Circ.* (1995). To appear.

[21] BERMAN, C. L., AND TREVILLYAN, L. H. Global flow optimization in automatic logic design. *IEEE Transactions on Computer-Aided Design of Integrated Circuits and Systems 10*, 5 (May 1991), 557–564.

[22] BERMAN, L., AND TREVILLYAN, L. A global approach to circuit size reduction. In *Advanced Research in VLSI, 5th MIT Conference* (1988), MIT Press, pp. 203–214.

[23] BERN, J., GERGOV, J., MEINEL, C., AND SLOBODOVA, A. Boolean manipulation with free BDD's. first experimental results. In *Proceedings of the European Conference on Design Automation* (Paris, France, Feb. 1994).

[24] BILLON, J. P., AND MADRE, J. C. Original concepts of PRIAM, an industrial tool for efficient formal verification of combinational circuits. In *The Fusion of Hardware Design and Verification*, G. J. Milne, Ed. Elsevier Science Publishers B.V.(North Holland), 1989, pp. 487–501.

[25] BLAKE, A. *Canonical Expressions in Boolean Algebra.* PhD thesis, Dept. of Mathematics, Univ. of Chicago, 1937.

[26] BOLLIG, B., LÖBBING, M., AND WEGENER, I. Simulated annealing to improve variable orderings for OBDDs. Presented at the International Workshop on Logic Synthesis, Granlibakken, CA, May 1995.

[27] BOOLE, G. *An Investigation of the Laws of Thought.* Walton, London, 1854. (Reprinted by Dover Books, New York, 1954).

[28] BOOTH, T. L. *Sequential Machines and Automata Theory.* Wiley, New York, 1967.

[29] BRACE, K. S., RUDELL, R. L., AND BRYANT, R. E. Efficient implementation of a BDD package. In *Proceedings of the 27th Design Automation Conference* (Orlando, FL, June 1990), pp. 40–45.

[30] BRAND, D. Redundancy and don't cares in logic synthesis. *IEEE Transactions on Computers C-32*, 10 (Oct. 1983), 947–952.

[31] BRAND, D. Logic synthesis. In *NATO ASI on Logic Synthesis and Silicon Compilation for VLSI*, P. Antognetti, G. D. Micheli, and A. Sangiovanni-Vincentelli, Eds. Kluwer, Dordrecht, The Netherlands, 1987.

[32] BRAYTON, R. Factoring logic functions. *IBM Journal of Research and Development 31* (Mar. 1987).

[33] BRAYTON, R., RUDELL, R., SANGIOVANNI-VINCENTELLI, A., AND WANG, A. Multilevel logic synthesis. Notes for Lectures at Oxford/Berkeley Summer Engineering Programme, July 1989.

[34] BRAYTON, R., SENTOVICH, E., AND SOMENZI, F. Don't-cares and global flow analysis of boolean networks. In *Proceedings of the IEEE International Conference on Computer Aided Design* (Santa Clara, CA, Nov. 1988), pp. 98–101.

[35] BRAYTON, R., AND SOMENZI, F. Minimization of boolean relations. In *Proc. Int. Symp. Circ. Syst. (ISCAS-89)* (Portland, OR, May 1989), pp. 738–743.

[36] BRAYTON, R. K. Algorithms for multilevel logic synthesis and optimization. In *NATO ASI on Logic Synthesis and Silicon Compilation for VLSI*, P. Antognetti, G. D. Micheli, and A. Sangiovanni-Vincentelli, Eds. Kluwer, Dordrecht, The Netherlands, 1987.

[37] BRAYTON, R. K., HACHTEL, G. D., MCMULLEN, C. T., AND SANGIOVANNI-VINCENTELLI, A. *Logic Minimization Algorithms for VLSI Synthesis.* Kluwer Academic Publishers, Boston, Massachusetts, 1984.

[38] BRAYTON, R. K., AND MCMULLEN, C. The decomposition and factorization of boolean expressions. In *Proceedings of the IEEE International Symposium on Circuits and Systems* (Rome, Italy, May 1982), pp. 49–54.

[39] BRAYTON, R. K., RUDELL, R., SANGIOVANNI-VINCENTELLI, A., AND WANG, A. R. MIS: A multiple-level interactive logic optimization system. *IEEE Transactions on Computer-Aided Design of Integrated Circuits and Systems CAD-6*, 6 (Nov. 1987), 1062–1081.

[40] BRAYTON, R. K., SENTOVICH, E. M., AND SOMENZI, F. Don't cares and global flow analysis of boolean circuits. In *Proceedings of the IEEE International Conference on Computer Aided Design* (1988), pp. 98–101.

[41] BRAYTON, R. K., AND SOMENZI, F. An exact minimizer for boolean relations. In *Proceedings of the IEEE International Conference on Computer Aided Design* (Santa Clara, CA, Nov. 1989), pp. 316–319.

[42] BREUER, M. A., AND FRIEDMAN, A. D. *Diagnosis and Reliable Design of Digital Systems.* Computer Science Press, Woodland Hills, CA, 1976.

[43] BROWN, D. A state-machine synthesizer—SMS. In *Proc. 18th Design Automation Conference* (June 1981), pp. 301–304.

[44] BROWN, F. M. *Boolean Reasoning: The Logic of Boolean Equations.* Kluwer, Boston, 1990.

[45] BROWNE, M. C., CLARKE, E. M., DILL, D. L., AND MISHRA, B. Automatic verification of sequential circuits using temporal logic. *IEEE Transactions on Computers C-35*, 12 (Dec. 1986), 1035–1044.

[46] BRYANT, R., AND CHEN, Y.-A. Verification of arithmetic circuits with binary moment diagrams. In *Proceedings of the Design Automation Conference* (San Francisco, CA, June 1995), pp. 535–541.

[47] BRYANT, R. E. Graph-based algorithms for boolean function manipulation. *IEEE Transactions on Computers C-35*, 8 (Aug. 1986), 677–691.

[48] BRYANT, R. E. On the complexity of VLSI implementations and graph representations of boolean functions with application to integer multiplication. *IEEE Transactions on Computers 40*, 2 (Feb. 1991), 205–213.

[49] BURCH, J. R., CLARKE, E. M., LONG, D. E., MCMILLAN, K. L., AND DILL, D. L. Symbolic model checking for sequential circuit verification. Tech. Rep. CMU-CS-93-211, School of Computer Science, Carnegie Mellon University, Pittsburgh, PA, 15213, July 1993.

[50] BURCH, J. R., CLARKE, E. M., LONG, D. E., MCMILLAN, K. L., AND DILL, D. L. Symbolic model checking for sequential circuit verification. *IEEE Transactions on Computer-Aided Design 13*, 4 (Apr. 1994), 401–424.

[51] BURCH, J. R., CLARKE, E. M., MCMILLAN, K. L., AND DILL, D. L. Sequential circuit verification using symbolic model checking. In *Proceedings of the Design Automation Conference* (June 1990), pp. 46–51.

[52] BURCH, J. R., CLARKE, E. M., MCMILLAN, K. L., DILL, D. L., AND HWANG, L. J. Symbolic model checking: 10^{20} states and beyond. In *Proceedings of the Fifth Annual Symposium on Logic in Computer Science* (June 1990).

[53] BUTLER, K. M., ROSS, D. E., KAPUR, R., AND MERCER, M. R. Heuristics to compute variable orderings for efficient manipulation of ordered binary decision diagrams. In *Proceedings of the Design Automation Conference* (San Francisco, CA, June 1991), pp. 417–420.

[54] CARLSON, S. *Introduction to HDL-Based Design Using VHDL*. Synopsys Inc., 1991.

[55] CERNY, E., AND MARIN, M. A. An approach to unified methodology of combinational switching circuits. *IEEE Transactions on Computers C-26*, 8 (Aug. 1977), 745–756.

[56] CHANG, H. Y., MANNING, E., AND METZE, G. *Fault Diagnosis of Digital Systems*. Wiley Interscience, New York, 1970.

[57] CHENG, K.-T. An ATPG-based approach to sequential logic optimization. In *Proceedings of the IEEE International Conference on Computer Aided Design* (Santa Clara, CA, Nov. 1991), pp. 372–375.

[58] CHENG, W.-T., AND CHAKRABORTY, T. Gentest—an automatic test-generation system for sequential circuits. *IEEE Computer 22*, 4 (Apr. 1989), 43–49.

[59] CHIODO, M., SHIPLE, T. R., SANGIOVANNI-VINCENTELLI, A., AND BRAYTON, R. K. Automatic reduction in CTL compositional model checking. In *Proceedings of the International Conference on Computer-Aided Design* (Santa Clara, CA, Nov. 1992), pp. 172–178.

[60] CHO, H., HACHTEL, G. D., JEONG, S.-W., PLESSIER, B., SCHWARZ, E., AND SOMENZI, F. ATPG aspects of FSM verification. In *Proceedings of the IEEE International Conference on Computer Aided Design* (Nov. 1990), pp. 134–137.

[61] CHO, H., HACHTEL, G. D., AND SOMENZI, F. Fast sequential ATPG based on implicit state enumeration. In *Proceedings of the International Test Conference* (Nashville, TN, Oct. 1991), pp. 67–74.

[62] CHO, H., HACHTEL, G. D., AND SOMENZI, F. Redundancy identification and removal based on implicit state enumeration. In *Proceedings of the International Conference on Computer Design* (Cambridge, MA, Oct. 1991), pp. 77–80.

[63] CHO, H., HACHTEL, G. D., AND SOMENZI, F. Redundancy identification/removal and test generation for sequential circuits using implicit state enumeration. *IEEE Transactions on Computer-Aided Design of Integrated Circuits and Systems 12*, 7 (July 1993), 935–945.

[64] CHO, H., JEONG, S.-W., SOMENZI, F., AND PIXLEY, C. Multiple observation time single reference test generation using synchronizing sequences. In *Proceedings of the European Conference on Design Automation* (Paris, France, Feb. 1993), pp. 494–498.

[65] CHO, H., AND SOMENZI, F. Sequential logic optimization based on state space decomposition. In *Proceedings of the European Conference on Design Automation* (Paris, France, Feb. 1993), pp. 200–204.

[66] CHOUEKA, Y. Theories of automata on ω-tapes: A simplified approach. *J. Comput. Syst. Sci. 8* (1974), 117–141.

[67] CLARKE, E. M., MCMILLAN, K. L., ZHAO, X., FUJITA, M., AND YANG, J. C.-Y. Spectral transforms for large boolean functions with applications to technology mapping. In *Proceedings of the Design Automation Conference* (Dallas, TX, June 1993), pp. 54–60.

[68] COLON-BONET, G., SCHWARZ, E. M., BOSTICK, D. G., HACHTEL, G. D., AND LIGHTNER, M. R. On optimal extraction of combinational logic and don't care sets from hardware description languages. In *Proceedings of the IEEE International Conference on Computer Aided Design* (Santa Clara, CA, Nov. 1989), pp. 308–311.

[69] CORMEN, T. H., LEISERSON, C. E., AND RIVEST, R. L. *An Introduction to Algorithms.* McGraw-Hill, New York, 1990.

[70] COUDERT, O., BERTHET, C., AND MADRE, J. C. Verification of sequential machines based on symbolic execution. In *Automatic Verification Methods for Finite State Systems, Lecture Notes in Computer Science 407*, J. Sifakis, Ed. Springer-Verlag, 1989, pp. 365–373.

[71] COUDERT, O., BERTHET, C., AND MADRE, J. C. Verification of sequential machines using boolean functional vectors. In *Proceedings IFIP International Workshop on Applied Formal Methods for Correct VLSI Design* (Leuven, Belgium, Nov. 1989), L. Claesen, Ed., pp. 111–128.

[72] COUDERT, O., BERTHET, C., AND MADRE, J. C. Formal boolean manipulations for the verification of sequential machines. In *Proceedings of the European Conference on Design Automation* (Mar. 1990), pp. 57–61.

[73] COUDERT, O., AND MADRE, J. C. A unified framework for the formal verification of sequential circuits. In *Proceedings of the IEEE International Conference on Computer Aided Design* (Nov. 1990), pp. 126–129.

[74] COUDERT, O., AND MADRE, J. C. Symbolic computation of the valid states of a sequential machine: Algorithms and discussion. In *1991 International Workshop on Formal Methods in VLSI Design* (Miami, FL, Jan. 1991).

[75] Coudert, O., and Madre, J. C. Implicit and incremental computation of primes and essential primes of boolean functions. In *Proceedings of the Design Automation Conference* (Anaheim, CA, June 1992), pp. 36–39.

[76] Coudert, O., and Madre, J. C. A new graph based prime computation technique. In *Logic Synthesis and Optimization*, T. Sasao, Ed. Kluwer Academic Publishers, Boston, MA, 1993, ch. 2, pp. 33–57.

[77] Coudert, O., and Madre, J. C. Towards a symbolic logic minimization algorithm. In *Proceedings of the 6th International Conference on VLSI Design* (Bombay, India, Jan. 1993), pp. 329–334.

[78] Coudert, O., and Madre, J. C. New ideas for solving covering problems. In *Proceedings of the Design Automation Conference* (San Francisco, CA, June 1995), pp. 641–646.

[79] Coudert, O., Madre, J. C., and Berthet, C. Verifying temporal properties of sequential machines without building their state diagrams. In *Computer-Aided Verification '90*, E. M. Clarke and R. P. Kurshan, Eds. American Mathematical Society – Association for Computing Machinery, 1991, pp. 75–84.

[80] Coudert, O., Madre, J. C., and Fraisse, H. A new viewpoint on two-level logic minimization. In *Proceedings of the Design Automation Conference* (Dallas, TX, June 1993), pp. 625–630.

[81] Coudert, O., Madre, J. C., Fraisse, H., and Touati, H. Implicit prime cover computation: An overview. In *SASIMI '93* (Nara, Japan, Oct. 1993), pp. 413–422.

[82] Darringer, J., Brand, D., Gerbi, J., Joyner, Jr., W., and Trevillyan, L. LSS: A system for production logic synthesis. *IBM Journal of Research and Development 28*, 5 (Sept. 1984), 537–545.

[83] Darringer, J., Joyner, W., Berman, L., and Trevillyan, L. Logic synthesis through local transformations. *IBM Journal of Research and Development 25*, 4 (July 1981), 272–280.

[84] De Micheli, G., Brayton, R. K., and Sangiovanni-Vincentelli, A. Optimal state assignment of finite state machines. *IEEE Transactions on Computer-Aided Design of Integrated Circuits and Systems CAD-4* (July 1985), 269–285.

[85] Detjens, E., Gannot, G., Rudell, R., Sangiovanni-Vincentelli, A., and Wang, A. Technology mapping in MIS. In *Proceedings of the IEEE International Conference on Computer Aided Design* (Nov. 1987), pp. 116–119.

[86] Devadas, S., and Keutzer, K. A unified approach to the synthesis of fully testable sequential machines. In *Hawaii International Conference on System Science* (Jan. 1990). also in IEEE Trans. on CAD, January, 1991.

[87] Devadas, S., Ma, H.-K. T., Newton, A. R., and Sangiovanni-Vincentelli, A. MUSTANG: State assignment of finite state machines for optimal multi-level logic implementations. *IEEE Transactions on Computer-Aided Design of Integrated Circuits and Systems CAD-7* (Dec. 1988), 1290–1300.

[88] Devadas, S., and Newton, A. R. Exact algorithms for output encoding, state assignment and four-level boolean minimization. In *Proceedings of the Hawaii International Conference on Systems Science* (Jan. 1990), pp. 387–396.

[89] Dolotta, T. A., and McCluskey, E. J. The coding of internal states of sequential machines. *IEEE Transactions on Electronic Computers EC-13* (Oct. 1964), 549–562.

[90] DRECHSLER, R., BECKER, B., AND GÖCKEL, N. A genetic algorithm for variable ordering of OBDDs. Presented at the International Workshop on Logic Synthesis, Granlibakken, CA, May 1995.

[91] DRECHSLER, R., SARABI, A., THEOBALD, M., BECKER, B., AND PERKOWSKI, M. A. Efficient representation and manipulation of switching functions based on ordered Kronecker functional decision diagrams. In *Proceedings of the Design Automation Conference* (San Diego, CA, June 1994), pp. 415–419.

[92] DU, X., HACHTEL, G. D., LIN, B., AND NEWTON, A. R. MUSE: A MUltilevel Symbolic Encoding algorithm for state assignment. *IEEE Transactions on Computer-Aided Design of Integrated Circuits and Systems CAD-10*, 1 (Jan. 1991), 28–38.

[93] DU, X., HACHTEL, G. D., AND MOCEYUNAS, P. H. MUSE: A MUltilevel Symbolic Encoding algorithm for state assignment. In *Proceedings of the Hawaii International Conference on Systems Science* (Jan. 1990), pp. 367–376.

[94] DUFF, C., AND SAUCIER, G. State assignment based on the reduced dependency theory and recent experimental results. In *Proceedings of the International Conference on Computer-Aided Design* (Santa Clara, CA, Nov. 1991), pp. 222–225.

[95] EHRICH, H. D. A note on state minimization of a special class of incomplete sequential machines. *IEEE Transactions on Computers C-21* (May 1972), 500–502.

[96] EICHELBERGER, E. B., LINDBLOOM, E., WAICUKAUSKI, J. A., AND WILLIAMS, T. W. *Structured Logic Testing.* Prentice Hall, Englewood Cliffs, 1991.

[97] EMERSON, E. A. Temporal and modal logic. In van Leeuwen [266], ch. 16, pp. 995–1072.

[98] EMERSON, E. A., AND CLARKE, E. M. Using branching time temporal logic to synthesize synchronization skeletons. *Science of Computer Programming 2* (1982), 241–266.

[99] EVEN, S. *Graph Algorithms.* Computer Science Press, Rockville, MD, 1979.

[100] FISHBURN, J., AND DUNLOP, A. TILOS: A posynomial programming approach to transistor sizing. In *Proceedings of the International Conference on Computer-Aided Design* (Santa Clara, CA, Nov. 1985), pp. 326–328.

[101] FLEISHER, H., AND MAISSEL, L. An introduction to array logic. *IBM Journal of Research and Development 19* (Mar. 1975), 98–109.

[102] FORTUNE, L., HOPCROFT, J., AND SCHMIDT, E. M. The complexity of equivalence and containment for free single variable program scheme. In *Lecture Notes in Computer Science 62*, Goos, Hartmanis, Ausiello, and Bohm, Eds. Springer-Verlag, 1978, pp. 227–240.

[103] FRANCIS, R. J., ROSE, J., AND VRANESIC, Z. Chortle-crf: Fast technology mapping for LUT based FPGAs. In *Proceedings of the Design Automation Conference* (San Francisco, CA, June 1991), pp. 613–619.

[104] FRIEDMAN, S. J., AND SUPOWIT, K. J. Finding the optimal variable ordering for binary decision diagrams. *IEEE Transactions on Computers 39*, 5 (May 1990), 710–713.

[105] FUJITA, M., FUJISAWA, H., AND KAWATO, N. Evaluation and improvements of boolean comparison method based on binary decision diagrams. In *Proceedings of the IEEE International Conference on Computer Aided Design* (Nov. 1988), pp. 2–5.

[106] Fujita, M., Matsunaga, Y., and Kakuda, T. On variable ordering of binary decision diagrams for the application of multi-level logic synthesis. In *Proceedings of the European Conference on Design Automation* (Amsterdam, Feb. 1991), pp. 50–54.

[107] Fujiwara, H. *Logic Testing and Design for Testability.* MIT Press, Cambridge, MA, 1985.

[108] Fujiwara, H., and Shimono, T. On the acceleration of test generation algorithms. *IEEE Transactions on Computers C-32*, 12 (Dec. 1983), 1137–1144.

[109] Gajski, D., Dutt, N., Wu, A., and Lin, S. *High-Level Synthesis, Introduction to Chip and System Design.* Kluwer Academic Publishers, 1992.

[110] Ghosh, A., Devadas, S., and Newton, A. R. Test generation for highly sequential circuits. In *Proceedings of the IEEE International Conference on Computer Aided Design* (Nov. 1989), pp. 362–365.

[111] Gilbert, E. N. Lattice theoretic properties of frontal switching functions. *Journal of Mathematics and Physics 33*, 1 (1954), 57-67.

[112] Gimpel, J. A reduction technique for prime implicant tables. *IEEE Transactions on Electronic Computers EC-14* (Aug. 1965), 535–541.

[113] Ginsburg, S. On the reduction of superfluous states in a sequential machine. *Journal of the Association for Computing Machinery 6* (Apr. 1959), 252–282.

[114] Ginsburg, S. Synthesis of minimal state machines. *IRE Trans. Electronic Computers EC-8* (Dec. 1959), 441–449.

[115] Ginsburg, S. A synthesis technique for minimal state sequential machines. *IRE Trans. Electronic Computers EC-8* (Mar. 1959), 13–24.

[116] Goel, P. An implicit enumeration algorithm to generate tests for combinational logic circuits. *IEEE Transactions on Computers C-30*, 3 (Mar. 1981), 215–222.

[117] Grasselli, A., and Luccio, F. A method for minimizing the number of internal states in incompletely specified sequential networks. *IEEE Transactions on Electronic Computers EC-14*, 3 (June 1965), 350–359.

[118] Grasselli, A., and Luccio, F. A method for minimizing the number of internal states in incompletely specified sequential networks. *IEEE Transactions on Electronic Computers EC-14* (June 1965), 350–359.

[119] Grasselli, A., and Luccio, F. Some covering problems in switching theory. In *Networks and Switching Theory*, G. Biorci, Ed. Academic Press, New York, 1968.

[120] Gregory, D., Bartlett, K., de Geus, A., and Hachtel, G. SOCRATES: A system for automatically synthesizing and optimizing combinational logic. In *Proceedings of the Design Automation Conference* (June 1986), pp. 79–85.

[121] Gross, D., Gu, B., and Soland, R. M. The biconjugate gradient method for obtaining the steady-state probability distributions of Markovian multiechelon repairable item inventory systems. In Stewart [255], pp. 473–489.

[122] Hachtel, G., Jacoby, R., Moceyunas, P., and Morrison, C. Performance enhancements in BOLD using "implications". In *Proceedings of the IEEE International Conference on Computer Aided Design* (1988).

[123] Hachtel, G., Lightner, M., Jacoby, R., Morrison, C., Moceyunas, P., and Bostick, D. BOLD: The Boulder Optimal Logic Design system. In *Hawaii International Conference on System Sciences* (1989).

[124] HACHTEL, G. D., HERMIDA, M., PARDO, A., PONCINO, M., AND SOMENZI, F. Re-encoding sequential circuits to reduce power dissipation. In *Proceedings of the International Conference on Computer-Aided Design* (San Jose, CA, Nov. 1994), pp. 70–73.

[125] HACHTEL, G. D., HERMIDA, M., PARDO, A., PONCINO, M., AND SOMENZI, F. Re-encoding sequential circuits to reduce power dissipation. Presented at the International Workshop on Low Power Design, Napa, CA, Apr. 1994.

[126] HACHTEL, G. D., AND JACOBY, R. M. Verification algorithms for VLSI synthesis. In *Design Systems for VLSI Circuits*. NATO ASI Series, 1986, pp. 264–300.

[127] HACHTEL, G. D., JACOBY, R. M., KEUTZER, K., AND MORRISON, C. R. On the relationship between area optimization and multifault testabilty of multilevel logic. In *Proceedings of the IEEE International Conference on Computer Aided Design* (Nov. 1989), pp. 422–425.

[128] HACHTEL, G. D., JACOBY, R. M., KEUTZER, K., AND MORRISON, C. R. On properties of algebraic transformations and the synthesis of multifault-irredundant circuits. *IEEE Transactions on Computer-Aided Design 11*, 3 (Mar. 1992), 313–321.

[129] HACHTEL, G. D., JACOBY, R. M., AND MORRISON, C. R. TECHMAP: Technology mapping with area and delay optimization. In *Proceedings of the International Workshop on Logic and Architecture Synthesis for Silicon Compilers* (Grenoble, France, May 1988).

[130] HACHTEL, G. D., RHO, J.-K., SOMENZI, F., AND JACOBY, R. Exact and heuristic algorithms for the minimization of incompletely specified state machines. In *Proceedings of the European Design Automation Conference* (Amsterdam, The Netherlands, Feb. 1991), pp. 184–191.

[131] HACHTEL, G. D., AND SOMENZI, F. A symbolic algorithm for maximum flow in 0-1 networks. In *Proceedings of the International Conference on Computer-Aided Design* (Santa Clara, CA, Nov. 1993), pp. 403–406.

[132] HACHTEL, G. D., AND SOMENZI, F. A symbolic algorithm for maximum flow in 0-1 networks. Presented at IWLS'93, May 1993.

[133] HALMOS, P. R. *Lectures on Boolean Algebras*. Van Nostrand Reinhold, London, 1963.

[134] HAMMER, P. L., AND RUDEANU, S. *Boolean Methods in Operations Research and Related Areas*. Springer-Verlag, Berlin, 1968.

[135] HARR, R. E. *Applications of VHDL to Circuit Design*. Kluwer Academic Publishers, 1991.

[136] HARRISON, M. A. *Introduction to Switching and Automata Theory*. McGraw-Hill, New York, 1965.

[137] HARTMANIS, J. On the state assignment problem for sequential machines, I. *IRE Transactions on Electronic Computers EC-10* (June 1961), 157–165.

[138] HARTMANIS, J., AND STEARNS, R. E. *Algebraic Structure Theory of Sequential Machines*. Prentice-Hall, Englewood Cliffs, NJ, 1966.

[139] HENNIE, F. C. Fault detecting experiments for sequential circuits. In *Proceedings of the 5th Annual Symposium on Switching Circuit Theory and Logical Design* (Princeton, NJ, Nov. 1964), pp. 95–110.

[140] HENNIE, F. C. *Finite-State Models for Logical Machines*. John Wiley, New York, 1968.

[141] HILL, F. J., AND PETERSON, G. R. *Computer Aided Logical Design*, fourth ed. John Wiley, New York, 1992.

[142] HOJATI, R., SHIPLE, T. R., BRAYTON, R., AND KURSHAN, R. A unified approach to language containment and fair CTL model checking. In *Proceedings of the Design Automation Conference* (June 1993), pp. 475–481.

[143] HOJATI, R., TOUATI, H., KURSHAN, R. P., AND BRAYTON, R. K. Efficient ω-regular language containment. In *Computer Aided Verification* (Montréal, Canada, June 1992), pp. 371–382.

[144] HONG, S. J., CAIN, R. G., AND OSTAPKO, D. L. MINI: A heuristic approach for logic minimization. *IBM Journal of Research and Development 18* (Sept. 1974), 443–458.

[145] HOPCROFT, J. A $n \log n$ algorithm for minimizing states in a finite automaton. In *Theory of Machines and Computation*, Z. Kohavi and A. Paz, Eds. Academic Press, New York, 1971, pp. 189–196.

[146] HOPCROFT, J. E., AND ULLMAN, J. D. *Introduction to Automata Theory, Languages, and Computation.* Addison-Wesley, Reading, MA, 1979.

[147] HOUSE, R. W., AND STEVENS, D. W. A new rule for reducing CC tables. *IEEE Transactions on Computers C-19* (Nov. 1970), 1108–1111.

[148] IBARAKI, T., AND MUROGA, S. Synthesis of networks with a minimum number of negative gates. *IEEE Transactions on Computers C-20* (Jan. 1971), 49–58.

[149] ISHIKAWA, J., SATO, H., HIRAMINE, M., ISHIDA, K., OGURI, S., KAZUMA, Y., AND MURAI, S. A rule based reorganization system LORES/EX. In *Proc. Int. Conf. Comp. Des. (ICCD-88)* (Oct. 1988), pp. 262–266.

[150] ISHIURA, N., SAWADA, H., AND YAJIMA, S. Minimization of binary decision diagrams based on exchanges of variables. In *Proceedings of the International Conference on Computer-Aided Design* (Santa Clara, CA, Nov. 1991), pp. 472–475.

[151] JACOBY, R., MOCEYUNAS, P., CHO, H., AND HACHTEL, G. New ATPG techniques for logic optimization. In *Proceedings of the IEEE International Conference on Computer Aided Design* (Nov. 1989), pp. 548–551.

[152] JEONG, S.-W., PLESSIER, B., HACHTEL, G. D., AND SOMENZI, F. Extended BDD's: Trading off canonicity for structure in verification algorithms. In *Proceedings of the IEEE International Conference on Computer Aided Design* (Santa Clara, CA, Nov. 1991), pp. 464–467.

[153] JEONG, S.-W., PLESSIER, B., HACHTEL, G. D., AND SOMENZI, F. Variable ordering and selection for FSM traversal. In *Proceedings of the IEEE International Conference on Computer Aided Design* (Santa Clara, CA, Nov. 1991), pp. 476–479.

[154] JEONG, S.-W., PLESSIER, B. F., HACHTEL, G. D., AND SOMENZI, F. Variable ordering for binary decision diagrams. In *Proceedings of the European Conference on Design Automation* (Brussels, Mar. 1992), pp. 447–451.

[155] JEONG, S.-W., AND SOMENZI, F. A new algorithm for the binate covering problem and its application to the minimization of boolean relations. In *Proceedings of the International Conference on Computer-Aided Design* (Santa Clara, CA, Nov. 1992), pp. 417–420.

[156] JI, Q., OH, Y.-S., LIGHTNER, M. R., AND SOMENZI, F. Technology independent estimation of area and delay in logic synthesis. In *SASIMI '92* (Kyoto, Japan, Apr. 1992), pp. 171–180.

[157] Kam, T., Villa, T., Brayton, R. K., and Sangiovanni-Vincentelli, A. A fully implicit algorithm for exact state minimization. In *Proceedings of the Design Automation Conference* (San Diego, CA, June 1994).

[158] Keutzer, K. DAGON: Technology binding and local optimization by DAG matching. In *Proceedings of the Design Automation Conference* (June 1987), pp. 341–347.

[159] Keutzer, K. Personal communication, Feb. 1989.

[160] Keutzer, K., Malik, S., and Saldanha, A. Is redundancy necessary to reduce delay? In *Proceedings of the Design Automation Conference* (June 1990), pp. 228–234.

[161] Knuth, D. Big omicron, big omega, and big theta. *SIGACT News, ACM* (Apr. 1976).

[162] Kohavi, Z. *Switching and Finite Automata Theory*, second ed. McGraw-Hill, New York, 1978.

[163] Kurshan, R. P. *Computer-Aided Verification of Coordinating Processes.* Princeton University Press, Princeton, NJ, 1994.

[164] Lai, H. C., and Muroga, S. Automated logic design of mos networks. *Advances in Information Systems Science 9* (1970), 287–335.

[165] Lai, Y.-T., and Sastry, S. Edge-valued binary decision diagrams for multi-level hierarchical verification. In *Proceedings of the Design Automation Conference* (Anaheim, CA, June 1992), pp. 608–613.

[166] Larrabee, T. Test pattern generation using boolean satisfiability. *IEEE Transactions on Computer-Aided Design 11*, 1 (Jan. 1992), 4–15.

[167] Lawler, E. *Combinatorial Optimization.* Holt Rinehart Winston, 1976.

[168] Lawler, E. L. An approach to multilevel boolean minimization. *Journal of the Association for Computing Machinery 11*, 3 (July 1964), 283–295.

[169] Lee, C. Y. Binary decision programs. *Bell System Technical Journal 38*, 4 (July 1959), 985–999.

[170] Lee, E. B., and Perkowski, M. Concurrent minimization and state assignment of finite state machines. In *IEEE Conference on Systems, Man and Cybernetics* (Halifax, Canada, Oct. 1984), pp. 248–260.

[171] Lee, S. C. *Modern Switching Theory and Digital Design.* Prentice-Hall, Englewood Cliffs, 1978.

[172] Lehman, E., Watanabe, Y., Grodstein, J., and Harkness, H. Logic decomposition during technology mapping. In *Proceedings of the International Conference on Computer-Aided Design* (San Jose, CA, Nov. 1995), pp. 264–271.

[173] Leiserson, C. E., Rose, F. M., and Saxe, J. B. Optimizing synchronous circuitry by retiming. In *Proceedings of the Caltech Conference on VLSI* (Mar. 1983).

[174] Leiserson, C. E., and Saxe, J. B. Optimizing synchronous systems. In *Proceedings of the Symposium on Foundations of Computer Science* (Oct. 1981), pp. 23–26.

[175] Lin, B., Coudert, O., and Madre, J. C. Symbolic prime generation for multiple-valued functions. In *Proceedings of the Design Automation Conference* (Anaheim, CA, June 1992), pp. 40–44.

[176] Lin, B., and Newton, A. R. Synthesis of multiple level logic from symbolic high-level description languages. In *Proceedings of the IFIP International Conference on VLSI* (Aug. 1989), pp. 187–196.

[177] LIN, B., AND NEWTON, A. R. Implicit manipulation of equivalence classes using binary decision diagrams. In *Proceedings of the International Conference on Computer Design* (Cambridge, MA, Oct. 1991), pp. 81–85.

[178] LIN, B., AND SOMENZI, F. Minimization of symbolic relations. In *Proceedings of the IEEE International Conference on Computer Aided Design* (Santa Clara, CA, Nov. 1990), pp. 88–91.

[179] LIN, B., WHITCOMB, G. S., AND NEWTON, A. R. Symbolic don't cares and equivalence in high-level synthesis. In *IFIP International Working Conference on Logic and Architecture Synthesis* (May 1990).

[180] MAILHOT, F., AND MICHELI, G. D. Technology mapping using boolean matching. In *Proceedings of the European Conference on Design Automation* (Glasgow, UK, Mar. 1990), pp. 180–185.

[181] MALIK, S., BRAYTON, R. K., AND SANGIOVANNI-VINCENTELLI, A. Encoding symbolic inputs for multi-level logic implementation. In *Proceedings of the IFIP International Conference on VLSI* (Munich, FRG, Aug. 1989), pp. 221–230.

[182] MALIK, S., SENTOVICH, E. M., AND BRAYTON, R. K. Retiming and resynthesis: Optimizing sequential networks with combinational techniques. In *Proceedings of the Hawaii International Conference on Systems Science* (Jan. 1990), pp. 397–406.

[183] MALIK, S., WANG, A., BRAYTON, R., AND SANGIOVANNI-VINCENTELLI, A. Logic verification using binary decision diagrams in a logic synthesis environment. In *Proceedings of the IEEE International Conference on Computer Aided Design* (Santa Clara, CA, Nov. 1988), pp. 6–9.

[184] MARCUS, M. P. Derivation of maximal compatibles using boolean algebra. *IBM Journal of Research and Development 8* (Nov. 1964), 537–538.

[185] MARLETT, R. A. EBT: A comprehensive test generation technique for highly sequential circuits. In *Proceedings of the Design Automation Conference* (June 1978), pp. 335–339.

[186] MCCLUSKEY, E. J. *Introduction to the Theory of Switching Circuits.* McGraw-Hill, New York, 1965.

[187] MCCLUSKEY, E. J. *Logic Design Principles.* Prentice-Hall, Englewood Cliffs, 1986.

[188] MCCLUSKEY, JR., E. J. Minimization of boolean functions. *Bell Syst. Technical Journal 35* (Nov. 1956), 1417–1444.

[189] MCCLUSKEY, JR., E. J., AND UNGER, S. H. A note on the internal variable assignments for sequential switching circuits. *IRE Transactions on Electronic Computers EC-8*, 4 (Dec. 1959), 439–440.

[190] MCELIECE, R. J., ASH, R. B., AND ASH, C. *Introduction to Discrete Mathematics.* Random House, New York, 1989.

[191] MCFARLAND, M. C. Using bottom-up design techniques in the synthesis of digital hardware from abstract behavioral descriptions. *Proceedings of the 22nd Design Automation Conference* (June 1986), 474–480.

[192] MCGEER, P. C., BRAYTON, R. K., AND SANGIOVANNI-VINCENTELLI, A. L. Performance enhancement through the generalized bypass transform. In *Proceedings of the International Conference on Computer-Aided Design* (Santa Clara, CA, Nov. 1991), pp. 184–187.

[193] McMillan, K. Class project on BDD-based verification. Private Communication, E. M. Clarke, 1987.

[194] McMillan, K. L. *Symbolic Model Checking.* Kluwer Academic Publishers, Boston, MA, 1994.

[195] McNaughton, R. Unate truth functions. *IRE Transactions on Electronic Computers EC-10* (Mar. 1961), 1–6.

[196] Mead, C., and Conway, L. *Introduction to VLSI Systems.* Addison-Wesley, Reading, MA, 1980.

[197] Mealy, G. H. A method for synthesizing sequential circuits. *Bell System Technical Journal 34* (Sept. 1955), 1045–1079.

[198] Micheli, G. D. Symbolic design of combinational and sequential logic circuits implemented by two-level macros. *IEEE Transactions on Computer-Aided Design of Integrated Circuits and Systems CAD-5,* 9 (Sept. 1986), 597–626.

[199] Miczo, A. *Digital Logic Testing and Simulation.* John Wiley, New York, 1986.

[200] Minato, S.-I. Zero-suppressed BDDs for set manipulation in combinatorial problems. In *Proceedings of the Design Automation Conference* (Dallas, TX, June 1993), pp. 272–277.

[201] Minato, S.-I., Ishiura, N., and Yajima, S. Shared binary decision diagram with attributed edges for efficient boolean function manipulation. In *Proceedings of the Design Automation Conference* (Orlando, FL, June 1990), pp. 52–57.

[202] Moore, E. F. Gedanken experiments on sequential machines. In *Automata Studies,* C. E. Shannon and J. McCarthy, Eds. Princeton University Press, 1956.

[203] Murgai, R., Nishizaki, Y., Shenoy, N., Brayton, R. K., and Sangiovanni-Vincentelli, A. Logic synthesis for programmable gate arrays. In *Proceedings of the Design Automation Conference* (Orlando, FL, June 1990), pp. 620–625.

[204] Muroga, S., Kambayashi, Y., Lai, H. C., and Culliney, J. N. The transduction method—design of logic networks based on permissible functions. *IEEE Transactions on Computers C-38,* 10 (Oct. 1989), 1404–1424.

[205] Niermann, T., and Patel, J. H. HITEC: A test generation package for sequential circuits. In *Proceedings of the European Conference on Design Automation* (Amsterdam, The Netherlands, Feb. 1991), pp. 214–218.

[206] Panda, S., and Somenzi, F. Who are the variables in your neighborhood. In *Proceedings of the International Conference on Computer-Aided Design* (San Jose, CA, Nov. 1995), pp. 74–77.

[207] Panda, S., Somenzi, F., and Plessier, B. F. Symmetry detection and dynamic variable ordering of decision diagrams. In *Proceedings of the International Conference on Computer-Aided Design* (San Jose, CA, Nov. 1994), pp. 628–631.

[208] Pareto, V. *Manual of Political Economy.* A. M. Kelley, New York, NY, 1971. English translation of "Manuale di economia politica." Translated by A. S. Schwier. Edited by A. S. Schwier and A. N. Page.

[209] Park, N., and Parker, A. C. Sehwa: A program for synthesis of pipelines. *Proceedings of the 23rd Design Automation Conference* (1986), 454–460.

[210] Parker, A. C., Pizarro, J., and Mlinar, M. Maha: A program for datapth synthesis. *Proceedings of the 23rd Design Automation Conference* (1986), 461–466.

[211] PAULL, M. C., AND UNGER, S. H. Minimizing the number of states in incompletely specified sequential switching functions. *IRE Trans. Electronic Computers EC-8* (Sept. 1959), 356–367.

[212] PETRICK, S. R. A direct determination of the irredundant forms of a boolean function from the set of prime implicants. Tech. Rep. AFCRC-TR-56-110, Air Force Cambridge Res. Center, Cambridge, MA, Apr. 1956.

[213] PIPPONZI, M., AND SOMENZI, F. An iterative approach to the binate covering problem. In *Proceedings of the European Conference on Design Automation* (Glasgow, UK, Mar. 1990), pp. 208–211.

[214] PIXLEY, C. A computational theory and implementation of sequential hardware equivalence. In *Computer-Aided Verification '90* (1991), E. M. Clarke and R. P. Kurshan, Eds., American Mathematical Society – Association for Computing Machinery, pp. 293–320.

[215] PIXLEY, C. A theory and implementation of sequential hardware equivalence. *IEEE Transactions on Computer-Aided Design 11*, 12 (Dec. 1992), 1469–1478.

[216] PIXLEY, C., BEIHL, G., AND PACAS-SKEWES, E. Automatic derivation of FSM specification to implementation encoding. In *Proceedings of the International Conference on Computer Design* (Cambridge, MA, Oct. 1991), pp. 245–249.

[217] PIXLEY, C., SINGHAL, V., AZIZ, A., AND BRAYTON, R. K. Multi-level synthesis for safe replaceability. In *Proceedings of the International Conference on Computer-Aided Design* (San Jose, CA, Nov. 1994), pp. 442–449.

[218] PLESSIER, B., HACHTEL, G., AND SOMENZI, F. Extended BDDs: Trading off canonicity for structure in verification algorithms. *Journal of Formal Methods in System Design 4*, 2 (Feb. 1994), 167–185.

[219] POAGE, J. F. Derivation of optimum tests to detect faults in combinational circuits. In *Proc. Symposium on Mathematical Theory of Automata* (Polytechnic Institute of Brooklyn, 1963), pp. 483–528.

[220] POTTOSIN, Y. V. Experimental evaluation of one method of minimizing the number of states of discrete automata. In *Synthesis of Digital Automata* (transl. from Russian), V. G. Lazarev and A. V. Zakrevskii, Eds. Consultants Bureau, New York, 1969, pp. 92–98.

[221] QUINE, W. The problem of simplifying truth functions. *Amer. Math. Monthly 59* (1952), 521–531.

[222] QUINE, W. A way to simplify truth functions. *Amer. Math. Monthly 62* (Nov. 1955), 627–631.

[223] QUINE, W. V. Two theorems about truth functions. *Boletin de la Sociedad Matematica Mexicana 10* (1953), 64–70.

[224] RABIN, M., AND SCOTT, D. Finite automata and their decision problems. *IBM Journal of Research and Development 3* (1959), 114–125.

[225] RANJAN, R. K., AZIZ, A., BRAYTON, R. K., PLESSIER, B. F., AND PIXLEY, C. Efficient BDD algorithms for FSM synthesis and verification. Presented at IWLS95, Lake Tahoe, CA., May 1995.

[226] RAVI, K., AND SOMENZI, F. High-density reachability analysis. In *Proceedings of the International Conference on Computer-Aided Design* (San Jose, CA, Nov. 1995), pp. 154–158.

[227] REUSCH, B., AND MERZENICH, W. Minimal coverings for incompletely specified sequential machines. *Acta Informatica 22* (1986), 663–678.

[228] RHO, J.-K., HACHTEL, G. D., SOMENZI, F., AND JACOBY, R. Exact and heuristic algorithms for the minimization of incompletely specified state machines. *IEEE Transactions on Computer-Aided Design of Integrated Circuits and Systems 13*, 2 (Feb. 1994), 167–177.

[229] ROBINSON III, S., AND HOUSE, R. Gimpel's reduction technique extended to the covering problem with costs. *IEEE Transactions on Electronic Computers EC-16* (Aug. 1967), 509–514.

[230] ROSS, D. E., BUTLER, K. M., KAPUR, R., AND MERCER, M. R. Fast functional evaluation of candidate OBDD variable ordering. In *Proceedings of the European Conference on Design Automation* (Amsterdam, Feb. 1991), pp. 4–10.

[231] ROSS, S. *A First Course in Probability*, third ed. Macmillan, New York, 1988.

[232] ROTH, J. P. Diagnosis of automata failures: A calculus and a method. *IBM Journal of Research and Development 10* (July 1966), 278–291.

[233] ROTH, J. P. *Computer Hardware Testing and Verification.* Computer Science Press, Potomac, Maryland, 1980.

[234] ROY, K., AND PRASAD, S. SYCLOP: Synthesis of CMOS logic for low power applications. In *Proceedings of the International Conference on Computer Design* (Cambridge, MA, Oct. 1992), pp. 464–467.

[235] RUDEANU, S. *Boolean Functions and Equations.* North-Holland, Amsterdam, 1974.

[236] RUDELL, R. *Logic Synthesis for VLSI Design.* PhD thesis, University of California, Berkeley, 1989.

[237] RUDELL, R. Dynamic variable ordering for ordered binary decision diagrams. In *Proceedings of the International Conference on Computer-Aided Design* (Santa Clara, CA, Nov. 1993), pp. 42–47.

[238] RUDELL, R., AND SANGIOVANNI-VINCENTELLI, A. Exact minimization of multiple-valued functions for PLA optimization. In *Proceedings of the IEEE International Conference on Computer Aided Design* (1986), pp. 352–355.

[239] RUDELL, R., AND SANGIOVANNI-VINCENTELLI, A. Multiple-valued minimization for PLA optimization. *IEEE Transactions on Computer-Aided Design of Integrated Circuits and Systems CAD-6*, 5 (Sept. 1987), 727–750.

[240] RUDELL, R., AND SEGAL, R. *BDSYN Users Manual*, Apr. 1986.

[241] SAHNI, S. *Concepts in Discrete Mathematics.* The Camelot Publishing Comapny, 1981.

[242] SALDANHA, A., BRAYTON, R. K., AND SANGIOVANNI-VINCENTELLI, A. L. Circuit structure relations to redundancy and delay: The KMS algorithm revisited. In *Proceedings of the Design Automation Conference* (Anaheim, CA, June 1992), pp. 245–248.

[243] SALDANHA, A., AND KATZ, R. PLA optimization using output encoding. In *Proceedings of the International Conference on Computer-Aided Design* (Nov. 1988).

[244] SASAO, T. Input variable assignment and output phase optimization of PLA's. *IEEE Transactions on Computers C-33* (Oct. 1984), 879–894.

[245] SASAO, T., Ed. *Logic Synthesis and Optimization.* Kluwer Academic Publishers, Boston, MA, 1993.

[246] SAVOJ, H., MALIK, A. A., AND BRAYTON, R. K. Fast two-level minimizers for multilevel logic synthesis. In *Proceedings of the IEEE International Conference on Computer Aided Design* (Nov. 1989), pp. 544–547.

[247] SCHULZ, M., TRISCHLER, E., AND SARFERT, T. SOCRATES: A highly efficient automatic test pattern generation system. *IEEE Transactions on Computer-Aided Design of Integrated Circuits and Systems CAD-7*, 1 (Jan. 1988), 126–137.

[248] SELLERS, JR., F. F., HSIAO, M. Y., AND BEARNSON, L. W. Analyzing errors with the boolean difference. *IEEE Transactions on Computers C-17*, 7 (July 1968), 676–683.

[249] SENTOVICH, E. M., SINGH, K. J., MOON, C., SAVOJ, H., BRAYTON, R. K., AND SANGIOVANNI-VINCENTELLI, A. Sequential circuit design using synthesis and optimization. In *Proceedings of the International Conference on Computer Design* (Cambridge, MA, Oct. 1992), pp. 328–333.

[250] SENTOVITCH, E., MALIK, S., AND BRAYTON, R. Peripheral retiming and resynthesis. In *hicss* (Jan. 1990), pp. 397–406.

[251] SIKORSKI, R. *Boolean Algebras*, second ed. Springer-Verlag, Berlin, 1964.

[252] SINGHAL, V., AND PIXLEY, C. The verification problem for safe replaceability. In *Sixth Conference on Computer Aided Verification (CAV'94)*, D. L. Dill, Ed. Springer-Verlag, Berlin, 1994, pp. 311–323. LNCS 818.

[253] SOMENZI, F. Gimpel's reduction technique extended to the binate covering problem. Unpublished Manuscript, Sept. 1989.

[254] SOMENZI, F., AND GAI, S. Fault detection in programmable logic arrays. *Proceedings of the IEEE 74* (May 1986), 655–668.

[255] STEWART, W. J., Ed. *Numerical Solutions of Markov Chains.* Marcel Dekker, New York, 1991.

[256] SWAMY, G. M., AND BRAYTON, R. K. Incremental formal design verification. In *Proceedings of the International Conference on Computer-Aided Design* (San Jose, CA, Nov. 1994), pp. 458–465.

[257] SWAMY, G. M., BRAYTON, R. K., AND MCGEER, P. A fully implicit Quine-McCluskey procedure using BDD's. Presented at IWLS'93, May 1993.

[258] TARJAN, R. Depth first search and linear graph algorithms. *SIAM Journal of Computing 1* (1972), 146–160.

[259] THISTLE, J. *Control of Infinite Behavior of Discrete Event Systems.* PhD thesis, University of Toronto, 1991.

[260] TJIANG, S. Twig reference manual. Tech. rep., AT&T Bell Laboratories, 1985.

[261] TOUATI, H. *Performance Oriented Technology Mapping.* PhD thesis, University of California, Berkeley, 1990.

[262] TOUATI, H., SAVOJ, H., LIN, B., BRAYTON, R. K., AND SANGIOVANNI-VINCENTELLI, A. Implicit enumeration of finite state machines using BDD's. In *Proceedings of the IEEE International Conference on Computer Aided Design* (Nov. 1990), pp. 130–133.

[263] TOUATI, H. J., BRAYTON, R. K., AND KURSHAN, R. P. Testing language containment for ω-automata using BDD's. In *1991 International Workshop on Formal Methods in VLSI Design* (Miami, FL, Jan. 1991).

[264] TRICKEY, H. *Compiling Pascal Programs into Silicon.* PhD thesis, Stanford University, 1985. Stanford Computer Science Report STAN-CS-85-1059.

[265] TRICKEY, H. Flamel: A high level hardware compiler. *IEEE Trans. on CAD CAD-6*, 2 (1986), 259–269.

[266] VAN LEEUWEN, J., Ed. *Handbook of Theoretical Computer Science.* The MIT Press/Elsevier, Amsterdam, 1990.

[267] VILLA, T. Constrained encoding in hypercubes: Algorithms and applications to logical synthesis. In *UC Berkeley Electronics Research Laboratory* (May 1987).

[268] WANG, A. *Algorithms for Multi-Level Logic Optimization.* PhD thesis, University of California, Berkeley, 1989.

[269] WEGENER, I. On the complexity of branching programs and decision trees for clique functions. *Journal of the Association for Computing Machinery 35*, 2 (Apr. 1988), 461–471.

[270] WEI, Y.-C., AND CHENG, C.-K. Toward efficient hierarchical designs by ratio cut partitioning. In *Proceedings of the IEEE International Conference on Computer Aided Design* (Santa Clara, CA, Nov. 1989), pp. 298–301.

[271] WHITCOMB, G. Exact factoring of incompletely specified functions. EE290ls class project report, UC Berkeley, May 1988.

Index